TULIPS

T. borszczowii Regel

TULIPS

Taxonomy, morphology, cytology, phytogeography and physiology

Z.P. BOTSCHANTZEVA

Translated and edited by

H.Q. VAREKAMP

Institute for Horticultural Plant Breeding, Wageningen

A.A. BALKEMA / ROTTERDAM / 1982

ISBN 90 6191 029 3

Translation of:
Z.P. Bočanceva, Tjul'pany, Morfologija, Citologija i Biologija
(Izd. Akad. Nauk Uzb.S.S.R.,Taškent, 1962)

Distributed in U.S.A. by M.B.S., 99 Main Street, Salem, N.H.03079

Printed in the Netherlands

Foreword

Zinaida Petrovna Botschantzeva was born in the city of Veriy (nowadays Alma-Ata, Kazakhstan, Central Asia) on 10 October 1907 as the eldest daughter of 7 children in the family of an employee of a local trade firm.

After preliminary training she entered the Central Asian State University (MASU) in 1925, taking biology courses and working part-time as a herbarium paster in the Institute of Soil Sciences and Geobotany in Tashkent. She took part in her first botanical expedition as an assistant in the summer of 1926. The event was decisive for her and enabled her to make her definite choice of study, which was to be botany. She applied for a position at the Botanic Garden of MASU in Tashkent to become an assistant to the garden staff in the autumn of 1926.

During 1927, she worked in the montane botanical station of the University in the Chimgan region, where she had the opportunity to become familiar with microscopy. She did some research on the leaf anatomy of several Chimgan trees and shrubs and took a special interest in the distribution of stomata on the leaf blades. During this initial period of study, her older friends P.A. Baranov, E.A. Mokeeva and I.A. Kultiasov helped her in many ways to set a firm base to her scientific interests.

In 1930 she left university with a qualification in morphology and cytology of plants. Shortly afterwards she was taken on to the staff of MASU Botanic Garden as a botanist. When the administration of the Botanic Garden was taken over by the Academy of Sciences of Uzbekistan in 1944, she became a senior scientific officer. She was involved in the layout and construction of the new botanic garden in Tashkent at the time and took an active part in many difficult and hazardous expeditions through montane Middle Asia. She collected material for scientific work and added further specimens to the existing botanical collections. These expeditions led her through the Zailiyskiy Alatau, the Chu-Iliyskye mountain range, the Kirgiz range, the Ferghan range, the Kuraminskiy massif, the western part of the Tien Shan and the Turkestan range. Her special field of study were the ornamental bulbous plants belonging to the Liliaceae and Iridaceae, both under natural conditions and in cultivation, in particular the Middle Asian species of *Tulipa, Allium, Eremurus* and *Iris.* Among the new species she described were: *Tulipa affinis, T. anadroma, T. butkovii, T. tschimganica,* and *T. vvedenskyi.* Other plants she discovered were *Eremurus zenaidae* Vved., *Iris zenaidae* Vved. and *Tulipa zenaidae* Vved., all named in her honour by professor A.I. Vvedensky of Tashkent.

Various scientific problems attracted her interest, such as the biological aspects of flowering, fertilization and remote hybridisation of some Papaveraceae, *Liriodendron tulipifera, Parthenium argentatum*, arborescent *Salsola* and others. She investigated the embryology, taxonomy and cytogenetics of these plants, but her favourites always remained the Middle Asian tulips.

The publication of her main work, the monograph "Tyul'pany" sums up the results of many years of investigations into the genus. The monograph was to become her doctoral thesis, which she successfully defended in June 1960.

Despite a grave illness during her later years, she pursued her work in the botanic garden with all the energy she could muster. She readily shared her knowledge and experience with younger students and helped many of them. In August 1973, after a long illness, she died in Tashkent, where she had worked for so many years, greatly missed by her many friends and family.

The present translation of her main work will enable botanists and both amateur and industrial bulb growers abroad to get to know the Middle Asian tulips Zinaida Botschantzeva loved so much. They represent a field of interest only vaguely known to a limited number of botanists outside the Soviet Union, mainly through older publications such as Komarov's Flora of the Soviet Union (1935), other local floras of these regions and lesser known publications that are virtually unobtainable in other countries. A great need exists for modern up-to-date floras and monographs on genera of the Central Asian republics and a preference is expressed for English translations so that knowledge can be shared by other botanists in the Western World. The translation of Zinaida Botschantzeva's book on Middle Asian tulips gives a better understanding of that part of the flora for the present period that separates us from modern taxonomic reviews or complete floras of these remote regions.

We gratefully acknowledge the help and support with this project by the Director, Academician N.V. Tsitsin and the Vice-Director P.I. Lapin of the Main Botanic Garden of the Academy of Sciences of the Soviet Union. They have given invaluable help and guidance through the many stages of the project, including official permission for translation and reproduction.

We record gratefully the financial help and moral support given by the Director of the Institute for Horticultural Plant Breeding, Ir. C. Dorsman, and by the Board of Directors of the Dutch Bulb Exporters Association, Hillegom, the Netherlands.

We are equally indebted to Mr. H.Q. Varekamp, the taxonomist of the Institute for Horticultural Plant Breeding, who translated the text from Russian into English and took care of the editing. Combining his knowledge of taxonomy with his experience of Slavic languages he has spared no effort to give an accurate rendering of the original text. He has added a map to chapter 4 as well as a badly needed index to the end of the book.

We will remember the lively conversation with professor Botschantzeva in the office of the Botanic Garden of Tashkent in June 1969 and the beautiful water-colours, made by Georgi Semenov, that were shown to us. They form a splendid background to the text of the book. We then became so interested that it was decided that the Tulip book should be translated into English. The same day we requested professor Botschantzeva's permission to allow translation, to which she immediately consented.

The translation has now made it possible that botanists and horticulturists, who do not have a command of the Russian language, can appreciate the importance of her work. It is a well-deserved tribute to one who devoted so much of her energy to the study of the genus Tulipa.

Utrecht, December 1980

Michael H. Hoog
Curator University Botanic Gardens

Table of contents

Editor's preface

In the editing of this book some changes were made in the arrangement of the subject matter of chapters 10 and 11. A small-scale map has been added to chapter 4, which may be useful for an orientation about the distributional areas of the treated species.

All coloured illustrations have been reproduced from the original water-colours made by G.A. Semenov and M.K. Lukanina. The figures, where necessary, were redrawn by ing. J. Baër, formerly of the Institute for Horticultural Plant Breeding and the graphs were reproduced to scale by Mr. G.C. Beekhof of the Office of Joint Services at Wageningen.

In transliteration of Cyrillic characters I have followed, as far as titles of publications are concerned, the *ISO Recommendation R9*, 2nd ed. 1968. A few exceptions were made for personal names of Soviet botanists where it was known that they followed a different convention. For geographical names, mentioned in the text, I have followed *The Times Atlas of the World, Fourth Comprehensive Edition*, London, 1972.

The translator takes all responsibility for any error of fact or interpretation that may have crept into the translation.

H.Q. Varekamp
Institute for Horticultural Plant Breeding, Wageningen

Introduction

From their very beginning the Botanical Gardens of the Academy of Sciences of the Uzbek Soviet Socialist Republic (formerly run by the Middle Asian State University), have been a centre for the study of local species and cultivars of tulips, many of which are highly prized on the world market and hold a prominent place in the catalogues of Dutch bulbgrowers. However, it was only in recent years that tulips, mainly foreign cultivars, began to appear in the parks and gardens of Moscow and Leningrad. Lately they have also been planted in the public parks of Tashkent.

Because of their bright colours, their distinctive shapes and the beauty and size of their flowers, many local tulip species hold their own with foreign cultivars, often even surpassing them. We mention *Tulipa fosteriana*, *T. greigii*, *T. micheliana*, *T. kaufmanniana*, *T. vvedenskyi* and *T. albertii*.

Truly amazing by their flaming beauty are *T. greigii* and *T. micheliana* and very striking is also the pink *T. korolkowii* with its charming medium-sized flowers, while *T. ostrowskiana* and the closely related species attract attention by their graceful lily-shaped flowers.

The colours of wild tulip species can be extremely variable. The widespread *T. schrenkii* presents in its populations the most exquisite shades of rare cultivars and all red-flowered tulips from the Tien Shan have forms which offer a variety of colours, e.g. *T. albertii*, *T. kaufmanniana*, *T. ostrowskiana* and others. Also in the shape, texture and colour of their leaves the species show a greater variety than the cultivars. Besides, wild species have an invaluable advantage over cultivars in being immune to mosaic virus, a disease to which all tulip cultivars are susceptible.

Middle Asia, with its varied climatic and geographical conditions and surrounded by mountain ranges and deserts differs markedly from other areas of the USSR. A large part of Middle Asia is a vast desert with enormous mountain ranges in the east and south: the Kopet Dag in the southwest, the Pamir-Alay in the centre, extending to the border of Afghanistan and Kashgar, and the Tien Shan in the east. These mountains are not only barriers, preventing the migration of certain plants, but form also the bridges between Middle Asia and the adjacent countries for others. This situation makes it difficult to find the correct answer to problems of the phylogenetic relationship in many groups of plants, in particular the genus *Tulipa*, which extends from Middle Asia into the surrounding geographical areas.

Vvedenskiĭ has given in the 'Flora of the USSR', Vol. IV (1935), detailed descriptions and a classification of the tulips, growing within the Soviet territory. Although the tulips of the USSR have been studied much more thoroughly since 1935, his classification is in general still valid.

Tulips were cultivated in the East long before they became known in Europe. The Persian poet, Musharrifu'd-din Sa'di, describing a garden in his poem 'Gulistan', in 1258, wrote:

> The murmur of a cool stream
> bird song, ripe fruit in plenty,
> bright multicoloured tulips and fragrant roses

We see that as early as the 13th century there must have been multi-coloured tulips in Persian gardens. The poet's words lend themselves to two interpretations: he meant either variegated tulips, or garden tulips raised from differently coloured species. As only cultivated tulips are known to be susceptible to the mosaic virus disease, causing variegated flowers, it could be concluded from the latter surmise, that already at that time tulips were cultivated in the East.

In the 15th century 'Ali-Shir Nawa'i writes in the poem 'Farhad u Shirin': 'wine in green bowls I can compare only to scarlet tulips in a meadow'. And in the poem 'Leili u Majnun': 'when the tulip came to know sorrow a black mark appeared on its heart'. In his early years Nawa'i lived in Samarkand and could have seen the scarlet tulips with dark basal blotches growing in the Samarkand mountains. They could have been *Tulipa fosteriana* or *T. micheliana* with the pointed black blotches on the tepals. Hence the comparison between poignant sorrow and the black mark on the tulip's heart.

Tulips are being mentioned quite frequently by this poet:

She loved the black-eyed gazelle
the vast colorful beauty of steppe
with the tulips in flaming bloom

and further in the poem 'Haft Peikar' ('Seven Effigies') he speaks of 'the tulip-speckled meadow'. There are in these poems indications that cultivated tulips were grown in the gardens of this period. In 'Leili u Majnun', there are the following lines:

Qais and Leili, together with others,
passed through the gates of Iram
and tulips there charmed the eyes of all,
but their bleeding hearts were like the scarlet tulips

'Ali-Shir Nawa'i shows his admiration of the tulips in the following enrapturing lines in 'Farhad u Shirin':

As slender as the Cypress tree
the charming tulip, tender as a pheasant,
as soothing to the sorrowful heart
as moonlight in a cloudless night

It appears that tulips were already in the 15th century a source of inspiration to poets in the territory which is now Uzbekistan.

In 'Gardens and Gardening' (1951) is an article by Rollo Meyer which begins as follows: 'In 1123 the tulips are already mentioned in Russian literature under their Russian name "Lola". Although this word is not of Russian but of Turkish origin, its use here indicates that also in Russia tulips had become known much earlier than in Western Europe, where they were introduced much later, in 1554. At that time many tulip cultivars were already known in Turkey. The first bulbs were received in Vienna. Later they reached Antwerp, then Holland in 1571, and still later the British Isles.

The cultivated tulips were brought to Russia from Western Europe by Dutch merchants in the 16th and early 17th centuries. An official reference to cultivated tulips in Russia dates back to the 18th century. According to the 1701 inventory of the Nizhni Park[1], there were 1.000 large and 3.000 medium- and small-sized tulips (Zabelin, 1872, cited by Migushova, 1951).

Because of their beauty and striking appearance the tulips spread rapidly all over Europe. The 'Tulipomania' of 1634, when in the course of three years tulip bulbs became the subject of fierce speculation, is well-known. An extensive literature on the history of tulip cultivation in Europe has built up since 1554. Several works which have appeared in recent years in the Russian language give brief surveys of the history of tulip cultivation (Silina, 1953; Migushova, 1954; Danilevskaya, 1955).

The exportation of bulbs of wild species from Middle Asia, organised by the Tashkent Botanic Gardens during the period 1929-1934, enabled Dutch bulbgrowers to bring on the market in the late nineteen-fifties a large number of entirely new cultivars resulting from selections of wild Middle Asian species and from crossing the latter with established cultivars. In particular the hybrids of *T. fosteriana* with Darwin tulips have become popular in recent years. The Middle Asian tulip species have been studied at the Botanic Gardens of the Academy of Sciences of the Uzbek Soviet Socialist Republic for many years and the results were summarized by Vvedenskiĭ in his monograph (1935). Vasilevskaya (1936) studied the anatomy of tulips, while their cytology occupied our attention.

For a more detailed morphological description we have used in this work, with a few exceptions, only observations on species, which we collected in natural habitats, or which were raised from bulbs or seeds introduced from known locations. Herbaria usually contain either plants in flower but without capsules, or plants with capsules but without flowers. We therefore described not only the flowers but also the capsules in order to give all the characters of the species. Moreover, it serves no purpose to collect bulbs for further cultivation during flowering, because at that time they are far too weak, and later, without sufficient knowledge of the capsules and their distinguishing features, it is easy to confuse different species.

This work includes cytological descriptions of species of known wild origin. Descriptions of plants grown in European botanic gardens are often given without any indication of their place of origin. In many cases they are from garden sources or hybrids and as a rule there is no guarantee that the material under investigation truly represents the species under whose name the taxon is cited in cytological or other works.

We also give the results of our researches on tulip seeds, including their germination and the growth and development of the seedlings. These data have contributed to an understanding of the evolution and origin of the various species and varieties.

In recent years some studies have appeared dealing with the annual cycle of development of the adult bulb of cultivars grown in the north of the USSR (Silina, 1953; Migushova, 1954; Danilevskaya, 1955 and 1956) and in the central regions of Russia (Timofeeva, 1956). These publications do not include wild species and are confined to the study of the bulb.

We conclude this work by describing certain modifications of the wild species under the influence of external conditions prevailing in cultivation.

After studying the Middle Asian tulip species for many years, we have obtained many new varieties through clonal selection, interspecific hybridization and crosses between wild species and cultivars.

1. The Nizhni Park dates from the 18th century and is located in the well-known Leningrad suburb Petrodvorets. (editor)

CHAPTER ONE

Short review of the study of Middle Asian tulips

Regel

E. Regel, in the article 'Enumeratio specierum hucusque cognitarum generis *Tulipa*' (1873), was the first to review the genus and to divide the by then known 26 tulip species into groups. He also provided a key for their identification. His treatment was as follows:

A.[1] Inner tepals and filaments hairy at the base
B. Leaves linear or linear-lanceolate or narrowly lanceolate
C. Bulb tunics inside towards the apex and sometimes at the base with adpressed hairs
1. *T. sylvestris* L.
CC. Bulb tunics arachnoid-woolly all over the inner surface
2. *T. biflora* Pall.
CCC. Bulb tunics glabrous inside
3. *T. celsiana* Redouté
BB. Lower leaves ovate or ovate-oblong
4. *T. saxatilis* Sieber
AA. Inner tepals and filaments naked at the base, or the latter in *T. sibthorpiana* hairy, and the former in *T. sogdiana* sparingly ciliate at the base
D. Tepals at the base with a blackish or blue-blackish blotch (in one form of *T. montana* there is no dark blotch)
E. Tepals ovate or elliptic-lanceolate, obtuse or abruptly acuminate
F. Pedicels glabrous
G. Lower leaves ovate or ovate-lanceolate; upper leaves narrower
5. *T. oculus-solis* St. Amans
6. *T. didieri* Jord.
7. *T. borszczowii* Regel
8. *T. julia* K. Koch
GG. Leaves linear
9. *T. boissieri* Regel
FF. Pedicels more or less hairy
10. *T. greigii* Regel
EE. Tepals ovate, the outer caudate-acuminate
11. *T. eunanthiae* Orph.
EEE. Tepals lanceolate, the outer attenuate-acuminate
12. *T. aleppensis* Boiss.
13. *T. boeotica* Boiss. et Heldr.
DD. Tepals at the base without a blackish blotch
H. Tepals blunt or abruptly acuminate
I. Pedicels glabrous. Flowers erect
K. Leaves not margined, the lower ovate or ovate-lanceolate, or lanceolate; the upper narrower
L. Bulb tunics glabrous inside
14. *T. gesneriana* L.
LL. Bulb tunics inside towards the apex with adpressed hairs
15. *T. schrenkii* Regel
KK. Leaves with transparent cartilaginous margins, the lower oblong-lanceolate, the upper narrower
16. *T. lehmanniana* Merckl.
KKK. Leaves not margined, the lower linear-lanceolate, the upper narrower or wholly linear
17. *T. sogdiana* Bunge
18. *T. armena* Boiss.
19. *T. clusiana* Vent.
20. *T. montana* Lindl.
II. Pedicels glabrous. Flowers nodding
21. *T. sibthorpiana* Sm.
III. Pedicels shortly pubescent
22. *T. suaveolens* Roth
23. *T. strangulata* Reboul
24. *T. altaica* Pall.
HH. Tepals caudate-acuminate
25. *T. turcica* Roth
26. *T. undulatifolia* Boiss.

Regel used all the criteria known at the time and divided the species into groups quite adequately, but his work received little attention from western botanists. The defects of Regel's classification are largely due to the scantiness of the material at his disposal, which

1. For the sake of completeness Regel's key is cited in extenso. (editor)

prevented him from forming a true picture of the wide range of variability.

Baker's synopsis of the genus appeared in the following year and became widely known. He divided the genus into two subgenera: *Eutulipa* with rudimentary styles and *Orithyia* with the style as long as the ovary. *Eutulipa* was divided into 5 sections: *Eriobulbi, Gesnerianae, Scabriscapae, Saxatiles,* and *Silvestres.*

Boissier (1882) grouped Baker's first three sections into *Leiostemones* and the other two into *Eriostemones.* The section *Orithyia* was not dealt with in his treatment.

Taliev

Taliev (1930) maintained Boissier's grouping of Baker's sections. In the *Silvestres* he distinguished two additional branches:

a) *T. turkestanica* and *T. biflora,* together with an intermediate, which he called *T. praebiflora,* form his branch *Bifloraeformis.*

b) *T. tricolor, T. biebersteiniana* and *T. sylvestris,* with the intermediates, called *T. praetricolor, T. tricolor polychroma* and *T. praebiebersteiniana,* form his branch *Silvestriformes.*

He regarded the multiflorous *T. turkestanica,* with its perceptible style and filaments pubescent along their whole length, as being nearest to the prototype of the genus. Referring to karyological data of Newton and Tischler he mentions that the chromosome number, n = 12, is found in the 'archaic group of Liliaceae'. This apparently shows that the genus *Lilium* and the concurrent genera *Fritillaria* and *Tulipa* are very close to the evolutionary origin of the Liliaceae. According to Taliev the typical uniflorous tulips developed through reduction from the more or less multiflorous inflorescence of remote ancestors. Some of the large-flowered tulips are no more than 'deprived relations' of some lilies and the genus obviously took shape through adaptation to certain ecological conditions. Probably, in the first place, to life in high-mountain conditions and then to the life-forms of spring xerophytes. For these reasons he concluded that primitive tulip species should be multiflorous, with large flowers and should grow under high-mountain conditions. He also contends that Middle Asia[2] leads the world in the number of species and is the centre of origin of the genus. We believe that the practical implications of his views are a large-scale programme to be launched in Middle Asia for the utilisation and conservation of this source of wild tulip species.

Vvedenskiĭ

Vvedenskiĭ, an expert on the genus *Tulipa,* writes: 'the now accepted system of Baker and Boissier is highly artificial, but it would be difficult to attempt its revision in view of the scarce information about the tulip flora of South-West Asia' (Flora of the USSR, 4, 1935). He introduced two new monotypic sections: *Spiranthera* Vved. with *T. kaufmanniana* Regel and *Lophophyllon* Vved. with *T. regelii* Krassn. In addition he transferred *T. sogdiana* Bunge from sect. *Leiostemones* to sect. *Eriostemones,* paying special attention to the order of the species within the sections. In the 'Flora of the USSR' Vvedenskiĭ arranged the Middle Asian species into groups according to their degree of affinity, apparently regarding sect. *Tulipanum* as the most primitive by placing it at the beginning of the genus. Within the section the species are likewise arranged in order of affinity: *T. tubergeniana* Hoog, *T. kuschkensis* B. Fedtsch., *T. hoogiana* B. Fedtsch., *T. julia* K. Koch. In 1946 he described a fifth species in this section, *T. subquinquefolia* Vved.

Within sect. *Leiostemones* the following species are grouped together: *T. florenskyi* Woron., *T. lanata* Regel, *T. ingens* Hoog, *T. fosteriana* Irv.,[3] *T. carinata* Vved. This first group comprises species which are related to those of sect. *Tulipanum* by the hairiness of the bulb tunics. The hairs however are less copious, thinner and rippled. Then follows a group of two species, *T. praestans* Hoog and *T. subpraestans* Vved., which are placed together because of a peculiar character, the absence of a blotch from the base of the tepals. Some forms of these species, however, show a diffuse, pale orangy tint at the base of the tepals, which in our opinion connects them with the next group. In the shape of the flowers they resemble species of the preceding group around *T. fosteriana* Hoog ex. W. Irving. The third group is very distinct in the shape of its flowers and most species have anthocyanin spots or stripes on the leaves. It consists of *T. albertii* Regel, *T. greigii* Regel, *T. mogoltavica* M. Pop. et Vved. and *T. micheliana* Hoog, which show affinities with sect. *Tulipanum* in flower structure. Vvedenskiĭ's fourth group includes two Caucasian species: *T. eichleri* Regel and *T. schmidtii* Fom.

2. In the Russian literature a distinction is made between 'Srednyaya Aziya' (lit. tr.: Middle Asia) and 'Tsentralnaya Aziya' (lit. tr.: Central Asia). The former comprises the Irano-Turanian steppes, semideserts and deserts east of the Caspian Sea, extending north of the Himalayas to the Altai and Pamir. Central Asia comprises Sinkiang, Tibet and Mongolia. An essential difference between these areas is the distribution of the annual rainfall. In Middle Asia the rains fall mainly in winter and early spring favouring the occurrence of geophytes such as *Tulpa,* whereas in central Asia winter and spring are extremely dry and the rains fall in summer (see Walter, H. 'Die Vegetation der Erde in ökophysiologischer Betrachtung', Jena 1973). In this book the terms have therefore been literally translated in the above sense. (editor)

3. *T. fosteriana* Hoog ex W. Irving. (editor)

Another most interesting group contains three species: *T. maximowiczii* Regel, *T. linifolia* Regel and *T. wilsoniana* Hoog, which are closely related, not only in their external morphology, but also in their cytology.

The sixth group comprises species with leathery bulb tunics: *T. lehmanniana* Merckl., *T. behmiana* Regel and *T. borszczowii* Regel, which in the structure of the flower are close to the species grouped around *T. greigii* Regel, and also possess prolonged bulb tunics extending to ground level. This feature may be absent in the species related to *T. greigii*, but is always present in species of the *T. lehmanniana* group.

The small species, *T. korolkowii* Regel, *T. nitida* Hoog and *T. rosea* Vved., form a closely related natural group. The group comprising *T. schrenkii* Regel, *T. ostrowskiana* Regel, *T. kolpakowskiana* Regel and *T. zenaidae* Vved. is also wholly natural, but in our opinion *T. schrenkii* gravitates in flower structure to the species grouped around *T. greigii* and *T. zenaidae* to those grouped around *T. korolkowii.* The next species in this sequence, *T. brachystemon* Regel, shows affinity with the *T. greigii* group in its general habit. After this the affinities between the species cited in 'Flora of the USSR' become less clear and they are no longer arranged in natural groups. The sequence is as follows: *T. platystemon* Vved., *T. altaica* Pall., *T. dubia* Vved., *T. iliensis* Regel, *T. ferganica* Vved., *T. anisophylla* Vved., *T. korshinskyi* Vved., *T. tianschanica* Regel, *T. tetraphylla* Regel, *T. hissarica* M. Pop. et Vved. and *T. krauseana* Regel. Further detailed investigation has shown that *T. dubia* Vved. should be transferred to the section *Spiranthera* in which Vvedenskiĭ originally placed only *T. kaufmanniana,* which he singled out on account of the mode of dehiscence of its anthers. Our own observations suggest that *T. dubia* should be placed in the same section as *T. kaufmanniana* because its anthers gradually split into longitudinal strands in the course of 2 or 3 days and not because the anthers coil spirally.

T. iliensis, T. ferganica and *T. tetraphylla* should be included in the same group as *T. ostrowskiana* and *T. kolpakowskiana. T. tetraphylla* shows affinities with *T. korolkowii* but is none the less highly distinctive. *T. hissarica* is a very distinct species with no close relation to the other *Leiostemones* species. We have not had the opportunity to see the other species of this section in the living condition and therefore cannot judge whether they were arranged correctly.

After the above mentioned sect. *Spiranthera* comes sect *Lophophyllon* with only *T. regelii* Krassn., characterised by 'longitudinal, twisted crests' along its solitary leaf and hairy filaments.

Vvedenskiĭ includes in *Eriostemones* the following species: *T. violacea* Boiss. et Buhse, *T. caucasica* Lipsky, *T. sylvestris* L., *T. biebersteiniana* Schult., *T. patens* Agardh. ex Schult., and *T. sogdiana* Bunge. The first five species form the 'branch' *Silvestriformes* (Taliev, 1930). *T. sogdiana* is distinguished by the peculiar shape of its flower and by the absence, in most cases, of hairs from the base of the filaments. The second group, *Bifloraeformes* (Taliev, 1930) is not mentioned as such in 'Flora of the USSR'; it consists of the species which Vvedenskiĭ arranged in the following order: *T. buhseana* Boiss., *T. turkestanica* Regel, *T. turcomanica* B. Fedtsch., *T. bifloriformis* Vved., *T. biflora* Pall., *T. polychroma* Stapf, *T. koktebelica* Junge, *T. callieri* Halacsy et Lev., *T. orithyioides* Vved., *T. dasystemonoides* Vved., *T. dasystemon* (Regel) Regel and *T. tarda* Stapf. The first five species form a natural group. It is now apparently natural to classify Caucasian and Crimean species together, but the last four species of this section are obviously interrelated and link sect. *Eriostemones* with *Orithyia.*

According to Vvedenskiĭ sect. *Orithyia* comprises three species: *T. heteropetala* Ledeb., *T. uniflora* (L.) Bess. ex Baker and *T. heterophylla* (Regel) Baker. Later Popov, mainly on the grounds of the peculiar shape of the seeds and the type of flower, separated *T. heterophylla* as an independent genus, *Eduardoregelia* M. Pop.

Vasilevskaya

Vasilevskaya (1936) divided *Tulipa* into nine groups according to the anatomical structure of the bulb tunics, but with the following reservation: 'It is natural to assume that the similarity of the species within these groups is due to their genetic affinity, but it would be methodologically wrong to establish phylogenetic connections, within and between these groups merely on the basis of anatomical characters. Her treatment of the Middle Asiatic tulip species resulted in the following groups:

I. *T. heterophylla* Baker and *T. tianschanica* Regel
II. *T. kuschkensis* B. Fedtsch., *T. tubergeniana* Hoog and *T. hoogiana* B. Fedtsch.
III. *T. lanata* Regel, *T. ingens* Hoog and *T. sp. nov.*
IV. *T. albertii* Regel, *T. carinata* Vved., *T. micheliana* Hoog, *T. kaufmanniana* Regel, *T. dubia* Vved., *T. greigii* Regel, *T. mogoltavica* M. Pop. et Vved. *T. schrenkii* Regel and *T. fosteriana* Hoog ex W. Irv. The last species occupies a special place in this group.
V. *T. korshinskyi* Vved., *T. ferganica* Vved., *T. hissarica* M. Pop. et Vved., *T. anisophylla* Vved. and *T. platystemon* Vved. Within this group *T. praestans* Hoog and *T. subpraestans* Vved. are treated as deviating species.
VI. *T. maximowiczii* Regel and *T. linifolia* Regel
VII. *T. turkestanica* Regel, *T. turcomanica* B.

Fedtsch., *T. dasystemon* (Regel) Regel, *T. dasystemonoides* Vved., *T. biflora* Pall., *T. bifloriformis* Vved., *T. tarda* Stapf, *T. buhseana* Boiss., *T. orithyioides* Vved., *T. paradasystemon* Vved., *T. binutans* Vved. and *T. orthopoda* Vved. *T. sogdiana* Bunge is placed on its own.

VIII. *T. kolpakowskiana* Regel, *T. zenaidae* Vved., *T. iliensis* Regel, *T. ostrowskiana* Regel, *T. tetraphylla* Regel, *T. nitida* Hoog, *T. korolkowii* Regel, *T. rosea* Vved., *T. brachystemon* Regel, *T. dykesiana* Vved., and *T. regelii* Krassn.

IX. *T. lehmanniana* Merckl., *T. behmiana* Regel and *T. borszczowii* Regel. In morphological structure *T. wilsoniana* Hoog differs markedly from these species.

Hall

Hall (1940) in his book on the genus Tulipa gives the following classification of the species known to him, placing the *Eriostemones* at the beginning of the genus:

Subgenus *Eriostemones* (Boiss. pro sect.)[4]

Section I — *Australes,* including the *T. australis* group consisting of: *T. australis, T. celsiana, T. grisebachiana, T. patens, T. primulina, T. urumiensis* and *T. sylvestris;* and the *T. hageri* group consisting of: *T. hageri, T. orphanidea* and *T. whittallii.*

Section II — *Saxatiles,* including the *T. humilis* group consisting of: *T. humilis, T. pulchella, T. violacea, T. aucheriana* and *T. lownei;* and the *T. saxatilis* group consisting of: *T. cretica, T. bakeri* and *T. saxatilis.*

Section III — *Biflores,* including *T. biflora, T. turkestanica, T. polychroma* and *T. tarda.*

In our opinion Hall's division of the *Eriostemones* into three sections is quite justified. One may, of course, argue with Taliev about which of the species – *T. australis, T. sylvestris* or maybe even *T. biebersteiniana* – should be regarded as primary in the first section, but the fact remains that this group is correctly subdivided into three clearly distinct sections, each containing more or less closely related species. These sections can be enlarged by the inclusion of species unknown to Hall. For instance, in 'Flora of the USSR' such species as *T. buhseana, T. turcomanica, T. bifloriformis* and others are included in the *Biflores.*

At the end of each chapter Hall mentions a group of species, some of which he qualifies as 'uncertain species', e.g. at the end of the Eriostemones: *T. androssowii, T. bessarabica, T. bifloriformis, T. buhseana, T. callieri, T. dasystemonoides, T. koktebelica, T. orithyioides, T. sogdiana* and *T. turcomanica.* He collects here both well-defined and doubtful species as well as synonyms.

Hall divides the *Leiostemones* into five subsections and discusses in a special chapter the 'solitary species' not included in any of them. In the first subsection, *Clusianae,* he treats *T. aitchisonii* as typical of the first group and then enumerates the species: *T. aitchisonii, T. aitchisonii* subsp. *cashmeriana, T. stellata, T. stellata* var. *chrysantha, T. chitralensis* and *T. clusiana.*

T. montana typifies the second group that includes *T. linifolia, T. maximowiczii, T. batalinii* and *T. montana.*

The second subsection *Gesnerianae* is also divided into two groups. The first is typified by *T. gesneriana* and includes *T. gesneriana, T. suaveolens, T. armena, T. galatica, T. hungarica, T. rhodopea* and *T. boeotica.* In the second group he enumerates 28 species of *Neotulipae.*

The third subsection of the *Leiostemones,* the *Oculus-solis* group, is again subdivided into two groups. The first is typified by *T. oculus-solis,* and in addition includes: *T. systola, T. julia, T. sharonensis, T. stapfii, T. cypria, T. undulatifolia, T. praecox, T. aleppensis* and *T. veneris.* He treats the second group under the heading 'Other species' and includes here: *T. kuschkensis, T. lanata, T. hoogiana, T. sovietica* and *T. amabilis.* This group contains species that other authors have placed in either *Leiostemones* or *Tulipanum.*

Hall calls the fourth subsection *Eichleres.* It consists of only one group comprising *T. eichleri, T. ingens, T. tubergeniana, T. fosteriana, T. micheliana, T. greigii, T. kaufmanniana* and *T. praestans.* In the 'Flora of the USSR' these species are distributed over the three sections *Leiostemones, Tulipanum* and *Spiranthera.* He placed *T. eichleri* next to *T. ingens* and *T. tubergeniana* between *T. ingens* and *T. fosteriana,* although it is not related to them. For some unknown reason *T. praestans* is placed next to *T. kaufmanniana.*

The *Kolpakowskianae,* as fifth subsection, include eight Middle Asian species: *T. kolpakowskiana, T. iliensis, T. altaica, T. kesselringii, T. corynestemon, T. ostrowskiana, T. korolkowii* and *T. lehmanniana.* Here too the species are arranged neither by relationship nor in alphabetical order. Vvedenskiĭ has long since established the synonymy of *T. kesselringii* and *T. tetraphylla. T. corynestemon* Hall is no doubt synonymous with the same species i.e. *T. tetraphylla.*

The 'solitary' species *T. schmidtii* Fomin. and *T. sprengeri* Baker are treated as a separate group. Furthermore he concludes from the literature that sect. *Orithyia* should comprise *T. uniflora, T. tianschanica, T. heteropetala* and *T. oxypetala.* He accepts a separate

4. Hall adopting Boissier's division into *Eriostemones* and *Leiostemones* raised the rank of these groups to subgenus. In the work cited the *Eriostemones* are divided into sections and the *Leiostemones* directly into subsections. (editor)

genus *Amana* for two species from China and Japan viz.: *T. edulis* and *T. graminifolia* and calls them *A. edulis* and *A. graminifolia.* At the end of the book there is a list of 83 species in alphabetical order, which the author has not seen in the living state.

Hall based his descriptions on cultivated plants of unknown or garden origin. For this reason, despite the author's thoroughness and conscientiousness, the descriptions are not always accurate. The division of species into groups does not reflect their phylogenetic connections or geographical distribution because many of the described species were from commercial sources. For instance, the species, which he treats under the name *T. sovietica*, is doubtlessly *T. hoogiana*, while the plants, which he described as *T. hoogiana*, represent one of the forms of *T. tubergeniana.* There was no reason for treating *T. amabilis* independently from *T. hoogiana* because Vvedenskiĭ has long since proved them to be conspecific.

Embryology

Highly interesting from the point of phylogenetics is the work of Romanov (1944), who drew attention to the wide variation of embryo sacs in the *Tulipeae.* No other genus of flowering plants has yet been found to be as rich in this respect as *Tulipa,* in which Romanov found five types of embryo sacs which he designated as Fritillaria-, Druza-, Adoxa-, Tulipa tetraphylla- and Eriostemones-type.

The Fritillaria-type is most common and was found in eighteen species: *T. heterophylla* (Tien Shan) of sect. *Orithya; T. tubergeniana* (Pamir-Alay) and *T. hoogiana* (Kopet Dagh) of sect. Tulipanum; *T. ingens, T. praestans, T. maximowiczii* (Pamir-Alay), *T. albertii, T. greigii* (Tien Shan), *T. micheliana* (Pamir-Alay), *T. borszczowii* (Syrdar'ya region), *T. lehmanniana* (Kyzylkum), *T. zenaidae* (Kirghiz Alatau), *T. ferganica* (Ferganskiy Mountains), *T. hissarica* (Pamir-Alay), and *T. gesneriana* and *T. praecox* (Mediterranean) of sect. *Leiostemones.*

The Druza-type embryo sac occurs in four species of sect. *Leiostemones: T. rosea* (Pamir-Alay), *T. korolkowii* (Syrdar'ya region), *T. wilsoniana* (montane Turkmenistan) and *T. chrysantha* (Iranian region).

Romanov found embryo sacs of Adoxa- and T. tetraphylla-types in the following *Leiostemones: T. ostrowskiana, T. kolpakowskiana* and *T. tetraphylla* from the northern and central Tien Shan. The Eriostemones-type is characteristic for the section bearing that name and is in its uniformity clearly different from the embryo sacs of species in the other sections. Even in *T. sogdiana,* which often lacks the typical Eriostemones feature (pubescent filaments) but is similar to these species in habit, Romanov found the embryo sac to be of the Eriostemones-type, as first described by Guignard (1900) for *T. sylvestris* and *T. celsiana.* Romanov also observed this type in *T. bifloriformis, T. tarda, T. sogdiana* and *T. dasystemon.* In 1939 Hall gave the list of species studied by Newton viz.: *T. australis, T. clusiana, T. dasystemon, T. hageri, T. humilis, T. pulchella, T. sylvestris* and *T. turkestanica.* In total ten species of *Eriostemones* were studied. In type of embryo sac *Eriostemones* species differ from those of other sections, although there is some slight similarity with *T. tetraphylla.* It should be mentioned here that in karyotype *T. tetraphylla* shows a closer affinity with the *Eriostemones* than with the *Leiostemones* (figs.11,17).

The Fritillaria-type embryo sac is most frequent in the other sections of the genus. Thus it follows that as far as the embryo sac is concerned the *Eriostemones* are more advanced than the species of other sections. Moreover, according to Romanov, the Fritillaria-type is most widely spread in those Liliaceae genera which come close to *Tulipa* such as *Lilium, Fritillaria* etc. Without being primitive for flowering plants in general, the Fritillaria-type is in this case to be regarded as the primary type in this group of genera.

Phylogeny

Buxbaum (1937) gave a scheme for the phylogenetic relations between the genera of the *Tulipeae.* According to this scheme the genus *Tulipa* branched off rather late from the main stem of the *Tulipeae,* which means that it is one of the youngest genera that have evolved in the flowering plants. Romanov refers to this scheme in his work (1944). Buxbaum regards the genus *Iphigenia* as the prototype of the entire subfamily. The genus *Gagea* with the haploid chromosome number, n = 12, was the first to branch off. From the *Platyospermum* branch of this genus separated at an early stage the ancestor of the genus *Lloydia,* which is also the progenitor of our contemporary *Lloydia* and *Girardiella* (*Lloydia serotina* has n = 12). The common ancestor of *Fritillaria* and *Lilium* (n = 12) was the first to branch off from the prototype of the genus *Lloydia;* the common ancestor of *Erythronium* and *Tulipa* (n = 12) followed later.

We will also consider Taliev's views on the evolution of the Liliaceae family in the light of Buxbaum's phylogenetic scheme. It is known that representatives of the genera *Gagea* and *Lloydia* never have large flowers and in no phylogenetic scheme known to us does the genus *Tulipa* derive as a direct descendant from the genus *Lilium.* For these reasons, Taliev's statement (1930) that *Tulipa* is *Lilium's* 'deprived relation' is quite unjustified. These genera developed side by side and quite independently and it was doubtless a progressive and not, as Taliev claimed, a regressive evolution. The size of the flowers increased more in the

Species known to occur in the USSR

Sect. *Tulipanum*

T. tubergeniana Hoog.
T. kuschkensis B. Fedtsch.
T. hoogiana B. Fedtsch.
T. julia K. Koch.
T. subquinquefolia Vved.*

Sect. *Leiostemones*

T. sosnowskyi Akhv. et Mirz.*
T. florenskyi Woron
T. kaghyzmanica Fom.*
T. lanata Regel
T. ingens Hoog
T. fosteriana Hoog ex W. Irving
T. carinata Vved.
T. affinis Z. Botsch.*
T. praestans Hoog.
T. subpraestans Vved.
T. butkovii Z. Botsch.*
T. albertii Regel
T. greigii Regel
T. mogoltavica M. Pop. et Vved.
T. vvedenskyi Z. Botsch.*
T. micheliana Hoog
T. eichleri Regel
T. karabachensis A. Grossh.*
T. mucronata Fom.*
T. schmidtii Fom.
T. maximowiczii Regel
T. linifolia Regel
T. wilsoniana Hoog
T. lehmanniana Merckl.
T. behmiana Regel
T. borszczowii Regel
T. korolkowii Regel
T. nitida Hoog
T. rosea Vved.
T. schrenkii Regel
T. ostrowskiana Regel
T. kolpakowskiana Regel
T. zenaidae Vved.
T. brachystemon Regel
T. platystemon Vved.
T. altaica Pall. ex Spreng.
T. iliensis Regel
T. ferganica Vved.
T. anisophylla Vved.
T. korshinskyi Vved.
T. tianschanica Regel
T. tetraphylla Regel
T. hissarica M. Pop. et Vved.
T. krauseana Regel

Sect. *Spiranthera*

T. kaufmanniana Regel
T. dubia Vved.
T. anadroma Z. Botsch.*
T. tschimganica Z. Botsch.*

Sect. *Lophophyllon*

T. regelii Krassn.

Sect. *Eriostemones*

T. violacea Boiss. et Buhse
T. caucasica Lipsky
T. lipskyi A. Grossh.*
T. sylvestris L.
T. biebersteiniana Schult.
T. ophiophylla Klok. et Zoz.*
T. scythica Klok. et Zoz.*
T. patens Agardh. ex Schult.
T. sogdiana Bunge
T. buhseana Boiss.
T. binutans Vved.*
T. talievii Vved.*
T. alaica Vved.*
T. turkestanica Regel
T. turcomanica B. Fedtsch.
T. bifloriformis Vved.
T. orthopoda Vved.*
T. biflora Pall.
T. polychroma Stapf
T. koktebelica Junge
T. callieri Halacsy et Lev.
T. orithyioides Vved.
T. dasystemonoides Vved.
T. dasystemon (Regel) Regel
T. neustruevae Pobed.*
T. tarda Stapf

Sect. *Orithyia*

T. heteropetala Ledeb.
T. uniflora (L.) Bess. ex Baker
T. heterophylla (Regel) Baker

Table 1

Distribution of species in the geographical regions of the USSR and adjacent countries, as given in the 'Flora of the USSR'

Geographical region	Number of species in sections							
	Tulipanum	*Leiostemones*	*Spiranthera*	*Lophophyllon*	*Eriostemones*	*Orithyia*	Total	%
Middle Asia	4	36	4	1	16	2	63	75.90
Caucasus	1	8	—	—	5	—	14	16.87
European part	—	1	—	—	8	—	9	10.84
West Siberia	—	2	—	—	3	2	7	8.43
East Siberia	—	—	—	—	—	1	1	1.21
Far East	—	—	—	—	—	—	—	—
Arctic region	—	—	—	—	—	—	—	—
Dzhungaria-Kashgaria	—	5	—	—	3	1	9	10.84
Iran and Afghanistan	1	4	—	—	4	—	9	10.84
Balkan Peninsula and Asia Minor	—	—	—	—	1	—	1	1.21
Lesser Armenia and Kurdistan	—	1	—	—	—	—	1	1.21
Mongolia	—	—	—	—	—	1	1	1.21
Western Europe	—	—	—	—	1	—	1	1.21

genus *Lilium,* but many contemporary tulips also possess large flowers, a feature which is observed in the youngest species of the genus now passing through their formative stages. For example, the group of recent species around *T. fosteriana,* growing in the Pamir-Alay (*T. affinis, T. victoris* etc). Further proof is furnished by the fact that the tulips of sect. *Orithyia,* which retained the ancient character of the family (the style) and under Buxbaum's scheme continue in this feature the ancestral line, have comparatively small flowers. Buxbaum believed that the small-flowered *Eriostemones* were so primitive that the *Leiostemones* could be derived from them.

In our opinion the sections *Orithyia* and *Eriostemones* should not be regarded as primitive. The only primitive feature of *Orithyia* is the ovary tapering into a style. The areas of both sections are actually lying beyond the area of the genus along its eastern border and their biology and karyology have not been studied. All this does not permit a conclusion as to their primitive stage of development. A branch of the *Eriostemones,* the *Bifloriformes* which converge in Middle Asia, contains among its many species a large number of polyploids scattered throughout the area of distribution. These facts emphasize the secondary nature of both the habitats and the species themselves.

Species and distribution

Since the publication of the fourth volume of the 'Flora of the USSR' (1935) nineteen new tulip species have been found, which were described by various authors:

Of sect. *Tulipanum: T. subquinquefolia* Vved.

Of sect. *Leiostemones: T. affinis* Z. Botsch., *T. butkovii* Z. Botsch., *T. kaghyzmanica* Fom., *T. karabachensis* A. Grossh., *T. mucronata* Fom., *T. sosnowskyi* Akhv. et Mirz. and *T. vvedenskyi* Z. Botsch.

Of sect. *Spiranthera: T. anadroma* Z. Botsch. and *T. tschimganica* Z. Botsch.

Of sect. *Eriostemones: T. alaica* Vved., *T. binutans* Vved., *T. lipskyi* A. Grossh., *T. neustruevae* Pob., *T. ophiophylla* Klok. et Zoz., *T. orthopoda* Vved., *T. paradasystemon* Vved., *T. scythica* Klok. et Zoz., and *T. talievii* Vved.

The 83 tulip species known to occur in the USSR are listed on p. 8. The 18 species marked with an asterisk are not cited in the 'Flora of the USSR', 4, 1935. Of the 83 species growing in the Soviet Union, 63 (75,9%) have a Middle Asian distribution. Four of them belong to *Tulipanum,* thirty six to *Leiostemones,* four to *Spiranthera,* one to *Lophophyllon,* sixteen to *Eriostemones* and two to the *Orithyia* section.

Fourteen species (16,9%) are found in the Caucasus: one *Tulipanum,* eight *Leiostemones* and five *Eriostemones;* together representing three sections of the genus.

The European part of the USSR follows with nine species (10,8%): one *Leiostemones* and eight *Eriostemones;* representing two sections. The relatively large number of *Eriostemones* in this case is explained by the fact that Klokov and others have divided one species, *T. biebersteiniana,* into three minor species.

West Siberia has seven species (8,4%): two *Leiostemones,* three *Eriostemones,* two *Orithyia.*

Only one species (1,2%) grows in East Siberia. It belongs to sect. *Orithyia.*

It appears that of all the geographical regions, into which the territory of the USSR is divided for convenience, Middle Asia is the richest as regards both the sections of the genus *Tulipa* and the number of species occurring there (table 1).

Below is a list of the species growing in the USSR and spreading to adjacent countries.

In all 22 of 83 species growing in the USSR (26,5%) spread beyond the boundaries of the country. There are, however, in the neighbouring countries numerous other species that do not occur in the USSR.

Of the 63 Middle Asian species 48 (76,2%) are endemic. Their distribution over the various zones and mountain belts is given in table 2. Two species which are not endemic inhabit the desert zone or the montane desert belt of the Caucasus.

List of the species growing in the USSR
and spreading to adjacent countries

Sect. *Tulipanum*

T. julia K. Koch

Sect. *Leiostemones*

T. florenskyi Woron.
T. micheliana Hoog
T. eichleri Regel
T. schrenkii Regel
T. lehmanniana Merckl.
T. altaica Pall. ex Spreng.
T. kolpakowskiana Regel
T. iliensis Regel
T. tetraphylla Regel
T. tianschanica Regel

Sect. *Eriostemones*

T. biebersteiniana Schult.
T. turcomanica B. Fedtsch.
T. patens Agardh. ex Schult.
T. biflora Pall.
T. buhseana Boiss.
T. sylvestris L.
T. polychroma Stapf
T. violacea Boiss. et Buhse
T. dasystemon (Regel) Regel

Sect. *Orithyia*

T. uniflora (L.) Bess. ex Baker
T. heterophylla (Regel) Baker

Table 2

Distribution of endemic tulip species in mountain belts and zones

Mountain belt or zone	Number of endemic species in each section						Total of		Endemic species, %	
	Tulipanum	*Leiostemones*	*Spiranthera*	*Lophophyllon*	*Eriostemones*	*Orithyia*	Endemic species	species per belt or zone	per belt or zone	of total number endemics in the region
Middle Asia										
Belt: upper	—	3	1	—	3	—	7	10	70.00	9.59
middle	4	17	3	—	7	—	31	34	91.18	42.47
lower	4	11	1	—	3	—	19	25	76.00	26.03
Zone: desert and semidesert	4	8	—	1	3	—	16	25	60.00	21.92
Caucasus										
Belt: upper	—	—	—	—	2	—	2	2	100.00	40.0*
middle	—	3	—	—	—	—	3	9	33.33	60.0
lower	—	1	—	—	—	—	1	3	33.33	20.0
Zone: desert or mountain desert	—	—	—	—	—	—	—	2	—	—
European part of the USSR										
Belt: upper	—	—	—	—	—	—	—	—	—	—
middle	—	—	—	—	—	—	—	—	—	—
lower	—	—	—	—	2	—	2	2	100.00	50.0
Zone: desert or mountain desert	—	—	—	—	—	—	—	—	—	—
steppe and meadow	—	—	—	—	2	—	2	7	28.58	50.0

* The Caucasus has 14 species, 5 of which are endemic.

CHAPTER TWO

General morphology of the tulip

The fully developed plant is composed of a bulb with roots and a flowering shoot, consisting of a stem, leaves and one or more flowers.

1. *The bulb*

The bulb, as described by Serebryakov (1952), is a specialized organ for vegetative reproduction serving at the same time as the plant's foodstore. The disc-shaped stem of the bulb is called the basal plate. The reserve material is stored in specialized fleshy leaf-bases or 'fleshy scales', varying in number from one in juvenile to six in adult plants. On the outside the bulb is covered by another specialized scale, the tunic, which in origin, life-span and morphology differs fundamentally from the fleshy scales. Dead scales of previous years clothe the tulip bulb, often lying at a considerable depth, and protect it from unfavourable external conditions. The apical bud of the adult bulb develops into the flowering shoot and the upper fleshy scale carries in its axil the bud for the main daughter bulb, which will replace the mother bulb the next year. At least one bud is initiated in the axil of each fleshy scale and the daughter bulbs or 'offset' bulbs which may develop from them are the means of vegetative reproduction. In many species they are usually suppressed when growing under natural conditions, but in cultivation they often lead to clonal propagation of certain forms of the wild species. The 'offsets' are carried a fair distance from the mother bulbs by the stolons. The initiation of stolons takes place during summer dormancy and is not directly connected with the depth at which the bulbs are seated.

The ever varying shape, size and colour of the bulbs are useful characters for separating species and groups of species. For instance mature bulbs of sections *Eriostemones* and *Leiostemones* can always be distinguished by their size, the number of scales and by the tunics. On the other hand, bulbs of *T. kaufmanniana* (sect. *Spiranthera*) as a rule cannot be distinguished from bulbs of *T. greigii* (sect. *Leiostemones*).

2. The stolon

The stolon is formed as a special outgrowth on the basal plate of the seedling. The long narrow part is spherical in cross-section and smooth; the wider lower part tubular, much thicker than the roots, hollow, and glabrous on the outside. The bud, which has been carried from the bulb to the bottom of the stolon cavity, subsequently develops into a new bulb. Similar stoloniferous outgrowths will be formed during the following years of the bulb's life, often till the first flowering. In many species the daughter bulbs will also form stolons. They can vary in length, e.g. in *T. fosteriana,* to nearly 50 cm long.

Stolon formation and subsequent dropping of the bulb ensure vegetative reproduction by removing the daughter bulb from the mother plant while maintaining the flow of nutrients between the two. Along the centre of the stolon runs an air cavity from the basal plate of the mother bulb to the bud, which is filled with stiff, erect, filamentous projections of the single-layered epidermis. The greater part of the stolon's outer surface is also covered by a single-layered, in this case glabrous, epidermis. It is only the lower, thickened part of the stolon (the new bulb) that is covered by two epidermal cell layers. On completion of the extension growth the two-layered part of the epidermis forms papillae, nipple-like cells with thick membranes. The papillae on the stolons of the bulbs were first noticed by Irmish as early as 1850 and later by Yanishevskiĭ in 1934. Their function is to anchor the bulb in the ground during shrinkage of the stolon and to protect it from becoming crushed by the soil. The thickness of the stolon wall varies. The thicker part, nearest to the basal plate of the bulb, has several vascular bundles which are arranged in a circle with the xylem turned inwards. However, at the other, thinner, end of the stolons there is only one row of vascular bundles. Like those in the leaf, they are collateral, with the xylem turned inwards.

The processes of protracted fusion of the sheath with the axial part of the stolon are remarkable in their morphogenesis and positive geotropic growth which carries the bud several centimetres downwards from the place of its initiation. In the first stages of growth the part of the basal plate next to the convex half of the bulb begins to develop laterally while the base of the attached half of the leaf sheath is growing vigorously. This proves that it is the node and not the internode that takes part in stolon formation. (Botschantzeva, 1956).

Certain morphological features of the stolon are useful characters to distinguish species. For instance the thin filamentous stolons, which easily disintegrate when withering, and the globose bulbs of juvenile and even adult plants of *T. praestans* are strikingly similar to the stolons and bulbs of *T. fosteriana*. The difference is that in the case of *T. praestans* the apical and axillary buds (or only the axillary buds) are carried away from their place of initiation by thin, filamentous stolons, leaving only dead scales, whereas in the case of *T. fosteriana* the daughter bulbs are often found in the same place as the old bulb. The main daughter bulb and all the other daughter bulbs of certain adult plants of *T. praestans* always retain the juvenile ability to migrate a little sideways and downwards by stolons, a phenomenon rarely observed in other species. It has also been found in *T. kaufmanniana* and in the variety 'King George V', one of the Darwin tulips, but their stolons, which develop into main daughter bulbs, differ morphologically from those of *T. praestans*. In *T. fosteriana* the main daughter bulb of an adult plant does not drop but remains where it is, inside the scales of the old bulb. In such a clump stolons develop from the apical buds of the other daughter bulbs, which often do not flower for many years but continue to reproduce vegetatively.

Whereas the stolons and bulbs of *T. fosteriana* and *T. praestans* are morphologically similar, the stolons and bulbs of *T. praestans* differ radically from those of *T. subpraestans,* although these two species are very much alike in flower structure and other morphological characters. The stolons of *T. subpraestans* are much thicker and stronger and grow vertically downwards instead of sideways as in *T. praestans*. *T. subpraestans* may often form a clump of bulbs which remain at the same level. Also the adult bulbs as a rule do not drop and are not globose, but radish-shaped.

The long, thick and soft stolon of *T. ingens* is somewhat constricted towards the ends and gradually develops into a long, narrow bulb which places this species close to *T. tubergenianum* and *T. kuschkensis* of sect. *Tulipanum.*

3. The roots

Only the seedling in its first year has a main root, which dies off towards the end of the first growing season and does not branch. During that stage of its life the plant has no other roots. As a rule, the roots of tulips of any age have no root hairs. The root epidermis does not form them and we only sporadically observed such hairs developing on the primary root of *T. ingens, T. maximowiczii, T. wilsoniana* and *T. dubia* in the first year of their life (Botschantzeva, 1956).

In the following years the whole root system consists of annually renewed adventituous roots, which are simultaneously initiated on the hoof-shaped lower part of the basal plate and on the side of the future main daughter bulb. As the roots cannot be regenerated, transplantation after the bulb has rooted will cause serious damage, because the plant can then only survive on the water and food reserves of the bulb.

Under Tashkent conditions the roots of tulip bulbs begin to grow and function with the advent of autumn. A fully developed root is a highly ephemeral organ and has a primary structure consisting of an epidermis, primary cortex and central cylinder. The primary cortex consits of a weakly developed exoderm, parenchymatous tissue of loosely arranged, thin-walled, large cells and a single-layered endoderm. The central cylinder with comparatively smaller cells than the primary cortex has a single-layered pericycle on its periphery. In the centre of the root is a vascular bundle with radially arranged xylem and phloem. It is the tip of the root that grows and if the growing point is torn off growth stops. Like the main root the adventitious roots cannot branch or be replaced.

4. The flowering shoot

The flowering shoot develops from the apical bud of the bulb. The development of an adult plant continues from the axillary bud of the innermost fleshy scale, i.e. sympodially. The monocarpic shoot which for many years, till first flowering, develops monopodially is the basic structural element of a tulip plant. This shoot consists of alternating long and short internodes.

5. The stem

The stem of the tulip appears in three main forms: the basal plate, the stem as a component part of the stolon and the stem of the flowering shoot with the peduncle. In a seedling, during its first year, the stem takes only two forms. The first, the basal plate, as a shortened and broadened modification of the stem, is characteristic of the bulb. The second form is that portion of the stem which takes part in the initiation, growth and development of the stolon. The third form, the stem of the flowering shoot, appears only after the tulip has passed through all the stages of its development and enters the adult phase. It consists of the section of the stem between the lowest leaf and the basal plate of the bulb, a centre part, bearing the assimilatory leaves, and an upper part, the peduncle, as a direct continuation of the main axis which is crowned by a flower.

The stem of many *Eriostemones* species branches above the upper leaf and the individual pedicels of the semi-umbellate inflorescence are each crowned by flowers of the same dimensions as those of uniflorous plants. The flowering shoot may also develop branches from the axils of all stem leaves. In some species in their natural environment and also in many plants, cultivated in Tashkent from large bulbs, we have observed the formation of first, second and third order branches on the stem of the reproductive shoot (Botschantzeva, 1953).

The stem is the first organ of the flowering shoot to be initiated in the bulb in the form of a meristematic dome. For a long time it grows laterally. In the initial stages this stem grows very much in the same way as the basal plate. It initiates more and more new parts of the shoot and consequently hardly grows in length. Thus the leaves on the reproductive shoot are initiated from below upwards on a shortened axis terminated by a flower. The internodes of the stem begin to grow only after the initiation of all organs, including the carpels of the flower, has been completed. In many species the stem of the flowering shoot, despite its changing functions (initiation of the leaves, shoot and flower), grows from September till March like the basal plate of the bulb. Only the nodes are functioning while there is hardly any growth of the internodes. We observed an interaction between the organs initiated on the shortened stem – the leaves of the shoot and the flower – and growth of the internodes. The removal of the green flower bud early in spring stops the growth of the internodes and the peduncle (Botschantzeva, 1952). The internode between the apex of the bulb and the lowest stem leaf is the longest. It carries the apical bud of the tulip to the surface of the ground. The intermediate internodes of the stem may differ in length from one species to another, but they are always longer than the internodes of the basal plate. The uppermost internode of the reproductive shoot, between the upper leaf and the flower, is of the same length as the lowest internode.

The stem of the flowering shoot has a primary structure. It consists of an epidermis, primary cortex and central cylinder. The epidermis is single-layered, covered with a cuticle, often with a waxy film and, in some species, with simple hairs. Directly below it is the primary cortex which has a collenchymatous ring consisting of one row of cells. The rather loose parenchymatous cells of the primary cortex have cellulose walls and contain chlorophyll grains. The peripheral cells of the central cylinder in the above-ground part of the stem are highly sclerified and form a rather strong ring of strengthening tissue separating the central cylinder from the primary cortex. The cells of the basic parenchyma vary in size and are larger in the centre next to the vascular bundles. Collateral vascular bundles are scattered in the central cylinder. On the periphery they are surrounded by strengthening tissue. Towards the centre their shape gradually changes and sometimes bicollateral vascular bundles may occur.

6. The basal plate

The basal plate (the base of the bulb) is a flattened disc-shaped stem. It has a strongly marked polarity. On its top develop the tunic, then the fleshy scales with buds in their axils and finally a flowering shoot in the centre. In the lower half of the basal plate at the side of the main daughter bulb lies a semi-circle of fine-celled tissue, the root meristem, which forms the adventitious root system. A large number of fibro-vascular bundles converge in the small basal plate and pass from the leaves and the stem of the flowering shoot into the bulb scales, the stolons and the roots. Migushova (1954) established that the collateral bundles of the flowering above-ground stem or leaves of juvenile plants on passing into the basal plate of the parent bulb are rearranged in concentric bundles of the amphivasal type with an internal phloem and are surrounded by a large-celled parenchyma. Nevertheless, on passing into the zone of root formation the vascular bundles are rearranged again and the concentric vascular bundles are transformed into vascular bundles of the root type.

Under Tashkent conditions the initiation of roots in the lower zone of the basal plate begins in many species after completion of the growing period of the parent plant. It is in evidence as early as mid-June. By the end of summer dormancy the roots are reaching the external epidermis of the basal plate, slightly lifting it and causing a swelling. Given sufficient humidity during that period the bulb could take root. Under Tashkent conditions tulip bulbs can only be grafted in the period between the beginning of summer dormancy and the middle of June, because later the whole meristem of the basal plate will be used up for root formation and the basal plate will not be capable of any healing, callus formation and concrescence.

7. The leaves

Tulips have five types of leaves: a) the single cotyledon; b) the sheathing green leaf of juvenile plants; c) the sessile green stem leaves of the flowering shoot; d) the modified basal leaf which forms the bulb's tunic; e) the modified basal leaves forming the fleshy scales of the bulb. In the first year there is only one cotyledon. Subsequently, up to the flowering phase, there appears every spring one 'petiolar' sheathing leaf above the ground, and it is only when the plant enters the adult phase that from 2 to 5 sessile stem leaves appear. As an exception there may be fewer or more leaves (*T. regelii* has one leaf and *T. schmidtii* has 12 leaves).

The cotyledon consists of a sheath, the cotyledon connective and the haustorium. During germination the upper part of the cotyledon remains in the seed. It is colourless, flat, and functions as haustorium; it never becomes green and soon withers after the seed has germinated. 'Cotyledon connective' is the name given by Vasil'chenko to the central green, usually cylindrical, part of the cotyledon. It is nearly circular in outline and constricted towards the top and the base; its vascular bundles are arranged in a circle. Before fixation in the soil, the cotyledon gets thinner, because the air-ducts in the parenchymatous tissue disappear. Underground the cotyledon retains its circular shape, has no chlorophyll grains and can be described as a 'petiole'. In many species the epidermis on one side of the cotyledon is flattened (Botschantzeva, 1956). In the lower part of the connective this flattened epidermis forms a groove, which is the remnant of the reduced upper plain of the connective (the blade of the cotyledon). Nearer to ground surface the groove in the cotyledon tissue becomes deeper and forms a closed sheath. The entrance to the sheath is often shielded by a small ligule. The upper portion of the sheath is for a large part covered with stiff, erect hairs which also shield the entrance. The bud of the seedling is hidden at the base of the cotyledonal sheath. This basal part which grows strongly dries up after completion of its growth and becomes the tunic of the first year's bulb.

The leaves of juvenile plants are arranged in biserial rosettes. The median vein of the first green leaf lies opposite the cotyledon because the first leaf produced after the cotyledon is a fleshy scale, which is a modified leaf with an undeveloped blade, serving as a reserve store and not as an assimilating organ. It is initiated in the first year's bulb almost simultaneously with the cotyledon, but withers when the development of the first green sheathing leaf of the second growing season is almost complete. From the second growing season up to flowering the seedling produces only one green leaf annually. Just above ground its leaf blade is constricted into a petiole which sheathes the secondary bulb.

The blade of the green leaf of a juvenile plant closely resembles the sessile leaves of adult plants. From the second year of the plant's life it gradually takes the shape which is typical of the basal leaf of the species or variety. In *T. regelii* one or two ridges appear on the upper surface of the leaf; in *T. fosteriana* a horny red or yellow edge; in *T. greigii* anthocyanin spots on the upper surface of the leaf; in *T. micheliana* stripes; in *T. maximowiczii* the long, narrow leaves develop red margins; in *T. vvedenskyi* the leaves become pubescent. It is therefore possible to identify many species by leaf characters already in the second year after sowing. In the following years these morphological characters will become more pronounced.

As in the case of the cotyledon a continuous, usually pubescent, channel runs along the modified petiole of the green leaf sheath of the juvenile bulb. The pubescence is particularly noticeable in the expanded lower part of the sheath serving as the tunic of the secondary bulb.

The structure of the tunic and its pubescence are important characters in the classification of tulips. For example, in many species the vascular bundles and cell walls of the sheath become lignified and the remnants of the tunics, accumulated during many years and not decomposed, reach up to soil surface. Tunics of this type may be found in *T. borszczowii, T. lehmanniana, T. behmiana* and some forms of *T. greigii.* Other species, however, have delicate, short tunics, e.g. *T. schrenkii* or coarse but not elongated tunics, e.g. *T. ostrowskiana, T. kolpakowskiana.* Of *T. korolkowii* and *T. greigii* there are plants both with short and elongated tunics. The type of indumentum of the tunics, being an important character in classification, led Vvedenskiĭ to exclude from section *Leiostemones* all species of which the tunics are densely lined with long, soft, strongly rippled hairs, like *T. tubergeniana, T. kuschkensis, T. hoogiana, T. julia* and *T. subquinquefolia.* He included them in section *Tulipanum.* The tunics of *T. ingens* on the other hand are lined with long, coarse, matted hairs; those of *T. wilsoniana* with thin, silky, long hairs, massed at the top of the bulb.

In most cases, when a plant reaches the adult phase, the sheathing leaf stops developing its blade and only the sheath, i.e. the tunic, continues to develop, but its consistence, indumentum and colour remain the same as in juvenile plants.

The colour of the tunic is also a very useful character. For example fresh tunics of *T. bifloriformis* are bright pink while those of *T. ostrowskiana* and *T. kolpakowskiana* are almost black.

When the adult phase is reached assimilation is carried on by the green sessile leaves of the flowering shoot. Their shape, particularly of the lowest leaf, very much resembles the blade of the sheathing leaves of juvenile plants. The higher up the stem the smaller the leaves become, their width diminishing more quickly than their length.

As a rule tulip leaves do not show much veining, but those of *T. praestans, T. carinata* and *T. anadroma* have thick midribs. A network of dark green cross-veins can be seen on the light green leaf blade of *T. kaufmanniana.* The leaves of the flowering shoot die off before the capsule is ripe.

The fleshy scales of the bulb are modified organs

which, unlike the tunic, never carry green blades. Their function is to accumulate food and moisture reserves in their thick parenchymatous tissues. In the first year the bulbs of most species have only one fleshy scale. The appearance of a second fleshy scale is a sign that these species will flower the following spring. Unlike that of the tunic the inner surface of fleshy scales is never pubescent, and when dried up they invariably become thin and papery. Adult bulbs may have up to six fleshy scales.

8. The flower

The flower of the tulip is cyclic, which means that all its parts are arranged in alternate whorls. This pentacyclic, trimerous flower consists of two whorls of tepals, two whorls of stamens and one whorl of carpels. The stamens may vary in length and their number equals that of the tepals. The brightly coloured, choripetalous perianth forms a double cup. The flower is actinomorphous, in most cases large, solitary and carried by a strong, usually erect, peduncle. In only a few species and varieties the flower buds and faded flowers droop for a short time, but before colouring of the bud the peduncle quickly straightens up as does the peduncle of the pollinated ovary later on. This was noticed in some species of sect. *Eriostemones* and in *T. tetraphylla* of sect. *Leiostemones* and also in hybrids from crosses with cultivated tulips of 'Duc van Tol Yellow' and 'Proserpine'. In other species, e.g. *T. ostrowskiana* and *T. sylvestris* and in cultivars of the lily-flowered group the large and heavy flower slightly bends the thin peduncle, lending the flowers a particular elegance.

Wild species show much variation in flower shape, which may be characteristic of certain species and even groups. In the case of *T. greigii* for instance, the inner tepals form a cylinder, slightly expanding towards the top, while the upper halves of the outer tepals reflex. The flower of *T. micheliana* has a similar shape. In the flower of *T. albertii* the inner tepals are hooded. The star-shaped flower of *T. maximowiczii* is very typical and resembles somewhat that of *T. linifolia.* Species of sect. *Tulipanum,* although being more closely related to those grouped around *T. fosteriana,* resemble in flower-shape those near *T. greigii,* etc.

9. The tepals

In structure, shape, and colouring the outer tepals differ somewhat from the inner tepals. For example, the outer tepals of *T. greigii* and *T. micheliana* are rounded to rhombic, while the inner ones are obtriangular with apiculate and pubescent tips. The outer tepals are much thicker than the inner ones, particularly at the base.

As regards colour intensity and pattern, the outer tepals differ noticeably from the inner ones. The former are on the outside, along the midrib, brightly coloured or covered with a glaucous, sometimes velvety bloom and often with a blotch at the base. The upper part of the flower is externally the same for both whorls and only the vein of the midrib and the blotch at the base of the inner tepals are of a different colour.

The inside of the outer and inner tepals are in the upper part often of the same colour, but the blotches at their bases always differ in shape, colour and margins. The hairy bases of the tepals are typical of sect. *Eriostemones.* When describing flowers of this section it is therefore necessary to indicate their shape as well as the shape, colouring and pubescence of the two whorls of tepals on both their inner and outer surfaces.

10. The stamens

Traditionally the presence or absence of pubescence has formed the basis of the division into two main sections. Species of section *Eriostemones* have pubescent filaments whereas those of section *Leiostemones* are glabrous. Later, also other characters were used to distinguish subsections in the genus. In 1935 Vvedenskiĭ used the mode of the anthers' dehiscence to separate *T. kaufmanniana* from sect. *Leiostemones* and placed it in the separate sect. *Spiranthera.* 'In *T. kaufmanniana* the anthers open slowly and gradually from apex to base and become contorted' (Vvedenskiĭ, 1935). The anthers of some new species we described from the Tien Shan also open gradually but without any specific contortion, e.g. *T. anadroma, T. tschimganica,* and *T. dubia.* The gradually dehiscing anthers of some typical forms of *T. kaufmanniana* do not always wind into a spiral either. Therefore, while maintaining the division of the genus into the sections recognized by Vvedenskiĭ, it was necessary to transfer *T. dubia* from sect. *Leiostemones* to sect. *Spiranthera* and to include two new species – *T. anadroma* and *T. tschimganica.*

In 1882 Regel described a tulip from the Dzungarskiy Alatau and named it *T. brachystemon,* i.e. with short stamens, and in 1935 Vvedenskiĭ describing a new species from the Pamir-Alay named it *T. platystemon,* i.e. with broad stamens. Finally, describing in 1940 a tulip received from us, Hall drew attention to the shape of the filaments of *T. tetraphylla* Regel and gave it the new name of *T. corynestemon.* The shape of the filaments varies widely from one species to another. In *T. micheliana* and *T. lanata* they are broadly triangular; in *T. borszczowii* like thin pegs; in *T. tetraphylla* club-shaped with a swollen top; in *T. platystemon* and *T. zenaidae* they are broader in the middle and narrowing towards both ends; in *T. greigii* very thick; and in *T. maximowiczii* quite thin.

The filaments of *Eriostemones* tulips are broadening to a quarter from the base, where they possess a small boss, clothed with short hairs and taper again towards the anthers. In cases where this pilosity is but weakly developed, the broadening is not very conspicuous.

The colour of the filaments is rather variable and may be typical of the species. As a rule the lower half of the filaments of *T. korolkowii* is black while the upper half is red; in *T. zenaidae* (the forms with a black blotch at the base of the tepals) the lower part of the filaments is black and the upper part orange; light-coloured forms of *T. kaufmanniana* never have dark filaments, whereas the orange filaments of *T. dubia* often have brown tips.

The four-celled anthers of all species are very similar in shape but differ widely in size and colour. Some species have a connective protruding beyond the anther. Sometimes it is stained and different in colour from the other parts of the stamen. For example, *T. buhseana* has yellow stamens but the apex of the anthers and the elongated tip of the connective is black. Certain species have rather large anthers while in others they are smaller and thinner. The mode of dehiscence is a good diagnostic character. The anthers of *Spiranthera* species dehisce gradually. The walls of their cells shrink transversely and the connective often winds spirally. Sometimes the anthers wind into a spiral longitudinally, in their upper part perpendicularly to the axis. The anthers of the species in this section are very narrow in comparison with their length and this main differentiating character is particularly noticeable in dried herbarium specimens.

11. The gynoecium

The superior ovary consists of three connate carpels and is trilocular. In each cell two rows of ovules are inserted on the placentae, formed by the incurved edges of the carpels.

Important features in the description of tulips are the size of the ovary in relation to the length of stamens and tepals, the shape of the ovary and the presence or absence of a style. For example, sect. *Orithiya* was excluded from *Tulipa* because the species of that section have long styles.

Important characters of ovaries with short styles or nearly sessile stigmas are their colour, the way in which the carpels merge into the style, and the shape of the stigma, e.g. the ovary of *T. ingens* has a wine-colour with a bluish tint; that of *T. fosteriana* is green and of *T. zenaidae* bluish, while the style and stigma are yellow.

The papillae on the stigma are frequently of a different colour from the seedless parts of the carpels (the style) just below the stigma. The stigma of *T. korolkowii* is yellow while the upper part of the style is red and the ribs of the ovary often have bright anthocyanin stripes.

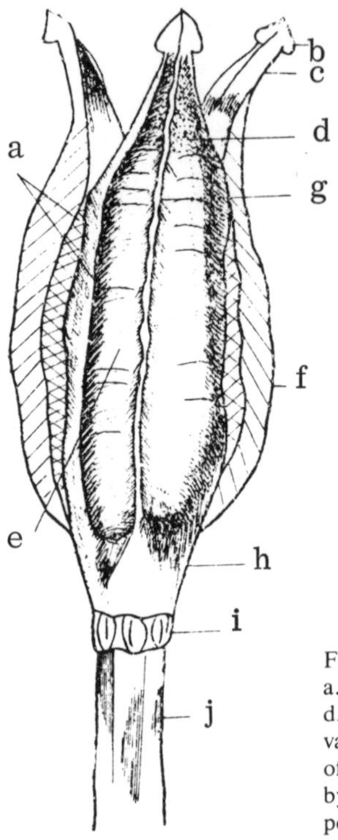

Figure 1. Tulip capsule
a. valves; b. stigma; c. style; d. upper seedless part of the valve; e. wing of the valve; f. ribs of the capsule; g. suture formed by fusion of the carpels; h. carpophore; i. tepal scar; j. pedicel.

12. The fruit

The fruit of the tulip is a trilocular capsule formed by the syncarpous and superior ovary (fig.1). In each locule two horizontal layers of flat seeds are attached to the incurved edges of the carpels. The pericarp is well developed and has characteristic lignified cells around the locular cavity along with some special tissue facilitating dehiscence. The capsule dehisces by longitudinal slits along the midrib of each carpel, forming typical fringes along the line of dehiscence. In many species the stigma grows rather large and remains attached to the 'style', the drawn-out upper part of the valves, which is seedless and varies in shape and size. In most species fusion of the carpels at the base of the capsule has resulted in a typically large and seedless carpophore. A large tepal scar separates the capsule from its pedicel.

Specific characters are length, diameter and shape of the capsule; size, shape and colour of the carpophore; the presence of a 'style', its size, shape and colour; the shape, colour and position of the stigma. A study of the valves of the capsule reveals a wealth of useful characters. Their shape, the suture formed by fusion of the carpels, its depth, texture of its surface, veining and colour are striking characters. The wings of the valves bulge on either side of the suture, which forms a median groove along the valves. They are as a rule convex but may also be flat, coarse or papery, transparent or opaque, wrinkled or smooth, often, especially in the upper part, transversely striped. More or less deep grooves run along the wings of the capsule close to its ribs; brown stripes running across the valves into these grooves often form an intricate lacy pattern. The outer edge of the valve along the rib is either brightly coloured or of the same colour as the rest of the valve.

When a capsule dehisces its inner septa also split along the strands of the pollen-conveying tissue. The shape of the capsule below the stigma, the length of the slits in the septa and the degree to which parts of the style and stigma reflex are typical of the species. Other distinguishing features are the tepal scars and the colour, thickness and relative length of the peduncle.

CHAPTER THREE

Morphological descriptions of wild tulips of the Soviet Union

In this book we have followed in principle the division into sections and the arrangement of the species within the sections of the genus Tulipa, as recognised by V.I. Vvedenskiĭ in the "Flora of the USSR", Vol. IV (1935).

SECTION *TULIPANUM* REB.

Vvedenskiĭ gives the following description of this section in volume IV of the "Flora of the USSR": 'Bulb tunics profusely arachnoid-shaggy over the whole inner surface. Flowers large, brightly coloured; filaments glabrous; ovary with a sessile stigma.'

Five species of this section grow in the USSR: *Tulipa tubergeniana* Hoog, *T. kuschkensis* B. Fedtsch., *T. hoogiana* B. Fedtsch., *T. julia* K. Koch and *T. subquinquefolia* Vved.. Four species are found in Middle Asia and one, *T. julia,* in the Caucasus.

Tulipa tubergeniana Hoog
plate 1, fig.2a

T. tubergeniana inhabits 'clayey slopes in mottled outcrop areas. Middle Asia: Pamir-Alay (SW). Endemic' (Flora of the USSR, Vol. IV).

The material studied originated from Uzbekistan, Denauskiĭ district, environs of Sina, mottled rocks on the slopes of a ravine near Derbent; Tadzhikistan, environs of Tutkaul on loess and red clay. On the white soils it was growing under *Artemisia* bushes.

Bulb profusely clothed with shaggy scales, seated at a depth of 10-15 cm, with the old tunic removed 3-5 cm in diameter, ovoid; with the tunic globose. Tunic grayish brown or brownish (very old tunics whitish grey), papery, on the inside thickly lined with gossamer hairs. Single hairs very long, thin, silky and strongly rippled.

Stem above ground together with the peduncle 5-30 cm long, covered with short, erect hairs; upper part of the peduncle reddish pink.

Leaves 3(4), falcate, reflexed, glaucous, more or less closely set, resembling a rosette. Only the upper leaf slightly higher up the stem. Forms from Derbent with glabrous leaves, red along the edges or with cartilaginous margins, ciliate, more or less downy, undulate, overtopping the flowers; basal leaf lanceolate, 18 × 2.8 cm; the upper linear 13.7 × 1 cm; successive leaves up the stem increasingly folded.

Flower red, solitary, scentless, resembling in shape *T. greigii,* but the outer tepals upright with the top part reflexed.

Outer tepals 2.5-10 cm long by 1.7-7 cm wide, rhombic with a long pointed, pubescent tip and slightly undulate margins; on the outside red-orange with a more pronounced whitish tint than on the inside; at the claw glaucous, with an elliptic, purple blotch, where the inner blotch shines through, becoming greenish towards its base (sometimes the outer blotch is green); on the inside red-orange with a large, elliptic, black blotch, 5.6 × 3 cm, slightly pointed at the top and with a well-defined whitish yellow margin (there may occur forms where this margin is hardly noticeable).

Inner tepals obovate or suboblong, 9-2.2 cm long, 6.3-1.5 cm wide, with an abruptly acute, pubescent tip; on the outside red-orange becoming somewhat more intensely orange towards the claw; near the claw with a rhombic, green or violet-claret blotch with its acute corner turned upwards and extending to one third the length of the tepal; on the inside red-orange, more glossy than outside and near the claw with an elliptic, black or black-claret blotch with an acuminate tip; this blotch, occupying the entire width of the tepal by its lower third part and extending to 3/5 the length of the tepal, is margined by a light yellow band, narrower than the one on the outer tepals.

Stamens 2.5-3 times shorter than the tepals, slightly longer than the ovary. Filaments thin, tapering, somewhat flattened, glabrous, black, a little lighter at both ends, three times shorter than the anthers. Anthers broad, dark purple. Pollen dirty yellow or brownish purple.

Ovary green with a light violet tinge, shaped like a short straight peg. Sitgma light green, sessile, broader than the ovary and with yellow papillae.

Capsule oblong, straw-coloured with brownish lines

and stripes, up to 4.6 cm long and up to 2.5 cm wide, obtuse at both ends; small plants bearing a nearly spherical capsule with a very short (0.7 cm) and broad, carpophore; in large capsules the carpophore wedged with a sharp prong into the groove of the valves; in small capsules a carpophore is absent. Style short, pronounced in the smaller capsules (e.g. a 2.2 cm long capsule with a 0.7 cm long style and a 7.7 cm long capsule with a 1 cm long style). Stigma brownish grey, reflexing, fragile.

The capsule opens by narrow, up to 0.2 cm wide slits, with almost parallel edges, reaching down to the carpophore and forming a 2 cm wide funnel where the upper, constricted parts of the valves reflex. Small capsules with a short, wide and wedge-shaped slit. The septa split to about one third of their length, i.e. to below the upper narrowed part of the valves. Below the stigma the slit is dull white on the inside, widens to the end of the thicker part of the valves, then narrows again, becomes straw-coloured at the curvature of the septum and retains the same width to the end of the slit.

Valves oblong with parallel edges, shortly pointed apex and obtuse base. Small capsules with ovate or narrowly ovate valves. Valves with in the middle a channeled suture running from the stigma to the prong of the carpophore. Suture brownish, somewhat darker towards the top. In many individuals an embossed vein runs along the suture and often bifurcates in the upper, narrow part of the valve.

Wings (convex parts of the valves) in smaller capsules more convex than in the larger ones; their solidity increasing with the size of the capsule: almost papery and semi-transparent in the smaller and thicker capsules and nearly opaque in the larger ones. Surface transversely wrinkled; the wrinkles finer in the upper part of the valves. Clear brown, thin veins in the grooves between the wrinkles connect the brown suture in the middle of the valve with the grooves along the ribs. On small capsules the wrinkles and lines are not so pronounced and are in part replaced by a mere roughness of the surface. Ribs along the line of dehiscence brown and separated from the wings by rather deep grooves in which the transverse, brown veins converge. The reflexed edges of the split ribs up to 0.5 cm wide and of a light straw-yellow colour. The upper, constricted part of the valve much darker than the lower part. Median groove below the stigma at the beginning of the widening valve depressed, forming a small pit.

Tepal scar narrow, whitish brown, on both sides bordered by a bright brown band and narrowing towards the peduncle. Small capsules with a whitish brown scar.

Peduncle up to 25 cm long, straw-coloured with a whitish tinge and sometimes also a brown sheen.

Seed light brown, flat, broadly ovate, 0.9 × 0.8 cm, with a somewhat lighter narrow margin. In the upper part of the endosperm, on its slanting side, a dark green crescent. Embryo straight, lying in the centre of the seed and clearly visible from both sides.

Under Tashkent conditions this species flowers around mid-April.

Tulipa kuschkensis B. Fedtsch.

plate 2, fig.2b

T. kuschkensis is found 'on clayey or rarely on stony slopes. Middle Asia: Montane Turkmenistan (Kushkinskiĭ River). Endemic' (Flora of the USSR, Vol. IV).

The plants studied came from the Kushkinskiĭ district of Turkmenia, near the village of Morgunovskiĭ (clayey slopes) and the Ashkhabad Botanical Garden.

Bulb ovoid with a grayish brown, papery tunic. Tunic on the inside densely lined with long, thin, strongly rippled, silky hairs.

Stem, including the peduncle, from 15-45 cm high (the tallest plants in this section), glabrous, green, glaucous. Our collections did not contain pubescent individuals, although the "Flora of the USSR" mentions that there may occur plants with in particular pubescent peduncles.

Leaves (3)4(5), widely spaced, the upper half falcately recurved with a narrow top, crispate, glabrous, glaucous, with ciliate margins, not exceeding the flower; the upper leaf, 18.5 × 1.65 cm, reaches somewhat beyond the flower in the early flowering stage; basal leaf lanceolate, 25 × 9 cm, high up the stem, 10 cm from ground level.

Flower solitary, red, scentless, beautifully shaped like that of *T. greigii* but differing in having the outer tepals folded along the midrib. When closed, both bud and flower conical with a broad, flat base.

Outer tepals 4.5 to 9 cm long by 2.25 to 5 cm broad, subrhombic or elliptic; midrib of the segments terminating in an acute, pubescent tip; claw near the basal blotch very fleshy; on the outside orange-red with a light raspberry-red tinge, on the inside glossy orange-red with a well-marked large, elliptic, black blotch (6 cm long tepals with a blotch of 2.5 × 1.2 cm), notched at the top and with a narrow, pale yellow margin.

Inner tepals obtriangular with the midrib terminating in an apiculate, pubescent tip; claw stout and incurved; median vein of the tepal deeply channeled; on the outside glossy orange-red, becoming more reddish towards the margins; on the claw with a sharply pointed, subrhombic yellow-green blotch (1.8 × 1.2 cm), with purple-violet streaks; on the inside glossy orange-red and a large (2.2 × 1.2 cm), truncated, oblong, black blotch with a yellow margin and somewhat constricted, notched apex. The inner tepals of the plants grown from bulbs which were brought from Kushka were a little broader than the outer ones, although according to the "Flora of the USSR" it should be the other way round.

Filaments short and stout, black, triangular at the base, (0.5 × 0.25 cm), gradually tapering towards the top. Anthers black or black-violet, when dry and dehisced 1.2 × 0.37 cm, more than twice as long as the filaments. Pollen violet-brown.

Ovary overtopping the stamens on the second day of flowering; formed like a pale green peg with a broadly overlapping stigma of the same colour.

Capsule straw-coloured with a light pinkish brown pattern, up to 4.8 cm long and 2.2 cm wide, oblong with an obtuse base and top. Carpophore short: at the ribs only 0.1 cm, at the place of insertion into the central groove 0.4 cm high.

Valves elliptic with an only slightly projecting carpophore and stigma. The deep groove running along the valve light brown all over or only in the upper half, but always darker than the wings. All along its length the valve has a narrow, crimson or brown band which becomes diffuse near the groove.

Wings convex, thin, papery, slightly wrinkled. Wrinkles often beginning already close to the base of the valves, but in some cases practically absent. The thin, transverse wrinkles run from the edge of the valve across the curved groove on to the ribs. Sometimes they may be coloured brown only near the edge of the valve with light brown veins extending from the edge to the median groove.

The capsule opens by wedge-like slits with very broad (0.6 cm), outward curving, felty edges of a dull yellow-orange colour. The slits extend to the base of the capsule and form a wide-mouthed funnel owing to the deeply split septa along the pollen-conducting tissue. Starting from the stigma as a broad band of the same width as the stigma itself, this tissue shrinks abruptly at 0.7 cm below the top of the stigma and reaches down to the curved part of the septum; the slit extending another 2 cm below the curvature.

Upper part of the valve attenuated, of a darker colour than the rest of the capsule; its central groove flattened and continuing like a vein or an embossment which bifurcates below (0.4 cm) the stigma and extends to its edges. Below the stigma the valve is abruptly constricted into a small broad style, 0.3 cm long, including the stigma. Stigmatic surface overlapping the style, towards the edges of a darker colour, but in general forming one single part with the valve.

Tepal scar brown, narrow (0.1 cm), but broader than the very long and slender peduncle, which is darker than the capsule.

Seed flat and triangular with rounded angles (0.8 × 0.55 cm), light brown with a 0.1 cm wide, lighter shaded margin, covered with minute wrinkles. Embryo clearly visible from both sides and of a lighter colour than the seed.

Under Tashkent conditions this species begins to flower around mid-April.

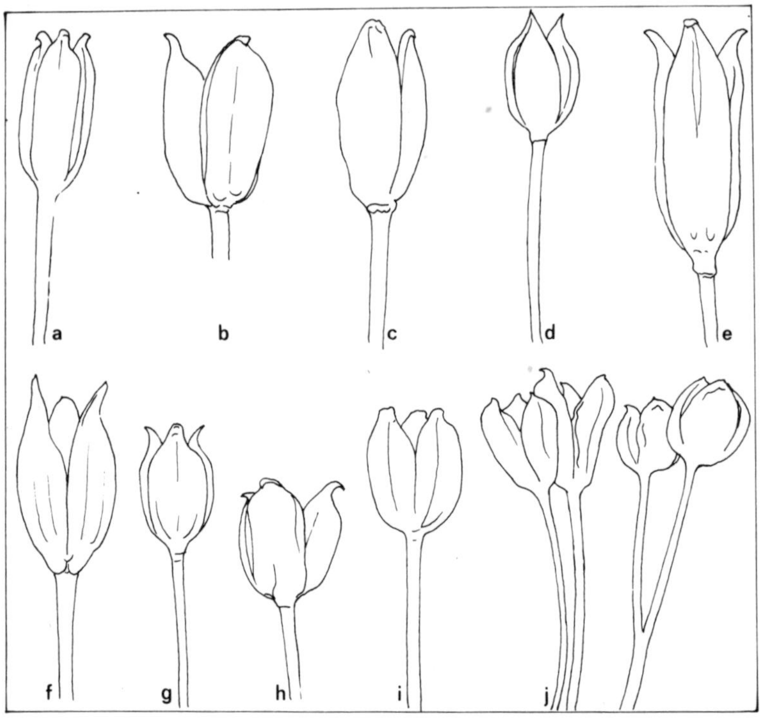

Figure 2. Mature tulip capsules
a. *T. tubergeniana* Hoog; b. *T. kuschkensis* B. Fedtsch.; c. *T. hoogiana* B. Fedtsch.; d. *T. julia* K. Koch; e. *T. ingens* Hoog; f. *T. fosteriana* Hoog ex Irv.; g. *T. carinata* Vved.; h. *T. affinis* Z. Botsch.; i. *T. praestans* Hoog; j. *T. subpraestans* Vved.

Tulipa hoogiana B. Fedtsch.

plate 3, fig.2c

T. hoogiana inhabits 'clayey or rarely gravelly slopes from the foothills to the mountain steppe zone. Middle Asia: Montane Turkmenistan, central and western Kopet-Dag. Endemic' (Flora of the USSR, Vol. IV).

We studied plants from Turkmenia, the Kopet-Dag range, environs of the Prokhladnoye settlement; from the Arvaz (eastern slope) and also plants obtained from the Ashkhabad Botanical Garden in 1950. We grew this species more than once from seed.

Bulb ovoid, up to 5 cm in diameter, coated with a thin, papery, dark brown tunic, densely lined with long, thin, silky, strongly rippled hairs.

Stem and pedicel glabrous, very glaucous, frequently with an intense anthocyanin pigmentation (in particular in the darker forms) resulting in a darkish claret-brown colour. Plants reaching to 40 cm above ground level.

Leaves 3-5, widely spaced, falcate, strongly reflexed, glaucous, glabrous but with cartilaginous and long-ciliate margins. We do not know of other species with similarly long cilia on the leaves, although there may be forms with nearly glabrous leaf margins. Some forms have almost linear shaped, folded, falcate or very undulate leaves. Leaves often reaching above the flower. Basal leaf lanceolate, at the base usually clasping the stem, 25.5 × 10 cm, sometimes tubular to a height of 3 cm. Quite often the basal leaf starts at about 10 cm from ground level giving these plants a leggy appearance. Upper leaf narrowly lanceolate, 20.5 × 2.5 cm.

Flower solitary, of a dark crimson or pinkish orange colour, beautifully shaped like a double cup with the outer tepals slightly reflexed and the inner ones upright, often with a slightly notched top. In the early stages with a rather strong, peculiar fragrance.

Outer tepals, 9.5 × 5.3 cm, gradually tapering into a pubescent tip, subrhombic or broadly elliptic; in the lower part rather abruptly narrowed into the claw; toothed along the edge of the curved part and often somewhat undulate. On the outside the overlying bloom resulting in a raspberry-red colour; the lighter forms with a silvery tint along the midrib and the raspberry-red of the edges often changed to light orange with red edges or a red shading filtering through the bluish bloom all over the tepal; the inner blotch may show through and become greenish towards its base, but more often it acquires a dark wine-colour along the centre of the fleshy part of the claw. On the inside glossy crimson often with an orange tinge filtering through. With pinkish orange and bright orange forms the inside colour is darker, which is apparently due to the absence of any glaucousness. Pinkish orange forms possess reddish streaks running all over the tepals. Near the claw on the inside an oblong (3.2 × 1.2 cm), obtuse or pointed blotch of widely varying colour from black to dark crimson and grayish green (in lighter forms). It may be of one colour throughout or possess a narrow light-green margin of varying width and intensity, frequently only in the form of lighter streaks surrounding the black blotch.

Inner tepals 9.2 × 5.3 cm, obovate or ovate, often somewhat spoon-shaped, with an apiculate, pubescent tip; on the outside and inside of the same colour, glossy dark crimson, orangy crimson, orangy pink or pure orange; the claw on the outside with a claret tinge spreading along the midrib, while the lighter forms become yellowish at the base; on the inside of the claw a narrowly elliptic or sharply triangular, margined blotch, 3.2 × 0.7 cm, of the same colour as the blotches on the outer tepals.

Stamens longer than the ovary and two, three or even more times shorter than the tepals. Filaments of the inner whorl the longest, 1-1.2 cm, widening towards the base, up to 0.4 cm, rather abruptly tapering off towards the anthers; the upper half or one third black or claret (in the lighter forms grayish green); in the lower half yellowish or reddish with gradual transitions from darker to lighter shades and quite light below the anthers. Anthers 1.8 × 0.4 cm, longer than the filaments; claret, pink, brownish or yellow, often coiled into spirals. Pollen grains brownish pink, brownish grey or yellow.

Ovary 2.2 × 0.6 cm, dark green with anthocyanin colouring spreading from the ribs over the carpels; of uniform thickness or just slightly constricted under the stigma. Stigmatic surface barely overlapping the constricted part of the ovary; the sides bright claret; papillae yellow.

Capsule 6.7 × 3.5 cm, pinkish or yellowish gray with a clear brownish purple pattern, more or less cylindric with rounded corners, obtuse at the base, acute at the apex. Carpophore short (0.6 cm), at the tepal scar 0.8 cm wide, near the valves without marked prongs, 1.5 cm wide. Upper part of the capsule with a small, sessile, reflexed but firmly fixed, brown stigma. The capsule splits right down to its base. The narrow, wedge-shaped slits with almost parallel edges form a 6 cm long funnel, 1.5 cm wide at the top and 0.2 cm wide near the curvature of the septa. Cleavage of the septa extending downwards to the end of the curvature. Inner surface beneath the stigma dull white with a ridge in the centre; in capsules with obtuse tops obconical, with acute tops subrhomboid in shape.

Valves narrowly elliptic or elliptic with an acuminate top. Central suture intensely brown with a purplish tint, forming a narrow, undeep groove in the lower 3/5 part of the valve but widening and almost flat in the upper part from where a raised vein in the groove runs down to the base of the suture.

Wings of the valves slightly convex in the lower half, more convex in the upper half, thick and coarse, lustreless, wrinkled only in the upper half or all over the surface. Transverse brownish purple veins running in the wrinkles form numerous anastomoses in the deep longitudinal grooves close to the curved edges of the valve. Edges of the ribs on the inside felt-like, 0.4 cm wide, straw-coloured with a pinkish tinge; on the outside with a dark brownish crimson band from which a purplish wash spreads to the very edge of the slits.

Tepal-scar narrow (0.1 cm), light brown, slightly raised, often with longitudinal crevices of a rhomboid shape.

Peduncle yellowish white with a pinkish tinge, sometimes brownish, short (up to 17 cm) and rather thick.

Seed broadly ovate, light brown, with a regular straight edge and near the slanting side a crescent of green tissue. Embryo small and short.

Tulipa julia K. Koch

plate 4, fig.2d

T. julia inhabits 'dry stony slopes. The Caucasus: Southern Transcaucasus. General distribution: Iranian Region' (Flora of the USSR, Vol. IV).

We have studied plants grown from bulbs received through Mr. Akhverdov, from Yerevan Botanical Garden. They were collected in 1953 in Armenia (Kitabskiy district near Yerevan); in the middle mountain belt on open southern slopes of marl sandstone covered by phrygana.[1] There were also plants grown from seed received from the same botanical garden.

Bulb 1.6-3.5 cm in diameter with a grey brown, papery or coarse (in the case of droppers) tunic, thickly lined with gossamer hairs, in particular at the apex.

Stem of plants grown in our conditions, rather short, 11.5-15 cm (on the first day of flowering), green with some brownish shades, glabrous.

Leaves 4, dark olive green, slightly glaucous, glabrous, channelled, crispate and usually reflexed, not reaching above the flower; leaf margins of a lighter colour, almost glabrous. Basal leaf, 13.5 × 3.5 cm, broadly lanceolate, set at ground level. Upper leaf, 11 × 1 cm, reflexed, narrowly lanceolate to linear, somewhat curled into a spiral, ascending.

Flower solitary with a faint pleasant scent at the onset of flowering, opening to a wide cup, dull crimson.

Outer tepals 5 × 3 cm, subrhombic with rounded angles and an apiculate, pubescent tip; on the outside crimson-claret with an orange-shaded base produced by the yellowish streaks radiating from the greenish yellow, 1.5 cm high, claw; inner blotch shining through on the claw as a diffuse greenish spot. On the inside reddish crimson; in some forms the outer tepals claret-orange with a brighter orange stripe along the midrib; at the base an elliptic, 1.1 × 0.7 cm, pointed or blunt, grayish black blotch with a broad yellow margin extending to the base of the tepal.

Inner tepals 4.4 × 2.8 cm, sub-obtriangular with rounded angles; top apiculate, notched; on both sides red-claret and more glossy than the outer tepals; on the outside on the claw a three-pointed, 1.4 cm high, greenish blotch, from which yellowish streaks radiate to the edges of the tepals; on the inside, against the bright yellow ground colour, a greenish black, obtriangular, blotch (1.1 × 0.8 cm) with rounded angles and a pointed top; margin of the blotch, produced by the ground colour of the tepal, yellow, up to 0.4 cm wide in the centre and broadening towards the edges by radiating streaks of increasing height.

Stamens as long as the ovary. Filaments 1.3 × 0.27 cm, long and tapering, black, at the base yellowish. Anthers oblong, 1 × 0.2 cm, black. Pollen brownish purple.

Ovary 2.1 × 0.4 cm, light green, constricted below the stigma, with a short style. The overlapping stigma and the part of the ovary just below it light yellow.

Capsule 4.8 × 2.3 cm, light straw-coloured, oblong, in some cases nearly ellipsoidal, at the base with a long (up to 1.4 cm) carpophore. The narrow carpophore sharply wedged into the valves while the ribs of the capsule continue as low ridges on the carpophore, disappearing half way down its length.

Valves elliptic, abruptly tapering at the top and drawn out into a 1 cm long style which is often darker than the rest of the capsule. On the top the remnants of the grayish-brown lobes of the small stigma divided into halves by the rather deep median groove are clearly noticeable. A thin brown vein running in the groove widens in the upper part of the wings and narrows again at the apex.

Wings of the valves convex, thin, papery, transparent, covered with light transverse wrinkles. Thin brown veins run in the transverse wrinkles of the upper half. Wings near the edges of the valves bordered by deep grooves formed by the reflexed ribs. The ribs all along the line of cleavage bordered by a brown stripe; edges of the ribs on the inside of a dull cream colour.

The capsule opens by narrow, wedge-shaped slits, terminating in a funnel formed by the reflexed parts of the style, while the septa split along the pollen-conducting tissue down to their curvature. Septa in the upper part more or less rhomboid, elongated towards the style and shortened at the curvature.

Tepal scar raised, very narrow, not wider than 0.05

1. Phrygana—a xerophytic formation with low shrubs and semi-shrubs, often consisting of thorn and cushion plants (Mediterranean region). (editor)

Plate 1. *T. tubergeniana* Hoog (75%)

Plate 2. *T. kuschkensis* B. Fedtsch. (70%)

Plate 3. *T. hoogiana* B. Fedtsch. (80%)

Plate 4. *T. julia* K. Koch

Plate 5. *T. lanata* Regel (75%)

Plate 6. *T. ingens* Hoog

Plate 7. *T. fosteriana* Hoog ex W. Irving (80%)

Plate 8. *T. carinata* Vved. (70%)

cm, light brown. Peduncle thin, of a lighter shade than the capsule.

Seed 0.7 × 0.65 cm, triangular with rounded angles, light brown, slightly oblique, with a narrow creased fringe. Embryo visible from both sides.

Under Tashkent conditions this species flowers in mid-April.

SECTION *LEIOSTEMONES* BOISS.

'Bulb tunics nearly glabrous on the inside, or hairy at the apex or all over, or with a thick woollen coating, or merely at the apex profusely arachnoid-shaggy; filaments glabrous; ovary with sessile stigma or with a rather long style but always shorter than the ovary' (Flora of the USSR, Vol. IV).

According to the 'Flora of the USSR', Vol. IV, 38 species of this section are found in the USSR, but we know of three more species not included there, which brings the total number of species classed as *Leiostemones* to 41.

Tulipa florenskyi Woron.

T. florenskyi inhabits 'dry stony slopes. The Caucasus: Southern Transcaucasia. General distribution: Iranian Region' (Flora of the USSR, Vol. IV). The description of this species is based on herbarium specimens preserved at the Botanical Institute of the USSR.

Bulb seated at a depth of up to 13 cm. Tunic elongated, but not beyond the height of the bulb itself, as a rule brown and thin, only when half-decayed grayish brown.

Leaves 3-4, strongly falcate, channelled, undulate, glaucous with cartilaginous margins; the upper leaf very narrow, erect.

Flower in shape resembling in our opinion somewhat that of *T. wilsoniana* Hoog.

Outer tepals obovate, rhombic or subrhombic, often tapering into a short, pubescent tip; basal blotch black, its shape, judging by herbarium material, oblong, 2.1 cm long on a tepal of 3.8 cm.

Inner tepals subobovate or subrhombic, in some cases with a short pubescent tip; basal blotch diffuse.

Filaments thin, slightly broadening towards the base, dark, but at the top and close to the base yellow. Anthers yellow or brown, when dry 1.5 times shorter than the filaments.

Ovary tapering to the stigma, reaching to 2/3 the height of the stamens. Stigma of a light colour, broadly overlapping the upper constricted part of the ovary.

Light yellow forms[2] exist at two localities in the area of distribution. These forms have darker bases of the tepals and also darker filaments and anthers; the filaments are very thin; the anthers broad, short, only half as long as the filaments or even less.

This species differs from *T. julia* in its bulb tunic which is rather densely lined with straight, coarse hairs; in the strongly falcate leaves (except the upper leaf), usually 4 in number, which are channelled more undulate and have cartilaginous edges. *T. julia* has broader leaves and apparently also differs in the shape of the basal blotch on the tepals.

Tulipa lanata Regel
plate 5

T. lanata is found 'on clayey, rarely stony, slopes in the lower mountain belt. Middle Asia: Pamir-Alay (Surkhan and Kafirnigan river basins). Endemic' (Flora of the USSR, Vol. IV).

We have studied plants collected by A.I. Vvedenskiĭ at Khadzha-Ipak, Denauskiĭ district in Uzbekistan; plants sent by T.I. Ryabova from the Dushanbe Botanical Garden and plants from the Shakhrinaus district, Solyamich canyon, eastern slope, altitude 1600 m.

Bulb up to 4 cm in diameter, ovoid. Tunic soft, thin leathery, brown, profusely covered with long, thin, woolly hairs.

Stem including the peduncle 13-30 cm, without any pigmentation and owing to the dense pubescence of a dull, pale green colour.

Leaves 4, in the first stages of flowering more or less closely set and overtopping the flower, later more widely spaced and not exceeding the flowers, reflexed, undulate and glaucous, in particular the upper leaf more or less pubescent; margins red, ciliate. Basal leaf, 16.5 × 5.5 cm, broadly lanceolate. Upper leaf, 18.2 × 2.2 cm, narrowly lanceolate.

Flower solitary, scentless, nicely shaped, resembling that of *T. greigii* Regel, on the outside bluish pink, on the inside bright red.

Outer tepals 6.5 × 3.2 cm, rhombic-subrhombic, tapering into a pubescent tip; pubescence on the outside extending from the tip to ca. halfway down the tepal, continuing along the edges as cilia, on the inside extending along the sides to ca. ¼ the height of the tepal; colour on the outside a silvery light pink becoming redder or more orange on the edges of the reflexed part; claw orangy at the top, greenish in the centre with light brown shadings also discernible in the other parts; blotch elliptic, shining through as a crimson or blackish shadow. Tepals on the inside bright scarlet, on the claw a large, elliptic or rhombic black blotch, 1.8 × 0.8 cm, with a well-defined pale yellow margin and an acute or blunt and notched top; pale yellow streaks radiate from both sides of the margin on the brightly coloured tepal.

2. The form described here has red flowers. (editor)

Inner tepals 6.7 × 3.1 cm, obovate or broadly elliptic, tapering into a large and broad, strongly pubescent tip; pubescence continuing as cilia along the edges of the reflexed part of the tepal; on the outside of the same colour as the outer tepals, silvery light pink, reddening towards the edges, on both sides of the median vein yellowish green; on the claw an elliptic, pointed blotch, 1.3 × 0.8 cm, of a greenish yellow, glaucescent colour; tepals on the inside scarlet red with an orange tint towards the tip and a more intense crimson near the lower half of the blotch; on the claw an elliptic, black blotch, 1.3 × 0.8 cm, with a very narrow, yellow margin and sharply pointed apex.

Stamens black, 1.8 cm long, 2.5-3 times shorter than the tepals. Filaments short, 0.4 × 0.25 cm, triangular, black, below the anthers white. Anthers 1.4 × 0.3 cm, dark claret. Pollen reddish purple.

Ovary 1.7 × 0.35 cm, cream-coloured, constricted below the cream-coloured stigma with wine-coloured papillae.

Capsule 4 × 1.6 cm, oblong, sitting on a short, 0.5 cm, carpophore which is 0.7 cm broad below the valves.

Valves straw-coloured, terminating in a 0.3 cm broad and 0.7 long style with from its centre a broad, brown strand of scar tissue running into the median groove of the valve. The style dark brown and the suture even darker; median groove brown with transverse wrinkles running across the wings.

Wings convex, the transverse wrinkles separated from the ribs by a deep groove. Ribs dark brown, cleavage surface dull white with whitish fringes entering the loculi of the capsule.

The part of the capsule from stigma to the curvature of the pollen-conducting tissue ellipsoid. Lobes of the stigma reflexed, forming a wide-mouthed funnel.

Peduncle of a lighter colour than the capsule; tepal scar broad.

Seed 0.8 × 0.8 cm, broadly triangular with rounded angles and a creased, narrow, margin; embryo straight, small, only half as long as the endosperm, showing through on both sides of the seed.

Under Tashkent conditions this species begins to flower in the fourth week of April.

Tulipa ingens Hoog
plate 6, fig.2e

T. ingens is found 'on clayey slopes, up to 2 500 m. Middle Asia: Pamir-Alay (Central Zeravshan, Kashkadar' ya river basin and south of Kugitang). Endemic' (Flora USSR, Vol IV).

We raised the plants from seed collected in the Botanical Garden. The original bulbs were brought from different locations in the area of distribution of the species.

Bulb elongated ovoid, in cultivated individuals up to 6 cm in diameter. Tunic blackish brown or brown, thin leathery and fragile with a typical pubescence consisting of sparse, rather long, adpressed hairs with their ends interwoven.

Plants at flowering 25 cm high. Peduncle pubescent, green with a wine-coloured shading in the upper part. Pubescence and degree of pigmentation vary widely in plants from the same locality and indeed from one seed number.

Leaves varying in number from 3-4 to 6, sitting rather low on the stem, in some cases almost forming a rosette; narrowly lanceolate, sometimes with slightly undulate margins, channelled, glaucous and sometimes densely pubescent.

Flower opening to a pointed star with a slightly depressed centre, resembling a typical lily flower, scentless. As a rule the flowers of *T.ingens* are trimerous, but sometimes also dimerous and tetramerous flowers occur.

Outer tepals from 7.6 to 15 cm long and from 2.8 to 6 cm broad (flowers from fully developed bulbs rarely have tepals smaller than 13 × 5 cm), broadly elliptic or subrhombic with rounded angles, tapering to a densely pubescent tip; the upper reflexed part often minutely dentate. On the outside along the midrib bluish crimson or pale yellow of varying intensity with a greenish shading becoming more pronounced towards the claw and changing gradually to scarlet-red at the margins; in some cases the glaucousness spreads to the margins of the reflexed part but the very edge of the tepal is invariably bright scarlet. On the inside pure scarlet, deepening in the centre of the tepal, but fading towards the top into a light crimson longitudinal stripe; on both sides, halfway across the blotch the red colour usually disappears and the ground colour of the mesophyll comes out as a whitish yellow on the lower margins of the tepal. The black blotch on the inside of the outer tepals 1.8 - 4.3 cm long and 1 cm broad, prolonged rhombic with smooth sides and a pointed or acute apex; it may also be ovate or obovate; in rare cases the blotch has a narrow, 1 mm wide, yellow margin.

Inner tepals somewhat shorter and wider than the outer ones, (7-)12(-14) cm long, (4.4-)5.7(-6.9) cm broad, narrowly obovate or rhombic with rounded angles, sometimes broadly ovate, always with an apiculate, pubescent tip and finely dentate reflexed part of the tepal. On the outside usually scarlet, in places crimson owing to an overlying bluish tint; basal blotch of a dull, yellow-green throughout. Forms exist with inner tepals orange-red on the outside; the deep orange along the midrib shading to purple towards the edges of the reflexed part. Claw on the outside often with a rhombic, purplish-brown blotch with a yellow margin, broadening towards the base. Tepals on the inside scarlet-red with purple shadings of varying intensity, deepening towards the centre above the blotch. On the claw always a pronounced black blotch, usually narrowly rhombic with an accuminate top; more rarely rounded or three-pointed with a slightly toothed margin; in some rare forms with a narrow yellow margin along the upper part of the blotch.

Stamens 2.5-4.5 cm long; in flowers that are just opening 1½ times longer than the ovary, but afterwards the ovary exceeding the dried and shrunken anthers. Filaments black, tapering towards the tip. Anthers black, brownish black or a dark wine-red. Connective in some cases slightly protruding beyond the anthers (0.3 cm). Pollen of various shades from nearly violet to dark brownish purple.

Ovary long (2.2-5.3 cm) and narrow (0.4-0.35 cm) green with purple or claret shadings deepening towards the stigma. Style green. Stigma overlapping, at the sides as a rule reddish-purple sometimes yellow; papillae pale yellow or yellow with red.

Capsule 11.5 × 2.3 cm, oblong and very narrow, pointed at both ends, straw-coloured. Carpophore 1.2 cm high, above the tepal scar 0.8 cm wide, at the base of the valves, where it meets the vein running in the central groove of the valve, 1.9 cm wide; ribs of the valves continuing on the carpophore as short ridges which flatten out and disappear halfway down. In capsules containing undeveloped seeds the lower part is very narrow, not broader than the carpophore. A true style is absent in this species, as the drawn-out, narrowly conical and seedless part merges gradually into the valves. Stigma small, light brown, easily broken off.

The capsule opens by long and narrow (0.1-0.2 cm) slits reaching down to the carpophore, while the valves curve outward to form a wide funnel (up to 3.5 cm). The septa split much deeper than their curvature which is in this species less pronounced than in other species. Their distended parts are between stigma and curvature thicker than below the curvature, in the upper part as broad as the stigma, lower down shaped like a more or less flat, gradually narrowing band with a median ridge and raised edges. In this part of the septa there is a strong resemblance between *T. ingens* and *T. fosteriana*.

Valves straw-coloured, narrowly lanceolate or oblong, parallel-sided, pointed at both ends. In the median groove often a distinct ridge running from the base to the narrow part of the valve. In other cases this ridge is not continuous, starts somewhat nearer to the carpophore and changes higher up into a raised vein, which bifurcates in the upper third of the capsule and extends to the base of the stigma as two parallel veins. Median groove and in particular the veins light brown, darkening beyond the bifurcation.

Wings thin and translucent, but firm in texture, rather convex, with only a few transverse wrinkles in the upper part. The whole surface covered with fine, light brown streaks (pale and barely visible near the base) anastomozing in the grooves near the ribs and forming an intricate polygon pattern. This pattern along with the transverse stripes and the brown-coloured ribs lends the top of the valve a much darker colour than the other parts. Edge of the ribs along the line of cleavage very narrow, 0.35 cm, pale yellow, of a felty texture and less reflexed than in *T.fosteriana.*

Tepal scar up to 0.2 cm high, brown. Peduncle up to 15 cm long, straw-coloured with a brownish tinge, rather thick at the tepal scar, abruptly thinning below.

Seed 0.8 × 0.3 cm, light brown, triangular with rounded angles, a little warped with a rather broad, slightly wrinkled margin, which broadens towards the top, in particular on the elongated side. Embryo straight, or comma-shaped, plainly visible from both sides.

Under Tashkent conditions this species flowers in the second week of April.

Plants of this species form rarely offsets, only very large bulbs do so. As far as we could observe *T.ingens* is clearly distinguished from other species and its essential characters only vary in the dimensions of certain parts. When comparing plants grown from seed with those brought as bulbs from wild habitats, we came to the conclusion that this species breeds true in cultivation and without artificial pollination does not hybridize with cultivars or wild species. Seedlings flowering for the first and the second time very much resemble the parent plant, the only difference being their somewhat smaller size in the first years of flowering. The capsules of *T.ingens* and *T.fosteriana* are very similar. The former are however narrower with a gradual transition of the valves into the carpophore and the seedless part below the stigma. The upper part of the vein in the median groove bifurcates. The wall of the valves near the tepal scar is much thinner and the peduncle is straw-coloured with a brownish tinge. In *T.fosteriana* the valves more or less abruptly broaden above the carpophore and below the stigma terminate in a blunt, style-like top. The vein in the median groove does not bifurcate and the pedicel is plain straw-coloured.

Tulipa fosteriana Hoog ex W. Irving

plate 7, fig.2f

T.fosteriana is found on 'fine-textured slopes and rock benches. Middle Asia: Pamir-Alay, Samarkand mountains. Endemic' (Flora of the USSR, Vol IV).

The samples studied were brought from the Pamir-Alay, (Samarkand mountains, the western end of the Zeravshanskiy range). In the prewar years and in 1949 they were collected more than once at Agalyk and Aman-Kutan.

Bulb 2-6 cm in diameter, ovoid. Dropper bulbs often perfectly spherical, sometimes flattened, turnip-shaped. Tunic blackish brown, leathery, on droppers often very tough but the skin of the stolon above the dropper fragile, decomposing towards the end of the growing season; on the inside densely lined with straight, coarse, golden hairs, in particular near the base and the top.

Stem 15-50 cm high, upper internodes successively longer. Stem and peduncle usually with a pinkish anthocyanin pigmentation, sometimes also pure green and covered with straight, erect hairs.

Leaves 3-5, rarely 6, widely spaced, declinate, slightly crispate, bluish or glossy green, on the upper side pubescent; often with red or cartilaginous whitish edges, ciliate; as a rule not exceeding the flower. Basal leaf set at ground level, oblong to broadly ovate, of widely varying width, but the length/width ratio always smaller than in the other leaves (from 7 to 30 cm long and from 3 to 16 cm wide). The second leaf often longer than the basal leaf, but much narrower (from 7 to 32 cm long and from 1 to 12 cm wide). The upper leaves much shorter and narrower (the third leaf 3-27 cm long and 0.4-10 cm wide).

Flower solitary (even in cultivation it is difficult to obtain multiflorous forms), glossy red, one of the largest and most beautiful among the wild species, widely variable, but in its double-cup shape typical of this species; only faintly fragrant.

Outer tepals 4.5-18 cm long and 2-8.5 cm broad, varying in shape from long ovate to narrowly rhombic and broadly lanceolate, terminating in a pubescent tip; pubescence often spreading over the upper half of the back of the tepals, the upper reflexed part finely dentate. On the outside bright red, often with an orange, bluish crimson or greenish shading of varying intensity along the midrib; on the claw a fan-shaped, greenish yellow blotch, in most cases single-pointed, less often convex and finely toothed. On the inside scarlet-red with varying orange tinges; in the centre a purplish shading deepening towards the claw. Blotch well-defined, fan-shaped, yellow margined, as a rule single pointed or convex with pointed or roundish edges; the width of the yellow margin varying from a 0.5 mm wide light halo to a broad distinct band. In most cases the blotch is deep black, but in some forms it is reduced to a brownish claret, in the darker parts streaked with yellow from the ground colour.

Inner tepals obovate, usually obtuse, with an apiculate and finely dentate top; on both sides bright red with the same shadings as on the outer tepals; on the outside of the claw a distinct yellow or yellow-green blotch, similar in shape to the three-pointed or notched inner blotch and sometimes partly covered by another narrow, triangular, anthocyanin purple, smooth or streaked blotch; on the inside of the claw a black or brown, usually three-pointed, sometimes two-pointed and more or less deeply notched, yellow-margined blotch (often also yellow throughout).

Stamens exceeding the ovary, with dehisced anthers 3 - 4 times shorter than the tepals. Filaments glabrous, triangular; in flowers with a dark centre black-claret with a yellow tip; in flowers with a yellow centre also yellow. Anthers black-violet, one and a half or two times longer than the filaments and splitting abruptly all along their length. Pollen purplish brown or yellow; abundant.

Ovary, long conical, constricted under the stigma, green. Stigma yellow, nearly sessile with large lobes overlapping the ovary.

Capsule straw-coloured, often with a light brown or brownish purple stripe along the median groove of the valves; usually 7.7 × 3 cm, in some cases longer up to 11 × 3 cm, very variable in shape, more or less oblong, but always with a pointed top which often resembles a style; at the base obtuse. Carpophore lighter in colour than the capsule and much narrower, 0.9 cm high and 0.9 cm wide at the tepal scar and 1.6 cm wide near the valves; ribs and grooves of the valves continuing on the carpophore as short ridges. Stigma small and broad, often with drooping and reflexed lobes.

The capsule splits right down to the base of the ribs, in most cases forming narrow wedge-shaped slits, their edges sometimes running parallel to the top of the seed-bearing part. The reflexed seedless part forms a wide-mouthed (up to 3 cm wide) funnel at the top. The septa split only to slightly below their curvature. The expanding pollen-conducting tissue below the stigma is narrowly lanceolate with three parallel ridges and ends at the curvature of the septum.

Valves broadly to narrowly lanceolate or broadly ovate with an acuminate top, straw-coloured, as a rule glossy, sometimes dull. All along their length an undeep median groove with a low-ridged vein in the centre (there are forms with no visible veins). The groove often widens in the upper third of the valve where its epidermis resembles scar-tissue. Colour of the groove brown or purplish brown of varying intensity and pattern, often light brown with a faintly coloured ridge. When the groove and ridge are dark purplish brown with a bluish shading; the edges of the groove are always dentate.

Wings thin but firm, transparent, convex, swollen in the lower half, wrinkled in the upper half. The whole capsule or only the upper third may be wrinkled. In the wrinkles thin transverse veins, often arched and with their ends slightly raised near the ribs. Ribs purple or light brown, bordered by a deep groove in which the transverse veines are joined by anastomoses. Reflexed edges of the split ribs up to 0.7 cm wide, straw-coloured changing to pale yellow at the edge; texture felty. Wings smoother below the stigma where the valves flatten out forming a more or less isosceles triangle over the seedless part of the capsule with smoothed out wrinkles and closely spaced veins.

Tepal scar up to 2 mm wide, raised, light brown often leaving a crenate pattern on the peduncle. Peduncle of the same colour as the capsule, sometimes slightly brownish in the upper part.

Seed 0.9 × 0.8 cm, irregularly triangular, somewhat creased and warped, light brown with a wide, bent margin. Embryo straight or comma-shaped, clearly visible.

In cultivation seedlings of this species take long to reach the adult phase. When juvenile individuals are planted they make abundantly offsets but only few bear flowers.

Under Tashkent conditions this species flowers in the first week of April.

In Tulipa the pubescent tips of the tepals are quite interesting in particular those of *T.fosteriana.* The top of the tepal is initiated and develops before the reflexed, lateral parts. The pubescence appears very early, already in the primordial stage and descends along the midrib on the outside of the tepal; with some plants for one third of its length. On the leaves however pubescence is in most cases only observed on the upper side i.e. the inside and the question arises why the tepals have it on the outside. The tepal as a modified leaf is comparable to the green leaf and the bulb tunic. The pubescence on the outside of the tepal is apparently a new development which has a protective function in the flower bud.

Tulipa carinata Vved.

plate 8, fig.2g

T.carinata is found 'on stony slopes in the middle and upper mountain belts. Middle Asia: Pamir-Alay (Chul'bair, Khodzha-Gurgur-Ata). Endemic' (Flora of the USSR, Vol. IV).

The plants we have studied were sent by V.P. Bochantsev and A.I. Vvedenskiĭ from the Denauskiy district, environs of the village of Sina and also raised from seeds brought by A.D. Pyataeva from the upper reaches of the Tupalang river.

Bulb ovoid, up to 4 cm in diameter; tunic black-brown, coriaceous, lined with woolly hairs; the hairiness varying considerably from one individual to another.

Stem up to 48 cm above ground level; the above-ground part, including the peduncle, pubescent.

Leaves 3-4, slightly declinate, the upper part erect, undulate, keeled, often twisted around the keel like a corkscrew. Basal leaf deflecting from the stem at 8 cm above ground level, 21 × 6.5 cm. broadly lanceolate. The following leaves widely spaced, bluish green, pubescent, with cartilaginous and ciliate, yellow or pinkish margins; towards the top successively narrower, but not shorter; upper leaf 28 × 3 cm. Leaves never exceeding the flower because of the rather long peduncle.

Flower red with a pink flush, solitary, scentless, resembling in shape that of *T.ingens* by the pointed tepals, but with a more pronounced arrangement of the tepals in two whorls. Inner tepals erect, the outer ones more or less reflexed. The very graceful flower keeps well after cutting.

Outer tepals 8.2 × 4.2 cm, obovate or broadly elliptic, tapering gradually to a downy tip; the pubescence extending down the midrib of the tepal; the outer tepals are frequently profusely pubescent on the outside near the top, the base and the edges of the reflexed part. Colour of the opening flower on the outside bluish brown with a crimson-pink flush subsequently becoming more red but retaining a light brownish tinge; towards the edges scarlet-red, on the claw greenish with sometimes the black part of the inner blotch shining through. On the inside scarlet-red, very similar to that of *T.fosteriana,* with at the base a small yellow blotch or in some cases a yellow-margined, black blotch. On the fan-shaped yellow blotch nearly always a black spot, 0.8 × 0.6 cm, which may be ovate with a notched or truncate top, broadly rhombic or subrhombic in shape; the yellow ground colour either clearly delimited or radiating on the red part of the tepal. The rather small blotch, in comparison with the size of the tepals, is typical of this species.

Inner tepals 8.8 × 4.1 cm, obovate to subrhombic, tapering to a long downy tip; on the outside pale or pinkish orange of varying intensity becoming red towards the edges of the outcurved part and on the claw a truncate or rhombic yellow blotch with often the lower part of the inner blotch shining through. On the inside bright orange-red, rather glossy (colour variants occur with deepening crimson and fading orange or the other way round); at the base of the claw, spreading over its entire width a smallish (up to 1.3 cm high), black rhombic spot, sometimes truncated and then obtriangular in shape, but usually with a pointed or rounded top. Forms with a yellow-margined blotch on the outer tepals also have a narrow margin around the blotch on the inner tepals, while the dark part of the blotch is reduced to brownish red streaks. The yellow blotch in this species is always clearly demarcated by its flat top.

Stamens 2.8 cm, somewhat longer than the ovary. Filaments 1 × 0.3 cm, gradually tapering to a more or less sharp point, black, at the base and below the anther yellow. Anthers 1.8 × 2.8 cm, black with a wine-red tinge; in rare cases yellow. Pollen light brownish, pinkish or yellow.

Ovary 2 × 0.4 cm, constricted below the stigma (to 0.2 cm), pale green, along the ribs often with cream and crimson shadings, or entirely pinkish blue. Stigma, at the sides pinkish blue, bright crimson with a purple tinge or cream-coloured, or even orange; papillae varying from cream to a dark wine-red.

Capsule 5.5 × 2.2 cm, carried by a tall peduncle, pale straw-coloured; shaped like a short cylinder, sometimes shortly and broadly obovoid with a long style; at the base abruptly narrowed into a rounded, light-coloured carpophore (0.7 × 0.6 cm); at the top first slightly broadening then above the wings narrowing into a 1.5 cm long style, which is at the base 1.7 cm and below the stigma 0.4 cm wide; this upper part of the capsule has a darker tinge, the median groove flattens out and is replaced by two parallel veins intersecting at the constricted part of the style and with a narrow, darker groove between them. The firmly attached grey-brown stigma lobes strongly reflexed.

Valves more or less elliptic with particularly in the upper part a darker coloured median groove.

Wings thin, semi-transparent, strongly convex, swollen in the upper parts, covered from top to bottom with a herringbone pattern of fine, light brown veins and with hardly any wrinkles; only the veins in the upper half of the wings running in undeep grooves; ribs pinkish brown.

The capsule dehisces along narrow, wedge-shaped slits, broadened noticeably at the top by the reflexed parts of the style. Septa strongly curved with in comparison to other species narrow strands of pollen-conducting tissue; they are split to about two thirds of their length, well below the curvature.

Tepal scar broad (0.2 cm), bright brown. Peduncle darker than the capsule, getting thinner below the scar (from 0.4 cm to 0.3 cm with a 30 cm long peduncle).

Seed large (1 × 0.9 cm), light brown, slightly warped with a wide undulate margin and a somewhat darker crescent in the upper long-sided corner of the endosperm. Embryo long, thin, visible from both sides.

Under Tashkent conditions this species begins to flower in the fourth week of April.

Tulipa affinis Z. Botsch.
plate 9, fig.2h

We obtained the samples studied from the Turkestanskiy range, the Uzungyr side ridge opposite the village of Tchurbek, where they were collected in the juniper zone at the upper boundary of a stony belt which runs as a wide bench all along the ridge. They were growing between large boulders in a fine-textured soil (melkozem). Endemic.

Bulb long ovoid, 2-4 cm in diameter, often with scarious, dark brown sheathes reaching halfway up the bulb; tunic sparsely lined with straight, long, adpressed, golden hairs tapering towards the base and the top. By its indumentum this species recalls *T.ingens.*

Stem including the peduncle up to 32 cm above ground level, green, in its upper part weakly pubescent.

Leaves 3, glaucous, pubescent on the upper surface, with cartilaginous, whitish edges, ciliate, a little undulate; basal leaf 21 × 6.2 cm, broadly lanceolate, set 2-3 cm above ground level; upper leaves 15 × 4.6 cm, lanceolate, more or less widely spaced.

Flower solitary, red, scentless, differing from all other known tulips in its rather unusual shape. Its basal part broad and rounded; the outer tepals with a slightly reflexed top and the edges of the curved part rolled outward; the inner tepals erect, slightly convex with a recurved top.

Outer tepals 7.5 × 3 cm, elliptic, gradually tapering to a reflexed, pubescent tip and with a finely-toothed margin. On the outside orange-red with a crimson tinge along the midrib; claw to a height of 1.3 cm green radiating on the ground colour; from the sides of the green blotch two more or less dense orangy green streaks run on both sides of the midrib and converge below the tip of the tepal. On the inside glossy, orange-red with a dark crimson tinge along the midrib which becomes deeper towards the blotch; on the claw a convex, black blotch (1.3 cm high) with a finely toothed top and a well-marked,bright yellow margin (0.15 cm wide).

Inner tepals 7.3 × 3.5 cm, subrhombic, more or less abruptly terminating in a rather large, pubescent tip; edges of the recurved part finely-toothed. On the outside glossy, orange-red; on the claw a large grayish violet blotch, 1.3 cm high, with a roundly scalloped top against a yellow-green fond, marking off the blotch by a ca. 0.3 cm wide margin. On the inside crimson-orange-red with a large black bloth, 1.6 cm high, as wide as the claw. This blotch has the same shape as the outer one, but it is flushed with red and has a narrow, 0.1-0.15 cm, bright yellow margin.

Stamens shorter than the ovary and 3-4 times shorter than the tepals. Filaments 0.7 cm long, 0.25 cm wide at the base, flattened, triangular, black, the constricted part below the anthers white. Anthers 1.3 × 0.25 cm, linear, black-purple. Pollen brownish violet.

Ovary 6.5 × 2.5 cm, long and narrow, a little broadening towards the base, of a deep anthocyanin purple colour, below the stigma lobes creamy green. Stigma large, its lobes projecting 4 mm beyond the ovary, cream-coloured.

Capsule on an average 6 × 2 cm, triangular in cross-section, oblong, at the top drawn out to a long, thin style (1.2 × 0.4 cm) with a rather large, slightly curved, light brown stigma. Carpophore rather long and narrow, 1.1 × 0.7 cm, lighter in colour than the capsule; inserted as a sharp, raised, wedge into the valves. On dehiscence the ribs split to the base of the valves from where they extend as short ridges halfway down the carpophore. As a rule the narrow, wedge-shaped splits terminate in a wide-mouthed funnel owing to the strongly reflexed upper third of the valves; the septa split down to the curvature, their pollen-conducting tissue running as a wide strand from the stigma down the narrow part of the valves to the loculi where it shrinks in width and merges gradually into the placenta.

Valves thin but strong with transparent wings divided by a rather deep median groove which in the upper part of the capsule flattens out into a group of dark, parallel veins which disappear near the true style. Upper third of the median groove brown, a little darker in the centre.

Wings creamy straw-coloured, convex, glabrous, with only at the top a few brown streaks or stripes. Ribs reflexed evenly cream-coloured, just a trifle brighter than the wings, with a narrow brown edge and separated from the wings by a narrow groove; upper part of the capsule completely brown owing to the merging colours of the median groove and the ribs. In particular the style evenly brown, but there are capsules with styles of the same colour.

Tepal scar of the same light colour as the peduncle, which is very long and has a slightly brownish tint, somewhat darker than the capsule.

Seed triangular with rounded angles, a little warped where the margin is widest. Embryo comma-shaped, rather long, clearly visible against the darker endosperm.

Under Tashkent conditions this species begins to flower in mid April.

Tulipa praestans Hoog

plate 10, fig.2i

T.praestans grows 'on stony and clayey slopes in the middle mountain belt. Middle-Asia: Pamir-Alay (southeastern part). Endemic' (Flora of the USSR, Vol. IV).

The plants studied were brought by V.P. Bochantsev in 1933 from the environs of Tutkaul and the Sebistan mountain in Tadzhikistan and have for many years been cultivated in the botanical garden. Other bulbs of this species were received in 1950 from Dushanbe Botanical Garden and in 1956 we lifted bulbs of *T.praestans* in the Varzob river valley and the Kandar area of the Gissarskiy mountain range. One individual was grown from seed received in 1950 from Lund in Sweden.

Bulb ovoid, 2 cm in diameter. Tunic thick, coriaceous, blackish brown; on the inside glossy, often at the top sparsely lined with thick, reddish brown, adpressed hairs and occasionally a few hairs at the base of the tunic as well.

Stem from 10 to 60 cm above ground level, green, often strongly pigmented with anthocyanin towards the top; the upper part, including the peduncle, covered with whitish, erect hairs.

Leaves 3 - 4, sometimes 5 or 6, even more in multiflorous individuals, depending on the number of secondary flowers; declinate, widely spaced, light grayish green; sometimes slightly undulate, often with red, ciliate edges; distinctly keeled and usually with a thick midrib; leaves of sterile plants semi-erect. Basal leaf set at ground level, oblong or lanceolate with a pointed or obtuse top.

Flower cupped, somewhat drawn out at the base, not opening to a large star like *T.subpraestans,* scentless.

Outer tepals broadly elliptic or subobovate, the top more or less acuminate or nearly obtuse with an apiculate tip. On the outside orangy red with a bluish crimson or yellowish shading along the midrib; the lower part of the tepal gradually becoming yellow and carrying a deeply notched blotch of varying size, which may either cover the lower quarter of the tepal or extend only a few millimeters from its base. On the inside orangy red throughout except for two narrow, colourless stripes (1.5 × 15 mm) running along the edges where they are covered by the inner tepals (there are forms without such stripes).

Inner tepals a little shorter than the outer ones (7.2 × 2.5 cm; 6.5 × 2.9 cm; 4.2 × 2.3 cm), lanceolate, elliptic or obovate; top obtuse with an apiculate and pubescent tip. On the outside orange-red, deepening towards the base to pure orange or becoming yellow. On the inside orangy red throughout and only at the place of attachment of a deeper orange tint.

Stamens with dehisced anthers 2.5 times shorter than the tepals. Filaments glabrous, somewhat swollen in the upper half, red, shading to yellow at the point of attachment. Anthers oblong, when dry two and a half times longer than the filaments; yellow, violet or claret. Pollen grayish brown, crimson or red.

Ovary violet to bluish, peg-shaped and carrying a sessile stigma with drooping lobes; shorter than the stamens. The sides of the stigma usually orange-red or pure red; its papillae always red but of varying intensity.

Capsule 4.3 × 2.8 cm, straw-coloured with an orange tint, shaped like an elongated cylinder with an obtuse top and base. Carpophore 0.6 cm high and 0.6 cm wide, inserted into the valve as a sharp prong. From the opened slits through which the seeds are liberated low ridges run onto this prong. The small sessile stigma drops off when the capsule opens. The capsule dehisces by wedge-shaped or nearly elliptic slits which are displaced a little sideways because of the twisted valves, but open widely right to the base. Valves 4 cm long, 1.8 cm wide in the centre of the capsule, 1.3 cm near the base of the upper seedless part and 2 cm next to the stigma. The septem splits to nearly three-quarters of its length; its width at the top is equal to that of the stigma and gradually diminishes towards the base, along with the flat strand of pollen-conducting tissue.

Valves elliptic, strongly convex with a deep median groove; orangy yellow, on the inside glossy and of a lighter shade than outside. In the upper part of the groove its diffuse orange colour spreads over the wrinkles on the valve, in some cases becoming brown.

Wings in the upper third of the capsule flattened with many orangy or in some cases brownish transverse wrinkles; below the stigma often of a more even and brighter colour than the rest of the capsule. Grooves along the suture deep, their edges strongly transversely wrinkled and of a deeper orange shade than the wings. Suture brownish, reflexed, its texture felty.

Tepal scar light brown, 1 mm thick. Peduncle straw-coloured with an orange tinge, 16 cm long.

Seed 0.7 × 0.6 cm, dark brown with a distinct margin, in particular at the upper, oblique part, of the seed where it is bent. Embryo straight, of a lighter shade than the rest of the seed.

Under Tashkent conditions this species begins to flower near the end of March. It retains its juvenile character of dropping the secondary bulb and also the other daughter bulbs are carried by stolons in different directions so that only empty tunics are left at the place of the parent bulb at the end of the growing season (in summer).

Tulipa subpraestans Vved.

plate 11 fig.2j

T.subpraestans is found 'on stony and clayey slopes in the middle mountain belt. Middle Asia: Pamir-Alay (Sebistan, Sarsaryak). Endemic' (Flora of the USSR, Vol. IV).

The bulbs studied were sent by V.P. Bochantsev from the environs of Tutkaul and later plants were more than once grown from seeds.

Bulb 1.5-4 cm in diameter, ovoid. Tunics coriaceous, brownish black, when half-decomposed becoming tawny black; on the inside glossy, rarely pubescent.

Stem and peduncle pubescent, glaucous. At flowering the plant is up to 40 cm high.

Leaves 3-4 on the main stem with a bracteal leaf on each branch; widely spaced, falcate, reflexed, coarsely undulate, glaucous, under Tashkent conditions often extending beyond the flowers; leaf margins cartilaginous. Lower leaves nearly glabrous, the upper leaves on both sides pubescent and strongly ciliate. Basal leaf often set high on the stem (up to 6 cm above ground), oblong-lanceolate, 43 × 12 cm; upper leaf lanceolate, 22.5 × 5.3 cm; bracteal leaf 5.5 × 0.6 cm.

Flowers as a rule 2 or 3 on first-order branches developing from the axils of the upper leaves. Weak individuals have only one flower, whereas vigourous plants can have from 5 to 6 flowers and very strong plants may develop flowers on second-order branches from leaf-axils on the first-order branches. Flower red with narrow and pointed, glossy tepals, when opened forming a simple, but very elegant star.

Outer tepals lanceolate tapering to a pubescent tip; on the outside bluish orange-red with a light brownish shading towards the top and only the very edge of the tepal pure brilliant red; claw to a height of 1 cm yellow shading to orange-red on the midrib (there are forms with darker coloured tepals and a leaden-coloured bloom along the midrib); on the inside bright and glossy orange-red (the orange tints are in this species less intense than in *T.greigii, T.fosteriana* a.o.); at the base, usually hidden by the filaments, a small, 0.3 cm wide, yellow blotch, 0.7 cm high at the edges, where it is covered by the inner tepals and 0.4 cm high in the notched centre where it becomes brownish and merges with the main colour (this blotch may be nearly absent in smaller plants, although the tepal invariably acquires a yellowish tint at the base of the claw; the hemisperical brownish spot in the centre of the blotch may also be absent).

Inner tepals 8.7 × 2.9 cm, lanceolate, slightly shorter and narrower than the outer ones, but tapering to a less pubescent tip. On the outside glossy, bluish orange-red, even slightly crimson (the bluish tinge is caused by a whitish admixture); at the base to a height of 0.9 cm yellow, merging with the main colour. On the inside very glossy red with barely noticeable orange tints, similar to those on the outer tepals; at the base to a height of 0.15 cm yellow with in the centre a notch with a brownish spot (in many forms absent). The place of attachment of the tepals to the pedicel is in this species always yellow.

Stamens 2.8 cm long, exceeding the ovary, 2.5-3 times shorter than the tepals. Filaments 0.7 × 0.35 cm, flat, tapering, red or wine-coloured, changing sometimes to yellow near the base. Anthers 1.8 × 0.25 cm, black or purplish black, coiling into a spiral on dehiscence. Pollen dark brown or grayish brown.

Ovary 1.7 × 0.6 cm, slightly constricted below the stigma, bluish green. Stigma broader than the ovary, crimson-red throughout including the upper seedless parts of the carpels below the stigma and the papillae of the lobes.

Capsule 1.5 × 2.8 cm, light straw-coloured, nearly spherical with an obtuse base and top; sometimes oblong, 6.8 × 2.3 cm. Carpophore of the same colour, smooth without any ridges, 0.6 cm high, at the base near the tepal scar 0.5 cm wide, at the top where it wedges in to the valves with three short prongs, 0.8 cm wide (the lower part of this capsule is very different from that of *T.praestans*). Style absent; stigma sessile, reflexed, very small. The capsule dehisces by wedge-like slits, usually opening widely right down to the carpophore. Slits 4 cm long and 2 cm wide near the stigma. The septum splits to approximately half the length of the capsule; its broadened part below the stigma, 1 × 0.7 cm, elliptic or narrowly rhombic with rounded corners.

Valves broadly elliptic, ivory-coloured. The wide suture, where the septum is attached broadened towards the top like a wedge with a sometimes bifurcating ridge. It is covered with a creased epidermis of the same colour as the wings or light brown with light purplish shadings.

Wings convex, the upper two thirds with a herringbone pattern of wrinkles; the lower part smooth and tumid. In all forms the median groove and the grooves between the wrinkles bluish brown. The wrinkles on the valves finer in the upper part, sometimes disappearing altogether, leaving only a rough surface. Ribs, strongly reflexed, separated from the valves by deep grooves; on the outside dark bluish crimson, sometimes brownish; on the inside, 0.35 cm wide, light orange or straw-coloured, texture felty.

Tepal scar rather broad (0.2 cm) brownish, scalloped along the upper edge. Peduncle up to 15 cm long, thinner than the carpophore, light straw-coloured.

Seed 1.2 × 0.9 cm, dark brown, angular, rounded at the top with a broad margin, bent at the long side of the upper corner. Embryo curved, barely visible.

Under Tashkent conditions this species begins to flower towards the end of the third week in April.

Tulipa butkovii Z. Botsch.

plate 12 fig.3g

The plants studied were brought by A.Ya. Butkov in 1954 from a sandy pass in Chimgan and in other years from various districts of the western Tien Shan: Bol'shoy Maydantal area on shingly slopes in fine earth, August 1955; Malyy Maydental area, among rocks in fine earth, August 1955; Akbulak-Karasay river valley, on shingly slope, 1 800 - 2 000 m, 1956; the river valley of the Akbulak near its confluence with the Serkesay, on shingly slope, 1956.

Bulb ovoid, up to 2.5 cm in diameter; tunic brown, papery, somewhat elongated, lined on the inside with short, adpressed, golden hairs, particularly abundant at the base and top.

Stem at flowering 15 cm high. Peduncle greenish brown, pubescent and glaucous (the waxy coating is easily rubbed off).

Leaves 3-4, closely set on the stem, reflexed, declinate, more or less pubescent, glaucous, with crispate, ciliate margins of a lighter shade; upper leaves channeled, at the beginning of flowering often exceeding the flower; basal leaf 12.7 × 4.5 cm, at flowering partly underground, broadly lanceolate; upper leaf 10.8 × 0.9 cm, narrowly lanceolate.

Flower brick-red, solitary, broadly campanulate or nearly stellate, scentless.

Outer tepals 5.5 × 2.2 cm, narrowly oblong or lanceolate, obtuse or rounded at the top, sometimes with a long pubescent tip and sparse pubescence descending down the midrib. On the outside light claret with crimson shadings in particular along the midrib, acquiring an orangy tint towards the margins and becoming yellow at the base; the overlying glaucousness lending the midrib an almost leaden colour. On the inside light claret-orange, glossy, in the lower half gradually darkening and merging with the diffuse margins of a long and narrow, beetroot-claret blotch, often deeply notched, reaching to 1/6 - 1/3 of the tepal.

Inner tepals 4.6 × 2.1 cm and 6.2 × 2.3 cm, oblong or narrowly ovate with rounded tops and more or less apiculate, pubescent tips. On the outside orange-claret, glossy, with a light crimson tint along the midrib and greenish orange claws. On the inside orange-claret, shading to a beetroot tint in the lower half of the tepal where it has a small, rectangular brownish claret blotch with unsharp margins. Tepals at the place of attachment for 1½-2 mm orange-coloured.

Stamens about half as long as the tepals, exceeding the ovary. Filaments 1.1 × 0.11 cm, often equaling the ovary, thin, broadened towards the base; in various shades of red. Anthers yellow or dark, gradually dehiscing from top to base (under Tashkent conditions often in the course of one day) but not coiling like those of *T.kaufmanniana.* Pollen yellow or claret.

Ovary 1.3 × 0.4 cm, broadening in the lower part but contracting abruptly at the base, tapering towards the stigma, cream-coloured or pale green with a shade of claret along the ribs and below the stigma. Stigma overlapping, on the sides claret, papillae yellow.

Capsule 7 × 2.2 cm, cylindrical, ovoid or conical but always broader at the base than near the style; grayish with a bright pattern of brown-claret spots and stripes. Carpophore of varying shape and dimension, as a rule rather thin, somewhat broadening towards the valves, separated from the valves by a more or less deep groove and entering the median groove by a sharp prong. The tops of the valves taper to a rather long style with a small, longitudinally folded, broader stigma at the top.

The valves may be of any shape from lanceolate, elliptic to broadly ovate and triangular. They are distinguished from those of other species growing in these regions by their fragility and by the seeds showing through the thin wings of the capsule, lending them a grayish tint. Median groove bright brownish claret, sometimes light brown; its brown stripe broadening towards the top and narrowing again on the style; the bright stripes running from this groove across the wings sometimes ramify before reaching the edge of the valve, but usually merge with the unbroken, brownish claret stripe along the rib.

Wings transparent, as a rule convex and transversely wrinkled, in the upper half presenting a herringbone pattern, and separated from the ribs by a deep groove. Sometimes the whole wing surface is covered with wrinkles. The edges of the ribs strongly reflexed, of the same colour as the groove or brownish claret or pure brown. Style darker than the rest of the valves.

The capsule opens by wedge-like slits reaching down to the base of the carpels and has a wide-mouthed funnel at the top, owing to the strongly reflexed style and upper seedless parts of the valves. The reflexed edges of the ribs, 0.35-0.40 cm wide, on the inside cream-, straw- or orange-coloured.

Tepal scar up to 0.2 cm wide, curved, brown. Peduncle orangy brown.

Seeds large, triangular with rounded angles, brown, with a membranous margin which is broader and warped on the slanted side. In the endosperm opposite the broadened part of the margin lies a narrow crescent of dark tissue. Embryo lying somewhat off-centre towards the broadened margin.

When raised from bulbs or seeds under Tashkent conditions this species starts growth later than others and begins to flower towards the end of April, which is also later than all other Middle Asian species of the *Leiostemones*.

Tulipa vvedenskyi Z. Botsch.
plates 13, 14, fig.4d

We collected the plants studied in the western Tiën Shan in the Angren river valley, 1 500 m above the village of Tyurk, where they were growing on slopes formed by outcrops of bedrock with a thin cover of fine earth. Endemic.

Bulb up to 3 cm in diameter, ovoid or in rare cases pyriform; tunic dark brown, papery, on the inside sparsely lined with adpressed hairs except for the apex which is thickly covered with hairs.

Stem 15-20 cm high, glaucous, sometimes with wine-coloured shadings; densely covered with short, white, squarrose hairs.

Leaves 4-5, closely set, reflexing, glaucous, crispate, sometimes pubescent with sparse, short hairs; margins slightly ciliate; leaves becoming smaller higher up the stem and as a rule not exceeding the flower. Basal leaf, up to 25 × 6 cm, lanceolate, set at ground level. Upper leaf, up to 23.5 × 2 cm, narrowly lanceolate.

Flower solitary, red, sometimes orange and exceptionally straw-yellow, finely shaped like a double cup somewhat resembling the flower of *T.greigii*, but with the inner tepals more and the outer tepals less reflexed. Also the bases of the tepals are less sturdy and the bottom of the flower is not so flat as in *T.greigii.*

Outer tepals up to 10.5 × 6.2 cm, rhombic-subrhombic with a rounded or acute top and gradually tapering to the base. In all forms the tepals terminate in a pubescent tip, the indumentum descending down the midrib in the form of sparse, short bristles. On the outside, along the midrib whitish orangy red with broad edges of bright scarlet-red; in the lower half to a height of 2.5 cm yellow with often a barely translucent dark blotch. On the inside scarlet-red, often with a yellow blotch near the base covering the whole tepal up to a height of 2 cm; blotch with a flat or convex top, its streaks merging with the main colour, often shading to a brownish tint in the upper part. On the yellow blotch usually a narrowly elliptic, dark or light brown spot, twice as high as the yellow blotch and with an acute tip. This spot may sometimes be much smaller in which case the background on the sides is claret with a wash of yellow above it. Frequently the basal blotch is wholly dark brown or nearly black, has a rhombic shape and extends like a long tooth with a rounded or acute top across the upper half of the tepal.

Inner tepals 11.5 × 5.2 cm obovate with an apiculate, pubescent tip and slightly undulate upper margin. On the outside orange-red shading to yellow in the lower part, often to a height of 3 cm; the yellow base obtriangular or rhombic often with an acute-angled or rhombic blotch formed by a confluence of dense claret streaks; sometimes this superimposed blotch completely covers the yellow base, leaving only a yellow halo near the top; in other cases the basal part is green instead of yellow. On the inside more intensely red and glossy than outside; blotch at the base variable, black-brown-claret with one of these shades predominating and lighter towards the edges. The dark blotches may be rhombic in shape with a pointed or rounded apex and lighter tinted sides or acutely triangular extending high up the tepal. The lighter coloured blotches may have a yellow halo and in some individuals this blotch was yellow throughout with a truncated or rounded top radiating on the main colour of the tepal. Often the blotch is only discernible as a light shadow.

Stamens at least three times shorter than the tepals, a little longer than the ovary. Filaments glabrous, broadening towards the base (up to 0.45 cm wide and 1.3 cm long), yellow or brown often becoming claret-brown in the upper half and yellow at the base; the tip below the anther yellow. Anthers, up to 2 × 0.4 cm, yellow or black-violet; in the lower part bright yellow, towards the tip brown; they may be as long as the filaments, but as a rule they are longer; dehiscing abruptly. Pollen yellow, black-violet, brownish or even grayish brown.

Ovary up to 2.3 × 0.6 cm, yellowish-blue or pale green, often with anthocyanin pigmentation in the centre and along the ribs; usually tapering to the stigma but it may also be cylindrical. Stigma overlapping, yellow, sometimes with red sides and papillae, but the latter mostly yellow.

Capsule 9 × 2.6 cm, narrowly oblong, triangular in cross section, with an acute top and base; straw-coloured with brown patterns. Carpophore 1 cm long at the ribs and 1.5 cm long below the suture; 1.4 cm wide, rounded. Top of the capsule gradually tapering to a rather long style crowned with large, reflexed, light brown stigma lobes. Stigma light grayish brown, overlapping the seedless part of the capsule. Dehiscense along wedge-shaped slits down to the carpophore. Edges of the ribs light brownish or buff, reflexed; the upper seedless part strongly reflexed forming a 3.5 cm wide funnel. The septa at first do not split deeper than the seedless part (up to 2.5 cm), afterwards they split for another 2.5 cm along the seams of the septa, in all 5 cm.

Valves compact and strong, translucent, oblong with an acuminate top; median groove with a low ridge in the centre and a 0.4 cm broad, dentate, light brown stripe running all along the valve becoming more pronounced in the upper part; from its teeth varying numbers of brown stripes branch off along the veins.

Wings convex with in general not many transverse wrinkles, increasing in numbers in the upper part; in the groove along the ribs the transverse veins form a network of acute-angled interlacements; edges of the ribs on the outside light brown, on the inside of a felty texture, buff. The seedless part below the stigma deepening in colour to brownish.

Tepal scar broad (0.4 cm), brown. Peduncle thick and sturdy, rather short (15-35 cm) with a brownish tint. Pedicels of very large capsules often split longitudinally when dry.

Seed dark brown, triangular with rounded angles and a narrow bent margin; embryo of a lighter tint, comma-shaped.

Under Tashkent conditions this species begins to flower in the third week of April.

Tulipa albertii Regel

plate 15

T.albertii grows 'on gravelly slopes. Middle Asia: Lake Balkhash area (Targyl, Kuyankuz); Tien Shan (Karatau, Chu-Iliyskiye Mts.). Endemic' (Flora of the USSR, Vol. IV).

We received the first samples of this species from the Karatau range through P.A. Gomolitskiĭ, then in 1949 from the Chokpar railway station through V.P. Bochantsev and in 1955 and 1956 samples collected near the Otar railway station were received from the Alma-Ata Botanical Gardens. In 1956 we crossed the Karatau range and brought back besides *T.greigii* fine samples of *T.albertii,* which we picked from the crevices between vertically standing slates where there was no other vegetation.

Bulb ovoid, up to 4 cm in diameter; tunic much elongated, black to tawny; the elongated part of the tunic densely and the rest of the inside sparsely lined with adpressed, stiff hairs.

Stem including the peduncle standing 20 cm above ground, thick, downy and glaucous with a light wine-coloured pigmentation.

Leaves 3-4, crispate in varying degrees, very glaucous almost blue with a slight greenish tint, pubescent; margins cartilaginous, ciliate. Basal leaf 14.1 × 6.2 cm, set at ground level, broadly lanceolate; the higher leaves set close to the basal leaf. Upper leaf 11 × 2.3 cm, lanceolate.

Flower solitary, scentless. It is often very beautiful, in particular that of the plants from the Karatau mountains which, although not as tall as those from Chokpar, are more distinct in colour and shape. The outer tepals are strongly reflexed, whereas the inner ones curve inwards. The flowers of this species vary widely in shape and colour; from orange with a reddish tint in plants of the Balkhash area to orange, orangy pink, golden- or silvery yellow and dark claret of varying intensity in plants from the Karatau. There follows a description of the form which was brought by V.P. Bochantsev from Chokpar:

Outer tepals 8 × 6 cm, rhombic with rounded angles, tapering gradually to a pubescent tip, always more or less reflexed. On the outside along the midrib purplish orange, towards the edges dark orange with a reddish tint, the purple shade getting darker towards the base and turning to a wine-colour; in some cases the base of the tepal next to the stem has a greenish colour. On the inside glossy, dark orange with a reddish tint deepening towards the centre of the tepal. Blotch more or less rounded with a notched or pointed top (1.1 cm high, 1.3 cm wide), darkish often black with a purplish shading, deepening towards the edges.

Inner tepals subrhombic with an acuminate base and broadly rounded, notched top which carries a small pubescent tip. Some forms have spoonshaped, more or less convex tepals with a cordate top. On the outside glossy, dark orange shading to yellow on the fleshy lower part of the tepal, where it has a narrow,acute-angled, heart-shaped blotch, 2.5 cm high, 1.6 cm wide, formed by claret streaks. On the inside glossy, dark orange; blotch of the same dimensions as that on the outside but more rounded and darker, brownish black with a claret shading becoming paler towards the base of the tepal. Tepals at the place of attachment yellow.

Stamens 2.9 cm long, 2-3 times shorter than the tepals; at the first stage of flowering longer than the ovary. Filaments 1.4 cm long, 0.4 cm wide at the base; although narrowing towards the top and terminating in a pointed tip they remain on the whole rather broad; yellow. Anthers as long as the filaments or a little longer, as a rule dark; in lighter forms sometimes yellow.

Ovary 2.2 × 0.7 cm, cylindrical, slightly constricted below the stigma, light green. Stigma broader than the ovary, cream-coloured.

Capsule 4.3 × 2.5 cm, nearly ovoid; at the base abruptly narrowed at nearly right angles to the tepal scar and separated from the short carpophore by a transversely wrinkled groove; at the top tapering to a short and rather broad style (0.5 × 0.6 cm). The lobes of the stigma are firmly attached to the valves and can only be broken off together with bits of the style. The rather solid capsule dehisces along narrow, longitudinal slits but the valves do not separate over the whole length of the slits, the distance between the upper parts being only 0.5 cm. Also the septa split only as far as the seedless part and deflect but slightly.

Valves broadly or narrowly ovate, rounded at the base and tapering to a short style at the top. Median groove bluish brown in its upper part; the uneven brown colour is unbroken and there are no transverse brown stripes branching off.

Wings of a light straw-colour, thin and smooth, convex, only across the seedless upper part of the capsule slightly wrinkled. Ribs a grayish brown, fading towards the groove between the wings and the ribs; a pattern of raised veins is noticeable against the diffuse grayish brown background. The edge of the valve does not curve. Style below the stigma of a darker shade than the rest of the capsule.

Tepal scar 0.15 cm wide, convex, light brown. Peduncle rather sturdy and short, not more than 12 cm, darker than the capsule, pigmented with brown and claret.

Seed 0.6 × 0.5 cm, subtriangular with rounded angles, bright brown with a dark green spot in the obtuse-angled corner; margin rather wide, warped, but of regular width all around.

Under Tashkent conditions this species begins to flower in the third week of April.

Tulipa brachystemon Regel

T.brachystemon grows 'on gravelly slopes. Middle Asia: Dzungaria-Tarbagatay, Dzungarian Alatau. Endemic' (Flora of the USSR, Vol. IV).

The present description is based on plants preserved at the botanical institute of the Academy of Sciences in Leningrad; they were collected in the Semirech'ye region in the foothills near Kapal (Maklakova farm) in April, 1908, Shnitnikov no. 15; also by Regel in April 1879 (nearly white); in Semirech'ye in the Taldy-Kurgan district at the headwaters of the Karatal river, 'Soldatskaya Shohel', by Shipchinskiĭ in July 1928 on a gravelly slope at an altitude of 1 750 m; by Regel in Chingil'dy; by Lidug in the environs of Kapal.

Bulb 2.5 cm in diameter, seated at a depth of 12 cm;

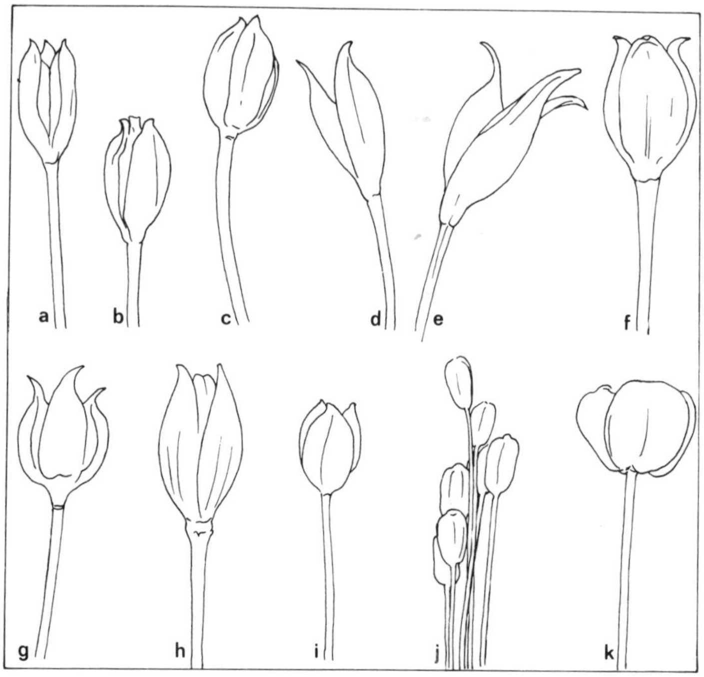

Figure 3. Mature tulip capsules
a. *T. kolpakowskiana* Regel; b. *T. brachystemon* Regel; c. *T. ferganica* Vved.; d. *T. kaufmanniana* Regel; e. *T. tschimganica* Z. Botsch.; f. *T. dubia* Vved.; g. *T. butkovii* Z. Botsch.; h. *T. anadroma* Z. Botsch.; i. *T. biebersteiniana* Schult.; j. *T. buhseana* Boiss.; k. *T. talievii* Vved.

tunic black, leathery, on the inside near the top coverd with adpressed, stiff hairs.

Stem 10-28 cm high; peduncle glabrous, pigmented by anthocyanin.

Leaves 4(3), widely spaced, ascending (even the basal leaf), with slightly undulate edges; margins cartilaginous, glabrous or slightly ciliate. Basal leaf 22 × 3 cm, linear-lanceolate; internode between basal and second leaf 5 cm. Upper leaf 7 × 0.3 cm, narrow, grass-like, attaining 14 cm above ground level.

Flower solitary, yellow or cream-coloured, as far as could be ascertained shaped like a simple star or a shallow cup.

Outer tepals oblong or lanceolate with an obtuse or slightly acuminate top, glabrous. On the outside with a purplish tint along the top of the midrib; on the inside evenly coloured wihout any blotches.

Inner tepals elliptic, subrhombic or obovate, of the same colour as the outer tepals, but without any purple shadings.

Stamens about 2 or 3 times shorter than the tepals. Filaments glabrous, contracted towards the base, but broadening markedly towards the anthers below which they taper abruptly to a sharp point. As to their colour we can only state with certainty that the anthers are of a lighter shade than the filaments, while the latter are darker than all the other parts of the flower. Anthers light brown or yellow, about as long as the filaments, deeply notched at the base where the sharp tip of the filaments is inserted.

Ovary with a well-pronounced style, bottle-shaped, constricted below the stigma, of a lighter shade than the anthers.

In its habitat this species flowers late in April.

Tulipa brachystemon Regel

plate 16, fig.3b

Different from the species described in Flora of the USSR.

The plants we have studied were sent by V.P. Goloskokov from the Dzhungarskiy Alatau.

Bulb up to 2.5 cm in diameter; tunic black, leathery, on the inside near the top covered with adpressed, stiff hairs.

Stem 10-22 cm high, glaucous. The peduncle of some plants slightly pigmented by anthocyanin, of others very pubescent; in some individuals glabrous.

Leaves 3, the upper surface pubescent, very glaucous, in particular the upper leaf crispate, deeply channeled, falcate; margins cartilaginous. The two lower leaves set close together with their bases sticking in the ground. Basal leaf, 22 × 7 cm, lanceolate. Upper leaf 12 × 1.5 cm, set at 4-8 cm above ground level.

Flower solitary, yellow. The upper third of the outer tepals slightly reflexed along the midrib, with its margin at the top incurved and below this reflexed; the tip of the inner tepals very slightly incurved. In these characters this flower resembles that of *T.albertii.*

Outer tepals 5.9 × 3.9 cm, elliptic, shortly acuminate with a pubescent tip; the upper third on the outside also pubescent; along the midrib, particularly near the base, thick and solid. On the outside a tapering claret stripe (up to 1.5 cm wide); at the base to a height of 1.3 cm and along the edges lemon-yellow. On the inside at the top lemon-yellow; in the centre more orange-tinted; at the base an elliptic, light brown spot (0.2 × 0.8 cm) with a pointed top through which the ground colour remains discernible; about 0.5 cm above this spot an obcordate, clear claret blotch (1.2 × 1.2 cm) with a dentate top.

Inner tepals 5.5 × 4.0 cm, obtriangular with a notched top and apiculate, pubescent tip. On both sides pure lemon-yellow, shading towards the base to a more orangy tint; in the centre over the convex part of the tepal at 0.9 cm from the base a narrowly oblong, barely visible blotch (1 × 0.3 cm) of light brown streaks.

Stamens 2.1 cm long, 2.5-3 times shorter than the tepals. Filaments 1.1 cm long, at the base 0.2 cm broad, higher up thin (0.1 cm) and rounded, terminating in a pointed tip, lemon-yellow. Anthers 1 × 0.4 cm, elliptic, pointed, cream-coloured.

Ovary 1.6 × 0.5 cm, nearly cylindrical, tapering only slightly towards the stigma; at the base dark green, pale yellow towards the stigma. Stigma flat, broader than the ovary, light yellow.

Capsule 4.5 × 2.0 cm, long conical, light straw-coloured with an overlying bluish tint; tapering to a short (0.4 cm), broad style with a slightly overhanging stigma of a lighter shade; carpophore absent.

Valves oblong with a rounded base and acute top. Median groove claret brown with a glaucous tint, broadening towards the stigma. The short, transverse, orange streaks which run from the median groove across the wings become more numerous and larger in the upper part of the capsule where they extend to the groove along the ribs. Edges of the valves cream-coloured, their outer rims dull brown.

The capsule opens only slightly on dehiscense and the valves do not reflex. The septa split for only a small part of their length below the stigma; their upper part narrowly rhombic, thick, tapering towards the curvature.

Tepal scar broad, grayish yellow, on both sides bordered by thin, brown stripes. Peduncle grayish, tapering off below the tepal scar.

Seed 0.8 × 0.55 cm, obliquely triangular, light brown with a lighter, even margin.

Under Tashkent conditions this species begins to flower in mid-April. The described *T.brachystemon* plants which were raised from bulbs received from V.P.

Goloskokov are not quite in conformity with the description of this species given in the 'Flora of the USSR'. They differ in the following characters mentioned above: the shape of the tepals, the pubescense of the plant and the shape of the filaments. Besides there is no mention of the red and dark spots on the tepals. Judging by the structure of the flower and the colour of the leaves the plants here described could be classed with *T. albertii*, but with respect to the distributin of the spots on the tepals and their shape they may be *T.greigii.*

Tulipa greigii Regel
plate 17, fig.4a

T.greigii grows 'on clayey slopes. Middle Asia: Aralo Caspian (Kzyl-Orda), Syr-Dar'ya; Tien Shan. Endemic' (Flora of the USSR, Vol. IV).

The samples we have studied were grown in the Botanical Garden and came from the Bostandykskiy district, the mountains near Karankul'say in the Chirchik river valley and from the Karatau range (collected long ago by P.A. Gomolitskiĭ and by us in 1956). Further from the Kichikemin river near Samsonovka village (collected by us); from the Darbazin district near the Stolovaya mountain; from the Fainbulak gorge in the Aksu-Dzhabaglinskiy reservation; from the environs of the Chokpar station (collected by Bochantsev); from the Alma-Ata Botanical Garden, the Otar district; from the Aktau mountain (collected by Sumarokov); from Angren, etc.

Bulb under natural conditions ovoid, when cultivated nearly turnip-shaped or spherical; in many cases with a considerably prolonged tunic of varying texture from coarse and leathery to delicate and papery. The bulbs with prolonged tunics are easily distinguishable from *T.kaufmanniana* bulbs. The tunic of that species has its apex and base densely covered with coarse, adpressed, golden hairs; the rest of its surface being much more sparsely covered.

Stem varying in height, from very short with a rosette of leaves to 40-45 cm high including the peduncle; thick and sturdy, often pigmented by anthocyanin in pink and brownish claret tints; densely pubescent with short, erect hairs.

Leaves (3-)4(-5) usually rather closely set with only the top leaf more widely spaced although there are high forms with widely spaced leaves; as a rule, reflexed and more or less crispate, glaucous or bluish, the upper surface with claret-coloured stains of varying intensity and shape which in some plants may disappear after flowering; the upper leaves pubescent above, ciliate. Basal leaf set at ground level, broadly lanceolate-oblong, in some forms from the Karatau nearly round, in most cases pointed, from 10 to 32 cm long and from 3 to 16 cm broad. Upper leaf lanceolate, often bracteal, from 4-25 cm long and from 1-4 cm broad.

Flower solitary, orange-red, but some forms have dark red, orange, yellow, cream-coloured and even multicoloured flowers. They are perfectly shaped with broad bases, the outer tepals strongly curved outwards for half their length, the inner tepals erect with rounded tops resembling a wineglass. They are also very elegant when not yet fully opened.

Outer tepals from 3-14 cm long and from 3-11 cm broad, rhombic with rounded angles or ovate, in both cases with an acute top and apiculate, pubescent tip; along the midrib, especially in the lower part, thick and fleshy, the edges and in particular the top part thin, undulate and finely toothed; in opened flowers reflexed. On the outside often pubescent along the midrib from the tip down to the base; in the orange-red and dark red forms along the midrib of a darker shade with a crimson tint of varying intensity; in the orange and all yellow forms, from bright yellow to pale lemon-yellow and cream-coloured, along the midrib pinkish crimson to red of varying intensity; towards the base either greenish or with an elliptic translucency from the inner blotch. The plants raised from the bulbs we had brought from the Karatau range had buff, reddish, bright yellow and cream-coloured flowers lacking along the midrib on the outside of the outer tepals the pinkish red tints that are characteristic of the lighter coloured individuals of this species. On the inside about the same colour as the edges on the outside, only brighter, owing to the more glossy inner surface of the tepals. Blotch on the lower half of the tepal, 2 cm high, 1,5 cm broad, rhombic or ovate, black and usually with an acute top; in some forms the main colour of the tepal, shading gradually to yellow on the claw, frames a smaller black blotch of varying intensity; in plants which were collected in the Chu valley and in the Chu-Iliyskiye mountains this yellow basal part of the tepal is larger and the black blotch is of a lighter shade, becoming brownish. In the yellow, light orange and cream-coloured forms the black blotch of the outer tepals is reduced to an ellipse consisting of faint brownish or greyish streaks, often not reaching to the base of the tepal, and may disappear altogether, as is the case with plants from the northern part of the area of distribution and in some populations from the Karatau range. The yellow, light orange and cream-coloured forms often have a red blotch above the black and yellow ones, which although very variable in shape, dimensions, and intensity, is always concave in its lower part; it tapers towards the top where it often fades away. There are plants in which this blotch extends over the whole central part of the tepal, from the yellow basal blotch to the pubescent tip, while other plants we brought back from the Karatau range with no black blotch on the lower part of the tepal, had a clearly marked, semicircular red blotch.

Inner tepals from 4 to 16 cm long and from 3 to 10 cm broad, oblong-obovate or obtriangular with rounded angles, abruptly tapering to a pubescent tip; along the midrib nearly as thick as the outer ones; the main colour similar to that of the outer tepals, but more brilliant and only the yellow forms a little less glossy than the orange and red ones. On the outside at the base often a diffuse, rhombic or fan-shaped yellow blotch; the red forms have a triangular or rhombic spot consisting of purple streaks overlying the yellow blotch or the main colour; in the yellow forms the yellow blotch has around the top part a halo of crimson red streaks of varying length and density, semicircular in shape or with an acute top. The central part of the midrib is red. On the inside of the same colour as on the outside, but the dark blotch on the lower part of the tepal much larger, elliptic, oblong, nearly round or subrhombic. In lighter coloured forms it is often only a faint oblong spot in the centre of the lower part, consisting of brownish streaks and resembling the cap of a mushroom. In the yellow and buff-orange forms the lower part of this blotch is usually colourless with above its yellow margin a red semicircle of varying intensity and width. In forms from the Karatau range it is small and well-defined; there are also plants with orange, yellow and cream-coloured flowers without any red spots on the inside of the tepals.

Stamens with dry anthers 3-4 times shorter than the tepals, about the same length as the ovary. Filaments short, 0.6 × 0.4 cm, sometimes narrower (0.6 × 0.25 cm), as a rule broadly triangular, glabrous, black or yellow darkening towards the top. Anthers very broad, 2.5 × 0.5 cm, yellow, violet or grayish; dehiscense abruptly and when dry coiled into a loose spiral.

Ovary 3.3 × 0.6 cm, usually cylindrical, in some forms tapering, pale green. Stigma large, much broader than the ovary, light yellow.

Capsule 11 × 4 cm, in some cases smaller, oblong to triangular with obtuse top and base; tetramerous forms are often found and sometimes dimerous forms. Carpophore short (up to 1 cm high), usually thick (2.2 cm), separated from the valves by a wrinkled groove with the seams of the ribs running as ridges on the carpophore. Upper seedless part of the valves only slightly tapering, but in some capsules elongated and contracting in a style much in the same manner as with *T.kaufmanniana,* the only difference being the shorter carpophore. The capsule dehisces along wedge-shaped slits, in particular in the seedless part with their rims wide apart because of the strongly reflexed edges of the ribs; the slits in the seed-bearing part remaining narrow.

The septa often split deeper than the upper, seedless part of the capsule (5cm). They may be as broad as the stigma and contract as a wedge towards the seed-bearing part or they may retain the same width as the stigma for half their length, then broaden and contract abruptly to continue as a narrow band in the central seed-bearing part; in other cases the septum right below the stigma is narrower and runs without expanding as a narrow band to the seed-bearing part of the capsule which is split deeper and wider than in the other types.

Valves oblong with an obtuse base and tip; median groove flat and broad, deepening towards the base and the top, minutely wrinkled, brown all along, darkening at the top, or brown only at the top or at both top and base with no colour in the centre.

Wings slightly convex, often entirely covered with fine wrinkles; in other cases more coarsely wrinkled in the upper part, where it sometimes shows a pattern of transverse brown stripes; towards the ribs brownish with narrow darker bands along the edges. Ribs separated from the wings by grooves in which run clearly-marked lines which converge below the stigma; reflexed inside part of the ribs broad, buff-coloured with a felty texture. Stigma large, reflexed, straw-coloured.

Tepal scar broad, 0.4 cm, with brown scallops. Peduncle sturdy, straw-coloured with a grayish tint, in very large plants ribbed, up to 16 cm long.

Seed 1.3 × 0.7 cm, subovate, light brown with a broad, bent margin and a comma-shaped embryo of a lighter shade. Both capsules and seeds can be much smaller than stated above and sometimes larger.

Under Tashkent conditions this species begins to flower around mid-April.

Tulipa mogoltavica M. Pop. et Vved.

T.mogoltavica grows on 'stony slopes in Middle Asia: Tien Shan (Mogoltau). Endemic' (Flora of the USSR, Vol. IV). The plants from Mogoltau described here were picked from between stones in meadows on the top of the lateral spurs of the mountain range. Besides some forms with *T.mogoltavica* characters were found in the Angren river valley (near the village of Sanginak) among typical *T.greigii.*

Bulb ovoid, up to 5 cm in diameter (the Angren forms), bulbs from Mogoltau smaller; tunic coriaceous, black to brown, lined with adpressed hairs, more densely at the base and the top.

Stem including the peduncle 19 cm above ground level. In its upper part, including the peduncle, pigmented with anthocyanin and shortly pubescent.

Leaves 4, the two lower ones set close together, reflexed, the upper widely spaced, erect, not exceeding the flower; glaucous, on the upper surface covered with numerous claret spots and rather pubescent; edges strongly crispate with cartilaginous, ciliate margins; in some forms the top leaves pubescent on both sides. Basal leaf 15 × 5.7 cm, lanceolate, set at ground level; upper leaf 10 × 1.0 cm, linear.

Flower solitary, bright red, resembling in shape *T.greigii*, faintly fragrant.

Outer tepals (3-)5(-10) cm long, (1.8-)3(-6) cm broad, subrhombic with an acuminate top tapering to a pubescent tip; upper half reflexed; on the outside pubescent all over, more densely at the top and the base. Colour on the outside orange red, along the midrib crimson, glaucescent and on the lower half a greenish elliptic blotch caused by the translucent black blotch on the inside. On the inside orange-red of varying intensity with an elliptic, black, blotch reaching to ⅓-¼ of the tepal (e.g. on a 5 cm long tepal 1.6 × 1 cm); blotch often bordered by a 1-1.5 cm wide pale yellow margin which often fades away at the top or is only discernible as a cream-coloured background at the sides of the blotch; some Mogoltau forms have blotches without any margin.

Inner tepals (3.1-)5.1(-10.2) cm long, (1.9-)3.1(-6.2) cm wide, obovate, rounded or subrhombic, closely set, erect; median vein terminating in a short, pubescent tip. On the outside brilliant orange-red; the large rhombic blotch with rounded angles greenish yellow, toothed at the top; in its centre a triangular or rhombic spot consisting of dark, purple streaks. On the inside of the same colour as the outside; on the claw a large, elliptic blotch reaching to ⅓-¼ the height of the tepal, always larger than on the outer tepals; its lower two-thirds intensely black, the upper part with crimson shadings, the top flat or sometimes emarginate. The blotch is quite often bordered by a well-defined yellow margin, wider than on the outer tepals, from 2-2.5 mm on a blotch of 2 × 0.9 cm, not counting the yellow margin.

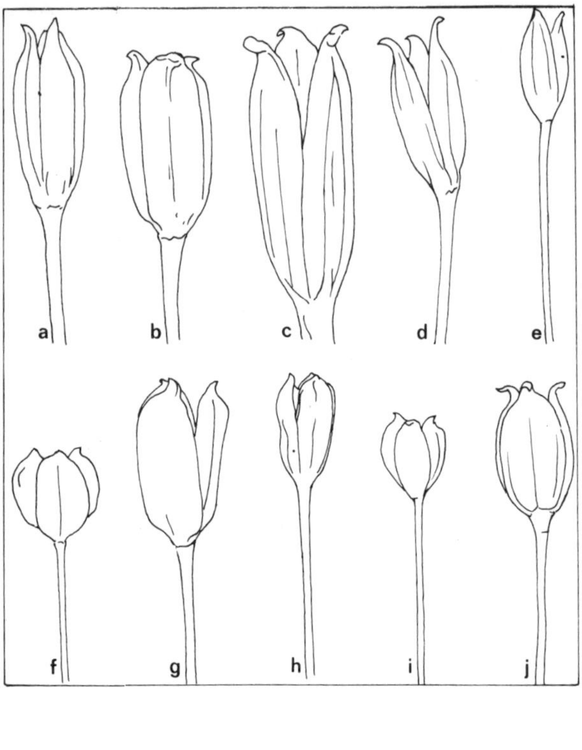

Figure 4. Mature tulip capsules
a. *T. greigii* Regel; b. *T. micheliana* Hoog; c. *T. eichleri* Regel; d. *T. vvedenskyi* Z. Botsch.; e. *T. maximowiczii* Regel; f. *T. wilsoniana* Hoog; g. *T. borszczowii* Regel; h. *T. korolkowii* Regel; i. *T. nitida* Hoog; j. *T. schrenkii* Regel.

Stamens 2.5-3 times shorter than the tepals, equalling the ovary or shorter. Filaments flat, triangular, 0.65 × 0.3 cm (at the base), glabrous, black with a red tip below the anthers. Anthers 0.9 × 0.25 cm, black or violet black, curled into one turn of a spiral. Pollen dark brownish violet.

Ovary 1.6 × 0.5 cm, wedge-shaped towards the stigma, pale greenish yellow, below the stigma lobes reddish. Stigma broadly overlapping the ovary, yellow with reddish shadings.

Capsule 4.2 × 2.2 cm, subtriangular with the top drawn out into a pointed beak, at the base contracting abruptly into the small carpophore. Carpophore up to 0.3 cm high and 0.7 cm wide, pale straw-coloured, narrowing towards the peduncle; the median vein and the ribs extending on the carpophore as ridges. Style rather thin, 1 × 0.4 cm, with a small light brown stigma. Stigma lobes strongly reflexed along the median axis. The capsule dehisces along the ribs but remains practically closed and although it can be seen through the opening in the style that the septum is split along its central part, it is not possible to ascertain how far.

Valves elliptic to ovate, slightly expanding towards the carpophore and narrowing towards the style. Median groove deep, flattening out on the seedless, style-shaped part of the capsule; brightly coloured by a herringbone pattern of dense brown streaks of varying length which extends over a part of the wings and disappears in the lower quarter of the groove leaving only a brownish coloured vein.

Wings strongly convex, rather thin but solid, faintly transparent, smooth or just slightly wrinkled at the top, from greyish to straw-coloured or nearly white (the greyish tint is caused by the seeds shining through); next to the ribs bright brown with in the grooves along the ribs a pattern of interlacing light brown veins. Edges of the valves barely separating on dehiscense. Ribs towards the top of the capsule darkening in colour which merges with that of the median groove, resulting in a very dark style-shaped part, sometimes with three dark stripes on each valve.

Tepal scar 2 mm broad, light brown. Peduncle short, up to 10 cm, light straw-coloured to nearly white.

Seed obliquely triangular with rounded angles, brown, with a broad and thin, slightly bent margin and a straight embryo of a lighter shade.

Under Tashkent conditions this species begins to flower towards mid-April.

Plate 9. *T. affinis* Z. Botsch. (70%)

Plate 10. *T. praestans* Hoog

Plate 11. *T. subpraestans* Vved.

Plate 12. *T. butkovii* Z. Botsch.

Plate 13. *T. vvedenskyi* Z. Botsch.

Plate 14. *T. vvedenskyi* Z. Botsch.

Plate 15. *T. albertii* Regel

Plate 16. *T. brachystemon* Regel

Tulipa micheliana Hoog

plate 18, fig.4b

T.micheliana grows 'on stony or clayey slopes. Middle Asia: Mountainous part of Turkmenistan, Pamir-Alay. General distribution: Iranian region' (Flora of the USSR, Vol. IV).

The plants we have studied were received from the Ura-Tyube railway station, Dragomirovo, Bakharden; from mountainous Turkmenia, Tumanovskiy settlement; from the Samarkand mountains, Aman-Kutan; through the Ashkhabad Botanical Garden from the Kopet-Dag range and the Koren-Dag range; from the Nuratau range. Apart from these we raised plants from material which was received from the area around Tehran in Iran. The progenies of most of these collections were raised from seed.

Bulb ovoid, from 2-8 cm in diameter; tunic coriaceous, often rather prolonged, black to tawny, lined, in particular at the top and the base, with adpressed hairs.

Stem including the peduncle 35 cm above ground level; pubescent.

Leaves 3-5, rather widely spaced, glaucous with on the upper side longitudinal, broken or unbroken, wine-coloured stripes of varying intensity; the stripes which are very brightly coloured in early spring fade towards the end of the vegetation period; in some cases there are no stripes at all and the leaves are more or less pubescent with cartilaginous, ciliate margins. When in full bloom they usually do not extend beyond the flower. Basal leaf set at ground level, lanceolate, up to 30 cm long and 20 cm broad; the second leaf often 1-2 cm longer than the first. Upper leaf, lanceolate, up to 21 cm long and 7.5 cm broad.

Flower solitary, bright red to dark crimson, very large, finely shaped, resembling *T.greigii*, but differing clearly from the latter by the crimson colour range and the conspicuous wedge-shaped, light margined black blotch on the outside of the inner tepals; scentless.

Outer tepals, rhombic-subrhombic, up to 16 cm long and 9.0 cm broad with an acute top often terminating in a 1 cm long pubescent tip, convex, solid at the base, in the lower part often with ciliate margins, near the top undulate with thin, finely toothed edges, along the midrib and the reflexed upper edges pubescent; in some cases the tepal is pubescent all over, in particular the top of the midrib, the reflexed edges and the thick part of the claw being densely covered with soft hairs. On the outside from orange red to dark crimson, glaucous except for a narrow edge; at the base an oblong green blotch extending to one quarter the length of the tepal and not clearly marked owing to the blotch on the inside shining through. On the inside of the same colour as on the outside, but very glossy, not glaucescent; at the base a large purplish black blotch of varying shape with a pale yellow or white margin (sometimes no margin) reaching about halfway up the tepal. The shape of the blotch may be rectangular, oblong, elliptic or acute-angled with an acute, blunt, or tridentate top, but always clearly marked e.g. on a 13.2 × 7.2 cm tepal the oblong black part of the blotch with a truncated, tridentate top may be 5.8 × 1.8 cm with a margin 0.15 cm wide at the sides and 0.9 cm at the top including the teeth.

Inner tepals obtriangular or obovate, up to 15 cm long and 9 cm broad, with an apiculate, frequently up to 1 cm long, pubescent tip; midrib and upper part of the tepal also pubescent. On the outside and inside of the same colour as the outer tepals, only on the inside very glossy, almost lacquered. On the outside at the base a subrhombic, greenish cream-coloured blotch with an acute top and an oblong, dark violet centre. On a 13.5 × 6.7 cm tepal it is up to 7 cm high; the blotch can also be violet throughout or partly violet with only a lighter coloured halo. On the inside at the base a similar oblong blotch with a pointed or truncated, tridentate top surrounded by a light yellow, often toothed, margin which broadens in the lower part; sometimes there is a broad dark blotch with only a margin around the top and sometimes there is no margin at all. The blotch on the inside of the tepal of the described individual is 8 cm high, has a 7 × 1.3 cm black centre and a light margin, 1 cm broad around the top and 0.2 cm broad on the sides.

Stamens 2.5-4 times shorter than the tepals, a little longer than the ovary, with very mobile anthers. Filaments short, flattened, triangular (0.4 cm broad at the base), glabrous, black, but near the base and at the tip yellowish or reddish. Anthers large and broad (in our plants 2.6 × 0.5 cm), oblong when opened, violet or sporadically yellow. Pollen brown in various shades, sometimes purple or even yellow.

Ovary 3.4 × 0.7 cm, cylindrical, sturdy, dark bluish green. Stigma overlapping, large, greener than the ovary; papillae yellow.

Capsule 9.0 × 3.0 cm, nearly cylindrical with an obtuse base and top; carpophore 0.5 cm high and 1.5 cm wide, narrowing towards the capsule and separated from the valves by a wrinkled constriction. Upper seedless part of the capsule very large, 3 cm long, 2.8 cm broad at the base, 1.5 cm broad below the stigma. Stigma sessile, large, 0.5 cm long and 1.3 cm broad, grayish brown, firmly fixed at nearly right angles to the valves and with a convex fold, 0.2 cm high, at the top.

The capsule dehisces throughout its length along narrow slits, up to 0.35 cm wide. The edges of the reflexed ribs and the upper seedless part of the valves form a 2 cm wide funnel. The septa split only as far as the seedless part i.e. 3 cm; the scar left by the suture of the septa 1.4 cm wide below the stigma and gradually narrowing to 0.2 cm near the curvature. Cleavage surface along the pollen-conducting tissue dull cream-coloured with in the centre a group of fibrous yellowish strands, particularly noticeable where the stigma merges into the seedless part of the valves.

Valves straw-coloured, very long with parallel edges, obtuse at both ends. Median groove light brown in the upper three quarters with dark brown veins merging in the lower part of the groove and bifurcating in the upper part where they form a mass of dark and lighter shaded anastomoses.

Wings slightly convex, smooth with alternately opaque and transparent straw-coloured stripes (not brown as in other species); groove along the ribs deep, rounded, covered with a network of brown veins. Ribs a dull brown which merges in the upper part with the colour of the stigma; seam of cleavage very broad (0.7 cm), reflexed, buff-coloured.

Tepal scar scalloped, light brown, protruding above the broadened base of the carpophore and the pedicel. Peduncle sturdy, up to 15 cm long, whitish straw-coloured, lighter than the capsule; in larger plants ribbed.

Seed broadly ovate, 1.1 × 0.9 cm, brown with a lighter shaded embryo; margin broad and flat.

Under Tashkent conditions this species begins to flower in the second week or early in the third week of April.

Tulipa karabachensis A. Grossh.

T. karabachensis grows in Azerbaydzhan, the Nagorno-Karabakhskaya autonomous region, Gadrut district, between the Kagnyjurd pasture lands and mount Ziarat. J.I. Kirpichnikov collected this species on southern stony slopes, where it flowered on 29th May 1948. According to Grossheim (1940) it grows only in the middle and the upper belt of the Karabakh range along the edge of the forest and it flowers in April-May. Endemic.

Bulb seated at a depth of 15 to 22 cm, up to 2.5 cm in diameter, nearly globose. It has three fleshy scales and is covered with many half-decayed, brown and light brown tunics. Tunic on the inside sparsely lined with coarse, straight, light-coloured hairs, somewhat denser towards the base and the apex.

Stem 40 cm above ground level, leaved part of the stem 12-15 cm, peduncle usually short, 7-9 cm; both parts glabrous.

Leaves 4, widely spaced, declinate with reflexed top, glabrous, lanceolate, not exceeding the flower; basal leaf 13.5 × 3.5 cm, broadly lanceolate, keeled in the lower part; upper leaf 7 × 1.3 cm, lanceolate with the top drawn out to a 1.5 cm long, narrow tip. The short and comparatively broad leaves are very typical of *T. karabachensis*.

Flower solitary. It is difficult to form an opinion on the shape of the flowers of the plants which are preserved in the herbarium of the Academy of Sciences of the USSR. Besides it is only the light yellow form of this species which is kept there, while Grossheim mentions that occasionally also red ones are found, both the yellow and the red individuals without blotches at the base of the tepals. Nothing definite can be said about the differences in colour between the outside and inside surfaces of the tepals, although it was clear that the lower half of the tepals is of a brighter yellow shade than the upper half.

Outer tepals 6.8 × 3.2 cm, rhombic with rounded angles and a rather obtuse top or with an attenuated top with a pointed purplish tip; on both sides yellow.

Inner tepals 6.5 × 3.2 cm, narrowly obovate or subrhombic with rounded angles, often with an emarginate top with a pointed midrib. Colour similar to that of the outer tepals, but more intensely yellow at the base.

Stamens exceeding the ovary. Filaments up to 1.5 cm long, rather thin, slightly broadening towards the base (to 0.12 cm), often brighter yellow than the tepals, sometimes white. Anthers in the dried state short, 0.7 × 0.2 cm, yellow.

Ovary long 2 × 0.3 cm, narrowing towards the stigma, gradually passing in a short style of a lighter shade than the ovary. Stigma rather broad (0.4 cm), of a light shade; papillae slightly overlapping the ovary.

Tulipa eichleri Regel
fig.4c

T.eichleri grows on 'dry slopes and among field crops. Caucasus: Eastern Transcaucasus. General distribution Iranian Region (?)' (Flora of the USSR, Vol. IV).

The plants studied were raised from bulbs and seeds which were sent by B.V. Serdyukov from the Tbilisi Botanical Garden.

Bulb ovoid, up to 4 cm in diameter, very much resembling bulbs of cultivated tulips; tunic rather prolonged, coriaceous, black to tawny, on the inside hairy, particularly at the apex and base.

Stem up to 39 cm high above ground level; the above-ground part of the stem and the peduncle densely pubescent, green.

Leaves 3-4, widely spaced, reflexed, glaucous, pubescent, slightly crispate; margins cartilaginous, ciliate. Basal leaf 19 × 5.5 cm, lanceolate, set at ground level. Upper leaf 10.5 × 1.1 cm, nearly linear.

Flower large and beautiful, cup-shaped, in the upper third tapering, sometimes broadly campanulate, base flat; colour red (medium madder); scentless.

Outer tepals 7 × 3.5 cm, broadly lanceolate gradually tapering to a broad tip. On the outside light wine-coloured, glaucescent; at the base, covering the claw, a tridentate, light green blotch, up to 1.6 cm high. On the inside glossy, light wine-coloured (medium madder) with an elliptic, black to purple or blackish brown blotch (1.7 × 0.8 cm), surrounded by a rather narrow (up to 0.2 cm), yellow margin which covers the edge of the base of the tepal.

Inner tepals 6.6 × 3.0 cm, obovate tapering rather abruptly to a pubescent tip. On the outside brilliant light claret with sometimes a crimson shading along the median vein and at the base a rhombic, tridentate, greenish yellow blotch (up to 1.8 cm high). On the inside also brilliant light claret with on the claw a rhombic, black to purple or black to brown blotch, covering the base of the tepal and bordered at the top by a well-defined, 0.15 cm wide, slightly dentate margin.

Stamens 2.5 cm long, a little shorter than the ovary. Filaments 0.9 × 0.25 cm, tapering towards the top, black; the base and the tip below the anthers white. Anthers 1.5 × 0.3 cm, black with a purplish tinge, sometimes yellow; dehiscing abruptly.

Ovary 2.7 × 0.5 cm, cylindrical without any constriction below the stigma, light green. Stigma and papillae cream-coloured, much larger and broader than the ovary.

Capsule up to 9.5 cm long and 2.4 cm wide, of an orangy straw-colour, very much resembling in texture, colour and shape the capsules of cultivated single early tulips and among the wild species that of *T.schrenkii*. Shape long and narrow cylindrical with an acute top and base. Carpophore rather long, 2 cm at the place of insertion in the valve, 1.5 cm near the ribs; at the tepal scar 0.9 and next to the valve 1.3 cm wide; evenly coloured, lighter than the upper part of the capsule, but darker than the lower part of the wings. Apex of the capsule bluntly pointed; just below the stigma the valves are constricted and drawn out to a 0.4 cm long style carrying the reflexed, but firmly attached, dark brown remnants of the stigma.

Valves nearly rectangular, acute at both ends; median groove shallow, sometimes absent leaving only a 0.3 cm wide strip of longitudinally creased orangy tissue, but more frequently with a slightly raised vein or ridge all along the groove.

Wings solid, convex, opaque, covered from top to bottom by numerous transverse wrinkles, with orangy light brown veins, larger and wider spaced in the lower part of the capsule, finer and denser towards the top. Upper pointed part of the valves covered with wrinkles which radiate from the top of the median groove to the edges of the wings. In the centre of the style below the stigma a short groove. Ribs separated from the wings by rather wide and deep grooves in which the veins of the valves converge. Edge next to the line of cleavage of a darker shade than the rest of the capsule. On dehiscense the capsule forms very narrow slits terminating in a wide-mouthed funnel owing to the strongly reflexed upper part of the valves and the stigma. Edges of the split ribs barely reflexed, up to 0.6 cm wide, buff with a pinkish orange tinge; texture felty. A typical feature of this species is the long fringe of hairs running from the inner edge of the cleavage surface straight to the centre of the capsule; the hairs are very thin and rippled, not straight and thick as in other species.

Tepal scar up to 0.15 cm high, usually somewhat thicker than the carpophore, grayish brown. Peduncle up to 27 cm long, 1.0 cm broad next to the tepal scar, narrowing to 0.5 cm below it, light straw-coloured.

Seed 0.8 × 0.65 cm, obliquely triangular with rounded corners, in various shades of brown, slightly warped and surrounded by a fragile narrow margin. Embryo comparatively small, only half as long as the seed and of a lighter shade.

Under Tashkent conditions this species begins to flower in the third week of April.

Tulipa schmidtii Fom.

T.schmidtii grows 'on slopes and among field crops. The Caucasus: Southern Transcaucasia, Tal. Endemic' (Flora of the USSR, Vol. IV).

The description is based on plants preserved in the herbarium of the botanical garden of the Academy of Sciences of the USSR.

Bulb 5.5 cm high and 2.5-5.0 cm in diameter, ovoid, seated at a depth of 25 cm; tunic thin, coriaceous, light brown turning to tawny black when decayed, on the inside lined with woolly hairs becoming denser at the base.

Stem, including the peduncle, up to 40 cm above ground level, both pubescent.

Leaves 6-12, set high on the stem, extending beyond the flower, which sometimes attains a height of 17 cm; declinate, slightly reflexed, keeled, glaucous; margins ciliate. Basal leaf 27-30 cm long and 2.0-4.5 cm broad, narrowly lanceolate with coarsely undulate margins. Upper leaf narrow, nearly grass-like, 17-20 cm long and 0.7 cm broad, erect.

Flower solitary, shaped like a deep cup, red. Outer tepals broadly lanceolate or rhombic with rounded angles, tapering to a pubescent tip. On the inside red and at the base an oblong, yellow-margined black blotch (5-10 cm long and 1.5-2.5 cm broad) with a notched top. Inner tepals oblong to obovate, somewhat drawn out at the top and terminating rather abruptly in a short tip. On the inside on the claw a broadly rhombic, yellow-margined black blotch with a notched and dentate top.

Stamens up to 2.7 cm long, 2-3 times shorter than the tepals. Filaments, 1.1 cm long, black to wine coloured, thin, tapering towards the tip. Anthers in the dried state

1.5 × 0.25 cm, of nearly the same dimensions as the filaments, apparently yellow. Pollen yellow.

Ovary 2 × 0.3 cm, a little shorter than the stamens, without any noticeable constriction below the stigma; in the upper part red. Stigma sessile, slightly broader than the ovary.

Capsule 6-7 cm long and 2-2.5 cm broad.

Tulipa maximowiczii Regel

plate 19, fig.4e

T.maximowiczii grows 'on mottled rock outcrops (always ?). Middle Asia: Pamir-Alay (Vakhsh, Kyzyl-Su). Endemic' (Flora of the USSR, Vol. IV).

The plants studied were collected by V.P. Bochantsev in Tadzhikistan near the village of Sangtud. We have also raised them from seeds harvested from these parent plants and from plants which were brought by F.N. Rusanov from the Sarbo Pass.

Bulb 1.0-2.0 cm in diameter, ovoid or globose; tunic scaly, stiff, black, at the top profusely lined with woolly hairs.

Stem at flowering 15-25 cm above ground level. Peduncle glabrous, green, below the flower pinkish brown.

Leaves (5)6(-8), widely spaced, exceeding the flower, making an acute angle with the stem and reflexing halfway up, linear (not crispate), glaucous; the edges ciliate, red. Basal leaf long, 29.5 × 2.5 cm and set 3-5 cm above ground level. Upper leaf 15.5 × 1.0 cm.

Flower solitary, elegant, red, stellate (opening widely to a nearly flat star), scentless.

Outer tepals up to 6.1 cm long and 3.2 cm broad, broadly lanceolate with a sharp pubescent tip. On the outside along the midrib bluish crimson, towards the edges scarlet, at the base greenish. On the inside brilliant scarlet showing a well-defined obtriangular, bluish black blotch with a notched top covering the lower part of the tepal to a height of 1.5 cm; blotch margined with white or cream-coloured stripes which become a little broader towards the top.

Inner tepals 6 × 3.5 cm, long obovate with a conspicuous, abruptly pointed tip. On the outside brilliant scarlet, at the base to a height of 1.5 cm yellowish green. On the inside likewise brilliant scarlet and at the base a large bluish black, obtriangular, blotch (1.5 cm high) with a notched top and narrow white stripes fading towards the base of the tepal.

Stamens of the individual described 1.8 cm long, i.e. three times shorter than the tepals, but in some smaller plants the stamens were half the length of the tepals. Filaments 0.8 × 0.3 cm, black, flat, broadening in the lower quarter and tapering towards the base and the tip; below the anther white. Anthers violet, 1.2 cm long and 0.25 cm wide in the individual described, but there are forms with anthers shorter than the filaments. Pollen dark violet.

Ovary 2.3 × 0.35 cm, peg-shaped, slightly tapering, exceeding the stamens, light green with red to violet ribs and red below the stigma. Stigma overlapping the ovary with crimson lobes.

Capsule from 2.0 to 4.5 cm long and from 1.0 to 1.4 cm broad, oblong and very narrow; the smaller capsules nearly globose, but all capsules with slender, drawn-out beaks. Since the valves leave only 2 mm between their bases and the tepal scar, there is practically no carpophore. The acuminate seedless part of the carpels, 1 cm long and 1 cm broad at the base, forms a narrow beak at the top. Stigma small, its grayish brown lobes placed at right angles to the style are easily broken off.

The capsule dehisces along rather wide wedge-shaped slits with parallel sides in the lower part and widening at the top but not forming a funnel since the seedless parts of the valves do not reflex as far as in other species. The septa split to approximately two-fifths of the length of the valve.

Valves fine and thin, straw-coloured, in small capsules in sideview like a double-convex lens, in larger capsules more rectangular with parallel ribs and acute base and top. Median groove shallow and of the same colour as the rest of the capsule. Upper seedless part of the valves on the whole more yellow than the rest of the valves; the edges often pinkish.

Wings nearly flat, only slightly bulging at the base and the top, sporadically covered with fine transverse wrinkles but these are usually only discernible as creases of the epidermis; in the upper third straw-coloured, along the ribs often pinkish. Cleavage surface of the ribs of a light buff colour; texture felty. Along the edge of the ribs a well-marked groove, the rim not reflexed.

Tepal scar 0.1 mm broad, flat, light brown. Peduncle up to 17 cm long, of a lighter colour than the capsule (the tallness of the plants is due to the elongated internodes).

Seed 0.5 × 0.5 cm, triangular with rounded corners, brown, with a flat margin, which is rather wide in comparison with its size.

Under Tashkent conditions this species begins to flower towards mid-April.

Tulipa linifolia Regel

T. linifolia grows 'on stony slopes. Middle Asia: Pamir-Alay (Pyandzh). Endemic' (Flora of the USSR, Vol. IV).

Bulb 1-2 cm in diameter and 1.5-2.5 cm high, ovoid, seated at a depth varying from 2 to 11 cm; tunic black or dark brown, thin but tough and coriaceous, profusely lined at the top with woolly hairs.

Stem with short internodes, not more than 2.5 cm long. Peduncle at flowering up to 27 cm long from base to top. Stem and peduncle glabrous.

Leaves (3-)6(-8), closely set, falcate, reflexed, crispate, usually not exceeding the flower. Basal leaf 8.0 × 0.8 cm; upper leaf 5.0 × 0.5 cm (sample from plants with seven leaves). Smaller plants have very narrow leaves (0.1-0.2 cm).

Flower solitary, stellate, red. Tepals 2-4 cm long and 1.2-3.5 cm broad, obtuse with a short pointed tip.

Outer tepals rhombic with rounded angles or subrhombic, inner tepals obovate or subobovate; both red and at the base a well-marked, bluish black, flatly truncated blotch with white-margined sides.

Stamens with dry anthers less than half the length of the tepals. Filaments broadening towards the base, glabrous. Anthers two-thirds the length of the filaments, grayish yellow.

Ovary flask-shaped, red. Style very short with an overlapping stigma of a lighter shade.

Under Taskent conditions this species flowers in mid-April.

Tulipa wilsoniana Hoog[3]

plate 20, fig. 4f

T.wilsoniana grows 'on stony slopes in the steppe zone of the mountains. Middle Asia: mountainous Turkmenistan. Endemic' (Flora of the USSR. Vol. IV).

The studied plants were collected in the foothills near the Prokhladnoye settlement and sent by the Ashkhabad Botanical Garden.

Bulb 1.5-2.5 cm in diameter; tunic prolonged, hard and coriaceous, black to tawny, near the top profusely lined with long soft, rippled hairs.

Stem 5-15 cm above ground level, glabrous. Peduncle pinkish or wine-coloured, glabrous.

Leaves 4-5, closely set, forming a rosette, at the beginning of flowering with their bases below ground; reflexed, channeled, glaucous; margins crispate with purplish edges. Basal leaf 12 × 1.5 cm, narrowly lanceolate. Upper leaf 7 × 0.4 cm, narrowly lanceolate to linear. The second leaf 14 × 1.2 cm, always longer and somewhat narrower than the basal leaf.

Flower resembling in shape that of *T.korolkowii;* shaped like a double cup, but with a cuneiform base instead of a flat one; scentless. Young flower buds pendulous, straightening up before opening.

Outer tepals 5 × 2.5 cm, subrhombic or broadly lanceolate with a small pointed tip. On the outside orange to crimson, the crimson along the midrib shading to purple, towards the claw slightly greenish. On the inside brilliant claret deepening towards the base; at the base a small brownish black or purplish black blotch (0.5 × 0.4 cm) broadening towards its notched top, the colour on its sides fading and turning to yellow near the base of the tepal.

Inner tepals 5 × 2.7 cm, narrowly obovate. On the outside brilliant claret to orange with a purple shade towards the base, at the point of attachment nearly black. On the inside of the same colour as on the outside and at the base extending over the whole width of the tepal a purplish black blotch, up to 0.4 cm high.

Stamens three times shorter than the tepals, 1.5 times longer than the ovary. Filaments 0.65 × 0.2 cm, conical, glabrous, for the lower third of their length black, above claret. Anthers 1.0 × 0.15 cm, narrow, yellow. Pollen yellow.

Ovary 1.3 × 0.35 cm, pale green, contracting towards the base and the broad and sessile, yellow stigma.

Capsule 2.7 × 1.7 cm, shortly cylindrical with an obtuse top and base; in smaller plants almost globose. A characteristic feature of this capsule is the carpophore as a continuation of the valves of which the ribs fuse at the base without any constriction between valves and carpophore. The capsule dehisces along wide wedge-shaped slits, 2.5 cm long and 1.4 cm wide at the top. The septa split further down than their upper expanded part. Below the stigma they are of the same width as the stigma, but become more than twice as wide in the cavity and contract again towards the seed bearing part of the capsule which does not split.

Valves elliptic, thin but firm, semitransparent, of a whithish straw-colour, often with a delicate pinkish shading; at the top rounded, carrying a small beak-like stigma, up to 1 mm high, with reflexed lobes; at the base terminating in the tepal scar; median groove shallow, but well-marked, not extending over the seedless part and not right down to the base of the capsule, bright brownish claret, broadening to a diffuse spot in the top part and with a small crimson halo spreading over the transverse wrinkles. The whole upper part of the capsule rather flat with transverse stripes and a light grayish crimson tint in the centre.

Wings slightly convex, transversely wrinkled in the upper two-thirds of the capsule, separated from the ribs by a groove with a pattern of crimson veins. Edges of the ribs reflexed, about 2 mm broad, of a pinkish buff colour, below the stigma light brownish; texture felty.

Tepal scar less than 1 mm thick, contracted to the 8 cm long, thin, whitish peduncle. At the end of the flowering stage the internodes stretch and attain a length of 6.8 cm.

Seed 0.65 × 0.50 cm, obliquely triangular with rounded angles, brown with a light brown straight embryo and surrounded by a flat margin of the same colour.

Under Tashkent conditions this species begins to flower in the second week of April.

3. The correct name of this taxon is *T. montana* Lindley. See Sealy, J.R. in: Curtis's Bot. Mag. 174 (4), t. 433, 1963. (editor)

Tulipa lehmanniana Merckl.

plate 20,21

T.lehmanniana grows in 'hamadas (stony desert) overblown with sand and on outcrops of mottled rock. Middle Asia: Kyzylkum, mountainous Turkmenistan, Syr-Dar'ya, Amu-Dar'ya. General distribution Iranian region (northern Afghanistan)' (Flora of the USSR, Vol. IV).

The plants studied were brought by A.Ya. Butkov from the environs of Kushka in 1930.

Bulb from 1.3 to 4.5 cm in diameter, ovoid; tunic prolonged to ground surface, very tough and thickened at the sides, almost scale-like, black to tawny, on the inside densely lined with curly woolly hairs.

Stem including the peduncle from 16 to 25 cm above ground level, glabrous, glaucous.

Leaves 4, not exceeding the flower, closely spaced, declinate, reflexed, crispate, glaucous, glabrous; margins without cilia. Basal leaf 18 × 2.7 cm, lanceolate, set at ground level. Upper leaf 8.0 × 0.8 cm, linear.

Flower solitary, yellow, orange or red, resembling the flower of *T.greigii* but narrower and much smaller. The following description is based on the yellow form from Kushka.

Outer tepals rhombic with rounded angles and an acute top with a slightly pubescent tip; 4 cm broad in the centre and on an average 5.6 cm long. On the outside quite smooth, yellow, paler along the broad midrib; the blotch from the inside faintly discernible. On the inside bright yellow deepening towards the base on which an elliptic, sharply indented, black to brown blotch (2.0 × 0.9 cm) with lighter shaded edges.

Inner tepals 6.0 × 4.0 cm, obovate, on both sides of the same yellow colour; the blotch on the inside showing through on the outside. On the inside the yellow ground colour deepens towards the oblong, black blotch (1.8 × 0.7 cm) with brownish edges and slightly tapering top.

Stamens overtopping the ovary by a third. Filaments half as long as the ovary, yellow; in some individuals the sharp tip of the filaments of a darker shade. Anthers and pollen yellow.

Ovary green; stigma light yellow.

Capsule 4.8 × 2.5 cm, long ovoid with an obtuse base and top, light straw-coloured. Stigma broad (0.6 cm) and short (0.3 cm), triangular, erect, purplish. Carpophore short and smooth, 0.6 cm broad near the peduncle and 1.3 cm broad near the valves; 0.3 cm long near the ribs and 0.5 cm long below the median groove of the valve.

Valves oblong, smooth, compact; median groove in the centre of the valve deep, often with a raised central vein and with a sunken pit below the stigma.

Wings thick and coarse, whitish, on the whole evenly convex, but bulging above the carpophore and below the stigma; surface smooth with only the transverse veins showing through from the inside; edges of the wings separated from the ribs by a groove. Ribs of a faintly purplish shade.

The capsule dehisces along short and wide, wedge-shaped slits narrowing towards the base; edges of the ribs rather thin, of a dull pinkish shade, with fine, fringy hairs. The split septa expose the rounded rhombic pollen-conducting tissue which is at the top of the same width as the stigma, then gradually broadens to contract again towards the curvature of the septum and extends further down as a thick even strand.

Tepal scar nearly completely hidden by the bulge of the carpophore. Peduncle usually pigmented by anthocyanin, pink or brown, but always of a darker shade than the capsule.

Seed 0.9 × 0.55 cm, sharp-angled, dark brown, with a narrow bent margin; embryo curved, shining through on both sides.

Under Tashkent conditions this species begins to flower in the beginning of April.

Tulipa borszczowii Regel

plate 23, fig.4g

T.borszcowii grows 'in sandy and sandy argillaceous deserts. Middle Asia: Aral-Caspian, Kyzyl-Kum desert (northern part). Endemic. Described from parts of the Kara-Kum desert near the Aral Sea'. (Flora of the USSR, Vol. IV).

The plants here described came from a sandy area near the town of Arys' close to the Syr-Dar'ya river, Kazakhstan.

Bulb 2-4 cm in diameter, ovoid; tunic prolonged to ground surface, very hard and coriaceous, black to tawny, densely lined with woolly hairs.

Stem, including the peduncle, up to 20 cm above ground level, glaucous, glabrous; in rare cases slightly pigmented with anthocyanin.

Leaves 3, widely spaced, lanceolate, glaucous, reflexed; margins undulate, with white cartilaginous edges, glabrous or sparsely ciliate (Syr-Dar'ya); not overtopping the flower. Basal leaf 18.0 × 4.0 cm, set at ground level. Upper leaf 7.0 × 0.8 cm, almost bracteal.

Flower solitary, either yellow, orange or orange-red (all plants from the sandy areas near Arys' close to the Syr-Dar'ya river had orange-red flowers, whereas those from Tyuratam railway station had yellow flowers with orange and orange-red only in exceptional cases). In structure the flower resembles *T.greigii,* but with the outer tepals more erect. The following description is based on the flowers of the orange to red form.

Outer tepals 6.0 × 4.5 cm, rhombic with rounded angles and an attenuated top terminating in a short pubescent tip; the lower third bent at right angles to the pedicel, higher up erect with only the tip reflexed and with toothed edges. On the outside orange to red, in the lower half more orange, the top red, along the midrib glaucous and at the base the blotch on the inside shining through. On the inside of the same colour; the blotch elliptic and consisting of purplish brown streaks.

Inner tepals 7.3 × 4.6 cm, long, obtriangular with a slightly rounded, notched top, carrying a small tooth on the midrib. On the outside orange-red, fading towards the claw, but near the blotch almost orange; blotch up to 1.9 cm high, as broad as the tepal, subrhombic, purple to brown. The inside of the tepals of the same colour as the outside but with a crimson shading along the midrib and the red colour extending lower down, leaving the orange ground colour only on the sides above the blotch; blotch rhombic, black with a light purple shading, up to 2 cm high and as broad as the tepal.

Stamens 1.9 cm high, black to purple. Filaments 1.1 cm long, black to purple, at the base of a lighter shade and broader (0.2 cm), the top pointed. Anthers long and rather narrow (0.8 × 0.3 cm), oblong, black to purple.

Ovary 2.2 × 0.5 cm, cylindrical, bluish green, the upper part below the stigma bright green. Stigma broader than the ovary, light greenish yellow.

Capsule 4.6 × 2.1 cm, cylindrical with a broadly rounded base and top without any trace of a style, light straw-coloured. Stigma short (0.6 cm) and broad, sitting directly on the obtuse top of the valves and of the same light colour. This species has no carpophore, the valves extending right down to the tepal scar even slightly overlapping it. The partly hidden tepal scar grayish below the valves and brown at the peduncle.

Valves coarse and compact, elliptic. Median groove shallow, gradually deeper towards the top with a rather broad and deep pit below the stigma (as a rule this part of the median groove is flattened in other species); along the groove, starting a few millimetres above the base, runs a raised vein.

Wings rather flat, only bulging in the upper part near the pit in the groove, the whole surface covered with fine transverse wrinkles which become denser in the top part and extend right across the grooves along the ribs to the edge of their cleavage surface. None of the indidviduals we examined showed brown or purplish shadings along the ribs.

The capsule dehisces along wedge-shaped slits, very narrow at the base and broadening to 1.5 cm below the stigma, forming only a narrow funnel. Edges of the ribs on the inside narrow (0.3 cm), of the same colour as the valves with a pinkish shade. Septa splitting for about three-fifth of their length; pollen-conducting tissue at the top as broad as the stigma, gradually contracting towards the curvature and extending as a thick even strand as far down as the split septa.

Peduncle short (6 cm), at the base and below the capsule of the same thickness, yellowish, darker than the capsule.

Seed very flat, subtriangular with rounded angles, dark brown, with a narrow, slightly bent, margin of a lighter shade; embryo three-quarters the length of the seed and also of a lighter shade.

Under Tashkent conditions this species flowers in the latter half of April.

Tulipa korolkowii Regel

plate 24. fig.4h

T.korolkowii grows 'on sandy and gravelly slopes in foothills. Middle Asia: Syr-Dar'ya. Endemic. Described from Farish' (Flora of the USSR, Vol. IV).

The plants studied were collected in the environs of the Darbaza station on the Middle-Asian railway.

Bulb 2-3 cm in diameter, ovoid; tunic tough almost scaly, somewhat prolonged (particularly in juvenile individuals), black, lined with woolly hairs near the top.

Stem, including the peduncle, 18-20 cm high, glabrous, greenish blue.

Leaves 3, closely set to more or less widely spaced, falcate, reflexed, channelled, very glaucous. Basal leaf 11 × 0.75 cm to 25 × 1.4 cm, linear to lanceolate or strap-shaped; upper leaf from 11 × 0.75 cm to 17 × 0.6 cm, narrowly linear.

Flower usually solitary (in cultivation there may be several), either red, yellow or multicoloured; there are no other shades of this species; scentless. Its shape is charasteristic and somewhat resembles a *T.greigii*-flower, but in some forms the inner tepals stand very erect.

Outer tepals 5.1 × 3.5 cm, rhombic with rounded angles or obovate, sometimes with a very small pubescent tip and often with very undulate margins. On the outside usually orange-red and glaucous, more red towards the edges of the reflexed part and more yellow towards the base; in yellow forms pale yellow with crimson streaks along the midrib and sparse green veins; in both forms the inner dark blotch at the base discernible on the outside. On the inside orange-red with a small (0.6 × 0.5 cm) acutely rhombic, black blotch and at the sides yellow stripes running from the truncated base to halfway the height of the blotch; in yellow forms orange-yellow with a large, broadly lanceolate orange-red blotch, half as broad as the tepal and three-quarters the height of the tepal with streaky edges and a small broadly rhombic, black spot at the base.

Inner tepals 5.2 × 3.4 cm, obtriangular, in rare cases

obovate, sometimes with a notched top and a pubescent tip on the midrib. On the outside red with an orange shading towards the midrib; at the base a small fan-shaped yellow blotch, shaded with anthocyanin and with an acute top; yellow forms have on the outside yellow edges with numerous pinkish red stripes all over the area of transition, from yellow to unbroken orange-red, and at the base a violet blotch of the same shape as the one on the inside of the tepal. In the red forms on the inside bright red with a well-marked, broadly rhombic, black blotch, truncated at the base. In yellow forms lemon-yellow along the upper edge of the tepal, lower down unbroken red which extends as a sharp prong over the yellow ground colour, sometimes right to the top of the tepal; at the base a small, fan-shaped, black spot, similar to that of the red forms.

Stamens placed higher than the ovary. Filaments 0.9 × 0.2 cm, broad and thick, glabrous, a little broadened in the central part, tapering to a sharp tip; the lower half yellow, the upper half black, the tip red; in multicoloured forms the black part is a little larger than in the red forms. Anthers 0.7 × 0.2 cm, yellow; pollen yellow.

Ovary one and a half times shorter than the stamens, cream-coloured with an anthocyanin shading. Stigma sessile, overlapping the ovary, pale yellow.

Capsule shortly cylindrical with an obtuse top and base and sometimes a small carpophore not more than 0.6 cm long; stigma small, sessile, or not more than 0.2 cm high; long capsules somewhat constricted in the middle. The capsule dehisces for two-thirds to half its length along wedge-shaped slits, 1.0 cm wide below the stigma and 2.3 cm long. The septa split to the same distance as the ribs; the irregularly rhombic part below the stigma and the split edges of the valves shaped like an anchor.

Valves elliptic or oblong, whitish straw-coloured, often with a yellowish shade in the upper part, very thin, transparent and where the seeds are shining through with a grayish pink tinge. Median groove sharply cut out, in the upper part brown and sometimes bifurcating.

Wings convex, covered all over with fine transverse wrinkles and alternating opaque and transparent stripes. Edges of the valves 1.5 mm wide, reflexed, buff to pinkish. Ribs in dry, closed capsules, rising above the wings as low, brown ridges and converging below the stigma as darker brown stripes with the grey to brown stripes of the bifurcated median groove. The pit below the stigma by the median groove is typical of this species.

Tepal scar narrow, light brown, partly hidden by the slightly attenuated carpophore. Peduncle up to 15 cm long, thin, straw-coloured with a light brown shading.

Under Tashkent conditions this species begins to flower in the beginning of April.

Tulipa nitida Hoog

plate 25, fig.4i

T.nitida grows 'on stony and gravelly slopes. Middle Asia: Pamir-Alay. Endemic. Description based on cultivated plants from Bukhara without precise indication of the place of origin' (Flora of the USSR, Vol. IV).

The plants we have studied were collected in the Denauskiy district of the Uzbekskaya SSR, 578 m left of the Termez highway on mottled rocks and in the lower mountain belt of the Baba Tag range, near Baysun, on outcrops of sandstone.

Bulb 1.5-2.5 cm in diameter, ovoid; tunic lined near the top with woolly hairs.

Stem, including the peduncle, up to 18 cm above ground level, glabrous, yellowish green, glaucous.

Leaves 3, as a rule very crispate but sometimes almost straight, rather widely spaced, falcate, reflexed, channeled, glabrous and very glaucous, nearly bluish green; the edges of a lighter shade. Basal leaf 15 × 2.0 cm, linear to lanceolate, set at ground level. Upper leaf 15 × 1.0 cm, linear.

Flower solitary, red, scentless; in shape resembling the flower of *T.korolkowii,* but with longer inner tepals and different proportions of the floral parts.

Outer tepals 5.2 × 3.3 cm, long rhombic with rounded angles; the edges of the reflexed part undulate, finely dentate; the top not pointed. On the outside red and glaucous, at the base darker and along the edges a narrow (up to 1 cm broad) yellowish green stripe (0.7 cm high). On the inside bright red, glossy, with an orange-crimson shading; at the base a well-marked, up to 0.5 cm high blotch with a notched top, solid black in the centre, fading to grayish at the edges.

Inner tepals 5.4 × 2.7 cm, obovate; at the top finely dentate. On the outside bright red with a deeply notched, blackish brown blotch up to 0.5 cm high. On the inside equally bright red, only more glossy, with a black blotch (up to 0.4 cm high) covering the base of the tepal and sharply delimited from the main colour.

Stamens longer than the ovary, 1.6 cm, i.e. more than three times shorter than the tepals. Filaments 0.6 × 0.2 cm, for three-quarters of their length black, of a lighter shade and slightly tapering towards the base, red and a little expanding in the upper part, pointed below the anthers. In some plants the filaments are entirely black; in others the red part is much larger than the black part. Anthers 0.8 × 0.2 cm, oblong, yellow. Pollen yellow.

Ovary 1.5 × 0.35 cm, tapering towards the style, green, below the style pigmented with anthocyanin; style up to 0.15 cm long, cream-coloured; stigma as broad as the ovary, broader than the style, yellow with red papillae.

Capsule up to 3.5 cm long and 1.5 cm broad, ovoid; top obtuse with a short style; at the base often at-

Plate 17. *T. greigii* Regel (60%)

Plate 18. *T. micheliana* Hoog (110%)

Plate 19. *T. maximowiczii* Regel

Plate 20. *T. wilsoniana* Hoog

Plate 21. *T. lehmanniana* Merckl. (60%)

Plate 22. *T. lehmanniana* Merckl. (60%

Plate 23. *T. korolkowii* Regel (75%)

Plate 24. *T. nitida* Hoog

tenuated; whitish straw-coloured. Carpophore often very short (0.2 to 0.5 cm), directly passing in the valves, slightly bulging over the tepal scar. Top of the valves terminating abruptly in a short style, 0.2 cm long and 1.2 cm broad, carrying a light brown stigma of the same width.

The capsule opens right up to the base along wedge-shaped slits, 3.2 cm long and 1.5 cm wide at the top. Although the septa split much below their curvature, the upper part of the valves draw so far apart that the septa crack transversely below their expanded part. The fused tissues of the septum below the stigma are strongly expanded and contract again towards the curvature, forming a 0.5 cm long and 0.35 cm broad 'scutellum'.

Valves elliptic or broadly lanceolate, with a truncated base when the carpophore is short and gradually broadening towards the middle when the carpophore is long. Median groove sharply cut, flattening out on the seedless part, of the same colour as the rest of the capsule, sometimes in the upper part purplish brown; the smooth part a little more yellow or brown than the rest of the capsule. The part of the valves below the stigma light brownish.

Wings very thin but firm, slightly convex, whitish straw-coloured, covered with very thin transverse wrinkles and without brown stripes. Rims of the valves whitish brown extending as thin stripes in the narrow groove along the ribs. Cleavage surface of the ribs narrow (0.15 cm), dull brownish turning to a buff colour.

Tepal scar narrow, not wider than 0.1 cm, raised. Peduncle strongly tapering below the scar, up to 13 cm long, brown or a whitish straw-colour.

Seed 0.65 × 0.5 cm, shaped like a scalene triangle, dark brown with a narrow, slightly bent margin.

Under Tashkent conditions this species begins to flower in mid-April.

Tulipa rosea Vved.
plate 26

T.rosea grows 'on outcrops of mottled rocks. Middle Asia: Pamir-Alay, the eastern part of the foothills of the Turkestanskiy mountain range. Endemic. The plants here described came from Kim'. (Flora of the USSR, Vol. IV).

The plants we studied came from mass collections made in the environs of the village of Kim in the Fergana Basin on the Belesnik upland (50 km south of Leninabad), in 1957.

Bulb 1.5-2.0 cm in diameter, ovoid, seated at a depth of 5-10 cm; tunic hard, coriaceous, black, often prolonged nearly to ground level, on the inside towards the apex and in the prolonged parts densely lined with long woolly hairs.

Stem, including the peduncle, short, at flowering up to 11 cm above ground level, glabrous, brownish green.

Leaves 3, rather closely spaced at the base of the stem, strongly reflexed, channeled and a little curled, glabrous, glaucous, usually overtopping the flower. Basal leaf 13 × 1.8 cm, linear to lanceolate; the top leaf 11.5 × 0.7 cm, narrowly linear or narrowly lanceolate.

Flower pink, of a paler shade in the eastern part of its area of distribution and bright pink in the western part; scentless. In shape of the *T.korolkowii* type with widely set, erect, inner tepals and the outer tepals slightly reflexed along the midrib.

Outer tepals 4.4 × 3.2 cm, rhombic with the reflexed parts finely dentate and slightly undulate. On the outside pinkish crimson with a glaucous tinge except for the very edge of the reflexed part; at the base greenish with the dark inner blotch shining through. On the inside orangy pink with the pink shade predominating on the lower half of the tepal and at the base a small (0.5 × 0.4 cm), well-defined, black blotch with a shortly acuminate top.

Inner tepals 4.2 × 2.7 cm, long obtriangular with a finely dentate and undulate top. On the outside pinkish crimson with a faint orange shading towards the top and at the base the inner blotch shining through as a purple blotch, up to 0.5 cm high and as broad as the tepal. On the inside more reddish with at the base a well-defined, obtuse or three-peaked black blotch, up to 0.6 cm high, covering the whole base of the tepal.

Filaments 0.75 × 0.2 cm, flat, tapering towards the top, glabrous, black becoming gradually reddish below the anthers and at the base. Anthers 0.7 × 0.2 cm, yellow.

Ovary 1.2 × 0.3 cm including the stigma, bright or pale pink, hardly constricted below the stigma. Upper part of the ovary (style) and stigma light yellow; lobes of the stigma slightly overlapping the top of the ovary.

Under Tashkent conditions this species begins to flower in mid-April.

Tulipa schrenkii Regel
plate 27, fig. 4j

T.schrenkii grows 'in steppes, deserts and semideserts. European part: Transvolga area (southern part), Black Sea area, Crimea, Lower Don, Lower Volga. Caucasus: Ciscaucasus, Dagestan, Western Transcaucasus (northern part) and Eastern Transcaucasus. West Siberia: Upper Tobol, Irtysh (western part), Middle Asia: Aral-Caspian (northern part). General distribution: Lesser Armenia and Kurdistan. Described from the environs of Ishim' (Flora of the USSR, Vol. IV).

In our collections were plants raised from bulbs brought from Berchogur railway station where they were dug up on mount Bokhtubai from among stones. Others were sent by Zubashko from Kamysh-Burun peninsula on the Crimea. Besides we raised them several times from seed collected in the natural habitats or from cultivated plants.

Bulb 1.5-4.0 cm in diameter, ovoid with a rather tender, coriaceous, black to tawny tunic, lined all over with adpressed hairs.

Stem at flowering 10-30 cm above ground level; the upper part including the peduncle pigmented by anthocyanin, glabrous (in the Crimean plants often shortly pubescent).

Leaves 3-4 (5); the lower ones more or less reflexed, falcate, the upper ones widely spaced, glaucous, glabrous or pubescent with undulate and ciliate edges. Basal leaf 16 × 0.6 cm, set at ground level, lanceolate; top leaf 11 × 2.7 cm, narrowly lanceolate. There are plants with larger leaves.

Flower usually solitary, very variable in shape from narrowly cup-shaped to a widely opened bowl very similar in type to a liliaceous flower and with a faint pleasant scent. In comparison with other species *T.schrenkii* offers the richest variety in shape and colour; there are claret-red, yellow, pink, white and multicoloured forms.

Outer tepals on an average 6.7 × 3.4 cm, long rhombic, at the top rather abruptly tapering to a pubescent tip. On the outside the widely differing shades softened by a glaucous bloom; along the midrib often greenish yellow or whitish and at the base a rhombic, greenish yellow blotch. On the inside of a little lighter colour than outside with no blotch or only a faint yellow shading at the base, or a rhombic blotch with an acute top which may be pure yellow, black, grayish and sometimes streaky with a yellow margin of varying width; these blotches do not cover more than one quarter of the length of the tepal.

Inner tepals on an average 6 × 3 cm, they may be broader, long obovate or rhombic with rounded angles or elliptic, tapering to a pubescent tip which is somewhat shorter than on the outer tepals, or with a notched top. On the outside brighter coloured than the outer tepals, often without any blotch but only paler towards the claw and sometimes with a fan-shaped, yellow or green blotch covering the base and extending to one-fifth of the height of the tepal. On the inside of the same bright colour as outside, with no blotch at all, a diffuse blotch or a well-marked, narrowly fan-shaped blotch with a dentate apex terminating in a large tooth; the blotch may be pure yellow, yellow with light brownish streaks or black with a yellow margin.

Stamens 2-3 times shorter than the tepals, a little longer than the ovary. Filaments 0.7 cm long, narrowly triangular, glabrous, yellow or black. Anthers nearly twice as long as the filaments and rather broad, yellow or darkish.

Ovary 2.2 × 0.4 cm, narrowly cylindrical, light green but darker along the ribs; stigma large, sessile, overlapping the ovary, cream-coloured.

Capsule 3.6 × 1.6 cm, in many cases much larger. Valves solid and hard, elliptic with a somewhat attenuate top and base, a short style and a seedless basal part (the carpophore); median groove running over the whole length of the valves.

Wings of a grayish straw-colour, covered with a dense network of transverse wrinkles. Along the ribs, the median groove and the wrinkles pinkish brown. Stigma often broader than the style, dark coloured.

Tepal scar dark. Peduncle grayish or pinkish brown, darker than the capsule.

Under Tashkent conditions this species begins to flower in the second week of April.

Tulipa ostrowskiana Regel

plate 28

T.ostrowskiana grows on 'clayey and gravelly slopes in the lower mountain belt. Middle Asia: Tien Shan (northern part). Endemic. Described from Alma-Ata' (Flora of the USSR, Vol. IV).

The plants studied were received from the Zailyskiy Alatau and had been collected in the environs of Alma-Ata and Talgar. We also raised this species several times from seed of various origins.

Bulb 1.5-3.0 cm in diameter, ovoid; tunic black, tough and coriaceous, close to the top densely lined with adpressed hairs.

Stem, including the peduncle, 18 cm high above ground level; peduncle thin, flexible, bluish green, in some forms with a claret-brown tinge.

Leaves 2-3, glabrous, glaucous, strongly deflecting from the stem and closely spaced, a character by which this species differs from *T.kolpakowskiana;* in typical forms lying nearly flat on the ground with only the top leaf sticking up; very crispate, somewhat reflexed and falcate; margins cartilaginous, white, ciliate. Basal leaf 15 × 3.0 cm, lanceolate, channeled. Top leaf 12 × 0.9 cm, linear-lanceolate, more strongly channeled.

Flower solitary, a very elegant star-shaped cup, orangy red, orange, yellow, multicoloured and in numerous subtle shades of these main colours; often fragrant, highly heliophilous; in bud and at the end of flowering drooping.

Outer tepals long rhombic or broadly lanceolate, somewhat undulate near the top and with a sharp tip. On the outside along the midrib bluish crimson; the margins of varying shades and intensity: sometimes very dark, orange to red, golden yellow or multicoloured; the red forms may show longitudinal yellow stripes along the edges of the lower third of the tepal. On the inside the same colour pattern as the margins on the outside, but with a yellow, often three-peaked blotch, covering the whole base to a height of 0.8-1.0 cm; on this yellow blotch in some cases an obtriangular dark spot of varying shape and shade. Some forms have only blotches on the inner tepals or, as with yellow forms, no blotches at all, while orange forms have pure yellow blotches.

Inner tepals narrowly rhombic with rounded angles, abruptly tapering to a sharp tip. On the outside yellow, orange or multicoloured, but in most cases orangy red with more orange along the midrib and towards the base; on both sides of the midrib two orange or sometimes crimson stripes of a brighter shade than the main colour; base of the tepal covered by a notched orange blotch (1-0.6 cm high in the indentation). On the inside more intensely coloured than outside; at the base covered with a yellow blotch delimited from the main colour by an almost straight line; there may be a flattened brownish spot with a streaky yellow margin of varying width on this yellow blotch.

Stamens longer than the ovary, two and a half to three times shorter than the tepals. Filaments 0.8 × 2.5 cm, narrowing towards the top and the base, but with a sharp tip; yellow, light brown, red or claret, but the tip below the anther always of a lighter shade. Anthers 1.2 × 0.2 cm, varying from yellow to black. Pollen yellow, black, brown or claret.

Ovary 1.6 × 0.4 cm, peg-shaped, varying from bluish green to a greenish cream-colour, along the ribs brighter, below the stigma yellow, green or red. Stigma broader than the ovary, red; the lobes sometimes brown, the papillae yellow.

Capsule 3.0 × 1.1 cm, narrowly ellipsoid with nearly parallel edges, only slightly narrowed towards the 3 mm long carpophore. Valves at the base and the top abruptly rounded. Seedless part of the capsule, the style, 0.5 cm high and 0.7 cm broad at the base, attached to the valves like a brownish prong. The rounded triangular dent on the prong below the stigma is typical of this species. Stigma small, sessile, and easily broken off.

The capsule dehisces right down to the carpophore along narrow slits, 2.3 cm long and 0.45 cm broad, with nearly parallel edges and only slightly broader in the middle. At the end of the slits the reflexed tops of the valves form a wide-mouthed funnel. The septa do not split very deep, 0.9 cm.

Valves subelliptic, of a whitish straw-colour; median groove shallow, bifurcating for 0.5 cm below the beak and at that place of a dull grayish brown colour. The colour of the valve's edges merging with the brownish colour of the beak.

Wings, slightly convex, delicate, papery (the seeds shining through), covered with fine transverse wrinkles; in many cases with very thin, barely visible, transverse brown stripes in the upper part below the beak. Edges of the ribs arched, not curving outwards, of a brownish to pinkish shade; along the ribs a clear-cut, shallow groove.

Tepal scar slightly raised. Peduncle very long, up to 35 cm, narrowing below the tepal scar, lower down thin and slender, whitish grey.

Seed small, 0.5 × 0.4 cm, light brown with hardly any margin.

Under Tashkent conditions this species begins to flower in mid-April.

Tulipa kolpakowskiana Regel

plate 29, fig.3a

T.kolpakowskiana grows on 'clayey and gravelly slopes in the foothills. Middle Asia: Dzungarskiy Alatau (southern part); Tien Shan (northern part). General distribution Dzungaria-Kashgar area. Described from Alma-Ata' (Flora of the USSR, Vol. IV).

The plants studied were received from the Chu-Iliyskiye mountains (the Sary-Bastau range in the environs of Androkai station on the Turkestan-Siberian railway) and from the environs of the Chokpar station; the Zailiyskiye Alatau (from the Talgar river valley near the Talgar railway station and Razdol'ye).

Bulb 1.5-3.5 cm in diameter, ovoid; tunic hard, coriaceous, black to brown, lined with rather thin and straight, adpressed hairs.

Stem, including the peduncle, 10-35 cm high above ground level, glaucous, glabrous.

Leaves 3, all three erect (never lying on the ground), deflecting at an angle of no more than 30-35° from the stem, slightly reflexed at the top and with 2-3 cm long internodes; the upper leaves successively smaller and narrower, at first bloom exceeding the flower, very glaucous, glabrous, somewhat undulate but deeply channeled; the edges ciliate. Basal leaf 18 × 2.2 cm, lanceolate. Top leaf 13 × 0.7 cm, seemingly twice as narrow by being flatly folded.

Flower solitary, yellow, in bud pendulous but straightening up before flowering, shaped like a double cup; with a faint honey scent.

Outer tepals 2-6 cm long and 0.7-3 cm broad, long rhombic tapering to a sharp tip. On the outside along the midrib greenish with the yellow ground colour showing through, softened by a bluish tinge and the green of the midrib extending to the tip; the edges yellow. Colour on the inside, in the centre, much the same as on the outside only more golden yellow; towards the edges much paler.

Inner tepals long obovate or narrowly subrhombic, elongated in the lower half, the top attenuated into a long tip. On the outside yellow, more golden towards the base and with a greenish midrib. On the inside also yellow, becoming brighter towards the base.

Stamens 2.5 to 3 times shorter than the tepals and a little longer than the ovary. Filaments 0.8 × 0.21 cm, of the same width throughout their length, at the top more or less rounded, glabrous, bright yellow. Anthers as long as the filaments but paler yellow; dehiscing abruptly along their whole length.

Ovary 1.35 cm long and 0.25-0.35 cm broad, shaped like a thick peg, bright green, darker along the ribs. Stigma almost sessile, its lobes clasping the ovary, cream-coloured.

Capsule 4.0 × 1.5 cm, when still closed in the upper part ellipsoid, almost cylindrical, somewhat expanding towards the top particularly when the capsule opens. This species has practically no carpophore. The obtuse top of the valve has a small pointed tip (0.2 cm long and 0.2 cm broad at the base) to which the small stigma lobes are attached which fall off when the capsule ripens.

The capsule dehisces along wedge-shaped slits to 1 mm above the tepal scar; the slits widen a little in the central part of the capsule (0.7 mm) owing to the strongly outcurving ribs; towards the top the slits narrow again (to 0.5 cm). The septa split only for 1.3 cm from the top and in the upper part of the carpels draw apart for 1 cm, which is less than in species with a large seedless part in the top of the capsule.

Valves oblong, obtuse at the top and at the base where they are somewhat narrower than above; straw-coloured; median groove rather shallow, but deeper in the central part, sometimes bifurcated below the stigma for a length of 0.5 cm; in some plants the top of the groove has a dull brown colour.

Wings slightly convex or, particularly in the central part, flattened; rather soft and thin, but thicker than in *T.ostrowskiana,* almost opaque; along their whole length with fine wrinkles and very thin, transverse brown stripes. Edges of the ribs along the cleavage surface of a light brown colour deepening towards the stigma. The opened capsule has a typical shape caused by the strongly reflexed edges of the valves.

Tepal scar rather narrow, brown. Peduncle thin, up to 0.25 cm in diameter, 28 cm long; with a stronger whitish sheen than the valves of which the yellow shade predominates. Capsule-bearing plants have long internodes.

Seed 0.45 × 0.3 cm, light brown with practically no margin.

Under Tashkent conditions *T.kolpakowskiana* begins to flower in the second week of April.

Tulipa zenaidae Vved.
plate 30

T.zenaidae grows 'on fine-textured and stony slopes in the lower mountain belt. Middle Asia: Tien Shan (Kirgizkiy Alatau). Endemic. Described from Shekule, Aleksandrovskiy range' (Flora of the USSR, Vol. IV).

We brought the material from Mount Shekule in the Kirgizkiy Alatau (Tien Shan) and we received from the Frunze Botanical Garden plants which had also been collected in the Kirgizkiy Alatau by R. Shpak in the Dzharly-Kainda river valley and on the southwestern slopes of the Mokraya-Burulyu gorge.

Bulb up to 3 cm in diameter, ovoid; tunic tough, black, becoming brownish red when old, near the top rather densely lined with straight, adpressed hairs.

Under our conditions the plants remain very small when flowering. The flower opens as soon as the bud appears above ground and at the beginning of flowering the bases of the leaves are underground.

Stem, including the peduncle, glabrous, slightly pigmented by anthocyanin; in the later stages of flowering continuing growth and reaching to 15 cm above ground level.

Leaves 3, at the beginning of flowering rather closely spaced and extending beyond the flower, later widely spaced not overtopping the flower; glabrous, light green with veins of a brighter shade, glaucous; the margins ciliate, not crispate; the tips often red. Basal leaf 19 × 6.0 cm, lanceolate. Top leaf 7.0 × 1.5 cm, lanceolate.

Flower solitary, very characteristic although the short peduncle detracts from its decorative value; shaped like an elegant double cup, the upper third of the outer tepals strongly curved outward and the inner tepals only slightly reflexed at the top; scentless. It has a tendency to develop laciniate tepals (Parrot-type), in particular the inner tepals.

Outer tepals 5.3 × 3.0 cm, broadly lanceolate or narrowly rhombic, pointed. On the outside along the midrib reddish pink on a yellow background which only shows through on the edges; at the base a lanceolate blotch (17 cm long, 0.4 cm broad) of the same colour as the edges. On the inside of the same yellow colour as the edges on the outside; at the base a small rectangular, black blotch with a tooth in the middle (0.5 cm high including the tooth and 0.5 cm broad). On the inside of the tepal, at the base, stomata can be clearly seen with the aid of a magnifying glass, particularly in the black area.

Inner tepals oblong or elliptic, abruptly tapering to a sharp point. On the outside yellow with a whitish tinge. On the inside a warm, golden yellow deepening towards the base, which is covered with a small black blotch up to 0.3-0.4 cm high with a straight top.

Stamens 1.6 cm long, two and half to three times shorter than the tepals. Filaments 0.6 × 0.15 cm, peg-shaped, with a sharp tip, glabrous, in the lower half black, the upper part yellow. Anthers 1.0 × 0.15 cm, oblong, yellow.

Ovary 1.3 × 3.25 cm, barrel-shaped, green or violet

to almost black with a bluish tinge. Stigma narrower than the ovary, bright yellow.

Some forms have no black blotches at the base of the tepals and the yellow of the base merges with the ground colour while also the filaments are entirely yellow. There are forms with orange and red flowers, but the plants we raised from old collections as well as the new ones we received were all yellow.

Capsule 3.6 × 1.5 cm, ellipsoid, nearly cylindrical in the seedbearing part; at the base attenuated to a short carpophore (0.3 × 0.5 cm); seedless part below the stigma brown. Stigma small, sessile, fragile. The capsule dehisces right down to the carpophore along three slits with nearly parallel edges and terminating in a funnel at the acuminate tips of the valves which may be reflexed after dehiscence; in some forms the slits are much wider at the base. The septa do not split very deep, only for 2 cm in the capsule described here.

Valves long rectangular with an acuminate top and base, straw-coloured with brownish shadings; median groove deep, often with one or two longitudinal brownish veins.

Wings flattish, coarse, strongly wrinkled, with about seven thin, brownish veins running from the median groove and forming a lacy pattern along the edges, but not extending to the grooves along the ribs; in the upper part of the capsule the pattern becomes denser and near the median groove darkens to brown and sometimes dirty pink. Edges of the valves strongly and evenly reflexed all along their length. Carpophore broader than the peduncle, bulging over the tepal scar.

Tepal scar brown. Peduncle of a much lighter shade than the capsule, often soft. By the time the capsule ripens it reaches to 32 cm above ground level with a distance of 6 cm between the internodes, the basal leaf remaining at ground level.

Seed 0.5 × 0.4 cm, dark brown with a small bent margin.

Under Tashkent conditions this species begins to flower in the latter half of April.

Tulipa ferganica Vved.

fig.3c

T.ferganica grows 'on stony slopes in fine-textured soil in the middle mountain belt. Middle Asia: Pamir-Alay (Alayskiy range); Tien Shan (Ferganskiy and Chatkal'skiy ranges). Endemic. Described from the Kugart river' (Flora of the USSR, Vol. IV).

We brought the material for our studies – bulbs and seeds – from the Ferganskiy mountain range, where it was collected at various places along the Kugart and Yassy rivers. P.A. Gomolitskiĭ collected plants at an altitude of 2 700 m on the Baubashata and Shamurat ridges in 1951 and T.I. Slavkina and P.M. Murzova from the Arkit Forest near lake Sarychelek.

Bulb up to 3,5 cm in diameter, ovoid; tunic black to brownish, tough, coriaceous, near the apex and the base lined with adpressed hairs.

Stem, including the peduncle, up to 25 cm above ground, pubescent, glaucous, sometimes pigmented by anthocyanin, sturdy. Although the flowers frequently open when still very close to the ground, the peduncle elongates considerably during flowering.

Leaves from 3 to 5, closely set (sporadically more widely spaced), reflexed; the lower leaves coarsely undulate, the upper more or less crispate, towards the end of flowering not overtopping the flower, glaucous, pubescent (the upper leaves more so than the lower ones which are nearly glabrous); edges cartilaginous, ciliate, light yellow. Basal leaf up to 16 cm long and 6.7 cm wide, broadly lanceolate, clasping the stem at the base. Top leaf up to 13.5 cm long and 2.0 cm broad, narrowly lanceolate.

Flowers one or two, in cultivation sometimes more, star-shaped with a somewhat concave base, yellow, on the outside bluish pink; scentless.

Outer tepals on an average 5.3 × 2.3 cm, lanceolate with an acuminate top and a pubescent tip. On the outside along the midrib pinkish brown of varying intensity, glaucescent; at the claw green extending like a wedge nearly halfway up the tepal and often even discernible at the top of the tepal; along the edges bright yellow bands up to 0.4 cm broad. On the inside bright yellow without any blotches at the base.

Inner tepals on an average 5.0 cm long and 2.0 cm broad, oblong with the acuminate top terminating in a slightly pubescent or naked tip. On the outside of the same yellow colour as inside and with only in the lower half along the midrib a rather broad greenish band, which sometimes extends as a narrow stripe to the top of the tepal; occasionally this band and the stripe have a reddish tinge. On the inside pure yellow without any blotch at the base.

Stamens two or three times shorter than the tepals, longer than the ovary. Filaments 0.7 × 0.2 cm, gradually tapering to the top, orange-yellow, paler towards the base. Anthers 0.9 × 0.2 cm, oblong, orange; dehiscing abruptly all along their length.

Ovary 0.9 × 0.35 cm, short and thick, yellowish green, of a darker shade towards the stigma and often along the ribs and in the middle of the carpels with anthocyanin streaks. Stigma sessile, broader than the ovary, the bright yellow lobes just overlapping.

Capsule 4.6 × 2.0 cm, long ovoid, of a dull, pale straw-colour. The flat, broad zone below the stigma, formed by coalescence of the tops of the wings, the median groove and the ribs, dull light brown. The carpophore has specific characters: it is whitish, nearly cylindrical, not very high, almost as thick as the peduncle and inserted like a sharp prong below the median groove in the protruding valves. On dehiscence the valves are forced widely apart forming rounded slits in the lower part of the capsule which slightly contracts in the centre to widen again at the top.

Valves compact, in opened capsules elliptic and owing to the strongly reflexed edges in the lower part, broader at the top than at the base; on the outside dull, on the inside glossy yellowish; below the stigma 0.7 cm wide, in some cases less, terminating in a blunt or indented top carrying the small and narrow, dark remnants of the stigma. Edges of the split ribs strongly reflexed, light brown, narrowing towards the top; texture felty. Median groove in the lower half deep and narrow (0.1 cm), flattening out in the upper half to 0.7 cm wide below the stigma, throughout its length covered with finely wrinkled light brown scar tissue and with a raised vein in the lower half of the groove which bifurcates in the top part.

Wings in the upper part with numerous transverse wrinkles; between the wrinkles light brown; only the whitish, convex lower part of the wings smooth. Along the ribs a gently sloping groove, strongly curved in the lower three-quarters of the capsule and of the same dull, light brown, colour as the ribs.

Tepal scar consisting of two raised parallel bands of a dull brown colour. Peduncle long, darker than the capsule, densely covered with short, whitish hairs, clearly visible to the naked eye.

Seed obliquely triangular, dark brown, with a rather broad, bent and somewhat wrinkled margin. Embryo of a lighter shade.

Under Tashkent conditions this species begins to flower in the third week of April.

Tulipa tetraphylla Regel
plate 31

T. tetraphylla grows 'on stony slopes. Middle Asia: Tien Shan (central Tien Shan). General distribution Dzungaria-Kashgar area. Described from the Kochkor river' (Flora of the USSR, Vol. IV).

We collected the plants for our studies in the Zailyskiy Alatau, near the village of Talgar, while others were brought by Ya.S. Vesnevskiĭ from the upper reaches of the Naryn river.

Bulb 1.5-3.0 cm in diameter, broadly ovoid; tunic tough and coriaceous, brownish black, lined at the apex with thick and straight, adpressed golden hairs.

Stem 3-25 cm above ground level, glabrous. Peduncle highly glaucous, near the flower evenly bluish, glabrous.

Leaves 3-7 (in vigorous plants), but as a rule 4-5, closely spaced, their bases still underground at flowering, reflexed, falcate, strap-shaped, with an acute top, channeled, often crispate, glabrous and highly glaucous; margins white, ciliate. They extend beyond the flower and are all of nearly equal length, e.g. the basal leaf 14.7 × 9.0 cm and the top leaf 13.5 × 5.5 cm; in some cases longer and broader.

Flowers 1-4, yellow, shaped like a double cup, in bud pendulous; scentless.

Outer tepals 2-6 cm long and 0.7-2.9 cm broad, broadly lanceolate or long rhombic, the top half reflexed. On the outside the yellow green along the midrib running parallel to the outline of the tepal and margined by a broad, claret-crimson band, consisting of dense streaks; the upper edge yellow. On the inside pure yellow.

Inner tepals 1.3-8.0 cm long and 0.8-3.5 cm broad, narrowly obovate with an attenuate top terminating in a tooth. On the outside yellow, the midrib green; on both sides a large, pointed, streaky green blotch, broadening and deepening in colour at the base, fading towards the top. On the inside pure yellow.

Stamens three times shorter than the tepals, one and a half times longer than the ovary. Filaments thickish, broadening in the upper half, pointed below the anthers, glabrous, yellowish green. Anthers equalling the filaments (0.7 cm), yellow; sometimes coiled into a spiral. Pollen dirty yellow.

Ovary 1.1 × 0.35 cm, a little shorter than the stamens, thick with rounded ribs, bright green. Stigma sessile, yellow.

Capsule 3.7 × 1.7 cm, broadly ellipsoid, more or less abruptly narrowed towards the base and the top. Carpophore up to 0.5 cm long and 0.4 cm broad. Style short, up to 0.4 cm; in many cases a style is absent and the capsule is not narrowed at the top because the valves terminate abruptly in the stigma, which is small and easily detached. The capsule dehisces along wide, wedge-shaped slits (1.1 cm wide at the top with a 2.1 cm long slit) with straight edges; the septa separate for more than half the length of the capsule (1.8 cm), which accounts for the wide slits.

Valves broadly lens-shaped, with a light constriction at the base and the top; median groove deep not continuing on the carpophore and flattening out below the stigma, brown all along its length; the brown stripe broader in the upper part and vanishing at the base of the capsule.

Wings, unlike those in other closely related species, strongly convex and slightly protruding above the carpophore; delicate, papyraceous, straw-yellow to white and transparent (seeds visible). In the upper third transversely wrinkled; wrinkles sparse, but rather deep with pale, lightbrown, transverse veins. Reflexed edges of the ribs separated from the wings by a groove of a light brown colour which darkens in the upper part; in some cases only the veins, forming a narrow and toothed margin along the edges, are coloured. The brown stripes of the median groove and the edges converge below the stigma lending a light brown colour to the upper narrowed part of the valves.

Tepal scar 1.5-2.0 mm broad, brown. Peduncle narrower than the seedless part of the capsule, light brown, darker than the capsule, much shorter than in other related species; under our conditions not longer than 15 cm. The leaves, which are rather closely set during flowering, become much more widely spaced after this period.

Under Tashkent conditions this species begins to flower in the second week of April.

Tulipa korshinskyi Vved.

T.korshinskyi grows 'on stony slopes in the middle and upper mountain belts. Middle Asia: Pamir-Alay. Endemic. Described from Sary-dash' (Flora of the USSR, Vol. IV).

The following description is based on plants from the original habitats which are kept in the herbarium of the Academy of Sciences of the USSR in Leningrad.

Bulb up to 3 cm in diameter, ovoid, seated at a depth of (7-) 10 (-16) cm and sometimes nearer to the surface; tunic dark brown almost black, lined at the apex and the base with adpressed, straight, orangy hairs. The large number of dead sheaths and the globose shape of the bulb suggest that it lies in a loose sandy or gravelly soil.

Stem during flowering underground together with the bases of the closely set leaves and the flower which as a rule opens quite close to ground level. Stem and peduncle glabrous. In the period between opening of the flower and ripening of the capsule the peduncle attains a length of 5.5. cm, sometimes more.

Leaves (2-) 4 (5), closely set on the stem, strongly declinate, often falcate, usually glabrous; the upper leaves sometimes ciliate; margins crispate; extending beyond the flower. Basal leaf lanceolate, at flowering from 5 to 14 cm long and from 1.5 to 3.0 cm broad; the top leaf from 3.0 to 13 cm long and from 0.5 to 0.9 cm broad.

Flower solitary, stellate with a thin peduncle. Tepals on the inside white with yellow basal blotches.

Outer tepals up to 3.3 cm long and 0.6 cm broad, linear-lanceolate, with a violet shading on the back.

Inner tepals up to 3.0 cm long and 0.7 cm broad, obovate to rhombic; the inner yellow blotch shining through on the outside which has a narrow wedge-shaped, grayish patch along the midrib.

Stamens a quarter shorter than the tepals and the anthers a third longer than the filaments. Filaments narrow, of uniform thickness, pointed below the anthers; yellow with an orange tip. Anthers bright yellow, the yellow connective protruding above the apex; dehiscing abruptly.

Ovary shorter than the anthers; the style half as long as the ovary and with a lighter shaded stigma.

Capsule 4.0-3.1 cm long and 1.6 cm broad, whitish, ellipsoid with an acuminate top and base. Carpophore 0.4 cm high, narrowing towards the peduncle, expanding towards the valves into which it fits like a sharp prong; the ribs of the capsule not extending on the carpophore. The valves gradually contract towards the top of the capsule and terminate in a sharp tooth (up to 0.3 cm long in dry capsules) which is often of a darker shade; the lobes of the stigma fall off.

Valves divided into halves by the well-marked median groove with dentate edges.

Wings more or less convex, thin, papyraceous, of a whitish straw-colour, in the upper part covered with fine wrinkles which sometimes have a brownish pink shade. Between the reflexed edges of the valves and the wings a shallow groove. Edges of the ribs along the line of dehiscense light brown; cleavage surface of the ribs a dull cream colour; very thin.

The capsule dehisces along wide wedge-shaped slits down to the carpophore, leaving practically no fringe along the cleavage. The septa split along the pollen-conducting tissue nearly to the base of the capsule, broaden in the upper part between the stigma and the curvature, but do not acquire the rhombic shape typical of other species; it is more long elliptic.

Tepal scar light brown, almost hidden by the bulging, white carpophore. Peduncle darker than the capsule, constricted below the tepal scar.

The two plants in the herbarium of the Botanical Institute which Vvedenskiĭ designated as the type are biflorous. In the plant on the left-hand portion of the sheet the subsidiary flower grows from the axil of the second leaf and has a small bracteal leaf on its stalk; the plant has three leaves besides the bracteal one. The biflorous plant in the right-hand portion of the herbarium sheet has four leaves and the subsidiary flower grows from the axil of the third leaf, its stalk for 1.5 cm coalescent with the pedicel of the main flower. There is no bract, apparently because it has intergrown with the top leaf. It is of interest that the filaments of the latter individual have on both sides long and narrow petalous wings which terminate at the apex below the anthers. The same could be observed in a uniflorous plant.

Tulipa hissarica M. Pop. et Vved.
plate 32

T.hissarica grows 'in the alpine mountain belt. Middle Asia: Pamir-Alay, Gissarskiy range. Endemic. Described from Mura Pass' (Flora of the USSR, Vol. IV).

The plants described here were received from the Dushanbe Botanical Garden, through T.I. Ryabova in 1950; from an expedition of the Institute of the Dry Subtropics in 1935 and from a collection by Z.P. Botschantzeva in the Varzob canyon (Gissarskiy range) in 1956, where they were growing in stony soil in the subalpine belt.

Bulb 1.0-1.9 cm in diameter, long ovoid; tunic papery, tawny to brown, prolonged and lined with adpressed hairs. After growing for several years in the same place the tunic is grayish and has become coarser with raised longitudinal veins, particularly when the bulbs have been transferred to conditions for cultivation. This happens also to the droppers which are initiated both in juvenile and adult plants and sink vertically.

Stem up to 7 cm high, glabrous, bluish green, at the beginning of flowering still hidden in the rosette, but afterwards elongating.

Leaves 3-4-5, very closely set, whorled, recurved, slightly crispate, short strap-shaped, bluish green with a reddish tinge which is more pronounced on the underside, along the edges and at the tip. In some cases the leaves are bright claret, glabrous or slightly ciliate and not broader than 15 mm. The upper leaves successively smaller and narrower and extending considerably beyond the flower; the top leaf sometimes petaloid.

Flower solitary, yellow, shaped like a flat star; with a pleasant scent faintly reminiscent of saffron. In cultivation this species may bear five flowers on the one stem.

Outer tepals 21-30 mm long and 7-8 mm broad, lanceolate, the edges of the upper half often notched. On the outside yellowish grey or claret-grey, in the lower third part greenish with very narrow yellow edges. On the inside golden yellow.

Inner tepals 22 to 26 mm long and from 8 to 13 mm broad, long lanceolate, often with an obtuse top, a narrow claw and sometimes notched edges. On the outside golden yellow, the base covered with green streaks which extend upward along the central part of the greenish midrib. On the inside golden yellow.

Stamens one and a half to two times shorter than the tepals; with dried anthers up to 18 mm long. Filaments yellow, glabrous, flat, tapering from the base to the top (up to 7.5 mm long), terminating in a very thin prong. Anthers linear-oblong, up to 9 mm long when dried, orange. Pollen orange.

Ovary 13 × 3.0 mm, shorter than the stamens, green, with a long (6 mm) style gradually broadening towards the ovary. Stigma light yellow, trilobate, strongly indented, almost sectioned, just a little broader than the style.

Capsule 3.2 × 2.1 cm, whitish straw-coloured, obtriangular with rounded corners, papery, very fragile. Carpophore small, 0.25 × 0.2 cm, in small capsules absent, a little thicker than the peduncle, white, very thin in comparison with the overhanging valves into which it fits like a sharp prong. Style up to 0.3 cm long, borne on the obtuse top of the valves, dark brown with a small brown stigma, its lobes easily dropping off.

Valves thin, papery, elliptic or obtriangular with rounded angles; median groove with a brown vein, often of a darker shade in the upper part and with transverse wrinkles separated by bright brown veins branching off in the upper two-thirds of the valve.

Wings convex with transverse wrinkles in the upper part and longitudinal creases in the lower part. The recurved edges of the valves separated from the wings by a groove of a darker brown shade than the rest of the capsule. Cleavage surface of the ribs very narrow (up to 0.1 cm), cream-coloured with sparse thin inward-pointing hairs.

The capsule opens along wedge-shaped slits down to the base of the valves; slits narrow below and wide at the top. The septa split right down to the base of the capsule and are not broadened below the stigma, the strand of pollen-conducting tissue being only a little thicker there.

Tepal scar dark brown, wider below the protruding capsule than above the thin, yellowish, peduncle.

Seed large, 0.7 × 0.35 cm, triangular with rounded angles or crescent-shaped, lightbrown, rough with a coarse, irregular, narrow margin; the embryo not shining through.

This species flowers from the middle till the end of March. In cultivation it is propagated vegetatively.

SECTION *SPIRANTHERA* VVED.

Introducing this section A.I. Vvedenskiĭ describes it as follows: 'Anthers opening gradually and slowly from apex to base, coiling. Filaments glabrous. Ovary with subsessile stigma. Bulb tunic more or less hairy on the inside'.

Tulipa kaufmanniana Regel

plate 33, fig.3d

T.kaufmanniana grows 'on stony slopes in the lower and middle mountain belts. Middle Asia: Tien Shan (Western Tien Shan). Endemic. Described from the Chrichik' (Flora of the USSR, Vol. IV).

The plants studied were collected in the Western Tien Shan in the mountains which form the basin of the river Chirchik, in the Pskem, Ugam and Chimgan valleys; along the Angren river and in the Aksu-Dzhabaglinskiy reservation and the Tavdy-Bulak gorge. This species inhabits the mountain shrub belt among bushes and between boulders.

In the Karankul'say gorge, on a mountain slope close to the river Chirchik, we came on a large rocky scree with fine earth between the rocks and overgrown with shrubs. The slopes with a deep layer of loess soil were covered with couch grass. Among the rocks and the bushes we found numerous capsules of *T.kaufmanniana* and on the couch meadows we found capsules of *T.greigii.*

Plate 25. *T. rosea* Vved.

Plate 26. *T. schrenkii* Reg

Plate 27. *T. ostrowskiana* Regel

Plate 28. *T. kolpakowskiana* Rege

Plate 29. *T. zenaidae* Vved.

Plate 30. *T. tetraphylla* Regel (80%

late 31. *T. hissarica* M. Pop. et Vved.

Plate 32. *T. kaufmanniana* Reg

Bulb. Although Vvedenskiĭ mentions in the 'Flora of the USSR' that the bulbs are ovoid, their shape varies widely: from very narrow, elongated like a radish to almost napiform. Bulbs of adult plants from 1.5 to 8.0 cm in diameter (in cultivation). Tunic variable in colour and texture, from tawny black and coriaceous to a delicate golden brown, papery, lined with adpressed hairs, the golden pubescence denser towards the apex and the basal plate.

Stem 10-50 cm above ground level; the upper part including the peduncle rather pubescent, sometimes glabrous, often pigmented by anthocyanin in a wine-colour of varying intensity.

Leaves from 2 to 5, closely set, glaucous with a dark network of veins against the lighter shaded leaf tissue, as a rule not crispate but sometimes slightly undulate, particularly in plants with narrow or rounded leaves, glabrous; margins ciliate. Upper leaves successively smaller not overtopping the flower, erect and slightly recurved; the lower ones declinate. Basal leaf from 2 to 20 cm broad, lanceolate or long elliptic; in some forms this leaf is almost round, particularly in plants with a stunted peduncle.

Flower, usually solitary (sturdy cultivated plants may have as many as five flowers), generally white or cream-coloured, but there are also plants with yellow, orange, flame-coloured, light red, brick-red and nearly claret-red flowers. The pollychrome forms are typical of this species and the experiments we have carried out led to the conclusion that the hypothesis about the origins of the polychrome forms being the result of interspecific hybridization is not correct. The shape of the flower varies from flatly stellate to lily-shaped or the reverse, cup-shaped, the most noble form of the cultivated tulips. Many forms have a pleasant, more or less strong, fresh scent.

Outer tepals 2.5-11 cm long and 0.8-4.5 cm broad, in most cases lanceolate or broadly lanceolate, sometimes elliptic and often recurved. On the outside along the midrib red, crimson, pink or bluish, sometimes greenish and in rare cases white; the edges of the recurved part of the same colour as the inside; at the base a deeply indented or elliptic, yellow blotch of varying size. On the inside white, cream-coloured, yellow, red etc. and a large, bright yellow blotch with diffuse edges, varying in shape from acute-angled to rounded or indented and two-fifths the length of the tepal; at a third from the top there are frequently symmetrical red streaks on the yellow ground colour of the blotch, sometimes merging to form a V-shaped or a solid large spot. There are forms in nature and in cultivation which do not possess these spots.

Inner tepals often erect, 2.5-10.5 cm long and 1.0-5.5 cm broad, long elliptic. oblong, or long rhombic with rounded angles; the top obtuse. The outside and inside of the same colour; white, cream-coloured, yellow, red or sometimes multicoloured; the inner yellow blotch shining through on the outside. In some lighter coloured forms the midrib is often green or red. On the inside at the base a fan-shaped, bright yellow blotch covering the whole claw and extending to approximately one-third the height of the tepal; in some cases it is notched near the midrib. From the blotch a yellowish wash spreads over the top half of the tepal, particularly in the cream-coloured and yellow forms.

Stamens one and a half to two times shorter than the tepals. Filaments glabrous, broadening towards the base, sharply pointed at the top, yellow to orange. Anthers yellowish, linear, shrinking transversely on discharging pollen, often 3-4 times longer than the filaments, opening gradually from apex to base in the course of several days and coiling into a spiral or becoming twisted.

Ovary much shorter than the stamens before dehiscence and exceeding the latter after coiling of the anthers; greenish yellow, along the ribs pigmented by anthocyanin. Stigma subsessile, some forms showing a trace of a style by a constriction of the ovary below the stigma. Stigma cream-coloured, its lobes protruding above the style.

Capsule 3-9 cm long and 1.5 to 4.0 cm broad, straw-coloured, triangular in cross section with a much attenuated top and base. Carpophore long, particularly in the larger capsules, up to 3 cm long and 1.5 cm across, thicker than the peduncle, but much thinner than the capsule, rounded and of a lighter shade than the capsule. Style often very long (up to 3 cm), variable in shape, i.e. with obtuse valves the style is short and with long drawn-out valves it is longer and forms an integral part of the valves. Style of the same light shade as the rest of the capsule. Stigma small, recurved along the brown median groove; the lobes firmly attached and only slightly broader than the style.

Valves lanceolate or broadly lanceolate; median groove shallow, varying in depth and colour from one form to another; its upper half often brown. In some cases the upper part of the valves shows a herring-bone pattern of small, usually colourless wrinkles, branching off from the median groove.

Wings coarse, compact, mostly flat, without transverse stripes. Groove along the ribs broad, gently sloping, transversely wrinkled, the rim of a darker shade. Cleavage surface of the ribs rather broad (up to 0.7 cm), pinkish yellow; texture felty.

The capsule dehisces along narrow wedge-shaped slits to the base of the valves, forming a rather wide-mouthed funnel at the top. The septa split to a little below their curvature. Where the valve broadens below the stigma the strand of pollen-conducting tissue contracts down to the curvature in the septa forming a long and narrow lens-shaped figure.

Tepal scar rather narrow (1-2 mm), brown, often

with irregular edges. Peduncle above thick, but narrower than the carpophore, becoming thinner over a distance of 4 cm and remaining of equal thickness below. It continues to grow after the flower has faded and may reach to 30 cm above the top leaf.

Seed large, e.g. 1.3 × 0.8 cm, triangular with rounded angles and of variable dimensions, light brown; margin broad, particularly in the upper part somewhat warped and bent. The endosperm shows a dark crescent in the top part of the seed against its longer side. Embryo long, clearly visible from both sides.

Under Tashkent conditions this species begins to flower in mid-February.

Tulipa anadroma Z. Botsch.

plate 34, fig.3h

The samples for our studies were brought by R.M. Murzova from the Chatkal'skiy mountain range, the Arkit Forest near lake Sarychelek, where it was found growing under walnut trees. Endemic.

Bulb up to 3.5 cm in diameter, ovoid; tunic brown, papery. Even in cultivation this species does not propagate vegetatively. In the open the plant attains a height of 30 cm above ground level, when shaded by trees it may be taller.

Stem in the upper part, including the peduncle, pubescent, green, without anthocyanin pigmentation.

Leaves 3-4, erect and only very little deflected from the stem, very straight, not crispate, keeled and folded along the well-pronounced midrib, green with a glaucous tinge, slightly pubescent; margins ciliate. Basal leaf set at 2-3 cm above ground, 21 × 7.4 cm, keeled, lanceolate, with an acute top. Upper leaves pressed close to the stem.

Flower solitary, as a rule overtopping the leaves, yellow, in the early stages shaped like a double cup which subsequently opens to a star with an elegant lily-shape; emitting a pleasant honey scent.

Outer tepals up to 7.5 cm long and 2.5 cm broad, lanceolate or long lanceolate, the recurved part laterally weakly developed. On the outside on both sides of the midrib a broad pinkish crimson band extending along the edges of the recurved part; its outer edges (1.5-2 mm) bright yellow; at the base to a height of 1.3 cm yellow which extends along the midrib as a narrow wedge, 2 cm high and 0.4 cm wide at its base. On the inside like the outside edges bright lemon-yellow with no blotch or orange shading at the base.

Inner tepals as large as the outer tepals, narrowly subrhombic, on both sides bright lemon yellow without any blotches at the base; only the midrib on the outside and the top on both sides of the recurved part acquire during flowering a reddish tinge, spreading in narrow streaks over the grooves along the midrib or covering the entire upper half of the tepal except the edges.

Stamens up to 2.5 cm long, ca. half the length of the tepals, very narrow; in bud with their tips joining above the ovary. Filaments long and thin, 1 × 0.15 cm, tapering to the top, light yellow, translucent. Anthers 1.5 × 0.15 cm, linear, dull yellow differing in shade from the filaments. As in *T.kaufmanniana* the anthers dehisce gradually from the top downward in the course of 2-3 days, but they do not coil.

Ovary 2.1 × 0.55 cm, somewhat tapering towards the top, pale greenish yellow, glaucous. Stigma overlapping, of a more yellow shade.

Capsule up to 4.8 cm long and 1.9 cm broad, narrowly conical with a narrowly drawn-out carpophore (0.9 cm long and 0.6 cm broad); in some plants the carpophore is hardly developed and the valves separate down to the tepal scar on dehiscense. The upper seedless part large, drawn out from one-third to two-fifth of the length of the valves without any constriction below the style. Lobes of the stigma coalesced with the valves, very small, barely noticeable, of the same colour as the valves.

Valves narrowly lanceolate to lanceolate. Median groove gently sloping, flattening out on the seedless part which is covered with a rough epidermis; in the groove a narrow vein or cicatrice, which in some cases extends like a ridge over the seedless part or bifurcates, the area between its branches covered with scar tissue.

Wings small in relation to the length of the capsule, more convex at the base than in the upper part where a close examination reveals fine wrinkles of the same colour as the valves. Edges of the valves recurved, separated from the wings by a gently sloping groove. This capsule differs from that of other species by the colour of the cleavage surface of the ribs being not brown and the part of the valves below the stigma having the same colour as the rest of the capsule.

The capsule dehisces down to the base along conical slits, narrow below and up to 3 cm wide above owing to the widely recurved upper parts of the valves which coil into a half spiral. Edges of the valves recurved, narrow, up to 0.2 cm broad, of a dull pale yellow colour.

Tepal scar somewhat sunken, of the same colour as the capsule, bordered below the carpophore and at the peduncle by a thin brown line. Peduncle constricted below the tepal scar, of the same colour as the capsule. In plants which are cultivated in the open up to 27 cm long.

Seed very variable in size and shape, from 0.8 × 0.8 cm to 0.4 × 0.3 cm and from nearly round to triangular with rounded angles, light or dark brown;the strongly plicate margins are typical. Embryo in most seeds barely visible.

Under Tashkent conditions this species begins to flower in the second week of April.

T.anadroma differs from *T.kaufmanniana* by its erect habit; the folded, carinate leaves with pronounced midribs; the absence of the orange-yellow blotch or shading at the base of the tepals; the shape and margin of the seeds. It differs from *T.ferganica* by its earlier flowering; the shape of the leaves, the flower and the capsule; and the mode of dehiscense of the anthers.

Tulipa dubia Vved.

fig. 3f, 5

T.dubia grows 'on slopes' in fine-textured soil with or without gravel; the upper mountain belt. Middle Asia: Tien Shan (western part). Endemic. Described from the Chotan river' (Flora of the USSR, Vol. IV).

The plants described here were collected in Chimgan (Western Tien Shan) at the top of a sandy pass on the eastern slope of a gorge in the Bol'shoy Chimgan range, close to a snow field; they were in the flowering stage. A.Ya. Butkov also brought plants from the Angren river and others were raised from seeds collected in natural habitats and in the botanical garden.

Bulb 1-3 cm in diameter; tunic papery, black or dark brown (old decayed tunics tawny), lined with adpressed hairs.

Stem. The flower of this species opens when it is still sitting in the rosette of the leaves, in rare cases when the stem has reached 10 cm above ground level. During flowering the peduncle grows strongly, up to 15 cm, and afterwards reaches 25 cm or even more above ground level. Peduncle strongly pubescent, glaucous, often coloured pink or brown by anthocyanin.

Figure 5. *T. dubia* Vved. (Reproduced from the original edition).

Leaves 2-4, very closely set, falcate, glaucous in particular along the midrib; the margins undulate and densely ciliate, at the tip villose. The two lower leaves with their bases underground when the flower opens. Basal leaf 14 × 6 cm, broadly lanceolate; top leaf 12 × 0.8 cm.

Flower solitary, yellow with on the outside bluish pink streaks, widely stellate; with a faint pleasant scent.

Outer tepals 4 × 1.5 cm, lanceolate to narrowly subrhombic with an acute top and rounded sides; terminating in a small, often pubescent, tip. The pubescence spreading from the tip downward along the midrib and from the peduncle upward over the base of the tepal. The lower third of the tepal often ciliate. On the outside along the midrib bluish pink, in the central part towards the base light yellow, the edges lemon-yellow. On the inside lemon-yellow; at the base a nearly rectangular (1.6 × 0.8 cm), tridentate orange blotch, which merges at the top with the ground colour of the tepal or extends as a broad orangy brownish band along the midrib to the red tip of the tepal, leaving only the edges lemon-yellow.

Inner tepals oblong with a drawn-out base or rhombic with rounded angles and an obtuse, sometimes acute top. On the outside lemon-yellow of varying intensity changing gradually to orange in the lower half. On the inside of the same shades as on the outside; for one-third to three-quarters of the length of the tepal orange which merges with the ground colour higher up.

Stamens ca. half as long as the tepals. Filaments sturdy, long triangular, broadening towards the base and with an abruptly pointed tip, glabrous, orange. Anthers yellow of the same shade as the tepals, oblong with a short acute tip, dehiscing gradually like those of *T.kaufmanniana* but not coiling into a spiral; on the inside of a brighter shade than on the outside. Pollen of the same colour as the anthers on the inside.

Ovary abruptly contracted at the base, broadening in the lower part, tapering towards the stigma, cream-coloured with a glaucous bloom. Stigma overlapping, yellow.

Capsule 6.0 × 2.4 cm, oblong with an attenuate base and top, whitish straw-coloured. Carpophore 1 cm long, 1.5 cm broad above and 1.0 cm broad next to the tepal scar, somewhat inflated, separated from the valves by a groove and bulging a little over the tepal scar. Upper seedless part of the capsule rather long (1.7 cm), formed by the attenuated tops of the valves (not style-shaped). Stigma lobes a little broader than the narrowed part of the valves, slightly reflexed, brown, easily falling off.

The capsule dehisces along deep but rather narrow (1 cm) slits, extending up to 3 cm along the fertile part of the capsule and above wedge-shaped owing to the seedless parts of the capsule sharply deflecting halfway up, forming a 3 cm wide funnel. The septa split to 3.5 cm from the base of the capsule.

Valves oblong to broadly lanceolate, of a yellowish straw-colour. Median groove broad and deep with a raised, oblong, ridge running along the centre and extending just a little on the carpophore; the groove flattening towards the top of the valves, where the ridge disappears; in the central part broad and brown, the colour not extending to the carpophore and only rarely over the upper seedless part. Numerous transverse, brown veins branch off from the groove across the wings, becoming denser in the upper half. Valves below the stigma of the same colour as their upper parts.

Wings rather coarse, convex; margins of the valves coarse, recurved, their edges brown spreading to the longitudinal groove between the edges and the wings. Cleavage surface of the ribs rather wide, pinkish yellow.

Tepal scar above broad, narrower at the thin peduncle. Peduncle of the same colour as the capsule, sometimes more grayish, much thinner than the carpophore, up to 20 cm long.

Seed 0.9 × 0.7 cm, with a broad margin which is broadest in the upper part and sharply bent.

Under Tashkent conditions this species begins to flower in the second or third week of April.

T.dubia differs from *T.kaufmanniana* by the more glaucous, nearly blue, narrow, pointed leaves with veins that are not clearly visible against the surrounding tissue; the later period of flowering and principally by the small size of its flowers with more narrow tepals; also by the anthers which open in the same way as those of *T.kaufmanniana,* but do not coil into a spiral. In Tashkent *T.dubia* starts to flower in mid-April when *T.kaufmanniana* and all its hybrids have already completed this phase and are setting seed.

Tulipa tschimganica Z. Botsch.

plate 35, 36, fig. 3e

The plants we have studied were collected in different areas in the Western Tien Shan: the Bol'shoy Chimgan gorge; the Chatkal and Ak-Bulak valleys and in the Tashkent reservation. The species grows on stony, gravelly or sandy soils at altitudes above 1 400 m. A habitat within this range is possible in deep valleys where snow accumulates in winter. It flowers usually next to snow patches on soils fed by water from the melting snow or on slopes which have been long under a snow cover.

Bulb up to 3.5 cm in diameter, ovoid; tunic papery, dark brown, sparsely lined with long and straight, adpressed hairs, more densely at the apex and the base.

Stem at flowering stage reaching up to 26 cm above ground level. Peduncle pubescent, covered with a glaucous bloom which is easily rubbed off.

Leaves 3-4, in the early stages of flowering often overtopping the flower, closely set, declinate, falcately recurved, channeled, pubescent, ciliate, glaucous; the edges of a lighter shade. Basal leaf up to 24 cm long and 7.5 cm broad, broadly lanceolate, during flowering with its base underground, at a later stage deflecting from the stem at ground level. Top leaf 20 × 2.2 cm, narrowly lanceolate.

Flower solitary, conical or star-shaped, red with a yellow base or yellow with red midribs on the outer tepals and red, inversely V-shaped, blotches on the inside; there are also multicoloured forms.

Outer tepals up to 7 cm long and 2.8 cm broad, narrowly to broadly lanceolate with an attenuated top. The red forms on the outside along the midrib bluish crimson, for 1/5 of their length downward to the base changing to yellow. On the inside light claret to orange, deepening in shade towards the base and on the base a yellow to orange fan-shaped blotch. The yellow forms on the outside along the midrib red-claret, at the base changing to yellow and covered with a glaucous bloom. On the inside at the base with an inversely V-shaped red blotch of varying dimensions, covering ca. ¼ of the length of the tepal.

Inner tepals up to 7 cm long and 3.2 cm broad, oblong or narrowly ovate with a more or less rounded, frequently notched, top. The red forms on the outside orange-claret, darker towards the orange blotch which covers the whole base. On the inside of the same colour as on the outside. The yellow forms on the inside bright yellow which changes to orange towards the base; at the place of transition from yellow to orange a V-shaped red spot of smaller dimensions than the one on the outer tepals.

Stamens nearly half as long as the tepals, overtopping the ovary. Filaments up to 1.2 cm long, on the whole thin but stightly broader at the base and pointed below the anthers, yellow with a brownish tip. Anthers nearly twice as long as the filaments, yellow or brownish; dehiscing gradually from the top downward, but not coiling as in *T.kaufmanniana.* Pollen yellow or brownish.

Ovary 2.3 cm long and 0.4 cm broad, narrowing at the base, drawn out into a style, yellowish green, often with a brownish shade along the ribs. Stigma overlapping, its lobes covered with yellow papillae, sometimes red on the sides.

Capsule up to 11 cm long and 2.7 cm broad, straw-yellow, narrowly conical, very elongated with a large seedless top, the style. Carpophore 1.5 × 1.5 cm, broadening and rounded towards the valves and fitting like a sharp prong into the valves from which it is separated by a rather deep groove. Style up to 3.5 cm long. Stigma small, folded longitudinally, brownish.

Valves lanceolate or long elliptic; median groove usually brownish claret which becomes more pronounced on the seedless part of the capsule.

Wings convex, slightly transversely wrinkled; along the edges of the valves a rather deep groove. The ribs of the valves in some cases brown, their edges recurved all along and parallel to the median groove. Cleavage surface of the ribs broad, up to 0.7 cm, dull cream-coloured or pale yellow. The capsule dehisces along wide wedge-shaped slits terminating in a wide-mouthed funnel.

Tepal scar brown, in many cases raised. Peduncle below the tepal scar broad, abruptly contracting lower down, of a darker shade than the capsule.

Seed large, triangular with rounded angles, brown; its scarious margin rather broad. Embryo shining through on one side of the seed.

Under Tashkent conditions this species begins to flower 20 days later than *T.kaufmannianna.* It is closely related to the latter, in particular the yellow and red forms which have similary coloured tepals. Besides the mode of dehiscence of the anthers differs from that of *T.kaufmanniana* only by their not coiling into spirals, but remaining straight as in all the other species. For many years *T.tschimganica* has been confounded with *T.kaufmanniana* from which it differs by ecological, phenological and some morphological characters.

Tulipa uzbekistanica Z. Botsch. et Scharipov

plate 11

T. uzbekistanica is described from southern Uzbekistan, the foothills south of the Dekhkanabad-Derbent highway, Mt. Kurukdagan, southern slope of the Tally Pass, in a gray saline clay soil.

Bulb ovoid, 1.0-2.5 cm in diameter, 4.5 cm long, seated at a depth of 5-20 cm. Tunic slightly prolonged, tough and leathery, on the inside only at the base and apex sparsely covered with short, straight, adpressed, yellowish hairs; old tunics dark brown, current year's tunics golden brown.

Stem at flowering up to 9 cm long, pale green, densely covered with short, white, squarrose hairs.

Leaves 3, broadly lanceolate, falcate, recurved, green with a glaucous tinge, densely pubescent, in particular on the upper side with short, white, squarrose hairs; margins cartilaginous, coarsely ciliate. Basal leaf 5.0 × 10 cm. second leaf 2.8 × 11 cm, upper leaf 8.7 × 1.0 cm.

Flower solitary, fragrant, 4.0-5.0 cm high and 5.0-7.0 cm in diameter, orange-red and usually in its centre a large dark brown blotch with a distinct, broad and dentate, bright orange, margin (in about 3% of the plants this blotch was absent).

Outer tepals 4.5 × 3.3 cm, broadly ovate with an attenuated, sharp, pubescent top; on the outside along the margins red-orange, on the midrib orange and at the claw with a rectangular green spot; on the inside red with an orange tinge and at the claw a dark brown, rectangular spot (0.6 × 0.6 cm) with a 0.5 cm wide, dentate, bright orange, margin.

Inner tepals smaller, 3.5 × 2.9 cm, broadly ovate with a denticulate, obtuse top; on the outside at the top red-orange, along the midrib orange, at the claw with a yellow spot on a green base; on the inside blood red with an orange tinge and on the claw, covering the entire width of the tepal, a 0.7 cm high, fan-shaped, dark brown spot with glaucous streaks, a sharply dentate top and a 0.6 cm wide, dentate, orange-yellow margin.

Stamens up to 2.2 cm long. Filaments cuneiform, yellow. Anthers light yellow, rarely claret, dehiscing gradually in 2 to 3 days and coiling. Pollen yellow.

Ovary 1.5 × 0.7 cm, tapering to the stigma, light violet with a glaucous tinge.

Capsule 5.0 × 2.2 cm (when ripe) oblong, tapering to both ends; the lower part white and smooth with decurrent ridges up to half its length; at the apex much elongated, almost without a discernible style; the ribs and sutures at first dark raspberry coloured, but later turning brown; the stigma does not fall off after maturation.

T. uzbekistanica should be included with the species of the section *Spiranthera* Vved., because of the mode of dehiscence of the anthers. It differs from the other species of this section in: plant habit; the narrower leaves with coarsely ciliate margins; the orange-red flowers with the distinctly margined brown blotch; its area of distribution that falls outside the distributional boundaries of the other *Spiranthera.*

T. uzbekistanica differs from *T. lanata* Regel, *T. fosteriana* Hoog ex Irv., *T. ingens* Hoog and *T. carinata* Vved. in the following characters: the tough and leathery, slightly prolonged tunic, that is on the inside only at the top and the base sparsely covered with short, straight, adpressed hairs; the low growth habit; the smaller number of broadly lanceolate, falcate and recurved leaves; the orange-red flower with at the base a dark brown blotch with a bright orange margin (flowers on the outside orange); the light yellow or claret anthers with the above-described mode of dehiscence; the light-violet, glaucously tinged, ovary; the mature capsule that is acute at both ends and has brown-coloured ribs.

SECTION *LOPHOPHYLLON* VVED.

'Leaf solitary with twisted longitudinal crests on the upper side; bulb tunic hairy on the inside; filaments hairy; ovary short-styled' (Flora of the USSR, Vol. IV).

Tulipa regelii Krassn.

plate 37 fig.6

T.regelii grows 'on stony slopes. Middle Asia: Lake Balkhash area. Endemic. Described from Kurtu, Andrakai' (Flora of the USSR, Vol. IV).

We studied plants raised from seed in 1951. The seed had been collected from plants grown from bulbs which were brought by V.P. Bochantsev from Chokpar station on the Turkistan-Siberian railway in Kazakhstan, where they were growing in a red clay soil. The seedlings flowered for the first time in 1956.

Bulb 1.5-3.0 cm in diameter, ovoid, seated at a depth of 10-15 cm; tunic prolonged, coriaceous, tawny, lined at the apex and the base with short, adpressed hairs.

Plant standing 5-8 cm above ground level. Stem pigmented by anthocyanin.

Leaf up to 11 cm long and 4 cm broad, lying flat on the ground, bluish green, usually solitary; sturdy plants may develop a small second leaf. The first leaf amplexicaul, overtopping the flower, oblong or broadly lanceolate, smooth and glabrous, glaucous, with longitudinal, 1-2 mm high crests on the upper side; along the edges and the top of the crests red anthocyanin stripes.

Flower as a rule solitary, white with a pinkish tinge and a faint pleasant scent (in cultivation we had plants with three flowers).

Outer tepals 2-3 cm long and 0.5-0.8 cm broad, lanceolate with an acute top. On the outside along the midrib cherry-coloured with a green shade, the edges pinkish and at the claw brownish. On the inside porcelain-white with a pinkish tinge; at the base a lanceolate (1 × 0.4 cm) golden yellow blotch.

Inner tepals much broader than the outer tepals, broadly triangular with rounded angles, an obtuse top and attenuate base; on both sides near the edges whitish villose. On the outside white with a dark green vein along the midrib and on both sides a brownish purple stripe. On the inside porcelain-white with at the base a large, rounded rhombic, golden yellow blotch (1 × 0.7 cm).

Stamens with dehisced anthers nearly half as long as the tepals and much longer than the ovary. Filaments almost of the same width throughout, terminating in a sharp tip; pubescent for ¾ of their length; at the base yellow, in the upper third orange. Anthers linear-oblong, yellow, 5-7 mm long after dehiscence. Pollen greyish yellow.

Ovary 10 × 3 mm, a little expanding towards the base, narrowing in a very short style below the stigma; greenish yellow, frequently cherry-coloured along the ribs. Stigma yellow, slightly broader than the style. No capsules were obtained during the period 1957-1959.

In Tashkent this species begins to flower towards the end of March.

SECTION *ERIOSTEMONES* BOISS.

'Bulb tunics on the inside glabrous or hairy or arachnoid-woolly; flowers in most species small, inconspicuous; filaments pubescent, in one species, *T.sogdiana,* glabrous; ovary with subsessile stigma or with a more or less developed style, which is never longer than the ovary' (Flora of the USSR, Vol. IV).

Tulipa sylvestris L.

T.sylvestris grows 'in woods; sometimes a weed. General distribution: Western Europe. Described from Southern Europe' (Flora of the USSR, Vol. IV).

The plants described here were received from the Institute of Plant Breeding of the Estonian Academy of Sciences through V.N. Esinovskaya, who informed us that they were growing in meadows in Estonia.

Bulb up to 2 cm in diameter; tunic coriaceous, black to tawny, densely lined at the apex and the base with adpressed hairs.

Plant reaching 46 cm above ground level. Stem, including the peduncle, glabrous, in the lower part pigmented by anthocyanin. Peduncle very slender, thin but strong, in the upper part brown.

Figure 6. *T. regelii* Krasn. (Reproduced from the original edition).

Leaves 3, widely spaced, more or less declinate, channeled, not exceeding the flower, dark green, glabrous, glaucous; the narrow reddish margins smooth, rarely ciliate. Basal leaf 24 × 2.4 cm, linear, with an acuminate top; deflecting from the stem at ground level. Top leaf 18.5 × 1.0 cm, linear with an acuminate top.

Flower shaped like an elegant golden star with pointed tepals.

Outer tepals 5.5 × 1.6 cm, lanceolate with an acuminate top; on the claw ciliate. On the outside over the midrib a broad band along the median vein and at the base of a bright greenish tinge, along the edges pinkish brown; the 0.2 cm broad edges of the tepal pure yellow. On the inside pure yellow without any blotches.

Inner tepals 5.5 × 2.6 cm, subrhombic with an acuminate top, convex and therefore giving the impression of being much shorter than the outer tepals. At the base drawn out into a long claw with densely ciliate edges. Also the tops of the tepals densely pubescent on the inside, whereas in other species this pubescence occurs at the tip on the outside and extends along the edges of the reflexed part of the tepal. On the outside bright yellow; along the midrib a thick green vein bordered by narrow greenish streaks and sometimes pigmented by anthocyanin; also the top of the tepal, in particular the tip of the midrib, may be pinkish. On the inside bright yellow.

Stamens 2.3 cm long, a little longer than the ovary. Filaments 1.4 × 0.3 cm, thick, flat, somewhat broadened and densely hirsute at the base, pointed at the top, yellow. Anthers 0.9 × 0.35 cm, orange, dehiscing abruptly. Pollen orange.

Ovary 1.6 × 0.4 cm, bright green, below the stigma tapering to a yellow style. Stigma small, narrower than the ovary, barely covering the style, yellow. Capsules did not develop in the material studied.

Under Tashkent conditions this species begins to flower in the last week of April.

Tulipa biebersteiniana Schult.
plate 38, fig.3i

T.biebersteiniana grows 'in steppes, on slopes, in shrubland communities on meadows in open woodlands and in river valleys. European part: Upper Volga (southern part), Middle Dnepr, the Volga-Don area, Transvolga area, the Black Sea area, Crimea, Lower Don, Lower Volga. Caucasus: Ciscaucasia, Dagestan, Eastern Transcaucasia. West Siberia: Upper Tobol, Irtysh (?). Middle Asia: Aral-Caspian, Lake Balkhash area (?). General distribution: The Balkans, Asia Minor (?)' (Flora of the USSR, Vol. IV).

The plants studied were collected by T.A. Ponomareva in the environs of Berchogur and received from B.V. Serdyukov (Tblisi Botanical Garden). We had also plants that were sent by chance with other plants by I.A. Zubashko from the Kerchenskiy peninsula (cape Kamysh-Burun) and plants received from P.A. Gomolitskiĭ from the environs of Aktyubinsk on 26th April 1957.

Bulb 1-2 cm in diameter, ovoid; tunic coriaceous, black to tawny, at the apex and the base lined with adpressed hairs or nearly glabrous and glossy on the inside, in particular the Caucasian forms.

Plant at the beginning of flowering not higher than 20 cm, but attaining 35 cm towards the end of this period. Stem, including the pedicel, glabrous, brownish claret.

Leaves 2-3-4, declinate or recurved, widely spaced (internodes 1.8-2.5 cm), bluish green, glabrous, strap-shaped and channeled with pointed beak-shaped tips; the margins sometimes ciliate and sometimes red (Crimean forms). Basal leaf 13 × 1.8 cm, broader than the upper leaves, at the beginning of flowering partly underground. Top leaf 15 × 1.2 cm.

Flowers 1-5, before opening and in cloudy weather always pendulous, when opened shaped like a broad golden star; with a faint pleasant honey scent.

Outer tepals 2.2 × 0.7 cm, lanceolate (in the Crimean forms smaller than in the plants from Berchogur and the Caucasus). On the outside along the midrib a broad brownish green band, often with a violet shading; the median vein brighter green and the ca. 1 mm broad edges of the tepal yellow. On the inside golden yellow, at the base greenish.

Inner tepals broadly lanceolate, sometimes long ovate; the top acuminate, the base abruptly narrowing in the claw, which is ciliate on the inside. On the outside yellow with a bright green vein broadening towards the base and becoming slightly reddish at the top; the yellow ground colour along the vein with a greenish shading. On the inside golden yellow, greenish at the base.

Stamens with dehisced anthers ca. half the length of the tepals, extending a little beyond the ovary. Filaments of varying length (1 and 3 cm long in the same flower) thickish, terminating in a sharp tip below the anthers, at the base light yellow, above the base yellow and at 1/4 from the top clothed with dense hairs. Anthers half or 1/3 the length of the filaments, oblong, broad, e.g. 0.3-0.4 cm long and 0.2 cm broad.

Ovary 1.3 × 0.3 cm, slightly narrowing towards the top, green. Stigma subsessile, yellow, narrow.

The plants from the Aktyubinsk area were 16 cm high, had much narrower leaves (the basal leaf 0.55, the top leaf 0.15 cm broad) and were on the whole of a smaller size.

Capsule 3.2 × 2.0 cm, ellipsoid, straw-yellow with a light orange tinge in the upper part and almost no carpophore; the valves extending to the tepal scar; the top slightly constricted and drawn out (up to 0.2 cm) like a very short style with a small brown stigma, which is in many capsules sessile.

The capsule dehisces down to the base along wedge-shaped slits (2.8 cm long and 1.2 cm wide at the top) without forming a funnel at the top. The septa split for one-third of their length to a little below their curvature with a flat rhombic expansion below the stigma of 0.7 × 0.7 cm; the edges remaining of even thickness lower down and along the curvature.

Valves coarse and compact, broadly lanceolate, of a dull colour, darker on the wrinkled part, lighter towards the base; median groove uneven, twisted, deeper in the middle, raised at the top and the base; along the groove one or two twisted ridges of the same colour as the valves.

Wings not well-pronounced owing to the valves being convex above the median groove and without grooves along the ribs. In the upper two-thirds of the valves transverse, light brown, wrinkles branch off from the median groove. Edges of the valves not arched, light brown, filtering through on the inside along the cleavage surface of the ribs, which has a dull orange colour; ribs not recurved. Top of the valves obtuse, darkest below the stigma from where two veins converge in the median groove with a narrow (0.7 × 0.1 cm) wedge-shaped depression between.

Tepal scar narrow, raised, dull light brown. Pedicel thin and long (26 cm), whitish with an orangy top.

Seed obliquely triangular with rounded angles and a narrow, bent margin. Embryo not visible.

Under Tashkent conditions this species begins to flower early in April.

Tulipa patens Agardh. ex Schult.

T.patens grows 'in steppes on stony slopes and solonchaks.[4] European part: Lower Don., Lower Volga. West Siberia: Upper Tobol, Irtysh, Altai (southwestern part). Endemic' (Flora of the USSR, Vol. IV).

The plants described were collected in the Semipalatinsk region, environs of Bobrovka, in 1933 and 1954, by Ermakov; one bulb from Z.M. Silina, Leningrad Botanical Gardens.

Bulb 1-2 cm in diameter; tunic prolonged to nearly twice the length of the bulb, thin coriaceous, brownish black to grey brown, at the apex and the base lined with adpressed hairs.

Stem 10-25 cm high, glabrous, pigmented by anthocyanin.

Leaves 2-3 (in our own plots 4), narrow, strap-shaped with a pointed top, recurved, glabrous; margins red, ciliate or glabrous. Basal leaf 24 × 1.7 cm, set at ground level. Top leaf 22 × 0.7 cm. The internodes between the leaves elongated (4.2 cm), extending over 15 cm of the stem, a feature which is uncommon in Middle-Asian tulips of this section. It may also occur in the related species *T.biebersteiniana* and *T.buhseana*.

Flower usually solitary, in cultivation there may be 2 or 3, in bud pendulous, somewhat straightening up during flowering; with the pleasant, delicate scent, typical of most species in this section growing within the territory of the USSR. The plants sent from Bobrovka had white flowers, while those from Leningrad were a delicate lilac-pink.

Outer tepals 4.5 × 0.9 cm, lanceolate with an acuminate top. On the outside along the midrib yellowish green, towards the base yellow; the midrib bright green; the broad edges lilac. On the inside lilac-pink with at the base a large (1.1 × 0.6 cm) elliptic, yellow blotch.

Inner tepals 3.9 × 1.8 cm, subrhombic with a long acuminate, recurved top and a short claw; the broadest part of the tepal at 1/3 from the base. On the outside bright pink lilac; covering the base a large, well-marked, bright yellow, fan-shaped blotch (1.1 cm high); along the median vein a green band bordered by dark brown stripes on the yellow blotch. On the inside of the same bright pink lilac colour and a similar yellow blotch (up to 1.3 cm high) at the base; the lower part of the tepal on both sides of the blotch sparsely ciliate.

Stamens half the length of the tepals, overtopping the ovary. Filaments 1.2 × 0.15 cm, flat, slightly broadened in the hirsute part at the base, tapering towards the top, yellow with an orange tip. Anthers 0.6 mm long after dehiscence, oblong.

Ovary 2 × 0.3 cm, peg-shaped, bright green with a subsessile, cream-coloured stigma.

Capsule 2.5 × 1.0 cm, oblong, straw-coloured with a brown tinge towards the top; towards the base tapering to a long, narrow seedless part, the carpophore; valves at the top terminating into a short, 0.25 cm long, style with small, brownish remnants of the stigma. The size of the carpophore varies greatly, up to 0.6 cm long and 0.2 cm broad; it is only distinct from the valves by the median groove continuing as a ridge.

The capsule opens down to the base along narrow wedge-shaped slits with practically no funnel at the top. The septa split down to their curvature with below the stigma long, narrowly rhombic edges.

Valves thinnish, broadly lanceolate, glossy or dull; median groove rather deep, continuing as a ridge on the carpophore and at the top which forms the style, often light brown, particularly in the upper ¾ of its length.

Wings slightly convex with along the ribs a narrow groove. In the upper part across the wings transverse wrinkles running from the grooves along the ribs to the median groove; the wrinkles denser near the top, par-

4. Solonchak—a wet saline soil with a high groundwater table, bearing a white salt crust in the dry season. (editor)

ticularly below the style. Outer edges of the valves slightly recurved and bordered by a narrow brown stripe. Cleavage surface of the ribs very narrow, of a dull colour, lighter than the valves; not turned outwards.

Tepal scar narrow, raised, corrugated. Pedicel up to 22 cm long, broadened towards the top, in the upper part darker than the capsule.

Under Tashkent conditions this species begins to flower early in April.

Tulipa sogdiana Bunge

T.sogdiana grows 'in sandy, rarely clayey, or stony deserts. Middle Asia: Kyzyl-Kum, Kara-Kum, Amu-Dar'ya. Endemic' (Flora of the USSR, Vol. IV).

The plants we have studied and described were received from A.D. Pyataev and came from the western Kopet-Dag, the Kyuren-Dag range, near Donat springs; growing in a sandy soil.

Bulb 1-1.5 cm in diameter, ovoid; tunic coarse, black to brown (young bulbs) or grayish (very old bulbs), somewhat prolonged, lined in particular at the base, the prolonged part and the top with adpressed hairs.

Stem 15 cm above ground level, glabrous, glaucous, with two widely spaced leaves; the basal leaf set at 1-2 cm above ground and 2-3 cm from the top leaf.

Leaves deflecting from the stem at an acute angle, slightly recurved, linear-lanceolate, glaucous, glabrous; the margins sparsely ciliate or glabrous.

Flower solitary, porcelain-white with pinkish edges, beautifully shaped like a cup with a somewhat narrowed top; with a rather strong and pleasant scent. Both the inner and outer tepals are very convex lending the flower its characteristic shape.

Outer tepals 2.6 × 0.8 cm, lanceolate, narrower than the inner tepals and with an acuminate top. On the outside along the midrib a broad brownish green band with a faint pinkish tinge; the margins porcelain-white with a touch of pink along the rims; at the base an elliptic, yellow blotch (0.7 × 0.3 cm). On the inside dull porcelain-white with a pinkish tip.

Inner tepals variable in shape, from long deltoid (2.8 × 1.2 cm) to broadly rounded (2.6 × 1.7 cm); the basal part more or less drawn out, ciliate; the top apiculate. On the outside dull porcelain-white; the midrib greenish yellow; at the base the greenish tinge spreading on both sides of the midrib, the yellow blotch on the inside shining through. On the inside dull porcelain-yellow with a trace of pink at the tip; the base covered with a well-marked, 1 cm high, bright yellow blotch with a convex, sometimes weakly indented, top.

Stamens 0.6 × 0.15 cm, overtopping the ovary, bright yellow, paler towards the broadened base, slightly tapering and with a sharp tip below the anthers. In many live plants from this habitat we found pale yellow hairs scattered on the broadened base of the filaments, particularly on the sides facing the tepals. It was the only character that distinguished these plants from *T.sogdiana* as described in 'Flora of the USSR' and obviously we witness in this species a transition from plants with glabrous to plants with hirsute filaments. Anthers 0.4 ×0.18 cm, oblong with short sharp tips, bright yellow. Pollen yellow.

Ovary 0.8 × 0.3 cm tapering to a very short style, pigmented pinkish by anthocyanin. Stigma subsessile.

Capsule small, 1.4 × 1.2 cm, subglobose or short ovoid, straw-coloured wiht a brown pattern and no carpophore. The valves curving abruptly to the tepal scar have obtuse tops from which protrudes a short, ca. 0.1 cm long, thin beak (the style) with a sessile, small brown stigma. The capsule dehisces along wedge-shaped slits while the septa split from the top to the base. There seems to be no thickened pollen-conducting tissue.

Valves rounded, papery and brittle; median groove rather deep, narrow, brown; the lateral wrinkles branching off from the groove brown. Edges of the valves deep brown. Cleavage surface of the ribs narrow, cream-coloured, without any fringe.

Tepal scar of a light shade, pressed against the capsule, broader than the peduncle. Peduncle narrowing abruptly right below the capsule, whitish straw-coloured.

Seeds of varying size, e.g. 0.5 × 0.4 cm, triangular with rounded angles, dull brown; margin narrow, irregular. Embryo not visible.

Under Tashkent conditions this species begins to flower in mid April.

Tulipa polychroma Stapf

fig. 7c

T.polychroma grows 'on dry slopes. The Caucasus: Ciscaucasia (eastern part), Eastern Transcaucasia. General distribution: Iran and Afghanistan. Described from Shurab' (Flor of the USSR, Vol. IV).

The 150 plants we have studied came from the Yerevan Botanical Garden and had been collected in the flowering stage by A. Akhverdov in the middle mountain belt at 700 m, Katav district, near the village of Avan, 31st March 1953. Habitat: stony semidesert overgrown with wormwood (*Artemisia* sp.), in dry clayey and stony places.

Bulb 1-1.8 cm in diameter, ovoid; tunic thin, coriaceous, light brown or tawny, a little prolonged, lined with arachnoid woolly hairs. The hairs very long, thin, rippled and dense in the upper third of the bulb. The bulb has two fleshy scales.

Stem 6-17 cm high, glabrous. Peduncle claret brown, glabrous.

Leaves 2, rather widely spaced, strap-shaped, channeled, tapering to a point, falcately recurved, glaucous, glabrous; the narrow red margins glabrous or sparsely ciliate.

Flowers 1-2, on the inside white with a yellow base, on the outside grayish purple, shaped like an elegant double cup; with a rather strong, pleasant scent.

Outer tepals 15-35 mm long and 6.5-8.5 mm broad, long ovate, recurved with straight edges; the base bearded. On the outside white; along the midrib brownish claret; towards the top and the base greenish. On the inside white; covering the base a well-marked, golden yellow blotch, up to 8 mm high, with a convex top.

Inner tepals 3-10 mm long, 20-14 mm wide, broadly lanceolate, convex, with an acuminate top and truncated claw; at the base densely ciliate, in particular the edges. On the outside white with a green-violet vein and often a narrow (up to 2 mm) blotch extending halfway up the tepal; on both sides of the blotch for 2/3 of the length of the tepal diffuse yellow. On the inside white; at the base a well-marked, fan-shaped, golden yellow blotch (up to 12 mm high) with a notched top.

Stamens overtopping the ovary, but with dehisced anthers half the length of the tepals. Filaments up to 11 mm long, thin, peg-shaped with a sharp tip; at the base a ring of golden yellow hairs. Anthers 2-7 mm long, oblong, yellow with a a short black tip.

Ovary 8-12 mm long, 1.5-2 mm broad, green, tapering to a short, barely developed, style with very small stigma lobes.

Capsule small, 1.7 × 1.3 cm, ellipsoid to nearly globose, straw-coloured with brownish patterns; at the base nearly right-angled and with no carpophore; protruding from the obtuse top a short, ca. 0.1 cm, thin beak with an inconspicuous, sessile stigma.

The capsule dehisces along wedge-shaped slits down to its base. The septa split to half their length and have a thick strand of pollen-conducting tissue along the edges. The septa hardly show any curvature and the valves recurve but little at the top.

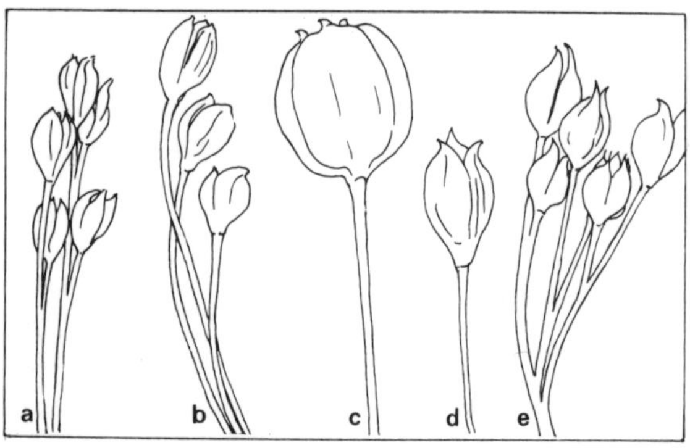

Figure 7. Mature tulip capsules
a. *T. bifloriformis* Vved.; b. *T. turkestanica* Regel; c. *T. polychroma* Stapf; d. *T. dasystemon* (Regel) Regel; e. *T. tarda* Stapf.

Valves elliptic, papery, brittle; median groove narrow, rather deep, brown, in the upper half a low ridge with narrow, brown stripes branching off along the lateral wrinkles and converging at the top in a brown-coloured spot. The valves are sometimes split below the stigma along the median groove.

Wings narrow, convex, straw-yellow; in the lower half smooth; in the upper half transversely wrinkled and striped, brown. Edges of the valves a little recurved, brown and practically no groove along the ribs. Cleavage surface of the ribs in these small capsules rather broad (0.15 cm), pale yellow.

Tepal scar pressed to the capsule, a little broader than the peduncle, brown. Peduncle tapering below the capsule over a distance of 6 cm, light brown.

Seed varying in size, e.g. 0.5 × 0.4 cm, rounded or triangular with rounded angles, dull brown; the margin narrow; the embryo barely visible.

Under Tashkent conditions this species begins to flower in the first half of March.

Tulipa buhseana Boiss.
fig.3j

T.buhseana grows 'on sandy and clayey soils. Middle Asia: Aral-Caspian, Lake Balkash area, Kyzyl-Kum, mountainous part of Turkmenistan, Tien Shan (foothills of the northern part). General distribution: Dzungaria-Kashgar area, Iran and Afghanistan. Described from Kara-Chokat' (Flora of the USSR, Vol. IV).

In an annotation to the description of this species Vvedenskiĭ mentions that it has a disjunct area of distribution and that the plants in the northern area are somewhat different from those in the southern area. He at one time proposed to treat the northern race as an independent species under the name *T.talievii* Vved. We had samples from both areas.

The plants described here came from the desert hills of Alymtau, northwest of Tashkent; from the western foothills of the Karatau range and from the Karoy plateau in the Karatau between two mountain chains.

Bulb 1-1.7 cm in diameter, ovoid; tunic somewhat prolonged, tawny to brownish, coriaceous, lined at the apex with thin, whitish, woolly hairs coiled into spirals.

Stem 10-40 cm high, erect, pubescent (including the peduncle), pinkish brown. The plant resembling a candelabrum.

Leaves 2, set at 5-7 cm above ground level, widely spaced, at the base making an acute angle with the stem, the upper part strongly inclined, narrowly strap-shaped, channeled, glabrous; the margins red and ciliate; they remain much below the flower. Basal leaf broadest, from 3 to 15 mm and even up to 20 mm in cultivation.

Flowers (1) 2 or 3 (-16) on long pedicels without bracteal leaves. The long peduncle lifting the whorl of pedicels 3-5 cm above the leaves is typical of this species. The flowers open widely above a narrowed base and the whole perianth curves a little inwards, i.e. the flowers are campanulate with the orifice turned upward. The tepals are cream-coloured.

Outer tepals from 1.1 to 3.0 cm long and from 0.2 to 0.6 cm broad, lanceolate, often bearded on the claw. On the outside along the midrib pinkish brown; towards the edges cream-coloured. On the inside cream-coloured with a small (7 × 3 mm), elliptic, yellow blotch on the lower half.

Inner tepals 1.0-2.9 cm long and 0.3-1 cm broad, broadly lanceolate, bearded on the claw. On the outside cream-coloured; the midrib green with the yellow blotch shining through. On the inside cream-coloured; covering the base a well-marked, fan-shaped, yellow blotch, extending to ¼ of the length of the tepal.

Stamens with dehisced anthers ca. half the length of the tepals. Filaments 5-10 mm long, golden yellow, rather thin with a sharp tip, narrowed at the place of attachment, broadened below the villose ring, above the ring often also hairy. Anthers 4-6 mm long, linear-oblong with short, sharp tips, yellow, sometimes dark at the tip. Pollen yellow or dirty yellow.

Ovary a little shorter than the stamens, yellow-green, tapering to a short style. Stigma pale green, slightly broader than the style.

Capsule ca. 2.1 cm long and 1.2 cm broad, cylindrical, in rare cases globose, whitish with a brown top. The base obtuse with the valves extending right down to the tepal scar; no carpophore. From the obtuse top protrudes a small, ca. 0.1 mm long style with a tiny brown stigma.

Valves rectangular with rounded edges; median groove well-marked, rather sharp, for ¾ of its length from the top light brown, below of the same colour as the capsule.

Wings, delicate, papery, convex, with no grooves along the ribs. The upper quarter of the capsule covered with light, transverse, brown wrinkles branching off from the median groove and below the style merging in a brown spot. The edges of the valves along the ribs brown. The capsule does not quite open on dehiscence and the ribs separate by narrow slits.

Tepal scar brown, closely pressed to the capsule, narrow but broader than the strongly tapering pedicel. Pedicels thin, darker than the capsule (yellowish or brownish), up to 8 cm long.

Seed 0.5 × 0.4 cm, light brown, triangular with one sharp and two rounded angles; the margin narrow. Embryo long, lying somewhat to one side of the seed.

Under Tashkent conditions this species begins to flower in the fourth week of March. *T.buhseana* differs from the other Middle-Asian species of this section by the slender habit, the candelabrum-shaped inflorescence, the shape of the flowers, the pinkish brown colour of the whole plant and the shape of the capsules which remain practically closed. We found in nature capsules of this species which remained closed until December.

Figure 8. *T. buhseana* Boiss.

Tulipa buhseana Boiss.

Plants of the form described here were found all along the railway line between Darbaza and Baygakum stations. They appeared to be all of a uniform type.

Leaves narrow, bluish green, glabrous, channeled, widely spaced, sharply recurved; the margins red, sparsely ciliate. Basal leaf 16 × 0.7 cm.

Stem including the peduncle, usually 25 cm above ground level; the peduncle well-developed, carrying many (up to 9 flowers) on long pedicels. The flower buds and faded flowers pendulous.

Flowers opening widely, stellate-campanulate.

Outer tepals 2.1 × 0.55 cm, lanceolate, at the base densely ciliate. On the outside grayish pink; towards the base yellow. On the inside cream-coloured and over the base a 4.5 mm high yellow blotch with a convex top.

Inner tepals 2.1 × 1.2 cm, triangular, in the lower half attenuated; at the base ciliate and bearded. On the outside cream-coloured with a grey midrib and the inner yellow blotch shining through in the lower part. On the inside cream-coloured and over the base a well-marked, 5.5 mm high, yellow blotch with a convex top.

Filaments of varying length in the same flower (0.7 and 0.5 cm), much broadened below the villose ring, sharply tapering above this part and remaining rather thin, the top pointed; at the base pale yellow, above orange yellow and pubescent. Anthers up to 3 mm long, oblong, yellow with a dark tip. Pollen yellow.

Ovary ca. two and a half times shorter than the tepals, a little shorter than the filaments, light green, peg-shaped, tapering to a short style. Style and stigma yellow.

Capsule 1.6 × 1.1 cm, straw-coloured, triangular to oblong with an obtuse top and base and no carpophore; at the top a short beak-shaped style (0.05 cm) with a small stigma of the same colour.

The capsule dehisces down to the base along narrow wedge-shaped slits, about 0.3 wide at the top owing to the upper parts of the valves not recurving. The septa do not split beyond the end of the broadened, wedge-shaped pollen-conducting tissue.

Valves rectangular with rounded angles; median groove rather narrow, deep, with a light brown vein. The wedge-shaped top of the valves brownish with a broad median ridge (0.4 cm).

Wings thin but strong, convex, smooth, sometimes with a few wrinkles below the beak; at the top flattened, brown. Along the ribs a narrow brown stripe and no groove, the edges flat and rigid; the cleavage surface from the ribs not visible from the outside, very narrow, not more than 1 mm in the central part, pale yellow, on the outside with a brown margin. The light yellow straw-colour of the rest of the capsule fading lower down and often broken by several narrow brown stripes.

Tepal scar very narrow, slightly raised, closely pressed to the capsule. Pedicel two to three times thinner than the scar, sturdy, of the same straw-colour as the capsule.

Seed 0.4 × 0.25 cm, obliquely triangular, dark brown with a thick, narrow margin; in the top part against the long side of the seed lies a narrow, black crescent of tissue, which contains chlorophyll after ripening of the seed. Embryo comparatively long for seed of this size (0.3 cm), clearly visible through the epidermis.

Under Tashkent conditions this form begins to flower in late March or early April.

Tulipa talievii Vved.

fig.3k

The plants for our studies were sent by V.P. Bochantsev from the environs of Chokpar station on the Turkestan-Siberian railway.

Bulb ovoid; tunic somewhat prolonged, coriaceous, reddish-tawny, lined with long adpressed, not interlaced hairs, rather sparse in the middle, dense at the apex and the base.

Stem including the peduncle 11.5 cm above ground level, green with a bluish pink shading; peduncle sparsely pubescent.

Leaves 2, widely spaced, linear, deflecting from the stem at a nearly right angle, bluish green, glabrous, channeled, overtopping the flower. Basal leaf 18 × 0.9 cm, falcate; top leaf 17 × 0.7 cm.

Flowers (1) 2 (3), widely campanulate.

Outer tepals 2.1 × 0.5 cm, lanceolate. On the outside grey with a pinkish tinge; towards the base yellow. On the inside cream-coloured with an elliptic yellow blotch (0.3 × 0.2 cm); along the base of the blotch a narrow band of dense hairs.

Inner tepals 2.0 × 0.7 cm, lanceolate, the lower half broadened and abruptly narrowed at the base. On the outside cream-coloured with a brownish green vein, the inner yellow blotch shining through in the lowest quarter. On the inside cream-coloured; over the base a well-marked yellow blotch (0.55 cm high) with a convex top; the base of the blotch, particularly on the sides, ciliate.

Stamens with closed anthers ca. twice as high as the ovary; after dehiscence contracting and extending only a little beyond the ovary. Filaments short (0.5 cm) with in the lower third a villose ring on the broadened basal part; above the ring conical, bright yellow, sparsely pubescent. Anthers oblong, yellow. Pollen yellow.

Ovary 0.8 cm long, narrowly peg-shaped, pale green with a short style and cream-coloured stigma.

Capsule 1.6 × 1.1 cm, whitish straw-coloured, long ellipsoid, with an obtuse top and base and no carpophore; on the indented top a small (0.1 cm long) brown remnant of the style with traces of a small dark stigma.

Valves rectangular with rounded angles; median groove rather deep, brown with transverse wrinkles branching off across the thin, convex wings, in particular in the upper half. Along the ribs no grooves; the edges of the valves along the cleavage line light brown. Cleavage surface of the ribs very thin, of a dull whitish straw-colour. Valves below the stigma indented.

The capsule dehisces to the base along deep wedge-shaped slits forming a wide-mouthed funnel. The septa split to ca. half the length of the capsule which is considerably beyond their curvature. The pollen-conducting tissue starting below the stigma as a broad band, abruptly tapers to the curvature where it continues as a mere thickening of the edges of the septa.

Tepal scar narrow, light brown, covered by the protruding bases of the valves. Peduncle of uniform thickness, more yellow than the capsule. The fruit stalks of some of our plants were 9 cm long, the peduncles of the flowering plants on an average 3.5 cm.

Seed small, triangular with rounded angles; the embryo not visible.

Under Tashkent conditions this species begins to flower in late March.

Tulipa turkestanica Regel

plate 39, fig.7b

T.turkestanica grows 'on clayey and stony slopes from the foothills ± to 2 500 m. Middle Asia: Pamir-Alay, Tien Shan (Mogoltau, Podshaata). Endemic' (Flora of the USSR, Vol. IV).

The plants described here were grown from bulbs which were brought from Aman-Kutan Forest lying in what is known as the Samarkand chain of the Zeravshanskiy mountain range.

Bulb up to 3.5 cm in diameter, broadly ovoid; tunic somewhat prolonged, black-tawny, coarse, coriaceous, lined at the apex with woolly hairs, close to the base sparsely pubescent. Hairs rather long, thick, coarse and only slightly rippled. Bulbs of adult plants have two fleshy scales and produce daughter bulbs.

Stem 10-30 cm high, brownish green, the upper part including the peduncle covered with short, erect, white hairs, all pedicels deflecting from the stem at the same height, like an umbrella.

Leaves 2-3, rarely 4, widely spaced, recurving and lying on the ground, linear, slightly channeled, bluish green with a claret tinge, the tips red, glabrous; margins red, ciliate. They are 5-25 mm broad and longer than the inflorescence.

Flowers 1-12, greenish cream-coloured, stellate but with the tepals a little incurved; with a strong rather unpleasant scent; in bud erect.

Outer tepals 1.2-2.5 cm long, at the base 0.3-0.6 cm broad, lanceolate, with a small ring of cilia near the base. On the outside green-violet, along the edges cream-coloured. On the inside cream-coloured with a greenish tinge; at the base an elliptic (8-11 × 4 mm) blotch with an acute top.

Inner tepals 1.1-2.4 cm long, 0.4-1.1 cm broad, lanceolate, concave; the edges near the base bearded, at the base ciliate. On the outside greenish cream-coloured with a large, well-marked, fan-shaped, yellow blotch covering the base and a green midrib. The yellow blotch (13 × 10 mm; 8 × 7 mm) rhombic, its height equalling 2/5 the length of the tepal.

Stamens in opened flowers half the length of the tepals. Filaments 7 mm long, yellow, thin with a densely villose ring around the broadened base and covered up to the constricted top with sparse hairs. Anthers 3-6 mm long, linear oblong, yellow with a violet top and terminating in a sharp tip.

Ovary 7 × 2 mm, yellowish green, broadened at the base and tapering to a short style (1 mm). Style pale yellow; stigma a little broader than the style.

Capsule up to 3 × 1.8 cm, whitish straw-coloured, oblong; at the base a short and rounded carpophore (0.7 cm long), fitted like a rhombic-shaped prong into the valves.

Valves broadly lens-shaped, pointed at the top and terminating in a short (ca. 0.1 cm) beak-shaped style with a tiny dark stigma; the edges not recurved; very thin and papery, covered from top to bottom with thin wrinkles radiating from the median groove, which is deepest in the central, coloured, part and at the top of the valve before it reaches the upper, flattened, zone below the stigma.

Wings flat, not clearly demarcated, white and glossy; along the edges with a shallow dent. Along the edges of the valves a thin, brown line which is also visible on the inside of the dull cream-coloured cleavage surface of the ribs.

The capsule dehisces down to the base along broad slits of even width, the valves separating by curving outward along the median grooves. The slits are screened by a thin web of hairs growing from the inner surface of the valves; the seeds being liberated through the deflected upper parts of the valves. Septa at the place of cleavage below the style rhombic with a ridge in the centre; they do not split beyond the beginning of the curvature.

Tepal scar very narrow, brown, raised. Pedicels thin, of a light shade.

Seed 0.7 × 0.65 cm, which is rather large for a species of this section; obliquely triangular with rounded angles and a narrow, wrinkled margin. Embryo visible.

Under Tashkent conditions this species begins to flower in mid March.

Tulipa turcomanica B. Fedtsch.

T.turcomanica grows 'on mountain slopes. Middle Asia: mountainous Turkmenistan. General distribution: Iran and Afghanistan' (Flora of the USSR, Vol. IV).

The plants described here were sent by A.D. Pyataeva from the Kyuren-Dag range, western Kopet-Dag, Turkmenia.

Bulb 1.5-3 cm in diameter, ovoid; tunic blackish or tawny, densely lined with thin, interlaced, web-like hairs.

Stem (15-) 19 (-30) cm high, pigmented by anthocyanin, glabrous.

Leaves 2, set at 2 cm and 4 cm above ground level, deflecting from the stem at nearly right angles, linear, glabrous, or slightly ciliate. Basal leaf 17 × 2.2 cm; top leaf 15.5 × 1.8 cm.

Flowers (1) 2 (3) on long pedicels branching off from a 2 cm long peduncle, broadly campanulate, very delicate with a pleasant scent. One of the best species of this section in Middle Asia.

Outer tepals 3.1 × 0.8 cm, lanceolate, pointed. On the outside reddish brown; along the midrib towards the base green. On the inside white with a diffuse, elliptic, greenish yellow blotch (up to 0.7 cm high); in the centre of the blotch a white densely pubescent band extending almost to the top.

Inner tepals 3.2 × 1.6 cm, ovate, pointed, convex, narrowing towards the base. On the outside white with a pinkish brown midrib and a fan-shaped, greenish blotch in the lower half, up to 0.6 cm high. On the inside white with a yellow blotch; white hairs scattered all over the blotch, particularly in the centre and often extending above the blotch. The edges of the tepals ciliate.

Stamens and filaments of varying length; the filaments linear, (0.6-0.8 cm), in the lower part light yellow and pubescent throughout, the upper part orange. Anthers yellow with black tips.

Ovary 1.5 cm long, a little shorter than the stamens and with a short style.

Under Tashkent conditions this species flowers early in April.

Tulipa bifloriformis Vved.

plate 40, fig.7a

T.bifloriformis grows 'on clayey and stony slopes in foothills, up to 1 900 m. Middle Asia: Syr-Dar'ya, Tien Shan (western), Pamir-Alay (?). Endemic. Described from Tashkent' (Flora of the USSR, Vol. IV).

The plants described here were collected in Tashkent on the banks of the Ankhor aryk;[5] close to the Darbaza station on the Turkistan-Siberian railway, in clayey hills in the Kyzkuduk area; in the Western Tien Shan at the foot of a sandy pass in Chimgan.

Bulb up to 2.0 cm in diameter, globose or ovoid; tunic often prolonged, coarse, reddish brown or pink (in old individuals tawny), lined with thin, arachnoid, woolly hairs. The plants frequently propagate vegetatively and develop a nest of daughter bulbs.

Stem 10-25 cm; at flowering often coloured claret by anthocyanin, softened by a dense pubescence consisting of short, unbranched, white hairs.

Leaves 2, widely spaced, linear-lanceolate, recurved, undulate or smooth; the margins red, ciliate; often surpassing the flower. Basal leaf broadest, 1.0-1.5-2.5 cm broad and 10-20-25 cm long.

Flowers 1-15, the inflorescence subtended by one bracteal leaf, stellate, the inner tepals slightly incurved and erect, the outer tepals deflecting at right angles from the pedicel; with a pleasant fresh smell of honey.

Outer tepals 2.5-3.0 cm long, 0.6-0.7 cm broad, narrowly lanceolate, bearded at the base. On the outside along the red-margined midrib brownish green, towards the edges cream-coloured, on the claw yellow; sometimes along the midrib brightly coloured by anthocyanin. On the inside cream-coloured, frequently with a greenish tinge; at the base a triangular golden yellow blotch (7 × 4 mm or 4 × 3 mm) with acute or rounded angles.

Inner tepals 2.4 × 3.0 cm long, 1.0-1.3 cm broad, lanceolate, abruptly contracted towards the spoon-shaped claw; bearded at the base. On the outside cream-coloured with a faint greenish tinge; the midrib green; on the claw yellow. On the inside also cream-coloured with at the base a well-marked, rhombic, golden yellow blotch (9 × 7 or 4 × 4 mm). When fading all tepals become pinkish.

Stamens ca. half the length of the tepals (10-17 mm). Filaments 7-12 mm long, yellow, thin, tapering; around the broadened base a densely villose ring and usually undulate hairs higher up. Anthers linear-oblong, after dehiscence 5-6 mm long, yellow, sometimes dark claret or violet; the connective terminating above the anthers in a sharp, dark tip.

Ovary 10-15 mm long, 2-3 mm broad, narrowly conical, pale green with a short pale yellow style (1.5-2 mm). Stigma villose, overlapping the style, but narrower than the ovary.

Capsule 4.7 × 1.7 cm, oblong sometimes globose, contracting gradually at the base and abruptly at the top where it terminates in a short beak. In unripe capsules the ribs and the median grooves of the valves are coloured red by anthocyanin stripes with in the upper third transverse stripes which become denser towards the top. In the ripe capsules this pattern of red stripes becomes brownish.

Valves thin, delicate, frequently transversely wrinkled near the top. Soon after ripening the capsule dehisces along the ribs down to the base, the valves often coiling into a spiral.

Seed 0.6 × 0.5 cm, dark brown with almost no margin; the embryo barely visible.

Under Tashkent conditions this species begins to flower early in March.

Tulipa orthopoda Vved.[6]

fig. 9

In an annotation to the description of *T.bifloriformis* Vvedenskiĭ writes: 'The plants from Karatau are very

5. Aryk—an irrigation canal. (editor)

6. The following names were not validly published: *Tulipa orthopoda* Vved. ex Poljak. 1958, Fl. Kazakhst. 2: 213, descr. ross. *Tulipa talievii* Klok. et Zoz, 1935, cit. soch.: 66, nom. nud. *Tulipa victoris* Vved. 1961, Tr. Tashk. Univ. 187, Bot.: 61, nom. nud. Czerepanov, S.K. 'Index nominum taxorum novorum Florae URSS et combinationem novarum in annis 1934-1970 post publicationem "Flora URSS" non valide editiorum. Nov. Syst. Plant. Vasc. 9: 327 (1972). (editor)

small and have a very hairy, usually prolonged tunic; leaves, closely set, as a rule surpassing the flower; buds and faded flowers erect' (Flora of the USSR, Vol. IV).

The plants described here came from the Dzhon plateau near the Baydzhansay gorge in the Karatau range (Kazakhstan). They were collected on a rocky slope in fine earth among stones, June 19, 1956. Endemic.

Bulb rounded with the tunic prolonged for 1-1.5 cm; tunic reddish brown or pinkish, coriaceous, profusely lined with long and thin, strongly rippled, webby hairs. This covering is so dense that the tunic does not show through.

Stem brownish claret, densely covered with unbranched, squarrose hairs; 2 cm high including the peduncle; the whole plant reaching only 3.7 cm above ground level when the first flower opens.

Leaves 2, closely set, rather short with an obtuse top, recurved, bluish green; the undersurface pigmented by anthocyanin; margins red. At the beginning of flowering the leaves remain with their bases underground (the bottom leaf for 2 cm), the blades lying flat on the ground.

Flowers 1-2, in cultivation more, broadly cup-shaped, the slightly recurved inner and outer tepals erect; faintly fragrant.

Outer tepals 2.5 × 0.5 cm, lanceolate; at the base on the inside girdled with delicate, long white hairs. On the outside along the midrib a broad green stripe, the pink ground colour gradually shading to white towards the edges. On the inside white with an elliptic, golden yellow, blotch (up to 0.8 cm high) at the base.

Inner tepals 1.8 × 1.0 cm, 1½-2 times broader than the outer tepals, triangular; the claw on the inside covered with white hairs becoming denser and longer towards the edges. On the outside white with a conspicuous green midrib and at the base a transversely rhombic, yellow blotch (7 × 8 mm). On the inside white with a similar blotch to that on the outside but more angular and orange-coloured.

Figure 9. *T. orthopoda* Vved. (Reproduced from the original edition).

Stamens of variable length (12 and 14 mm), with dehisced anthers overtopping the ovary; the stamens opposite the inner tepals with the longer filaments (7 mm), those opposite the outer tepals shorter (5 mm) and dehiscing later than the former. Filaments yellow, at the base cream-coloured, towards the top tapering to a very thin and sharp orange tip; in the cream-coloured part covered with a ring of long, whitish hairs, in particular dense along the edges and extending for 2/3 the length of the filament above this ring. Anthers linear-oblong, yellow, 4-5 mm long after dehiscence; connective yellow, protruding above the anther.

Ovary 11 × 1.5 mm, slightly narrowing towards the style; stigma subsessile, cream-coloured, not broader than the style.

Capsule 2.3 × 1.8 cm, whitish straw-coloured with light brown margins, subglobose-globose; without a carpophore; the valves elliptic with rounded top and base. As distinct from *T.talievii* the valves of this species are not notched at the top. They terminate in a short dry style, up to 1.5 mm long, carrying a tiny, narrow, yellow stigma. Wings are not demarcated and the valves do not show any grooves along the ribs.

Valves thin but compact, nearly flat, slightly convex in the upper part. Median groove in the upper half up to the seedless, flattened and often brown coloured part of the valves, bright brown; the lower half not coloured. In the coloured part of the groove a 2 mm wide strip resembling scar tissue and fanning out from this part of the groove many small and larger wrinkles with light brown veins in between. On the edges along the ribs a brown stripe with a lighter stripe running next to it all along the valves; the brown stripe also visible on the inside of the valves. Cleavage surface of the ribs, except for the stripes, dull white. The valves recurved along the median groove.

The capsule dehisces along wide, wedge-shaped, slits extending nearly to the tepal scar, the interlaced hairs on the inside edges of the ribs preventing further widening of the slits. The septa split to their curvature; the edge of the septa below the stigma and the pollen-conducting tissue obovate with the latter extending in a strand to the curvature.

Tepal scar brown, narrow and closely pressed to the capsule, but the valves not protruding over the tepal scar as they do in *T.talievii*. Pedicel(s) up to 5.5 cm long, thin, more yellowish than the capsule. Peduncle hardly developed, not more than 1 cm long.

Seed small, obliquely triangular, dark brown with a rather broad margin. Embryo not visible.

Under Tashkent conditions this species begins to flower in mid-March.

Tulipa binutans Vved.

plate 41

In an annotation to the description of *T.bifloriformis* Vvedenskiĭ refers to this taxon as follows: 'Plants from the Aleksandrovskiy mountain range have bulbs with less hairy tunics; the buds and the faded flowers are pendulous' (Flora of the USSR, Vol. IV). Later in 'Botanicheskikh materialakh AN UzSSR' 13 (1952) he describes it as an independent species.

The plants described here were grown from seed collected on mountain slopes in the environs of Tyul'kubas station on the Turkistan-Siberian railway.

Bulb ovate or globose; tunic slightly prolonged, reddish brown, lined only at the apex and also a little at the base with rather coarse, adpressed hairs. The innermost daughter bulb of flowering plants is stoloniferous and sinks to a lower level.

Stem up to 12 cm above ground level; stem and pedicels profusely covered with short, whitish, erect hairs and with a reddish brown anthocyanin pigmentation.

Leaves 2, rather widely set, strap-shaped, falcately curved, channeled, glabrous; along the margins and the acute top red.

Flowers stellate; the buds and the faded flowers pendulous, which is also typical of other related species, e.g. *T.buhseana.*

Outer tepals 2.6 × 0.6 cm, narrowly lanceolate. On the outside a broad green band with reddish brown shadings along the midrib and narrow cream-coloured edges. On the inside cream-coloured with on the lower half a small, elliptic, yellow blotch (0.7 × 0.3 cm); at the base of the blotch a strip of whitish cilia and along the edges in the lower half of the blotch also sparsely ciliate.

Inner tepals 2.5 × 0.65 cm, lanceolate, pointed, a little broader than the outer tepals, abruptly tapering into the claw. On the outside cream-coloured with a brownish green midrib and at the base, up to a height of 0.6 cm, yellow. On the inside cream-coloured with a well-marked yellow blotch covering the base and extending to the broadened part of the tepal (up to 0.7 cm). The base of the tepal bearded, the edges on both sides ciliate; the inside surface of the yellow blotch also sparsely covered with fine hairs.

Stamens of variable length, overtopping the ovary. Filaments 8.5 and 6 mm long with a thick villose ring in the lower third, thin (1 mm thick above the ring); between the ring and the sharp tip pubescent. Anthers up to 5 mm long, linear-oblong, yellow. Pollen yellow.

Ovary long and narrow, green with a cream-coloured, subsessile, overlapping stigma.

Capsule 2.8 × 1.6 cm, whitish straw-coloured with light brown ribs, ovoid with in most cases a very small (3 mm) and narrow carpophore. The top of the valves obtuse, not forming a style, the stigma being inserted in an indentation or on top of a very small protuberance.

Valves elliptic; the median groove in the central part deep, light brown, in the lower part flattened and almost colourless, in the upper part strongly wrinkled with the wrinkles extending across the valves; some capsules with thin brown stripes in the grooves between the wrinkles.

Wings convex, very delicate, papery, glossy, separated from the ribs by deep grooves; the rims of the ribs brownish, the brownish stripes converging below the stigma.

The capsule dehisces along slits right down to the tepal scar, forming a wide-mouthed funnel at the top. The septa split along the pollen-conducting tissue to the end of their convex part; the edges below the stigma broadened and with a longitudinal ridge which continues below the broadened part.

Tepal scar narrow (0.5 mm), raised. Pedicel thin, with a yellow tinge.

Under Taskent conditions this species begins to flower towards the end of March.

Tulipa biflora Pall.

T. biflora grows 'in solonets[7] steppes, solonchaks[8] and on clayey slopes. European part: Black Sea area (??), Crimea (??), Lower Volga. West Siberia: Upper Tobol (southern part), Irtysh (southern part). Middle Asia: Aral-Caspian. General distribution: Dzungaria-Kashgar area' (Flora of the USSR, Vol. IV).

Bulb up to 2 cm in diameter, ovoid; tunic brownish, a little prolonged, papery, lined with thin, arachnoid, woolly hairs.

Plant 13 cm high, stem including the peduncle glabrous, brownish green and particularly the pedicel(s) glaucous.

Leaves 2, widely spaced, linear, channeled, declinate, glaucous, overtopping the flowers; the margins claret-red, sometimes sparsely ciliate; the tips claret-red. Basal leaf 15.5 × 0.9 cm, top leaf 12 × 0.5 cm.

Flowers 1-2, broadly campanulate. Outer tepals, 1.9 × 0.4 cm, recurved at a right angle to the stem, lanceolate. On the outside grayish green or grayish violet with narrow white margins. On the inside cream-coloured to white and in the lower half a 0.6 cm high, sharply angular, yellow blotch with a fringe of yellow hairs.

Inner tepals 1.9 × 0.7 cm, narrowly rhombic with a sharp top and an abruptly attenuated base; along the

7. Solonets—a saline soil typical for parts of eastern Europe where eolian salt has been deposited in southern chernozem and chestnut soils. (editor)

8. See footnote on p. 62.

Plate 33. *T. anadroma* Z. Botsch.

Plate 34. *T. tschimganica* Z. Botsch. (80%)

Plate 35. *T. tschimganica* Z. Botsch. (80%)

Plate 36. *T. uzbekistanica* Z. Botsch. et Scharipov

Plate 37. *T. regelii* Krassn.

Plate 38. *T. biebersteiniana* Schult.

Plate 39. *T. turkestanica* Regel

Plate 40. *T. bifloriformis* Vved.

Plate 41. *T. binutans* Vved.

Plate 42. *T. neustruevae* Pob.

edges ciliate. On the outside white with a green midrib and the blotch on the inside shining through. On the inside white with a green midrib and a 0.6 cm high, sharply angular, yellow blotch with a fringe of hairs covering the base.

Stamens 0.9 cm, half the length of the tepals and nearly equalling the ovary. Filaments yellow, thin, with a sharp tip and broadened below the densely hairy ring; below the ring and at the sharp tip below the anther paler yellow. Anthers yellow, 0.25 cm long, oblong, with a small and sharp, yellow tip.

Ovary 0.75 × 0.2 cm, green, tapering to a small yellow style; stigma yellow, narrower than the ovary.

Capsule 1.3 × 1.3 cm, subglobose, without a carpophore; style short (1 mm) thin, brown and with the remnants of a small, flat, brown stigma protruding from the rounded top. The capsule dehisces down to the base along wedge-shaped slits, while the septa split to their curvature. The strand of pollen-conducting tissue broadens towards the base of the style and contracts towards the curvature.

Valves nearly round, glossy, semitransparent. Median groove rather deep, brown and frequently with a raised, longitudinal vein.

Wings transversely wrinkled, the rather flat wrinkles denser in the upper third of the capsule. Along the ribs of unripe capsules often a broad purplish stripe which becomes brown when they ripen. There are no grooves along the ribs and the edges, which are thin and on the inside lusterless, do not curve outwards. The capsule is darker below the stigma.

Tepal scar narrow, uneven, forming a ridge around the base of the capsule. Pedicel tapering from below the tepal scar, pinkish.

Under Taskent conditions this species begins to flower in late March.

Tulipa dasystemonoides Vved.

T.dasystemonoides grows 'in the alpine mountain belt. Middle Asia: Tien Shan. Endemic. Described from the Talasskiy Alatau: Maydantal' (Flora of the USSR, Vol. IV).

We have studied plants from Angren which were sent by A. Ya. Butkov in 1954. Besides, we studied plants grown from material which Butkov brought from the Pervomayskiy settlement.

Bulb 1-1.8 cm in diameter, ovoid; tunic black to tawny, thin, coriaceous, densely lined with woolly hairs.

Stem, including the peduncle, 5-8 cm high above ground level. Peduncle pigmented by anthocyanin, pinkish brown, densely covered with erect hairs.

Leaves 2, emerging almost simultaneously with the flowers, broadly lanceolate, green, extending a little beyond the flower; along the margins red, ciliate. At flowering stage the basal leaf 4.7-8.5 cm long, and 1.3-2.8 cm broad; the top leaf (3.4-) 5 (-7.5) cm long and (0.9-) 1.5 (-1.8) cm broad.

Flowers 1-2-3 on one peduncle, the main stem remaining underground during flowering; stellate with narrow segments; when just opened pale yellow becoming brighter on the second day; with a faint, pleasant scent suggesting saffron.

Outer tepals 1.9-2.7 cm long, 0.3-0.5 cm broad, narrowly lanceolate, terminating in a sharp tooth; on the claw ciliate. On the outside along the midrib greenish, towards the top and the edges claret-red, along the edges yellowish. On the inside yellow.

Inner tepals 1.1-2.7 cm long, 0.4-1.0 cm broad, elliptic-lanceolate, 1½-2 times broader than the outer tepals; the base acuminate, bearded and ciliate. On the outside yellow; the midrib bright green with two parallel narrow green stripes; the midrib and the stripes becoming pinkish towards the top and brown at the claw. On the inside evenly yellow and only the top of a much lighter shade.

Stamens with thread-like, yellow filaments (1 × 0.1 cm); the lower quarter paler, hairy and broadened, like a typical *Eriostemones.* Anthers 0.6 cm long, linear-oblong, yellow with a dark tip.

Ovary pale green, half the length of the anthers, constricted below the rather long and thick style, which is greener than the ovary. Stigma cream-coloured, overlapping.

Under the conditions prevailing in the spring of 1957 capsules failed to develop in spite of additional artificial pollination.

Tulipa neustruevae Pob.

plate 42

With the description of *T. dasystemon* Regel in the "Flora of the USSR", Vol. IV goes the following annotation: 'Another tulip occurring in the Ferganskiy range, in addition to *T.dasystemon,* grows apparently at lower altitudes and has firmer coriaceous bulb tunics. Cultivated specimens have displayed bright yellow flowers (*T.paradasystemon* m. in herb)'. Endemic.

The bulbs of the plants described here were collected by P.M. Murzova in 1953, in a walnut forest near lake Sarychelek in the Chatkalskiy mountain range. They were propagated in the Botanical Garden of the Academy of Sciences of the Uz.SSR.

Bulb subglobose-ovoid, 1-2 cm in diameter, 2 cm high; tunic blackish tawny to tawny, papery. There is often a daughter bulb in the axil of the first fleshy scale which develops a narrow green leaf, up to 9 cm long, when the first flower opens. Adult bulbs have two fleshy scales.

Stem (10-) 20 (-30) cm high, glabrous; pedicels brownish-green, glabrous.

Leaves 2, rather widely spaced, declinate at an angle of 30° (not lying on the ground), linear, sometimes slightly undulate, glabrous; the margins red, smooth or sparsely ciliate. Basal leaf broadest (0.5-2.4 cm).

Flowers 1-2-3, broadly stellate, golden yellow, with a pleasant smell of honey.

Outer tepals 1.5-3.0 cm long, 0.6-0.8 cm broad, lanceolate, pointed, at the base glabrous. On the outside along the midrib brownish-violet-greenish, the outer edges golden yellow. On the inside bright golden yellow.

Inner tepals as long as the outer ones but broader, (1.2–1.5 cm), broadly lanceolate; the abruptly attenuated base covered with white cilia, which are denser towards the edges. On the outside yellow; along the midrib green with on both sides a longitudinal, green stripe from the base to ca. 1/3 of the length of the tepal. On the inside bright golden yellow.

Stamens with dehisced anthers nearly half as long as the tepals. Filaments flat, gradually broadening in the lower third, above this part tapering and terminating in a sharp tip; for the lower 2/3 from the base covered with thin cilia. Anthers up to 5 mm long, oblong, yellow, with a very short yellow tip. Pollen yellow.

Ovary 14 × 3 mm, a little shorter than the dehisced stamens, gradually broadening towards the base, greenish yellow, with a rather long (2 mm), pale yellow style and slightly overlapping stigma lobes.

Capsule 2.7 × 1.2 cm, whitish straw-coloured; the valves irregularly rhombic with rounded lateral angles. The capsule is broadest at 1/3 from the top where it gradually contracts to a rather long style (5 mm) or, when the latter is absent, to the stigma. The carpophore only 1-2 mm long.

Valves delicate, papery; along the median groove and the ribs with rather thin brownish red stripes; the transverse wrinkles small and denser in the upper part.

Under Tashkent conditions this species begins to flower in mid-March.

Taliev in 1930 and afterwards Vvedenskiĭ in 1935 intended to describe *T.neustruevae* but this was not done until 1949, by E.G. Pobedimova, who based the description on a plant raised from a bulb in a greenhouse which failed to develop a peduncle. Plants with weakly developed peduncles occur both in cultivation and in nature. The species was described from a single plant which was dissected for sketching and no authentic specimen has been preserved.

Tulipa dasystemon (Regel) Regel

fig.7d

T.dasystemon grows 'in the subalpine and alpine belts. Middle Asia: Pamir-Alay, Tien Shan. General distribution: Dzungaria-Kashgar area' (Flora of the USSR, Vol. IV).

The plants described here were received from Yu. Mironova in July 1955 and came from the Zailiyskiy Alatau in the Tien Shan: Alma-Ata region above lake Issyk-Kul'. They were collected on a steep northeastern slope above a talus where they grew in fine earth among shingle with a thin grass cover. The year 1956 brought high plant mortality and flowering began only in April 1957. Apart from these plants, we had previously studied and described specimens of this species collected in the same area, in 1935.

Bulb 0.9-1.5 cm in diameter; tunic usually light brownish, sometimes black-tawny, papery, usually with only a scanty cover of straight, adpressed, white hairs in the upper half.

Stem at flowering 5 cm above ground level; both stem and peduncle glabrous; the peduncle in the upper part strongly pigmented, brownish claret.

Leaves 2, at flowering half underground, bluish green, glabrous; the margins with red edges. Basal leaf 10 × 1.2 cm, oblong with a rather obtuse top; top leaf narrower and more channeled, 8.0 × 0.7 cm, narrowly lanceolate, when still underground set at 2 cm from the basal leaf.

Flower solitary, rather bright yellow (not pale yellow, as stated in "Flora of the USSR", Vol. IV). On the Bol'shoy Chimgan mountain we collected white specimens with grayish midribs and a pleasant scent.

Outer tepals 1.9 × 0.4 cm, narrowly lanceolate. On the outside along the midrib a broad brownish or greenish claret band, only the curved edges yellow. On the inside pure yellow with a band of very delicate cilia at the base of the claw (below this band often a drop of nectar).

Inner tepals 1.7 × 0.7 cm, long rhombic or narrowly lanceolate; on the edges of the attenuated claw long and dense cilia which on the inside spread over the base. On the outside yellow; the midrib brownish green with in the lower half of the tepal on both sides a brownish green streak. On the inside yellow and the midrib faintly showing through at the claw.

Stamens 1.6–1.1 cm, yellow; with closed anthers just a little shorter than the tepals; with dehisced anthers 2/3 the length of the tepals. The stamens of the outer whorl dehiscing on the first day of flowering, the longer stamens of the inner whorl on the second day. Filaments of nearly uniform thickness throughout, with an apiculate tip, pubescent over the entire surface, yellow, paler in the lower third. Anthers 0.3-0.8 mm long, yellow; the short dark tip very conspicuous when the anthers are still closed.

Ovary green, constricted below the style, 1 cm long including the 0.5 cm long style; stigma pale yellow, narrower than the style.

Capsule 2.4 × 0.9 cm, of a grayish to pinkish straw-colour, long obovoid, in cross section triangular, strongly drawn-out at the base to a narrow carpophore,

0,5 cm long, 0,2 cm wide at the base and 0.35 cm wide below the valves; between carpophore and valves no constriction. The capsule terminates at the top in a long, narrow style (0.5 × 0.1 cm); the tiny remnants of the stigma falling off on dehiscence.

Valves grayish to pinkish, very delicate, papery, flat; median groove narrow, not extending on the carpophore, slightly pigmented in the upper part and with 5-6 thin brown wrinkles branching off across the upper third. Edges of the valves along the cleavage surface of the ribs narrow, brownish. Grooves along the valves and the lower seedless part of the capsule are absent and separate wings are not discernible since they form an integral part of the valves.

The capsule dehisces along narrow wedge-shaped slits, the outward curving edges of the upper seedless part of the valves forming a funnel at the top. The septa split to ca. 1/4 of the capsule and have along their edges below the stigma narrow strands of pollen-conducting tissue. The part of the valve forming the style dark brown near the top.

Tepal scar 0.5 mm wide, dark brown. Peduncle 7.5 cm long, thin, darker than the capsule.

Seed triangular with two rounded and one sharp angle near the tiny embryo; margin very small.

Under Tashkent conditions this species begins to flower in the second week of April.

Tulipa tarda Stapf
fig.7e

T.tarda grows 'on rocky and shingly slopes. Middle Asia: Tien Shan. Endemic. Described from Middle Asia: 'probably Turkistan'. With the notation: 'This very interesting species is known from a few locations in close proximity to one another, Kastek, Dzhol-Bulak and Dzhil'-aryk' (Flora of the USSR, Vol. IV).

The plants studied were brought from Kirgiziya, the Kastek river canyon, 10 km from the Mikhaylovskiy settlement. They were growing in the shrubland belt among stones and rocks.

Bulb 1.5-3.0 cm in diameter, ovoid; tunic black to brown, coriaceous, glabrous.

Stem, including the pedicels, at flowering from 5-11 cm above ground level; the very short peduncle giving the impression of the flower cluster emerging directly from the soil.

Leaves 3-7, closely set, linear, weakly channeled, more or less recurved, bright green, glabrous; the edges claret, often ciliate; slightly overtopping the flowers. Basal leaf 12 × 1.5 cm, top leaf 11.5 × 0.9 cm.

Flowers up to 15, white, broadly stellate; with a pleasant delicate scent.

Outer tepals 3.4 × 1.3 cm, broadly lanceolate, pointed. On the outside along the midrib a broad green stripe, towards the top with purple stripes, towards the claw yellowish and the narrow edges white. On the inside a large (2.5 × 1.3 cm), yellow, broadly lanceolate blotch, covering the larger part of the segment.

Inner tepals 3.5 × 1.7 cm, long rhombic, pointed, spoon-shaped; the base attenuated, covered with long, dense cilia. On the outside for 2/3 from the base yellow with a gradual transition to the white top part. On the inside for 3/4 from the base yellow, the top white. There are forms with entirely yellow tepals.

Filaments 0.8 cm long, yellow, from the broadened base upwards conical; at the base a ring of dense, pale yellow, long, villous hairs; above the ring pubescent. Anthers 0.5 × 0.15 cm, yellow with grey margins. Pollen yellow or yellowish green.

Ovary bottle-shaped with a rather long style and yellow, overlapping stigma.

Capsule 3.8 × 1.8 cm, globose or fusiform, rounded at the base; tapering to the top in a style (up to 1 cm long) with a small stigma not broader than the style.

Valves whitish straw-coloured, delicate, transparent; in this species quite flat; median groove in the upper half brown with thin brown stripes branching off along the transverse wrinkles and becoming denser towards the top of the valves. Wings only discernible by the slightly convex part of the capsule over the seed cavity, the typical grooves along the ribs being absent. The two halves of the valves recurve after dehiscence along the median groove, leaving the areas below the stigma and at the base as a continuation of the edges, which are of a duller colour than the wings. Cleavage surface of the ribs strongly recurved, broad (3.5 mm), dull straw-coloured, the scar along its edge brown.

Slits between the valves narrowly wedge-shaped, broadened near the stigma owing to the strongly deflected narrowed part (the style) of the capsule. The style ridged on the inside. The septa begin as a broad protuberance at the top of the seed cavity, which contracts lower down, and do not split beyond the wedge-shaped contraction.

Tepal scar raised, light brown, narrowing to the pedicel. Pedicels thin straw-coloured to white. Peduncle 11 cm long.

Seed 0.6 × 0.5 cm, obliquely triangular or broadly ovate, very dark, brownish, almost without a margin; the embryo not shining through.

Under Tashkent conditions this species begins to flower in mid-April.

CHAPTER FOUR

The Middle Asian centres of diversity of the genus Tulipa

V.I. Taliev (1930) attempted to interpret species formation in the genus *Tulipa* by applying the geographical method. He regarded the species of the section *Eriostemones* as the ancestral forms of the genus and gave a fairly accurate analysis of the western branch of this section, the *Silvestriformes,* but failed to understand the *Bifloraeformis* tending to Turkistan and the Aral-Caspian lowland. The latter include a large number of good, morphogeographically well-defined, species: *T.bifloriformis, T.binutans, T.buhseana, T.regelii, T.sogdiana, T.tarda, T.turcomanica, T.turkestanica* and some other species still awaiting their geographical recognition, viz. *T.orthopoda* and *T.talievii,* described by Vvedenskiĭ. Apart from these, there remains a related group of recent formae which occur on the northern slopes of the Pamir-Alay. Since Taliev did not know all these species and has not been able to examine living specimens, he could not give a true picture of the evolution in this section, although he dealt with the evolution of the genus as a whole.

In 'The process of species formation in the genus *Tulipa*', Taliev (1930) outlined his concept of the hypothetical ancestor of *Tulipa.* According to his description the primitive common ancestor of the *Liliaceae* must have been multiflorous, polyphyllous, polychrome, with large flowers, nectaries, a sessile stigma on a long style and leaves of widely different shape, for instance even broadly cordate leaves like those of *Lilium cardiocrinum.*[1] Taliev supports the theory of regressive evolution in the genus *Tulipa,* which he visualizes as the breaking up of a Central Asian prototype into several forms which spread in western and northeastern directions. 'Selection has led to a loss of characters and the appearance of those combinations of characters which, although morphologically and physiologically less luxuriant, are more specialized with respect to adaptability' (p. 117). Discussing the size of the flowers, which are rather large in the *Tulipanum* section and small in the *Eriostemones,* Taliev claims that they should all be conceived as 'deprived relatives' of the large lily flowers. There is no special objection to this claim although it cannot serve as a convincing argument for a more ancient origin of the large flowers in the genus *Lilium,* because in the genus *Fritillaria,* in the broad Linnean sense, the flowers are not larger than those of *Tulipa.* Nevertheless the author believes the genus *Fritillaria* to be more ancient than *Tulipa.* Some species of the section *Leiostemones* have larger flowers than species of the section *Tulipanum,* e.g. *T.greigii, T.fosteriana, T.micheliana,* their flowers becoming even larger in cultivation. Wild tulips are highly responsive to improved growing conditions both in nature and in cultivation. The flowers of *T.greigii* and *T.micheliana* are always larger on ploughed land than on virgin soil and in cultivation their size often increases, while both wild species and cultivars become multiflorous (Botschantzeva, 1953, 1954).

Our experiments have shown that the smaller number of leaves in *Tulipa* is indeed an adaptive reaction to the environmental conditions. By improving these conditions when growing plants from bulbs which had been collected in their natural habitat, we obtained an increase in the number of leaves. From *T.regelii,* which has as a rule one leaf and one flower, both in nature and in cultivation, we could raise plants with two leaves and three flowers. In *T.greigii, T.fosteriana, T.kaufmanniana, T.vvedenskyi* and many other species we have observed the initiation and development of axillary buds on the green stem, which developed into shoots with one or two leaves and a flower. Under less favourable conditions in following years the same plants did not develop subsidiary flowers and leaves, showing that they easily reverted to their former state.

The multiflorous character of *Eriostemones* under natural conditions, as mentioned by Taliev, is in our opinion not a primitive character, since it is caused by branching of the apical bud, a highly specialized trend in flower initiation. It can therefore not be regarded as a primitive, common, character of the whole genus and it is hardly ever observed in representatives of the *Leiostemones.* The multiflorous character in species of the latter section is different. In nature *T.praestans* often produces up to four terminal flowers on branches

1. *Lilium giganteum* Wallich., section *Cardiocrinum* (Baker) pro subgenus; see V.I. Taliev, 1930. (editor)

Figure 10. The Fergana basin separates the Western and Central Tien Shan from the Pamir-Alay mountain system.

Mountain ranges	*Mountain systems*
1. Tarbagatay range	Eastern Tien Shan
2. Dzhungarskyi Alatau	Northern Tien Shan
3. Ketmen range	,, ,, ,,
4. Zailiyskiy Alatau	,, ,, ,,
5. Terskey Alatau	Central Tien Shan
6. Chu-Iliyskiye mts	Northern Tien Shan
7. Kirgizskiy range	,, ,, ,,
8. Karatau range	,, ,, ,,
9. Talasskiy Alatau	Western Tien Shan
10. Chatkal'skiy range	,, ,, ,,

Mountain ranges	*Mountain systems*
11. Kuraminskiy range	Western Tien Shan
12. Turkestanskiy range	Pamir-Alay
13. Alayskiy range	,, ,,
14. Ferganskiy range	Central Tien Shan
15. Zeravshanskiy range	Pamir-Alay
16. Gissarskiy range	,, ,,
17. Darvazskiyi range	,, ,,
18. Nuratau range	
19. Kopet Dag	
20. Mugodzhary hills	

growing from the axils of green leaves and consequently retains to some extent the ancient character of the genus.

When recording flower initiation of our lilies in the botanical garden, we found that in *L.philippinense* Baker, *L.martagon* L. and *L.tenuifolium* Fisch. the flowers appeared on the branches growing from leaf axils and that almost every flowering branch had one or two leaves, i.e. the lily flowers showed the same type of branching as multiflorous *Leiostemones.*

The *Eriostemones* and *Leiostemones* sections of the genus *Tulipa* evolved in different ways. They separated very long ago, although one has to agree with Taliev that 'the uniflorous character typical of tulips has doubtlessly developed through reduction from a more or less multiflorous inflorescence of distant ancestors'. However, since by better feeding and watering it is comparatively easy to render tulips multiflorous and more leafy – a tendency retained by many species of the genus and which is to some extent a reversion – these characters can be regarded as having been lost by the genus, apparently during a xerothermic epoch before our time.

Without attempting to solve the whole problem of species formation in the genus *Tulipa,* we shall try to trace on the basis of a few characters the phylogenetic relationships in their distribution pattern in Middle Asia, proceeding from east to west (see the above map for the main geographical features).

It is not yet clear which representatives of the genus

occur in the Dzhungarskiy Alatau, but according to the information on hand, the tulips there have yellow and no red flowers.

T.patens with white flowers, on the outside along the midrib bluish, grows in the environs of Bobrovka not far from Semipalatinsk and like all other tulips from this section has a yellow blotch at the base of the tepals. *T.brachystemon,* yellow with pinkish backs of the outer tepals and V-shaped red blotches on the inside, is found in the Dzhungarskyi Alatau; *T.iliensis* with a yellow perianth and a white *Eriostemones* species inhabit the Ketmen range. Farther to the west on the sands of the Ili river valley we find yellow *T.behmiana* and white *T.talievii.*

South of the river Ili stretches from east to west the Zailiyskiy Alatau, where in the foothills, below rocky benches, grow *T.kolpakowskiana* with a yellow and *T.talievii* with a white perianth. The mountain scrub belt at higher altitudes is the habitat of *T.ostrowskiana* with orange-red, orange and yellow flowers and still higher up, between rocks, grows lemon-yellow *T.tetraphylla.* On the tops of lateral ridges we find a pale yellow *T.dasystemon* often accompanied by *Leontopodium* sp., while *T.heterophylla,* now called *Eduardoregelia* M. Pop., also yellow with bluish backs, grows in the alpine meadows.

In the mountains surrounding lake Issyk-Kul' as far as the upper reaches of the river Naryn, yellow *T. tetraphylla* is widely spread and yellow *T. dasystemon,* yellowish *T. tianschanica* and *T. heterophylla* are also found there. The bulbs which were collected in the spruce forest belt of the Terskey Alatau, in 1956, bore yellow flowers. It appears that all tulips growing in the mountains surrounding lake Issyk-Kul' at an average altitude of 1600 m are yellow or yellowish.

The first orange-red tulip in the eastern centre of diversity of the genus was found in the Zailiyskiy Alatau. Further westwards much brighter colour shades occur: *T. albertii* with its orange-red flowers inhabits the Chu-Iliyskiye mountain range and the Betpak-Dala hills. In the same area, somewhat east of the Chok-Par Pass, we find *T. greigii* with a predominance of orange shades in the perianth colour and large, diffuse, yellow blotches on the claw of the tepals. The dark blotches are not black, but brown becoming paler towards the edges. Yellow *T. kolpakowskiana,* cream-coloured, almost white, *T. talievii* and white *T. regelii* also grow in this area.

In the Alatau hills, 80 km from Frunze in the direction of Alma-Ata, grows orange-red *T. ostrowskiana* and a similarly coloured *T. greigii,* together with one small white tulip of the *Eriostemones* section. In the valley of the river Kichikemin, discharging into the Chu, an orange-red tulip resembling *T. ostrowskiana* was found, and also *T. tarda* with a white perianth. On the Kastek Pass in the same mountain system we collected white and yellow *T. tarda.* In the Chu valley grows a reddish orange *T. greigii.*

In the Kirgizskiy mountain range from Frunze to Dzhambul (Kaindy-Say) occurs an orangy red tulip, closely resembling *T. ostrowskiana,* and at lower altitudes also orange-red *T. greigii.* On screes and in fine earth between stones yellow *T. zenaidae,* with or without the small, well-marked, dark blotch at the base of the tepals, was found more often than the red form. White tulips of the *Eriostemones* section also grow there.

In the Karatau range, running southeast — northwest, we noted *T. albertii* and *T. greigii*, varying in colour from dark reddish orange and lemon-yellow to nearly cream-coloured. The light coloured forms of *T. greigii* have a bright red blotch in the centre on the inside of the tepals; the outer segments are red on the reverse. The colour of these two species varies considerably in the habitats on the gravelly soil of the central plateau, which runs parallel to the two ridges bordering this mountain system on the eastern and the western side. Two creamy *Eriostemones* are also present in the Karatau, viz. *T. orthopoda* on rocky slopes (granite), in cracks filled with fine earth, and *T. buhseana* on slates and shingle, in the central and eastern parts of the range.

In the lower mountain belt of the northern part of the Talasskiy Alatau occurs an orange-red *T. greigii,* which is a little redder than the forms from the Chu valley and the Ak-Tas mountains. Cream-coloured *T. kaufmanniana* with pink backs inhabits the western part of this range with at lower altitudes *T. binutans* and much higher a light yellow *T. dasystemon.*

Along the banks of the Syr-Dar'ya river in subsaline soils grows *T. buhseana.* In the Alkakol'kum and other sands on both sides of this river we found dark orange forms of *T. borszczowii* tinged with red. The area of *T. greigii* extends all along the eastern bank of the Syr-Dar'ya, between the Karatau and the river to Kyzyl-Orda and in the south to all the valleys of the western Tien Shan, the lower parts of the Karzhantau, Ugamskiy, Pskemskiy, Chatkal'skiy and Kuraminskiy chains. There is no clarity yet about the distribution of this species in the Kuraminskiy chain, in particular on the slopes facing the Fergana valley.

T.mogoltavica inhabits the Mogoltau ranges. It has the same bright red colour as *T.greigii* of the Western Tien Shan but not red-orange, like those from the eastern and northern part of the area, with clearly marked yellow margined, black blotches. A similar tulip is found on the eastern bank of the Angren river together with typical *T.greigii* from the western Tien Shan.

On all the ranges of the western Tien Shan grows *T.kaufmanniana* with cream-coloured, almost white, tepals, yellow towards the base. Also red, orange and yellow forms are quite frequent and sometimes one comes across multicoloured forms with orange or red stripes of varying intensity on the yellow ground colour. In the same mountains at a somewhat higher altitude on talus and rock benches we found another species, *T.butkovii,* with lanceolate glaucous leaves and claret-red flowers which, when cultivated in Tashkent, flowers one month later than *T.kaufmanniana.* There also, close to the snow line, grows *T.tschimganica* of the *Spiranthera* section with yellow, orange and red flowers. So far we have not found a form of *T.tschimganica* with creamy or almost white flowers. *T.dubia,* yellow, grows in loose substrate at the same altitude, not below 1 800 m. Everywhere and at varying altitudes one can find *T.bifloriformis* with almost white flowers. In certain locations of the Chatkal'skiy chain it occurs in slightly different forms, depending on the altitude of its habitat. According to the literature the *T.dasystemon* with white flowers grows at higher altitudes than any other species. We saw living specimens of this form in Chimgan, in 1959. No doubt the tulip which is called abroad *T.dasystemon* is our *T.tarda* from the mountains surrounding the valley of the river Chu. It is indeed often white with a large yellow blotch, but yellow forms also occur. Butkov collected yellowish *T.dasystemonoides* with a very delicate scent on stony talus in the upper reaches of the Angren river. Earlier this species had been brought from Kyzykurt.

In the Angren basin grow bright red *T.vvedenskyi,* red *T.greigii* and polychrome *T.kaufmanniana.* In the low hills of the western extremity of the Tien Shan occurs *T.korolkowii,* red with a black base and frequently also the orange-yellow form of this species along with bright red *T.greigii* and cream-coloured *T.bifloriformis.*

We have never found any red tulips in the Fergana basin. On the surrounding southern slopes of the Chatkal'skiy range grow: *T.neustruevae,* bright yellow; *T.ferganica,* yellow; *T.anadroma* (section *Spiranthera*), yellow. In the Ferganskiy range: *T.neustruevae,* yellow; *T.dasystemon,* yellowish and a white *Eriostemones.* On the northern slopes of the Alayskiy range grow the white *Eriostemones, T.turkestanica* and *T.dasystemon* and a yellow, narrow-leaved, very pubescent species, closely related to *T.ferganica.* In the foothills of the Fergana basin, environs of Namangan, occur a yellow *Tulipa* sp. and *T.rosea,* the pink variant of *T.korolkowii,* or its paler form which has lost its bright red colour that is found near the village of Kim.

A large orange-red tulip was brought in 1958 from an area above the Fergana basin, near the upper reaches of the river Naryn (a right-hand tributary of the Kekelemansay). This specimen had a yellow blotch at the base of the tepals and streaky brown spots on the yellow ground colour in the lower part of the inner tepals. The occurrence there of bright coloured tulips is very interesting. We have no information on other tulips in the Fergana basin, but on the Belesna upland at the entrance grow the same pink *T.rosea,* yellow *T.lehmanniana* and cream to white *T.buhseana.*

As regards species composition and colour of the perianth the Pamir-Alay mountain system strongly differs from the earlier mentioned regions,[2] which gravitate towards the Tien Shan. In the Uzungyr (the northern chain of the Turkestanskiy mountains) we found at an altitude of 2 000 m a red tulip resembling *T.carinata* from the *T.fosteriana* group, named *T.affinis.* In the same area, but at a somewhat lower altitude, grew bright cream-coloured *T.turkestanica.* On the way to Shakristan we found yellow *T.dasystemon* and in the Guralash reservation we collected two species, yellow *T.dasystemon* and an almost white *Eriostemones* tulip. Besides, staff members of this reservation told us that two different red species could be found 18 km north of the Kul'say experimental base. From the Mal'guzar range they brought us cream-coloured *T.turkestanica* and it is quite possible that the red *T.micheliana* is also found there.

In the Nuratinskiy range grow red *T.micheliana* and red *T.victoris,* a species close to *T.fosteriana.* It is the most eastern area of distribution of the whole series of species grouped around *T.fosteriana.* Of all the species inhabiting the Tien Shan mountain system, *T.vvedenskyi* seems at first sight the nearest species to this group, but a more thorough study reveals that it differs clearly from them by the slightly diffuse blotches on the claws of the tepals and by other characters.

In the Nuratau range are found cream-coloured *T.turkestanica* and in the sands of the foothills and farther into the desert yellow and orange-red *T.lehmanniana;* near Dzhizak red *T.korolkowii, T. nitida* and *T.micheliana.*

A wide variety of tulips grow in the western part of the Zeravshanskiy range, also known as the Samarkand mountains. They are all red: *T.fosteriana, T.ingens* and *T.micheliana. T.ingens* is light claret red without any trace of orange, like madder. *T.micheliana* can be scarlet with crimson shades. There is also *T.turkestanica,* cream-coloured.

The Kashka-Dar'ya area supports: red *T.ingens,* red *T.nitida,* red *T.tubergeniana* and cream-coloured *T.turkestanica.* The Gissarskiy range has *T.carinata, T. lanata, T.nitida, T.tubergeniana,* all of them red and *T.turkestanica,* cream-coloured. In the central part of this mountain range appear red *T.praestans* and yellow *T.hissarica,* while farther westwards one finds crimson *T.subpraestans* and white *T.orithyioides.* In the Darvazskiy mountains occurs red *T.linifolia* and near the

2. By 'regions' are meant here the vegetation regions of the USSR as delimited in Flora of the USSR, Vol. IV, 1935. (editor)

village of Sangtud on the river Vakhsh *T.maximowiczii*, also red.

Two yellow species were discovered in the Pamir-Alay: *T.platystemon* (Sary-Biya river) and *T.anisophylla* (Darvaz) and also the white *T.korshinskyi*. It is quite possible that still other species occur there since this mountain system has been poorly explored. In the Alayskaya valley grow yellow *T.dasystemon* and *T.alaica*, an *Eriostemones* with cream-coloured flowers.

Hence we find in the Pamir-Alay system the following species possessing red perianths: *T.affinis, T.carinata, T.fosteriana, T.lanata, T.victoris, T.tubergiana, T.micheliana, T.nitida, T.praestans, T.subpraestans, T.linifolia, T.maximowiczii*. At the present time we know in all thirteen species from that region that are red with different crimson shades. All the forms retain the basic colour of the species and only of *T.linifolia* a yellow form has been described, known from literature as *T.batalinii*, and of *T.nitida* the pink form *T.rosea*. Besides, there occur in the Pamir-Alay six species with yellow perianths: *T.anisophylla, T.dasystemon, T.ferganica, T.hissarica, T.platystemon, T.sp.* (northern Pamir-Alay) and also the following white ones: *T.alaica, T.bifloriformis, T.korshinskyi, T.turkestanica* and T.sp. (northern Pamir-Alay).

In the Kushka area a very beautiful red tulip was found — *T.kuschkensis*. The bulb has a shaggy tunic, a character for the first time recorded in the Pamir-Alay. Its area of distribution lies close to the Afghan border and very likely extends across it.

Farther west along the Kopet-Dag we find *T. micheliana*, red with crimson shades and more in the eastern Kopet-Dag *T. hoogiana* with a crimson-red perianth and a densely pubescent tunic. These two species are spread all along this mountain range, where in the steppe belt grow *T. wilsoniana*, crimson-red, and *T. turcomanica*, cream to white. Thus we see that mountainous Turkemenistan includes four species with red flowers: *T. hoogiana, T. micheliana, T. kuschkensis* and *T. wilsoniana*, besides a yellow form of *T. lehmanniana (T. ariana)* and the white *T. buhseana* and *T. turcomanica*. Typical of most western Middle-Asian species, inhabiting the Kopet-Dag, is the dominance of crimson shades in the red tepals instead of the orange shades of the species which inhabit the eastern parts of Middle Asia, e.g. the Zailiyskiy Alatau near Alma-Ata, the Chu river area near Frunze and further to the Karatau and in the western Tien Shan.

Our impression that the crimson shades predominate in red tulips occurring in the western part of the areas of distribution is to some extent supported by the Caucasian tulips. We had no chance to see living specimens of *T. florenskyi* and *T. schmidtii* and therefore found it difficult to form an opinion about their colours on the basis of herbarium specimens and descriptions from literature, but the colour of *T. florenskyi* is without doubt very similar to that of *T. julia*, while *T. schmidtii* is a little different. Besides two other Caucasian species with red flowers have been described which resemble *T. florenskyi*, viz. *T. kaghizmanica* and *T. sosnowskyi*. *T. eichleri* and *T. julia* are both red with bright crimson shades.

In the semidesert foothills of the northern Caucasus occur wine-coloured, orange and yellow forms of *T. schrenkii*.

The Caucasus is also the area of two *Eriostemones* with violet tepals, *T. violacea* from the Talysh region and *T. caucasica* from the upper Kuban. No similarly coloured tulips are known to exist in other vegetation regions of the USSR. There remain to be mentioned here the marvelous flowers of *T. polychroma* of which the white tepals acquire a pinkish crimson tint before fading and the only two yellow Caucasian tulips, *T. biebersteiniana* and *T. karabachensis*. We see that, although the areas have no species in common, the Caucasian tulips gravitate to the Pamir-Alay and mountainous Turkmenistan by their red-flowered species exhibiting similar crimson shades.

The Crimean tulips are somewhat different from the Caucasian ones and clearly show the influence of species from the Transvolga steppes. *T. schrenkii*, inhabiting the steppe zone of the Crimea, presents a wealth of colours, e.g. wine-coloured, pink of varying intensity, orange, yellowish flame-coloured and even pure white forms, but in none of its habitats does it show the flamboyant red colours of the Middle-Asian tulips. The other Crimean tulips are either yellow, like *T. biebersteiniana* and *T. oxypetala*, or white, like *T. callieri* and *T. koktebelica*.

In the European floristic region, as delimited in the Flora of the USSR including the Crimea, *T. schrenkii* also occurs in colours ranging from white, pink, yellow, orange to claret. Of the remaining species *T. sylvestris, T. biebersteiniana, T. callieri and T. koktebelica* are yellow and *T. patens* and *T. biflora* are white.[3]

The Syr-Dar'ya region has three red tulips: in semideserts *T. greigii* of which also yellow forms occur; in deserts *T. lehmanniana*, red-orange, orange and yellow; in semideserts *T. korolkowii*, red and yellow and also *T. bifloriformis* cream-coloured, almost white.

In the lake Balkhash area, in deserts and semideserts, one finds *T.albertii*, red-orange; in sandy

3. The author includes here the Crimean endemics, *T. callieri* and *T. koktebelica* as yellow species, presumably in error, since they are cited above as white-flowered species. It may be mentioned here that recent studies have shown both taxa to be identical with *T. biflora* Pall. (see: Mordak, E.V., 'O krymskikh tjul' panakh *Tulipa callieri* Halácsy et Levier i *T. koktebelica* Junge' in: Nov. Syst. Plant. Vasc. 12:132-134 (1975). (editor)

desert *T.behmiana,* yellow; in semideserts *T.altaica, T.biebersteiniana* and *T.kolpakowskiana,* all of them yellow, and *T.regelii* and *T.buhseana,* white.

In the Aral-Caspian region occurs in semideserts *T.greigii,* orange-red; in deserts *T.borzczowii* orange and red; in deserts and semideserts *T.schrenkii,* white, pink, red, yellow and orange. It is also the area of *T.biflora* and *T.buhseana,* both of which are white.

The Kyzyl-Kum has in the southwestern desert area *T.lehmanniana,* yellow, orange and red-orange; in the northeastern *T.borszczowii* with the same colour variants, and also two small, white tulips, *T.buhseana* and *T.sogdiana.*

In the Middle-Asia deserts, extending over an area from the Syr-Dar'ya river in the east nearly to the Caspian Sea in the west, yellow, orange and frequently also red-flowered forms of *T.lehmanniana* are widely distributed. In the Betpak-Dala and on the sands along the Ili river grows *T.behmiana,* yellow, accompanied by *T.talievii* (the eastern form of *T.buhseana).* A yellow *T.lehmanniana (T.ariana)* is found in the environs of Kushka. The white *T.sogdiana* ranges over nearly the same area as *T.lehmanniana,* except the eastern part. *T.ariana* is found there together with *T.turkestanica.*

The Karakumy desert is inhabited by only one species, the white-flowered *T.sogdiana.*

In the deserts of the Amu-Dar'ya region grows the already mentioned polychrome *T.lehmanniana* in yellow, orange and red forms.

In the Aral-Caspian lowland, in the Priaral'skiye Karakumy and on the sands along the Syr-Dar'ya river in the south one finds *T.borszczowii*, also polychrome, with yellow, orange and orange-red flowers. The semideserts in the northern part of this area are inhabited by *T.schrenkii,* another very polychrome species with a red, pink, orange, yellow or white perianth. This species is also widely distributed in the European part of the USSR. In the northern part of its area only yellow forms are found.

In the south and in the north, in the Mugodzhary hills, grows white *T.biflora,* spreading far southward with gradual transitions to *T.buhseana* and extending also to West Siberia and the southern European part of the USSR. In the same area, but more widely scattered, occurs *T.biebersteiniana* yellow, while *T.patens,* white, has been found only in the European part of the USSR and is hardly ever encountered in Middle Asia. The distributional pattern of these two species has been closely outlined by Taliev. Most interesting is, however, that within the territory of the USSR they extend farther north than any other species and that one has yellow and the other white flowers.

Western Siberia includes three yellow species: *T.heterophylla, T.altaica, T.uniflora* and two white ones, *T.biflora* and *T.patens.* Eastern Siberia has only one species, yellow *T.uniflora.*

Apparently the most widely distributed Middle-Asian species inhabited the ancient coastal desert of the Tethys and, after the bottom of the Tethys sea had risen, spread over vast desert areas. They include the desert inhabiting *T.lehmanniana* with polychrome flowers and the closely related *T.borszczowii.* Also *T.schrenkii* of semideserts and steppes, highly polychrome and occupying vast areas, extending, like those of the former species, beyond the boundaries of the USSR. The wide colour range, the ancient distribution pattern and the similarity of certain characters place the desert species *T.albertii* close to this group and to some extent also *T.greigii,* growing in foothills and semideserts. Both species hold their own with *T.lehmanniana* and *T.schrenkii* as regards the variation in flower colours, particularly in the Karatau mountain area.

Typical of the red tulips from the Tien Shan are their orange shades and the diffuse, indistinctly margined, blotches at the base of the perianth. Farther westward the Pamir-Alay tulips show an increasing tendency towards crimson instead of orange shades and where a yellow or a dark blotch occurs, it is either well-marked, or the dark blotch has a well-defined lighter coloured margin. Other species in the Pamir-Alay have densely hirsute bulb tunics, viz. *T.tubergeniana, T.subquinquefolia.* There is a close resemblance in the colour of the perianth between tulips from the Kopet-Dag and the Pamir-Alay, e.g. *T.hoogiana* with a densely hirsute bulb tunic. Caucasian tulips also show crimson shades, e.g. *T.julia,* likewise with a densely hirsute bulb tunic.

T.micheliana and *T.greigii* are vicarious species, *T.greigii* being typical of the Tien Shan foothills, while *T.micheliana* has spread in the Pamir-Alay, the Kopet-Dag and to Iran. There is, however, much more variation in the flower colour of *T.greigii* than in that of *T.micheliana,* which varies only from orange-red to crimson, while *T.greigii,* in the Karatau, ranges in colour from creamy white to dark red. In the western Tien Shan *T.greigii* varies from dark orange to dark red, the shades differing from those in the Karatau; in the Chu valley and in the Chu-Iliyskiye mountains from dark orange to orange.

As regards flower colour *T.ostrowskiana,* in particular the forms from the Kirgizskiy Alatau, comes close to *T.schrenkii* although it has a narrower colour range.

T.korolkowii appears to be typical of the Pamir-Alay, where it is widely distributed and was described as an independent species, *T.nitida.* In the eastern part of its area, in the environs of the Darbaza railway station, occur both red and yellow forms of *T.korolkowii.* It is the only red species found both in the Pamir-Alay and the western Tien Shan, where it grows in the lower mountain belt, being in particular adapted to habitats on sedimentary rock or disintegrated bedrock. In the Fergana valley this form has pink flowers. It was also described as an independent species, *T.rosea.*

Farther east in the Kirgizskiy Alatau occurs *T.zenaidae* with yellow flowers which is different from *T.korolkowii* but resembles it in the clearly marked black blotch at the base of the perianth.

The tulips of the eastern part of the distributional area of the genus, *T.albertii, T.greigii* and *T.korolkowii*, occupy a narrow strip of the deserts and foothills of the Tien Shan and are to some extent primary species. The proper mountain-inhabiting species of the Tien Shan: *T.kaufmanniana, T.ostrowskiana* and *T.tschimganica*, are younger derivatives which spread to the new territories of the raised Tien Shan ridges and have undergone some changes in adapting themselves to new conditions while retaining in some measure the ancient variability in flower colour, viz. the polychrome shades of different individuals in these populations.

Summarizing, we find in the region of the Tien Shan and the adjacent deserts six species with a high variability in flower colour: *T.albertii, T.greigii, T.kaufmanniana, T.ostrowskiana, T.tschimganica* and *T.korolkowii.* If the differently coloured bases of the tepals in the other species are disregarded, the first three are the most variable in flower colour. They even produce flowers that are completely white. Red-orange forms prevail in the first two species, while cream-coloured forms are predominant in *T.kaufmanniana.* The remaining species vary from light yellow to dark red.

In comparison with the above-mentioned species the Tien Shan tulips *T.butkovii* and *T.vvedenskyi* vary less in the basic colours of the perianth. Only light orange variations are found, but these are merely paler shades of the usual orange-red flowers of these species. In this case we see a similar variability to that shown by the Pamir-Alay tulips in which the basic colour hardly ever varies.

All the other species in that region belonging to the sections *Leiostemones* and *Spiranthera* have yellow flowers. They are: *T.anadroma, T.dubia, T.ferganica, T.iliensis, T.kolpakowskiana, T.tetraphylla, T.tian schanica* and *T. zenaidae.* The section *Eriostemones* includes cream-coloured or nearly white: *T.biflori-formis, T.buhseana, T.tarda, T.turkestanica* (only in the Mogoltau) and yellow *T.dasystemon, T.dasystemonoides* and *T.heterophylla,* the name of the last now being *Eduardoregelia heterophylla* M. Pop.

The Pamir-Alay is the third centre of diversity of the genus. There, and nowhere else, the adult bulbs of- *Leiostemones* species have retained the typical feature of the stolons growing horizontally instead of downwards. This habit is most noticeable in *T.carinata, T.fosteriana, T.praestans, T.lanata* and in many other species belonging to the section *Eriostemones.* The formation of stolons, typical of juvenile plants of both wild species and cultivars, disappears at various stages of maturity. With only a few exceptions it is lost in the adult phase (Botschantzeva, 1956). It should therefore be mentioned that adult bulbs of several Pamir-Alay species, belonging to the *Leiostemones,* form stolons which carry bulbs horizontally away from the parent bulb, a habit which is rather widespread among the *Eriostemones* (*T.bifloriformis* and others).

The type of pubescence on the bulb tunic of Pamir-Alay tulips varies widely. In *T.ingens* the tunic is sparsely lined with strands consisting of long straight hairs. In individuals of *T.carinata* and *T.lanata* the indumentum varies from very dense, nearly like that of *T.tubergeniana* of section *Tulipanum* which resembles cotton wool, to only weakly pubescent at the base and the apex of the tunic.

Thus, on the basis of colour patterns and certain other characters, it is possible to divide the area of the genus in the USSR into four large centres. The first includes the deserts and steppes comprising the Syr-Dar'ya, Lake Balkhash, Aral-Caspian, Kyzyl-Kum and Kara-Kum regions. The European areas of distribution are connected to this centre, while Siberia holds more or less a place of its own. The Tien Shan and the Pamir-Alay mountain systems form the second and third large centres. The fourth centre, mountainous Turkmenistan and the Caucasus, has connections with the Pamir-Alay. With only a few exceptions the polychrome *Leiostemones* species are absent in the third centre where crimson predominates among the red shades of the flowers and in a number of species also the profuse pubescence of the bulb tunic.

In the high mountain belt of Middle Asia grow only tulips with yellow, stellate, flowers and a limited number of strap-shaped leaves. These highland tulips are often polyploid, an indication of their being relatively young species which occupied these habitats rather recently.

There is very little variation in flower colour in the tulips close to the eastern boundary of the West-Siberian area, where species of the *Orithyia* section are found, and in those close to the eastern boundary of the area in Semirech'ye. Also the Fergana valley offers but a small variety of forms and colours. One centre, where diversification takes place in our time, lies in the deserts of Middle Asia, a second in the western Tien Shan and a third, an entirely independent one, in the Pamir-Alay.

Our studies of the species and forms of *Tulipa,* as they occur in nature, have shown the floristic isolation of the Pamir-Alay from the Tien Shan. The species occurring in the Pamir-Alay are in all cases closely related to those with distributional areas in the Kopet-Dag, in Iran and in the Caucasus and are isolated from the Tien Shan species. The areas of distribution and the characters of the latter show connections with the first centre of diversity in the Middle-Asian deserts.

The ancestors of *Tulipa* evidently did not inhabit the high mountain areas where tulips are rare. Moreover the presence of polyploids in those areas points to a much later occupation of these habitats.

The richest variety in the genus *Tulipa* is found in the middle and lower mountain belts, where also the basic chromosome number is strictly preserved. It appears that there, in the different mountain systems at various altitudes, but no higher than 3 000 m, lie the present centres of species formation of the genus *Tulipa*.

CHAPTER FIVE

Cytogenetic investigations of Middle Asian tulips

Nearly all species and forms of *Tulipa,* which over the last 30 years have formed part of the collection at the Botanical Garden of the Academy of Sciences of the Uzbek SSR, have been the subject of cytogenetic studies. The publications of the following authors have shown that the use of cytological data as taxonomic evidence and for phylogenetic interpretation can give very good results: Sakamura (1918), Nikolaeva (1922), Täckholm (1922), Karpechenko (1922), Kihara (1924), Delaunay (1915, 1922, 1925, 1928), Levitskiĭ and Kuz'mina (1927), Levitskiĭ (1931), Sveshnikova (1927), Avdulov (1931), Senyaninova-Korchagina (1932) and many others. The method found wide application after the publication of Levitskiĭ's work (1931) on suitable fixatives for studying karyotype morphology.

The chromosome number can be characteristic for certain groups, for example in *Triticum* (Sakamura, 1918; Nikolaeva, 1922). In other cases it remains unchanged in several related genera of a family, e.g. the somatic number of 24 in the genera *Erythronium, Fritillaria, Lilium* and *Tulipa* (Tischler, 1926; Darlington and Wylie, 1955). For comparative cytogenetic studies of taxa with a uniform chromosome number within a section or other category, it will be necessary to use also evidence from karyotype morphology. The basic morphological differences of karyotypes were first described by Navashin (1911) and Sakamura (1915, 1920) for plants and by Agar for the animal kingdom. In his studies of *Fritillaria tenella,* Navashin distinguished three types of chromosomes: one with a median centromere, one with a submedian centromere and another with a terminal position of the spindle attachment. He found that in chromosomes with a terminal spindle attachment there is always a kinetic constriction similar to that in the other types (the term 'kinetic constriction' was later introduced by Delaunay, 1928). Navashin (1911), who may be regarded as the founder of chromosome morphology, also discovered 'satellite' chromosomes, which play an important role in the construction of idiograms. Later on Chernoyarova (1914) and Delaunay (1915) observed segmented chromosomes, which besides the centromeric constriction, have secondary constrictions for which Delaunay (1928) introduced the term 'akinetic constriction'.

Levitskiĭ (1924-1931) systematised all the data on karyotype morphology and established a uniform terminology for its classification based on factual material. He presented the general structure of a chromosome as consisting of two arms joined by a kinetic constriction and gave special terms for chromosomes with median, submedian, subterminal and terminal attachments of the spindle threads. This structure is a primary character of normal chromosomes in both the plant and animal kingdoms (Levitskiĭ, 1931; Navashin, 1936). Satellites and akinetic constrictions are secondary developments.

The combination of karyosystematics with the more conventional methods has been found quite effective in solving taxonomic and phylogenetic problems (Levitskiĭ and Kuz'mina, 1927; Delaunay, 1928; Avdulov, 1930; Levitskiĭ, 1931). Delaunay (1928), having established the connection between karyology and systematics when studying the genera *Bellevalia, Muscari, Ornithogalum* and *Iris,* arrived at the conclusion that one may have to recognize the 'genus' as the most natural taxonomic unit after the 'elementary species'. For this genus, sometimes referred to as 'minor genus', we would prefer the term 'karyotype genus'. Levitskiĭ even claims that 'although chromosome morphology provides us in this respect[1] with very valuable combinations of characters, their taxonomic significance is in general just as relative as that of the external morphological characters'. This statement is supported by studies demonstrating the variability of the karyotype when for example two closely related species have so widely different karyotypes that, according to Delaunay, they should belong to different genera (Sveshnikova, 1927; Senyanikova-Korchagina, 1931). We also observed this in the case of *T. maximowiczii* and *T. linifolia,* morphologically very similar species, which judging by their karyotypes could be assigned to different genera. There is also the opposite case, viz. species belonging to different sections of the genus *Tulipa* which are so similar in karyotype that, according

1. Phylogenetic (Z.P. Botschantzeva).

to Delaunay, they should be classed together in the same 'minor genus'. In view of his treatment of the genus *Iris* the 'minor genus' should be conceived as a section of a genus. However, he was the only one who made an attempt to apply karyological data as a criterion of generic status. As a rule cytologists use their information only to obtain an idea about the closer or more remote systematic or phylogenetic relationships between taxa without concerning themselves with categories. Cytotaxonomists concentrate more on the definite taxonomic significance of karyosystematic characters and try to distinguish on this basis characters of families, genera, species and varieties. Among such characters Levitskiĭ includes chromosome size, basic number, segmentation and also polyploidy and states: 'to some extent every type of karyological change may be related to a certain taxonomic category, but this connection is only relative and permits numerous exceptions from the highest to the lowest categories' and in conclusion he again stresses: 'thus, karyological characters carry the same relative weight as the conventional morphological characters' (Levitskiĭ, 1931 pp. 208-209).

The first chromosome numbers, determined in *Tulipa,* were published by Guignard (1900), who studied the embryo-sac and the process of fertilization, followed by Ernst (1901), Strasburger (1908) and Newton (1927). With the exception of Newton, these researchers did not pay particular attention to chromosomes. Guignard dealing with four species in his embryological studies found the haploid number 12 for: *T. australis* Link, *T. clusiana* (probably), *T. gesneriana* L. and *T. sylvestris* L.

Newton investigated 26 species and 5 garden forms which were supplied by Dykes (1930). Unfortunately he did not describe any of the species he had studied, nor did he give an indication of the localities where they had originated. This has in some cases created insurmountable difficulties in comparing his results with those of our studies. Following Boissier, he divided the genus into two sections: *Leiostemones,* characterized by their glabrous filaments, and *Eriostemones,* with hirsute bases of the filaments. He studied the chromosomes of the following species in both the somatic and reduction division:

Section *Leiostemones* 2n(n);

T.batalinii Regel* 24; *T.linifolia* Regel* 24(12); *T.maximowiczii* Regel* 24(12); *T.chrysantha* Boiss. 48(24); *T.clusiana* DC. in Red. 60(24,12); *T.stellata* Hooker 48; *T.kolpakowskiana* Regel* 24(12); *T.greigii* Regel* 24; *T.eichleri* Regel 24; *T.kaufmanniana* Regel* 24(12); *T.praestans* Hoog* 24(12); *T.sprengeri* Baker 24; *T.armena* Boiss. 24; *T.galatica* Freyn 32; *T.viridiflora* Baker 24(12); *'Copper Colour' (hort.)* 24; *'Keizerskroon' (hort.)* 36; *'Massenet' (hort.)* 36; *'Murillo' (hort.)* 24(12);

Sect. *Eriostemones*

T.dasystemon Regel* 24(12); *T.humilis* Herbert 24(12); *T.pulchella* Fenzl. 24; *T.biflora* Pall.* 24; *T.turkestanica* Regel* 24; *T.primulina* Baker 24; *T.australis* Link 24(12); *T.sylvestris* L. 48(24); *T.orphanidea* Boiss. 24(12); *T.whittallii* (Dykes) A.D. Hall 48(24); *T.hageri* Heldr. 24(12).

The ten species marked with an asterisk grow in Middle Asia and for all of them Newton gives the somatic number 24. Besides counting the chromosomes, he also studied their morphology and distinguished three types: I. long chromosomes with a submedian centromere, II. medium sized chromosomes with a subterminal centromere and III. short chromosomes with a terminal centromere. He notes that in *Tulipa* the chromosomes of types I and II are almost equally long, the only exception being *T.armena* and *T.galatica* in which the chromosomes of type I are much longer than type II, as is also found in *Lilium* and *Fritillaria.* The short chromosomes of type III have a subterminal centromere in *Tulipa* and *Lilium* and a terminal one in *Fritillaria. Lilium, Fritillaria* and *Tulipa* (with the exception of five species and the tetraploid series) have two pairs of chromosomes with a submedian centromere.

Of *T.batalinii, T.linifolia* and *T.maximowiczii,* all chromosomes have a terminal centromere. These 3 species constitute a natural group differing both morphologically and cytologically from the other tulip species.

T.kolpakowskiana has neither chromosomes with a submedian centromere nor the very small chromosomes typical of the *T.linifolia*-group; they are of an intermediate type between those of *T.linifolia* and the normal complement.

T.armena and *T.galatica,* two closely related species differing only in flower colour, have chromosomes of type I, which are more isobrachial than usual, but differ in chromosome number: *T.armena* has 2n = 24 and *T.galatica* 2n = 32. The four extra pairs of chromosomes are very small and had not yet been observed in *Tulipa.* According to Newton they are the result of fragmentation and did not arise through doubling or hybridization. However, this was not confirmed by an examination of the complements of these species since there could not be observed any shortened chromosomes in *T.galatica* in comparison with those of *T.armena.*

Newton also studied several polyploids such as the tetraploid forms *T.stellata* and *T.chrysantha,* which gave rise to the hybrid *T.clusiana. T.whittallii* appears to

Figure 11. Metaphase plates

a. *T. kaufmanniana* Regel; b. *T. wilsoniana* Hoog; c. *T. subpraestans* Vved.; d. *T. kolpakowskiana* Regel; e. *T. maximowiczii* Regel; f. *T. ferganica* Vved.; g. *T. tetraphylla* Regel; h. *T. greigii* Regel; i. *T. linifolia* Regel; j. *T. wilsoniana* Hoog; k. *T. zenaidae* Vved. l. *T. maximowiczii* Regel; m. *T. praestans* Hoog; n. *T. fosteriana* Hoog ex Irv.

be a stable tetraploid, which evidently originated from the forms grouped under the name of *T.orphanidea.*

In his list of chromosome numbers of higher plants Tischler (1938), basing himself on Upcott (1935, 1936, 1937), Upcott and La Cour (1936), Hall (1934, 1937), Woods and Bamford (1937) and Darlington and Wylie (1955), gives the haploid numbers for the following tulip species:

T.altaica 12; *T.amabilis* 12; *T.armena* 12; *T.aucheriana* 12; *T.australis* 12; *T.batalinii* 12; *T.biflora* 12; *T.boeotica* 12; *T.borszczowii* 12, 12 + frag.; *T.chrysantha* 12; *T.cretica* 12; *T.cuspidata* 12; *T.cypria* 12; *T.dasystemon* 12; *T.eichleri* 12; *T.fosteriana* 12; *T.galatica* 12 + frag.; *T.gesneriana* 12; *T.greigii* 12; *T.grisebachiana* 12; *T.hageri* 12; *T.hoogiana* 12; *T.humilis* 12; *T.ingens* 12; *T.kaufmanniana* 12; *T.kesselringii* 12; *T.kolpakowskiana* 12, 24; *T.korolkowii* 12, 24; *T.kuschkensis* 12; *T.lanata* 12, 18; *T.linifolia* 12; *T.marjolettii* 12; *T.micheliana* 12; *T.montana* 12, 12 + frag.; *T.oculus-solis* 12; *T.orphanidea* 12, 18, 30; *T.ostrowskiana* 12, 24; *T.persica* 12; *T.planifolia* 12; *T.polychroma* 12; *T.praecox* 12, 18; *T.praestans* 12; *T.primulina* 12; *T.pulchella* 12; *T.rhodopea* 12; *T.saxatilis* 12, 18; *T.schmidtii* 12; *T.schrenkii* 12; *T.sprengeri* 12; *T.stapfii* 12; *T.stellata* 12; *T.suaveolens* 12; *T.tubergeniana* 12; *T.undulatifolia* 12; *T.urumiensis* 12; *T.violacea* 12; *T.wilsoniana* 12; *T.-aleppensis* 18; *T.clusiana* 18, 30, 12; *T.edulis* 24; *T.sylvestris* 24; *T.turkestanica* 24; *T.whittallii* 24.

Of the tulips growing in the USSR only 27 species are included in this list. The exact origin of many of them is not known.

It is only recently that a comprehensive study of the genus was started and it will therefore be clear that many evolutionary problems are still awaiting solution.

MATERIALS AND METHODS

The root-tips to be studied were obtained from species that had been collected in the course of many years in different regions of Middle Asia by expeditions of the Botanical Garden of the Academy of Sciences of the Uzbek SSR. The bulbs were planted out in the nursery and the plants were thoroughly studied. From each habitat 10 plants were taken and only in the case of some species, of which there were only a few individuals, did this number have to be reduced. With bulbs of Middle Asian tulips all roots are initiated at the same time, which was the main difficulty we had to overcome in our investigations (Botschantzeva, 1937). Removal of all the roots will in most cases kill the bulb, and even if only part of the roots is cut off, the bulb is weakened and will not flower for many years. However, in studying chromosome number and karyotype morphology, it will be necessary to preserve a normally developing plant. We used also smears from young anthers, according to the method of Levan (1931, 1932), for which a comparatively large number of buds are needed in the preliminary search for the suitable stages of division of the primary nucleus. It follows that the combined methods, root-tips and smears, can only be applied if sufficient material is available. For these reasons we only used pollen tubes at the stages of nuclear division when there was not enough material or when the plates obtained from root sections were unsatisfactory, and sometimes as a final check. Some species were therefore studied both in sections from root-tips and in smears from anthers and pollen tubes.

The root-tips were taken from bulbs and not from seedlings, because during the first year the latter make only one root which soon stops growth. Besides, our tulip seeds have a dormancy period of nearly one year and will only germinate after a cold period during winter, whereas bulbs start growth in October. From one adult bulb of a larger tulip nearly 100 root-tips can be obtained, from the smaller bulbs 10-15.

The bulbs were planted in pots filled with moist soil or in vessels containing just enough water to reach the basal plate, otherwise the bulbs will start to rot. To prevent rotting of the bulbs and excessive twisting of the roots, the tunics were removed.

In trying to find the optimal fixative for the tulip root sections we used several variants of chromacetformal (Navashin, 1914) and chromformol (Levitskiĭ, 1931). With chromformol fixation the best results were obtained by using equal parts (5 + 5) of 1% chromic acid and 10% formalin. Larger dosages of formalin cause considerable discoloration of the chromosome arms next to the kinetic constriction, increasing its size at the expense of the arms. Small dosages produce a similar picture to that obtained with chromacetformol. With most tulip species chromacetformol reveals the kinetic constrictions rather clearly, but tulip chromosomes being large and thick, this fixative makes it difficult to locate conveniently situated plates. At first we took root sections with a thickness of 10μ but later we cut them thicker, 24μ, because it appeared to be difficult to find in the thin sections intact chromosomes at the metaphase stage. When chromformol in the 5 + 5 combination is used for fixation of root sections the plasma is sufficiently transparent even in the 24μ sections. The microtomic preparations were stained with ferrohaematoxylin after Heidenhain.

The anthers of Tulipa species ripen at different stages of bud development. In making smears of for instance *T.praestans* very young buds must be used before their emergence from the soil, whereas buds of *T.maximowiczii* should be taken after the pedicel has

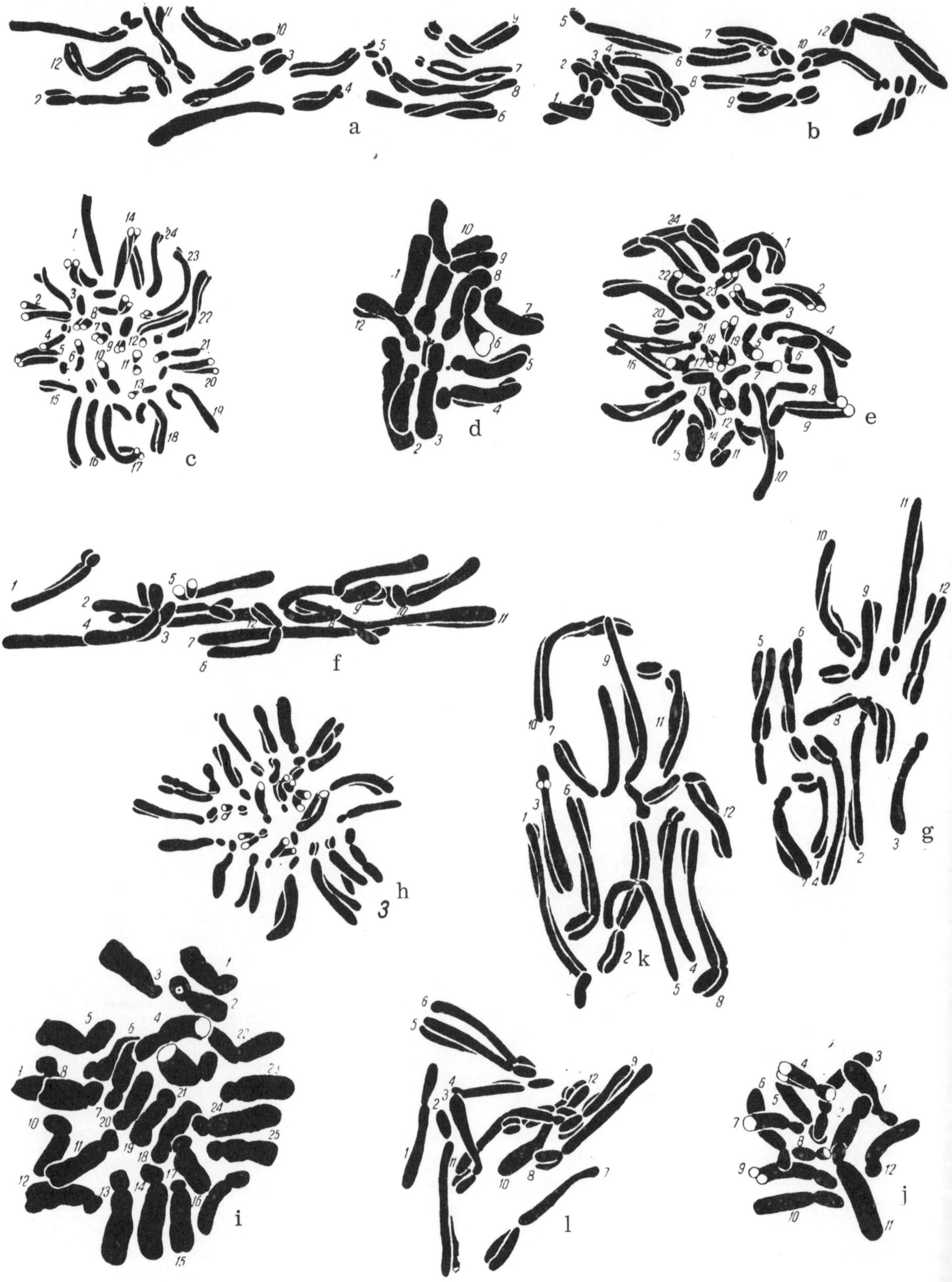

Figure 12. Metaphase plates (enlarged 1000 times)
a. *T. bifloriformis* Vved.; b. *T. bifloriformis* Vved.; c. *T. binutans* Vved.; d. *T. binutans* Vved.; e. *T. buhseana* Boiss.; f. *T. turkestanic* Regel; g. *T. buhseana* Boiss.; h. *T. bifloriformis* Vved.; i. *T. bifloriformis* Vved.; j. *T. turkestanica* Regel; k. *T. buhseana* Boiss.; l. *T* *binutans* Vved.

reached its normal length, 2 or 3 days before flowering. The preliminary treatment with aceto-carmine for determining the suitable stages of division, in particular of species with early-ripening pollen, damages the bud and the flower will be misshapen or killed. Nevertheless, it is worthwhile sacrificing several buds if there are more plants, since the bulbs will be saved. The ease with which the chromosomes can be counted, the clear image of their morphological characters (fig.12), the practically unaltered size and the simple technique in comparison with the microtomic sections are all in favour of a wider application of this method.

The anthers taken from a bud at the correct stage of development are crushed and smeared between two slides and immediately placed in a fixiative of the following composition:

1st solution — picric acid (sat. sol.), 75 ml; urea, 22 ml; chromic anhydride 1.5 ml.

2nd solution – 'glacial' acetic acid, 1.5 ml; formalin (commercial), 7 ml.

Heat the first solution before use and mix 21 ml with the second solution (8.5 ml) just before fixation, which takes from 2 to 6 hrs.; then rinse in water and stain in the following way: place one drop of 1% gentian-violet on the slide; rinse with distilled water; drop I-KI on the preparation and shake; rinse with distilled water; then pass it quickly through increasing concentrations of 10%, 20%, 35%, 50%, 75% alcohol and abs. alc; after checking the colour of the preparation in the abs. alc. under the microscope add 2-3 drops of a mixture of equal parts of clove oil and abs. alc. and pour off; then add one drop of pure clove oil and allow to stand for half an hour; pour off the clove oil; add 2-3 drops of a mixture of equal parts xylol and clove oil; pour off; rinse in xylol; add one drop of a mixture of two parts xylol and one part thin cedarwood oil; pour off; add one drop of a mixture of 1 part xylol and two parts thin cedarwood oil; pour off; place in pure cedarwood oil and cover with cover slip. If Canada balsam is used instead of cedarwood oil the pollen grains will shrink and their contents be deformed, which is highly undesirable.

The limited quantity of material of certain species compelled us to study also chromosomes at the stages of division of the generative nucleus in the pollen tube. As regards distinctness of the chromosomes, their morphology and size this method is in no way inferior to Levan's method (Botschantzeva, 1934).

When tulip pollen germinates in vitro in a suspended drop medium of 2% gelatin and 20% cane sugar spermatogenesis in the pollen tubes commences within 20 minutes and continues for 21 hours. The pollen tubes are wide enough for the chromosomes to arrange themselves freely and their length is but rarely affected. As a fixative chromformol gives quite satisfactory results in studying chromosome morphology. It is sufficient to keep the mounts in it for two hours and then to rinse them in water for one hour (Botschantzeva, 1937). After this the samples are stained with gentian violet in the same manner as the smears, using Levan's method but without the preliminary treatments with alcohol. The sections are sealed in cedarwood oil. Mounting samples from pollen tubes takes less time than making microtomic sections and the reduction to the haploid number of the chromosomes, the absence of deformations, and the clear image of their morphology all recommend the application of this method. Besides, once the interval between bringing the pollen in its medium and the division of the generative nucleus has been determined for a given species, one can always fix the pollen tubes after this time interval to obtain tens or even hundreds of metaphase plates to choose from.

All drawings were made with the aid of a drawing apparatus. In comparing chromosome size of the different species the measurements were taken according to the projection method, worked out by Levitskiĭ (1931).

Root-tips of *Tulipa* may sometimes show, besides normal chromosomes, very much shortened chromosomes (fig. 10m). As this shortening, although to a lesser degree, occurs quite frequently, we first divided the chromosomes into main groups according to the length ratios of the arms: less than 2:1, more than 2:1, more than 3:1, more than 4:1 etc. Then these divisions were subdivided according to size, absence or presence of akinetic constrictions and satellites.

SECTION *TULIPANUM* REB.

Data about chromosome numbers and karyotypes of species belonging to this section were first published by Upcott and La Cour (1936) for *T.hoogiana, T.kuschkensis* and *T.tubergeniana* and later confirmed by Woods and Bamford (1937). We used mainly root-tip sections for measuring and other morphological observations. Spermatogenesis was studied in pollen tubes in vitro.

The five species in this section growing in the USSR are: *T.hoogiana* B. Fedtsch., *T.kuschkensis* B. Fedtsch., *T.subquinquefolia* Vved., *T.tubergeniana* Hoog from Middle Asia and *T.julia* K. Koch from the Caucasus. *T.subquinquefolia* was described by Vvedenskiĭ as a new species in 1946.

Tulipa tubergeniana Hoog. The individuals studied were collected in Uzbekistan, the Denauskiy district in the environs of the village of Sina; in Tadzhikistan, the environs of Tutkaul. Hillcrests on loess and red clay.

In the idiograms, which were derived from the chromosomes at the stage of nuclear division in pollen tubes, the chromosomes are stretched to 1½ times the length of chromosomes of mitotic preparations, the relative length of the arms remaining unaltered. The length of the chromosomes in this species ranges from 9.25 to 21.25μ.

The 24 chromosomes of the diploid complement fall into six main groups (fig.13a, 14a).

Group I, one pair: The shortest of the complement.

Group II, two pairs: The 1st pair very large with the long arms 14.25μ and in the lower third of the short arms an akinetic constriction; the 2nd pair much shorter with the long arms 9.5μ.

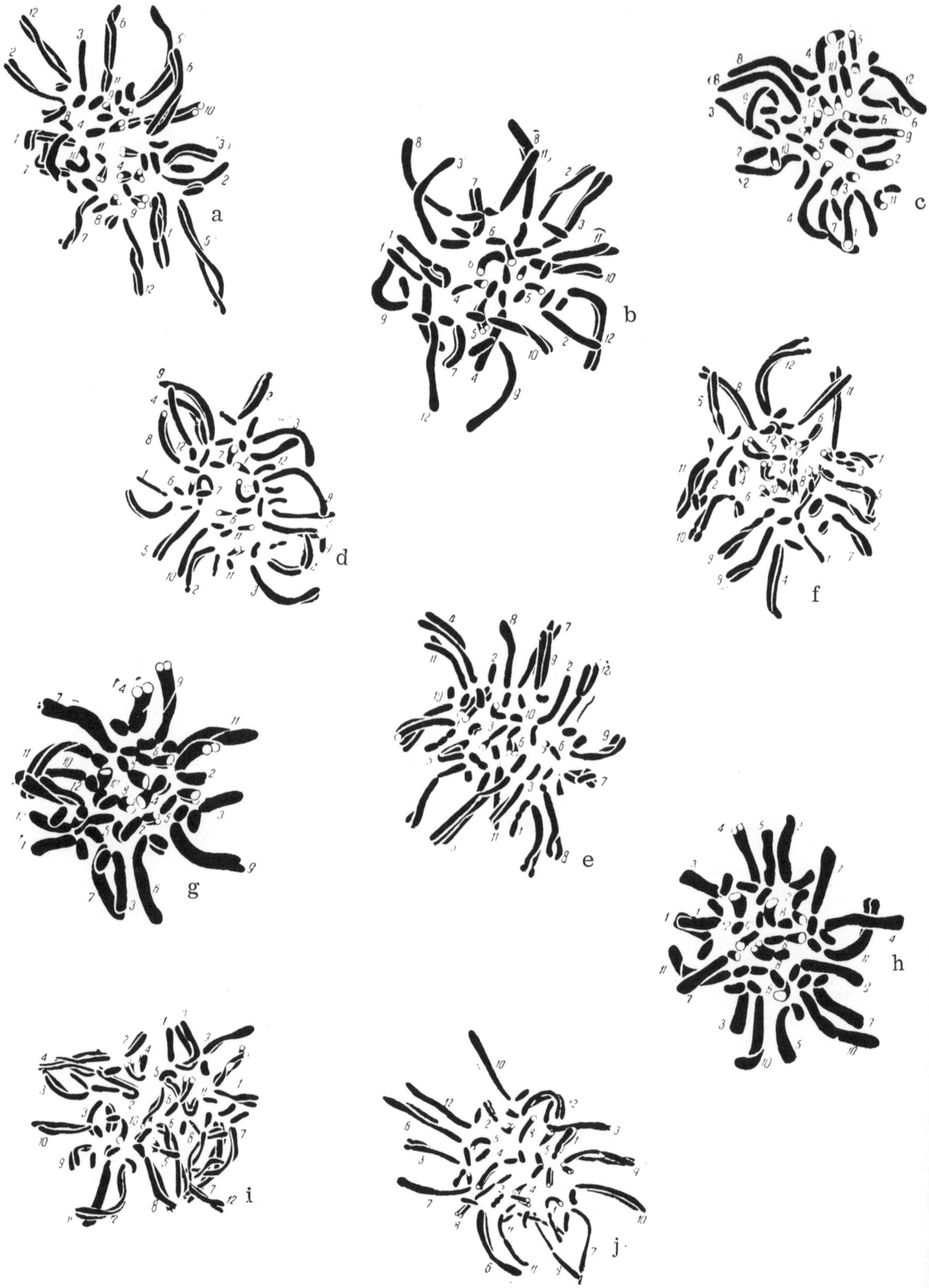

Figure 13. Metaphase plates from root-tip sections (enlarged 1000 times)
a. *T. tubergeniana* Hoog; b. *T. fosteriana* Hoog ex Irv.; c. *T. lanata* Regel; d. *T. victoris* Vved.; e. *T. kuschkensis* B. Fedtsch.; f. *T. inger* Hoog; g. *T. subpraestans* Vved.; h. *T. praestans* Hoog; i. *T. hoogiana* B. Fedtsch.; j. *T. carinata* Vved.

Group III, one pair: The long arms almost as long as those of the 2nd pair of group II, the short arms shortened.

Group IV, five pairs: The 1st pair with the longest arms of the set, 18.75μ, the short arms 3.75μ; the following pairs successively shorter with the long arms diminishing more radically than the short arms; the long arms of the 2nd and 3rd pairs, 14.0 and 13.25μ resp.; the 4th pair only slightly larger than the 5th; the long arms of the 5th pair 10.0μ, the short arms 2.5μ.

Group V, two pairs: The 1st pair with the long arms 10.62μ, the short arms 2.25μ; the 2nd pair 10.0 and 2.0μ resp.

Group VI, one pair: The long arms 13.25μ, the short arms 2.0μ.

The total length of the haploid complement is 175.50μ, of the long arms 140.0μ and of the short arms 35.50μ. Total index: 3.94.

Tulipa kuschkensis B. Fedtsch. The individuals studied were collected in Turkmenia, Kushkinskiy district, near the village of Morgunovskiy and in the environs of Kushka.

The 24 chromosomes are generally smaller than in *T.tubergeniana*, but more segmented. They fall into six main groups (fig.13e, 14b).

Group I, one pair: The long arms 7.88μ and an akinetic constriction below a spherical segment at the distal end; the short arms 5.5μ with two constrictions dividing the arms into three unequal parts.

Group II, two pairs: The 1st pair with the long arms 8.5μ, the short arms 3.0μ; the 2nd pair 8.13μ and 3.0μ resp.

Group III, one pair: Of nearly the same size as the corresponding pair in *T.tubergeniana;* the long arms 10.0μ, the short arms 2.75μ with in the middle an akinetic constriction.

Group IV, five pairs: the 1st pair with the long arms 10.0μ and an a. con. at ca. 3.0μ from the centromere, the short arms 2.13μ; the 2nd pair with the long arms 9.0μ, the short arms 2.0μ; the 3rd pair with the long arms 8.13μ and about halfway an a. con., the short arms 2.0μ; the 4th pair 7.5 and 1.9μ; the 5th pair 6.5 and 1.5μ resp.

Group V, two pairs: Of nearly equal size; the long arms of the 1st pair 8.0μ, the short arms 1.5μ.

Group VI, one pair: The long arms 9.38μ, the short arms 1.5μ; in the long arms an a. con. below a spherical segment close to the centromere.

The total length of the haploid complement is 128.76μ; of the long arms 101.00μ and of the short arms 27.75μ. Total index: 3.64.

Tulipa hoogiana B. Fedtsch. The individuals studied were collected in Turkmenia, environs of the Prokhladnoye settlement, the Arvaz river (eastern slope).

The 24 chromosomes of the diploid complement fall into five main groups (fig.14c).

Group I, one pair: The long arms 5.5μ, the short arms 2.75μ.

Group II, two pairs: The 1st pair with an akinetic constriction in the lower third of the short arms; the 2nd pair with the long arms 6.26μ, the short arms 2.25μ and an a. con. ca. halfway.

Group III, three pairs: The first pair with the long arms 8.5μ, the short arms 2.5μ and an a.con. at ca. 1.0μ from the centromere; the 2nd pair with the long arms 6.63μ, the short arms 2.0μ and an a. con. at the distal end of the long arms below a spherical segment; the 3rd pair with the long arms 6.25μ, the short arms 1.5μ and a similar constriction to that found in the 2nd pair at the distal end of the long arms.

Group IV, four pairs: The 1st pair with the long arms 8.5μ, the short arms 1.88μ and an a.con. in the middle; the 2nd pair with the long arms 7.5μ, the short arms 1.75μ; in the long arms an a.con. at 2.0μ from the centromere, in the short arms also an a.con. at ca. 0.7μ from the centromere; the 3rd pair with the long arms 7.0μ, the short arms 1.63μ; the 4th pair with the long arms 6.5μ, the short arms 1.63μ resp.; no secondary constrictions were observed in the 3rd and 4th pairs.

Group V: Absent.

Group VI, two pairs: The 1st pair with the long arms 10.5μ, the short arms 1.75μ, at the distal end of the long arms an a.con., separating a segment 1.25μ long; the 2nd pair with the long arms 9.63μ, the short arms 1.5μ and no secondary constrictions.

The total length of the haploid complement is 110.67μ; of the long arms 87.75μ and of the short arms 22.92μ. Total index 3.8.

It appears that in this section *T. tubergeniana* and *T. kuschkensis* are very close to each other in karyotype.

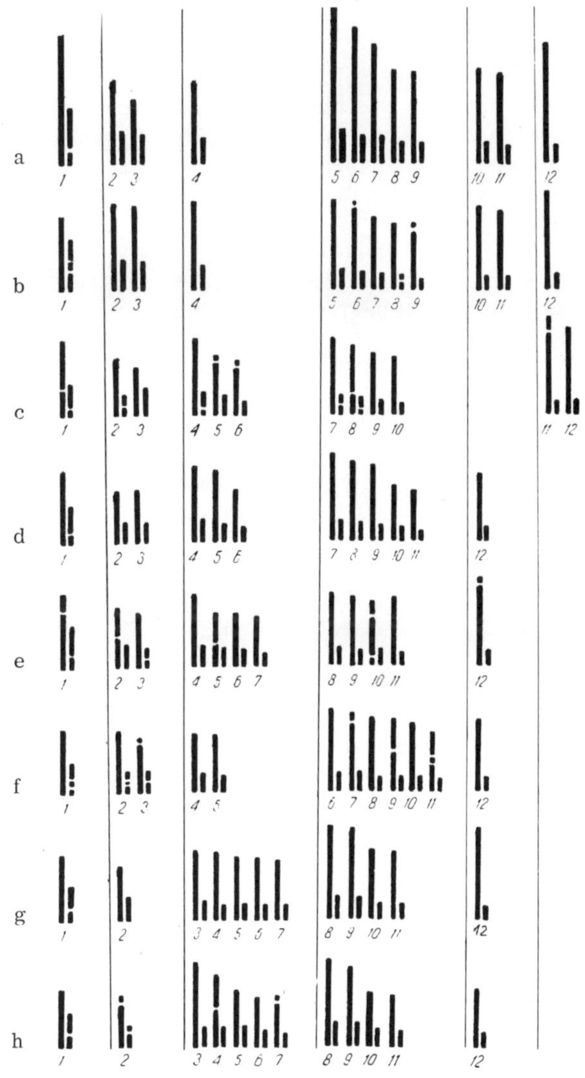

Figure 14. Idiograms of measured plates (enlarged 600 times) a. *T. tubergeniana* Hoog; b. *T. kuschkensis* B. Fedtsch; c. *T. hoogiana* B. Fedtsch.; d. *T. lanata* Regel; e. *T. ingens* Hoog; f. *T. carinata* Vved.; g. *T. fosteriana* Hoog ex Irv.; h. *T. victoris* Vved. (In figure 14a chromosome no 3 should be classed in group I, while no 1 should take the place of no 2 and no 2 that of no 3 in group II. In figure 14c chromosome no 3 should be classed in group I, while no 1 should take the place of no 2 and no 2 that of no 3 in group II.) (editor)

The shorter chromosomes of *T. kuschkensis* and their segmentation suggest a more recent origin of that species. *T. tubergeniana*, apparently, is a direct relative of *T. kuschkensis*.

Despite the great similarity in karyotypes of *T. hoogiana* and the two above-mentioned species no clear idea can be formed about the origin of the former, although the influence of some other karyotype seems certain. A better understanding of the phylogeny in the section *Tulipanum* may be obtained after the tulips of the neighbouring countries, Iran and Afghanistan, have been studied.

SECTION *LEIOSTEMONES* BOISS.

The Flora of the USSR, Vol IV, 1935, lists 38 species belonging to this section. Five more were described after its publication, which brings the number of *Leiostemones* growing in the Soviet Union to 43, of which we studied 28, all with areas of distribution in Middle Asia.

To facilitate the analyses of the karyotypes the species were divided into groups according to their relationship by external morphological characters:

1) *T. lanata, T. ingens, T. carinata, T. fosteriana, T. victoris*
2) *T. praestans, T. subpraestans, T. butkovii*
3) *T. albertii, T. mogoltavica, T. greigii, T. micheliana*
4) *T. maximowiczii, T. linifolia, T. wilsoniana*
5) *T. lehmanniana, T. sp.* from Kushka, *T. behmiana, T. borszczowii*
6) *T. korolkowii, T. nitida, T. rosea*
7) *T. schrenkii, T. ferganica, T. zenaidae, T. ostrowskiana, T. kolpakowskiana*
8) *T. tetraphylla*

Newton studied six Middle Asian *Leiostemones* species: *T. batalinii, T. maximowiczii, T. linifolia, T. greigii, T. kolpakowskiana* and *T. praestans*. Vvedenskiĭ in 'Flora of the USSR' mentions *T. batalinii* as a synonym of *T. maximowiczii* and consequently Newton determined the chromosome number of five Middle Asian species of this section. He found for all five 24 as the number of their somatic complement, but gave no morphological description, nor an indication of their origin. For instance for *T. maximowiczii* was found 2n=24. However, in studying plants of *T. maximowiczii*, collected by an expedition of the Botanical Garden of the Tashkent State University in the environs of Sangtud, we have several times found the number 2n = 22. Newton also draws attention to all chromosomes of this species having terminally attached spindle threads, but we observed both in somatic cells and in the smears (Levan's method) one pair of chromosomes with an approximately median attachment of the spindle threads. No doubt the name *T. maximowiczii* had been used for an entirely different taxon from the one to which we attribute this name and which was described by Regel. Our plants were collected from the original habitat and it is not clear where the material studied by Newton and his successors came from. The publications of La Cour and Upcott (1935, 1936, 1937) and of Woods and Bamford (1937) show the same deficiency, because these botanists studied progenies which had been grown in various botanical gardens in Europe and the USA, instead of plants which were collected directly in the natural areas of distribution. Moreover, large-scale experiments have shown that under favourable conditions wild tulip species interbreed both with other species and cultivars. The doubtful species mentioned by Newton are: *T. batalinii, T. linifolia* and *T. maximowiczii*. In publications by Upcott and La Cour: *T. borszczowii* hort., *T. kolpakowskiana* hort. and *T. ostrowskiana* hort. It is not very well possible to confuse *T. hoogiana* with *T. lanata* and yet in Woods and Bamford (1937) one finds in table 1 '*T. hoogiana*' with in brackets '*T. lanata*'. For these reasons only the data on chromosome numbers can be relied on in this case, because the somatic number, 2n=24, is typical for the genus *Tulipa* and a number of closely related genera, while minor deviations and polyploidy are comparatively rare.

It will be clear that without proper descriptions and information on the areas of distribution and habitats of the individuals studied, it is impossible to obtain reliable data which could be used for phylogenetic or other schemes. The names appearing in cytological and embryological publications are in general not trustworthy because the authors are in most cases not sufficiently acquainted with their materials and, when dealing with individuals of doubtful origin, attribute them to definite species.

Leiostemones, first group

Tulipa lanata Regel. The individuals studied were raised from bulbs which originated from Khadza-Ipak in the Denauskiy district, Uzbekistan (coll. A. I. Vvedenskiĭ and V.P. Bochantsev).

We examined chromosomes in somatic cells from root-tips and in pollen tubes during spermatogenesis. The 24 chromosomes of the diploid complement fall into five main groups (fig. 13c, 14d).

Group I, one pair: The long arms 7.5 the short arms 4.0μ with an akinetic constriction at approximately 1.0μ from the centromere.

Group II, two pairs: The 1st pair with the long arms 5.5μ, the short arms 2.25μ; the 2nd pair with the long arms about as long as those of the 1st pair, the short arms ca. 2.0μ.

Group III, three pairs: The 1st pair with long arms 8.25μ, the short arms 2.60μ; the 2nd pair 8.0 and 2.25μ; the 3rd pair 5.5 and 1.5μ resp.

Group IV, five pairs: The 1st pair with the long arms 9.5μ, the short arms 2.3μ; the 2nd pair 8.75 and 2.0μ; the 3rd pair 8.5 and

1.75μ; the 4th pair 6.0 and 1.25μ; the 5th pair 5.6 and 1.24μ resp.

Group V, one pair: The long arms 7.5μ, the short arms 1.5μ; the only secondary constriction was observed in the one pair of group I.

The total length of the haploid complement is 110.95μ; of the long arms 96.1μ and of the short arms 24.85μ. Total index: 3.87.

When studying nuclear division in pollen tubes we found, besides the normal number of 12 chromosomes, also pollen tubes with 11 chromosomes where in the majority of cases the chromosome of group I was missing and only in one case did this appear to be a chromosome from another group. Pollen tubes with 13 chromosomes were not observed. It is possible that pollen grains with n + 1 chromosomes do not germinate. The unbalanced chromosome complement and consequently the gametes with different chromosome numbers indicate that the given species is comparatively recent, although it may have existed for a long time without changing because of its means of vegetative reproduction, in which case its relative age is uncertain. The fact that among the 15 plants studied we did not find any with somatic cells containing 22 or 23 chromosomes, may be explained by the limited number of individuals we examined or by the lack of vitality of such forms.

Tulipa ingens Hoog. The individuals studied were collected in the Samarkand mountains (Satualdybaev); in the environs of the village of Agalyk, on gentle slopes (A. Ya. Butkov); Aman-Kutan (Z.P. Botschantzeva); Kugitang (A.F. Gnezdilo).

The chromosomes were studied in root-tip sections taken from adult bulbs; the diploid number is 24; they can be divided into five groups (fig.13f, 14e).

Group I, one pair: The long arms 8.0μ with an akinetic constriction about 3.0μ from its distal end, the short arms 4.5μ with an a.con. 1.25μ from the centromere.

Group II, two pairs: The 1st pair with the long arms 7.0μ and an a.con. approximately in the middle, the short arms 2.5μ; the 2nd pair with the long arms 5.6μ, the short arms 2.0μ and an a.con. at three quarters from the distal end.

Group III, four pairs: The 1st pair with the long arms 7.75μ, the short arms 2.25μ; the 2nd pair with the long arms 6.25μ and an a.con. at ca. 2.5μ from the distal end, the short arms 2.0μ; the 3rd pair 5.75 and 1.6μ; the 4th pair 6.0 and 1.7μ resp.

Group IV, four pairs: The first pair with the long arms 8.0μ, the short arms 2.0μ; the 2nd pair 7.5 and 1.75μ; the 3rd pair 7.0 and 1.75μ resp. and in the long arms two a.cons., one at ca. 1.5μ from the centromere, the other at ca. ¾ of its length from the centromere; the 4th pair with the long arms 7.25μ and the short arms 1.6μ.

Group V, one pair: the long arms 9.75μ and with a small, ca. 1.0μ long, segment at the distal end which in some cases, where the secondary constriction is elongated, could be taken for a satellite; the short arms 1.75μ.

The total length of the haploid complement is 109.45μ; of the long arms 85.35μ and of the short arms 24.1μ. Total index 3.54.

Tulipa carinata Vved. The individuals studied came from the Denauskiy district, environs of the village of Sina (coll. V.P. Bochantsev and A.I. Vvedenskiĭ). The flowers of *T.carinata* are smaller than those of *T.fosteriana.*

In this case we used only root-tips. The somatic complement can be divided into five main groups (fig.13j, 14f).

Group I, one pair: The long arms 7.25, the short arms 3.75μ; in the short arms two akinetic constrictions, the 1st at 0.75 and the 2nd at 1.75μ from the centromere.

Group II, two pairs: The 1st pair with the long arms 6.5μ, the short arms 2.3μ and two a.cons., one at 0.25μ from the centromere, the other at 0.63μ from the distal end; the 2nd pair with the long arms 6.0μ and at the distal end a spherical segment, 0.4μ in diameter, which may be a satellite; the short arms 2.25μ and an a.con. at ca. 1.0μ from the centromere.

Group III, two pairs: The 1st pair with the long arms 6.25μ, the short arms 2.0μ; the 2nd pair 6.25 and 1.75μ resp.

Group IV, six pairs: The 1st pair with the long arms 9.0μ, the short arms 2.0μ; the 2nd pair with the long arms 8.5μ and a spherical segment 0.75μ long at the distal end (the constriction separating this segment is unstable), the short arms 2.0μ; the 3rd pair with the long arms 8.25μ, the short arms 1.75μ; the 4th pair with the long arms 8.0μ and an a.con. approximately in the middle, the short arms 1.75μ; the 5th pair 7.6 and 1.6μ resp.; the 6th pair with the long arms 6.25μ and two a.cons. at ca. 2.5 and 3.0μ from the centromere, these arms appearing like two longish segments joined by an ellipsoid body, the short arms 1.4μ.

Group V, one pair: The long arms 7.75μ, the short arms 1.4μ.

The total length of the haploid complement is 111.45μ; of the long arms 87.6μ, of the short arms 23.95μ. Total index: 3.64.

Tulipa fosteriana Hoog ex Irv. The individuals studied were collected at Agalyk and Aman-Kutan in the Samarkand mountains (Zeravshanskiy range). In the nursery of the botanical garden we selected six forms varying in the colour of perianths and foliage.

The somatic number counted in root-tips of adult bulbs was always 24 (fig.13b). Numbers and karyotype morphology were checked by Levan's method in smears from young anthers (fig.11n). The diploid complement of 24 chromosomes can be divided into five main groups (fig.14g).

Group I, one pair: The long arms 7.4μ, the short arms 5.0μ with an akinetic constriction at ca. 1.25μ from the centromere.

Group II, one pair: The long arms 5.75μ, the short arms 2.5μ.

Group III, five pairs: Of rather uniform length and shape; the 1st pair with the long arms 7.6μ, the short arms 2.25μ; the 2nd pair 7.6 and 2.13μ; the 3rd pair 7.0 and 1.9μ; the 4th pair 7.0 and 1.9μ; the 5th pair 6.75 and 1.75μ resp.

Group IV, four pairs: The 1st pair with the long arms 10.13μ, the short arms 2.6μ; the 2nd pair 10.0 and 2.5μ; the 3rd pair 7.6 and 1.75μ; the 4th pair 7.5 and 1.75μ resp. The first two pairs of this group are the largest in the idiogram. The other two pairs come close to the two pairs of group III.

Group V, one pair: The long arms 10.0μ, the short arms 2.0μ.

The total length of the haploid complement is 122.35μ; of the long arms 94.32μ and of the short arms 28.03μ. Total index: 3.36.

Tulipa victoris Vved.[2] This new endemic species, described by Vvedenskiĭ, is morphologically very similar to *T. fosteriana.* The difference between them is not clear to us. The individuals studied were sent from Nuratau in the Koytash range (coll. V.P. Bochantsev).

2. This name was not validly published. See Czerepanov, S.K. in Nov. Syst. Vasc. 9: 327 (1972): *Tulipa victoris* Vved. 1961, Tr.Tashk. Univ. 187, Bot: 61, nom. nud. (editor)

We used root-tips of adult bulbs (fig.13d). The 24 chromosomes of the diploid complement fall into five main groups (fig.14h).

Group I, one pair: The long arms 6.26, the short arms 3.57μ with an akinetic constriction at 1.0μ from the centromere.

Group II, one pair: The long arms 5.5μ with an a.con. below a spherical segment at ca. 0.5μ from the distal end; the short arms 2.25μ with a similar spherical segment.

Group III, five pairs: The 1st pair with the long arms 8.75μ and the short arms 2.25μ; the 2nd pair with the long arms 7.75μ and an a.con. ca. halfway, the short arms 2.12μ; the 3rd pair 6.12 and 2.0μ; the 4th pair 5.5 and 1.8μ resp.; the 5th pair with the long arms 5.5μ and at the distal end a spherical segment ca. 0.5μ long, the short arms 1.75μ.

Group IV, four pairs; The 1st pair with the long arms 9.5μ, the short arms 2.0μ; the 2nd pair 8.75μ and 2.0μ; the 3rd pair 5.75μ and 1.4μ; the 4th pair 5.5μ and 1.25μ resp.

Group V, one pair: The long arms 6.5μ, the short arms 1.25μ.

The total length of the haploid complement is 105.2μ; of the long arms 81.37μ of the short arms 23.82. Total index: 3.34.

The karyotypes of this group of closely related species around *T. fosteriana,* concentrated in the Pamir-Alay, all have the diploid number 2n=24 and only in *T. lanata* was a deviation observed for the gametes. There are in this group no species with chromosomes with a terminal attachment of the spindle threads and all species have only one pair of chromosomes with a submedian attachment of the spindle threads; in all the other chromosomes it is subterminal.

With respect to the length-ratios of the arms, the species in this group have the following chromosomes in common: In group I, one pair with an akinetic constriction in the short arms; in group II, one pair; in group III, two pairs; in group IV, four pairs; in group V, one pair.

The second pair of chromosomes in group II is a character which *T. lanata, T. ingens* and *T. carinata* have in common, while *T. lanata* and *T. ingens* have more chromosomes in group IV and fewer in group III.

The difference in karyotypes between *T. ingens* and *T. fosteriana* is that the former has one extra pair of chromosomes in group II and one pair less in group III. The distributional pattern of the chromosomes in the length-ratio groups of *T. victoris* is the same as that of *T. fosteriana,* the only difference being the shorter and more segmented chromosomes of *T. victoris.*

In our opinion *T. fosteriana* is the primary species in this series. *T. victoris,* perhaps an incipient species, is quite recent and very close to *T. fosteriana;* then follow *T. ingens, T. lanata* and *T. carinata.* This group of species, in general but little differentiated and specialised, is evidently in the process of speciation.

Leiostemones, second group.

This second group includes *T. praestans, T. subpraestans* and *T. butkovii.* The first two species have thick, leathery, black to tawny coloured tunics, which are glossy and almost glabrous on the inside. A simple analysis shows that they differ from each other, although they have characters in common proving beyond doubt that there is a close relationship. Among our large-flowered wild tulips these are the only species that are as a rule multiflorous and in both the tendency to vegetative reproduction is well developed.

Tulipa praestans Hoog. The individuals studied were collected in Tadzhikistan: the Sarbo Pass (F.N. Rusanov, 1932); the environs of Tutkaul and in the Sebistan mountains (V.P. Bochatsev, 1933); the Varzob river valley (Z.P. Botschantzeva, 1956).

The chromosomes were studied in root-tips and also in pollen tubes during spermatogenesis. It was not possible to use Levan's smear method because of the early mitosis in the young pollen, which takes place when there is not yet any part of the plant showing above ground. At this stage the flower is tightly swathed in leaves and extracting the anthers will kill it and seriously damage the leaves. Newton (1927) found the somatic chromosome number 2n=24, which is confirmed by our investigations on 30 plants from three different habitats. The diploid complement can be divided into five main groups (fig. 13h, 15a).

Group I, one pair: The long arms 4.25, the short arms 2.25μ.

Group II, two pairs: The 1st pair with the long arms 6.6μ, the short arms 3.0μ; the 2nd pair 4.75μ and 2.25μ resp.

Group III, two pairs: The 1st pair with the long arms 8.75μ, the short arms 2.25μ; the 2nd pair 6.25 and 2.0μ resp.

Group IV, four pairs: The 1st pair with the long arms 9.4μ, the short arms 2.0μ; the 2nd pair 7.15 and 1.5μ; the 3rd pair 6.0 and 1.4μ; the 4th pair 5.75 and 1.25μ resp.

Group V, three pairs: The 1st pair with the long arms 9.25μ, the short arms 1.75μ; the 2nd pair 8.25 and 1.5μ; the 3rd pair 7.5 and 1.5μ resp.

The total length of the haploid complement is 108.55μ; of the long arms 83.9 and of the short arms 24.65μ. Total index: 3.40.

The chromosomes of *T. praestans* are morphologically of surprising uniformity. We did not observe any signs of secondary segmentation and they are rather thick and shortened, the longer arms even more shortened than the short ones. An attempt to divide such chromosomes into groups according to the length-ratio of the arms gives an entirely different picture from the usual complements.

Tulipa subpraestans Vved. The individuals studied were collected in the environs of Tutkaul (coll. V.P. Bochantsev).

We examined chromosomes in root-tips of adult bulbs and smears from anthers (Levan's method). The division in the anthers into a vegetative and a generative nucleus takes place in a large flower bud when this is still below ground surface. The diploid complement consists of 24 chromosomes (fig.13g), which fall into six main groups (fig.11c, 15b).

Group I, one pair: Small chromosomes; the long arms 3.25, the short arms 1.9μ.

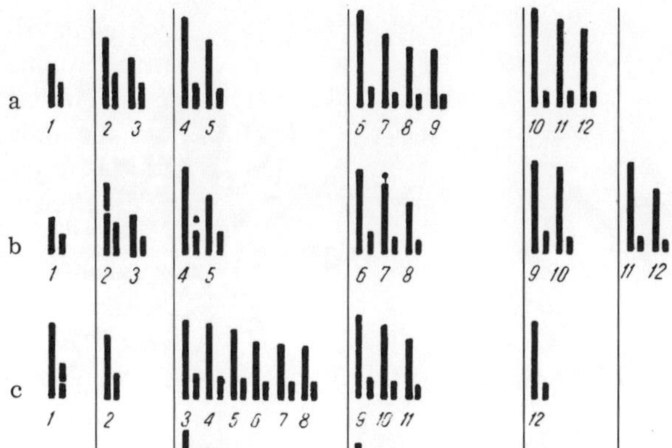

Figure 15. Idiograms of measured plates (enlarged 600 times) a. *T. praestans* Hoog; b. *T. subpraestans* Vved.; c. *T. albertii* Regel; d. *T. mogoltavica* M. Pop. et Vved.; e. *T. greigii* Regel; f. *T. micheliana* Hoog; g. *T. lehmanniana* Merckl.; h. *T. behmiana* Regel. (Judging by the dimensions and figure 15c, chromosome no 1 of *T. albertii* should fall into group II.) (editor)

Group II, two pairs: The 1st pair with the long arms 6.75μ and an akinetic constriction at 3.6μ from the centromere, the short arms 3.0μ; the 2nd pair, the smallest of the complement, with the long arms 3.5μ and the short arms 1.25μ.

Group III, two pairs: The 1st pair with the long arms 8.25μ, the short arms 2.25μ and a rather stable satellite which can be observed at mitosis in root-tip sections and young pollen; the arms of the 2nd pair 5.5 and 1.75μ.

Group IV, three pairs: The 1st pair with the long arms 8.0μ, the short arms 1.75μ; the 2nd pair with the long arms 6.75μ and a satellite at the distal end, its thread may be drawn into the chromosome body with the satellite sitting directly on the arm, the short arms 1.5μ; the 3rd pair 5.0 and 1.25μ resp.

Group V, two pairs: The 1st pair with the long arms 8.75μ, the short arms 1.75μ; the 2nd pair 8.3 and 1.5μ resp.

Group VI, two pairs: The 1st pair with the long arms 8.25μ, the short arms 1.25μ; the 2nd pair 6.0μ and 1.0μ resp.

The total length of the haploid complement is 98.45μ; of the long arms 78.3μ and of the short arms 20.15μ. Total index: 3.88.

A comparison of the karyotypes of *T.praestans* and *T.subpraestans* shows corresponding numbers and proportions for the chromosomes of group I, II and III. Both species have in group IV three and in group V two similar chromosomes. *T.praestans* has one extra pair in groups IV and V which are in *T.subpraestans* compensated for by the two pairs of group VI. The chromosomes of *T.subpraestans* are in general a little shorter and more segmented than those of *T.praestans*.

There can be observed certain similarities in the karyotypes of *T.carinata, T.praestans* and *T.subpraestans*, viz. the corresponding numbers of chromosomes in groups I, II and III. On the other hand the chromosomes of the last two species, particularly those of group I, are considerably reduced in size. A comparison of the chromosomes of groups IV and V of these three species with those of other species again shows a reduction of the chromosomes of *T.carinata* as against those of the closely related species *T.lanata, T.ingens* and *T. fosteriana*. A similar reduction is found when comparing *T.praestans* with *T.carinata* and *T.subpraestans* with *T.praestans*. The karyotypical relationship of *T.carinata* and *T.praestans* is supported by their external morphological characters and there exists without any doubt also a karyotypical and macro-morphological relationship between *T.praestans* and *T.subpraestans*.

Leiostemones, third group.

The three Middle Asian species with the characteristically shaped *T. greigii*-type flowers and anthocyanin spots and streaks on the leaves, *T. greigii, T. mogoltavica* and *T. micheliana*, form a natural group. *T. albertii* with a somewhat differently shaped flower and no anthocyanin spots on the leaves holds more or less a place of its own. Newton found for *T. greigii* 2n=24 and it was much later that other cytological data on representatives of this group appeared in the literature and then only for *T. micheliana* (Tischler 1940).

T. albertii Regel. The individuals studied were collected in the Karatau range (coll. P.A. Gomolitskiĭ).

The chromosomes were studied in root-tip sections taken from adult bulbs. The diploid chromosome number is 24. The chromosomes fall into five main groups (fig. 15c, 16a).

Group I, one pair: The long arms 7.0, the short arms 3.25μ with an akinetic constriction at 1.0μ distance from the centromere.

Group II, one pair: The long arms 6.0μ, the short arms 2.25μ.

Group III, six pairs: The 1st pair with the long arms 7.5μ, the short arms 2.5μ; the 2nd pair 7.25 and 2.0μ; the 3rd pair 6.5 and 2.0μ; the 4th pair 5.5 and 1.75μ; the 5th pair 5.25 and 1.75μ; the 6th pair 5.25 and 1.75μ resp.

Group IV, three pairs: The 1st pair with the long arms 8.0μ, the short arms 2.0μ; the 2nd pair 7.0 and 1.62μ; the 3rd pair 6.0 and 1.5μ.

Group V, one pair: The long arms 7.5μ, the short arms 1.5μ.

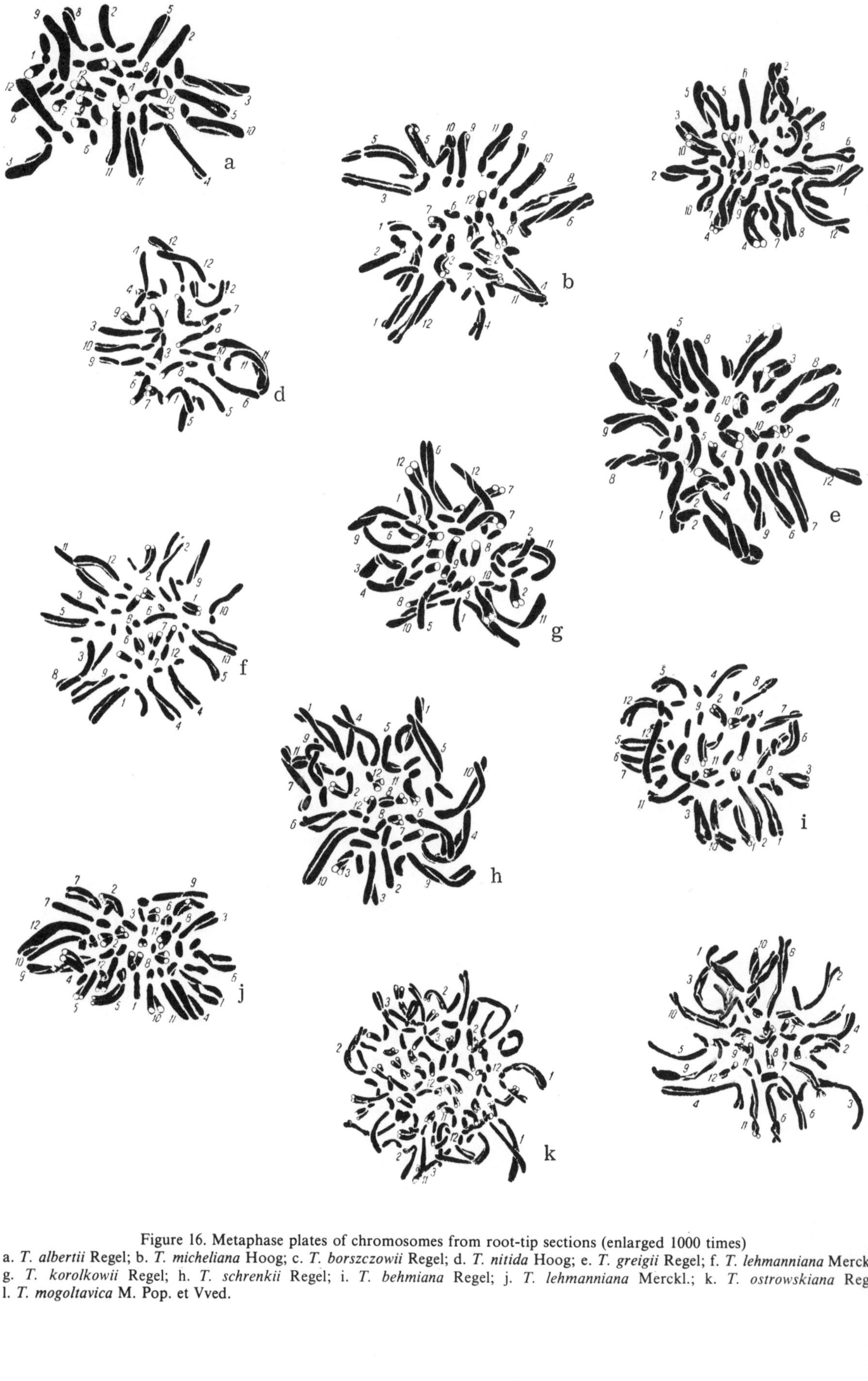

Figure 16. Metaphase plates of chromosomes from root-tip sections (enlarged 1000 times)
a. *T. albertii* Regel; b. *T. micheliana* Hoog; c. *T. borszczowii* Regel; d. *T. nitida* Hoog; e. *T. greigii* Regel; f. *T. lehmanniana* Merckl.; g. *T. korolkowii* Regel; h. *T. schrenkii* Regel; i. *T. behmiana* Regel; j. *T. lehmanniana* Merckl.; k. *T. ostrowskiana* Regel; l. *T. mogoltavica* M. Pop. et Vved.

The chromosomes of this species are thick and very little segmented. Despite the availability of ample material we found no akinetic constrictions beyond the one in the first chromosome pair. The total length of the haploid complement is 102.63μ; of the long arms 78.75μ and of the short arms 23.88μ. Total index 3.30. Index chrom. Al: 2.15.

Tulipa mogoltavica M. Pop. et Vved. The individuals studied were collected in the Mogoltau range (coll. I. I. Granitov).

The chromosomes were studied in root-tip sections taken from adult bulbs. The diploid chromosome number is 2n=24; they fall into five main groups (fig.15d, 16*l*).

Group I, one pair: The long arms 8.3μ, the short arms 5.5μ with an akinetic constriction at ca. 1.75μ from the centromere.

Group II, one pair: The long arms 6.0μ, the short arms 2.6μ.

Group III, five pairs: The 1st pair with the long arms 11.0μ, the short arms 3.25μ; the 2nd pair 9.0 and 3.0μ; the 3rd pair 7.5 and 2.4μ; the 4th pair 6.5 and 2.15μ; the 5th pair 6.24 and 2.0μ resp.; besides in the long arms of the 1st pair an a.con. at 4.5μ from the centromere; in the long arms of the 2nd pair an a.con. at 2.0μ from the centromere; in the long arms of the 3rd pair an a.con. at ca. 1.0μ from its distal end and in its short arms a similar constriction below a 0.3μ long spherical segment.

Group IV, four pairs: The 1st pair with the long arms 9.75, the short arms 2.0μ; the 2nd pair 8.0 and 1.8μ; the 3rd pair 6.5 and 1.62μ; the 4th pair 6.0 and 1.5μ resp.; besides, in the long arms of the 2nd pair one a.con. at ca. 3.0μ from the centromere and another at ca. 1.0μ of the distal end.

Group V, one pair: The long arms 7.5μ, the short arms 1.5μ.

The total length of the haploid complement is 121.93μ; of the long arms 92.3μ and of the short arms 29.63μ. Total index: 3.11. Index chrom. AI: 1.50.

Tulipa greigii Regel. The individuals studied were taken from mass plantations in the Botanical Garden of Tashkent State University. Others originated from the Bostandykskiy district (coll. Satualdybaev); from the Karatau range (coll. Gomolitskiĭ); from the Kichikemin river valley, environs of Semenovka (coll. Z.P. Botschantzeva); from Tyul'kubas station, the Fainbulak gorge, (coll. A. Ya. Butkov and V.P. Bochantsev).

The chromosomes were studied in root-tip sections of adult bulbs and in smears from anthers (fig. 11h). The idiogram was obtained by the last method and although the chromosomes are in this case somewhat shorter than those obtained from root-tips a check proved that the arms ratios had remained unaltered. In all cases the somatic number 24 found by Newton (1927) was confirmed by our own observations. The complement can be divided into five main groups (fig.15e, 16e).

Group I, one pair: The long arms 6.5μ, the short arms 3.75μ and in the lower third the usual akinetic constriction.

Group II, one pair: The long arms 6.75μ, the short arms 3.0μ.

Group III, four pairs: The 1st pair with the long arms 7.0μ, the short arms 2.0μ; the 2nd pair 5.75 and 1.9μ; the 3rd pair 5.5 and 1.8μ; the 4th pair 5.0 and 1.5μ resp.

Group IV, five pairs: The 1st pair with the long arms 7.25μ, the short arms 1.75μ; the 2nd pair 6.5 and 1.5μ; the 3rd pair 6.0 and 1.5μ; the 5th pair 5.4 and 1.25μ resp.

Group V, one pair: The long arms 7.5μ, the short arms 1.5μ.

The total length of the haploid complement is 97.85μ; of the long arms 75.15μ and of the short arms 22.7μ. Total index 3.31. Index chrom. A: 1.73.

Chromosome morphology was studied in the *T.greigii* plants from Tashkent. Of the other provenances we only counted the chromosomes. The chromosomes of *T.greigii* are very thick and without any segmentations, except in the 1st pair.

Tulipa micheliana Hoog. The individuals studied originated from Karakal, near the Ura-Tyube, Dragomirovo and Bakharden railway stations (coll. N.B. Nikiforova) and from the environs of the Tumanovskiy settlement (coll. A. Ya. Gnezdilo).

The chromosomes were studied in root-tip sections taken from adult bulbs. The number counted for the diploid complement is 2n=24; they fall into six main groups (fig.15f, 16b).

Group I, one pair: The long arms 8.0, the short arms 4.25μ and an akinetic constriction at ca. 1.5μ from the centromere.

Group II, one pair: The long arms 5.25μ and an a.con. at ca. 2.75μ from the distal end, the short arms 2.25μ an an a.con. at ca. 1.0μ from its distal end.

Group III, two pairs: The 1st pair with the long arms 8.0μ and two constrictions, one at ca. 2.25μ and the other at ca. 3.0μ from its distal end, the short arms 2.6μ and also an a.con. at ca. 1.5μ from the centromere; the 2nd pair with the long arms 7.25μ and an a.con. at 2.0μ from the centromere, the short arms 2.0μ.

Group IV, six pairs: The 1st pair with the long arms 9.5μ and a.con. at ca. 2.5μ from the distal end, the short arms 2.25μ and an a.con. at 1.0μ from the distal end; the 2nd pair with the long arms 8.74μ and two a.cons., one at ca. 0.75μ from the centromere and the other at ca. 1.0μ from its distal end, the short arms 1.9μ and an a.con. at ca. 0.74μ from the centromere; the 3rd pair 7.5 and 1.63μ resp.; the 4th pair with the long arms 7.0μ and two a.cons. at 0.75 and 2.5μ from the distal end, the short arms 1.5μ; the 5th pair 6.0 and 1.5μ; the 6th pair 5.75 and 1.4μ resp. and no indication of secondary constrictions.

Group V, one pair: The long arms 6.25μ and an a.con. at 1.75μ from the centromere, the short arms 1.25μ.

Group VI, one pair: The long arms 9.5μ and an a.con. below a 1.5μ long segment at the distal end, the short arms 1.5μ.

The total length of the haploid complement is 123.45μ; of the long arms 98.75μ and of the short arms 24.7μ. Total index 4.00. Index chrom. Al: 1.88.

Plants of this species were also collected in the area around Tehran in Iran. Preliminary studies of this material showed chromosome types which were somewhat different from those described here. There is a tendency towards a lengthening of the shorter arms which gives an entirely different idiogram. However, the material still needs thorough checking. As these chromosomes were examined in smears and those of the plants from our country in root-tip sections, it is difficult to make a comparison of the karyotypes. On the other hand we believe that this should be possible since we have found that in other species the arms ratios remain unaltered when examined by both methods and that only the absolute lengths change. The species which on taxonomic grounds are grouped around *T. greigii* have the following karyotype characters in common: one pair of chromosomes in group I, one pair in group II, two pairs in group III, three pairs in group IV and one pair in group V.

T. mogoltavica differs from *T. greigii* in the more segmented chromosomes, an extra pair of chromosomes in group III and one pair of chromosomes less in group IV. The chromosomes of *T. greigii* are twice as thick as those of *T. mogoltavica.*

T. micheliana has only two pairs of chromosomes in group III, but six pairs in group IV, which is one pair more than *T. greigii* and two pairs more than *T. mogoltavica.* Moreover *T. micheliana* has an extra group VI with one pair of chromosomes which is absent in the other species.

T. albertii has in group III one pair of chromosomes more than *T. mogoltavica,* two pairs more than *T. greigii* and four pairs more than *T. micheliana,* but in group IV *T. albertii* has one pair of chromosomes less than *T. mogoltavica,* two pairs less than *T. greigii* and three pairs less than *T. micheliana.* Thus we see that the differences in karyotypes within this group of related species are greatest between *T. micheliana* and *T. albertii,* which is confirmed by the external morphological characters of these species.

Leiostemones, fourth group

This fourth group formed by the species *T. maximowiczii* Regel, *T. linifolia* Regel and *T. wilsoniana* Hoog is a natural group. All three species have small bulbs with black, leathery tunics. Plants of *T. maximowiczii* may attain 30 cm in height, whereas the flowers of *T. wilsoniana* are sometimes borne just above ground level and the plants have to be dug out to show the stems. All three species have narrow leaves and red tepals with a not very high and obtuse, black blotch which is as broad as the tepals. A comparison of characters and observations in nature leads to the conclusion that *T. maximowiczii* and *T. linifolia* are closely related, while *T. wilsoniana* stands rather by itself within this group.

Newton (1927) investigated *T. maximowiczii, T. linifolia* and also *T. batalinii* which Vvedenskiĭ mentions as a synonym of *T. maximowiczii.* Newton found for all three species the same type of chromosomes, viz. with a terminal attachment of the spindle and a small spherical second arm. He also notes the one very short pair of chromosomes present in these three species. We had no opportunity to examine *T. batalinii,* but some conclusions may be drawn from fig. 2 in table 7 of his publication. However, we will first give the results of our studies of *T. maximowiczii, T. linifolia* and *T. wilsoniana.*

Tulipa maximowiczii Regel. The individuals studied were collected in Tadzhikistan near the village of Sangtud (coll. V.P. Bochantsev, 1933).

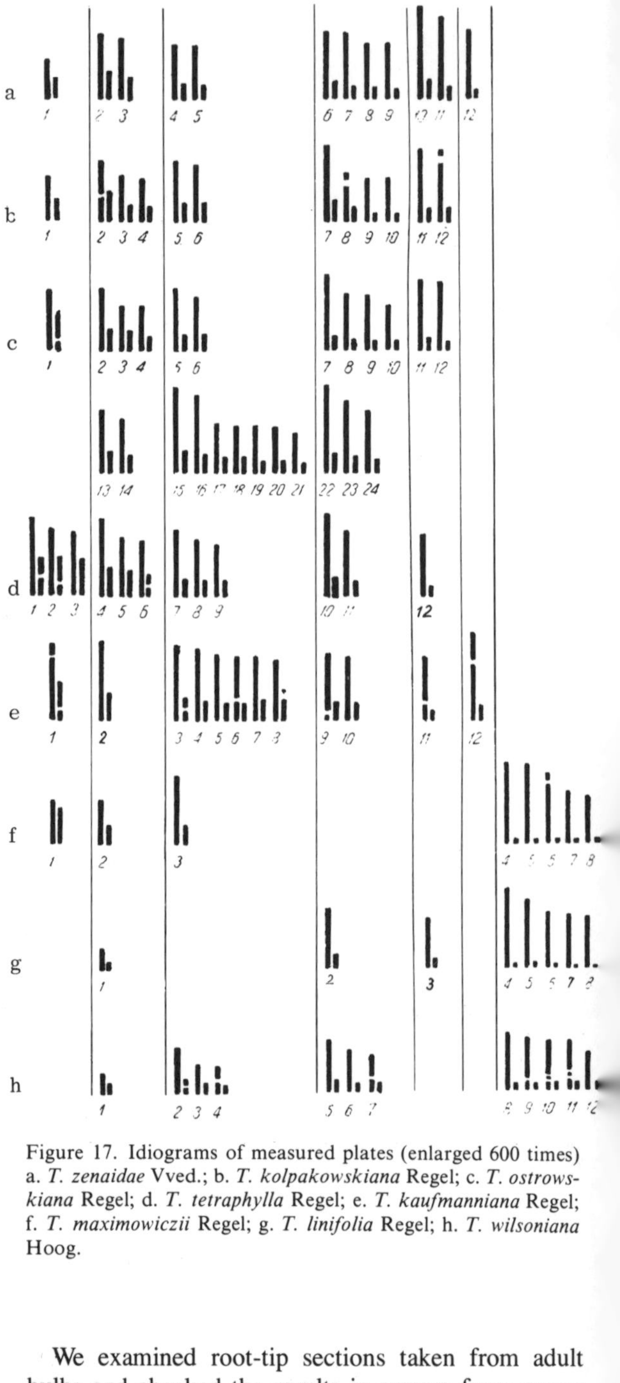

Figure 17. Idiograms of measured plates (enlarged 600 times) a. *T. zenaidae* Vved.; b. *T. kolpakowskiana* Regel; c. *T. ostrowskiana* Regel; d. *T. tetraphylla* Regel; e. *T. kaufmanniana* Regel; f. *T. maximowiczii* Regel; g. *T. linifolia* Regel; h. *T. wilsoniana* Hoog.

We examined root-tip sections taken from adult bulbs and checked the results in smears from young anthers (fig.11e). As distinct from other *Tulipa* species, the number of chromosomes in the root cells is 2n=22. This number has not been observed in any other species of the genus we have studied. Newton found 2n=24, but the examination of root-tip sections of thirty plants confirmed 2n=22 for their chromosome numbers (fig.11b). In the spring of 1935 we studied smears from young anthers during division of the primary nucleus and again obtained the same results: n=11 chromosomes.

The chromosomes can be divided by the length ratios of the arms into four main groups (fig.17f).

Group I, one pair: The long arms 4.75, the short arms 4.0μ; almost isobrachial chromosomes, or in Newton's terminology for other species 'with a median attachment of the spindle threads'.

Group II, one pair: The long arms 4.5μ, the short arms 2.0μ.

Group III, one pair: The long arms 7.75μ, the short arms 2.0μ.

The remaining eight pairs of chromosomes all have short arms, shaped like a spherical segment with a diameter of barely 0.5μ. They are placed in a special group IX, consisting of 'capitate' chromosomes, which were not observed in the previously studied species.

Group IX, eight pairs, Capitate chromosomes: With the long arms 8.75, 8.75, 7.5, 5.25, 5.0, 4.25, 4.25 and 3.5μ resp.; the 3rd pair has an akinetic constriction below a segment ca. 1.0μ long at the distal end of the long arms; besides the short arms of the 5th pair carry a spherical satellite attached to threads (the latter could only be observed in smears).

The total length of the haploid complement is 79.25μ; of the long arms 67.25μ and of the short arms 12.0μ. Total index: 5.6.

T. maximowiczii has three types of chromosomes, group I, II and III, which Newton does not mention, and among the capitate chromosomes is one pair with akinetic constrictions and one pair with satellites.

Tulipa linifolia Regel. The individuals studied originate from the Kaivak Pass, Kala-i-Khumb (coll. P.A. Baranov, 1934).

The chromosomes were examined in root-tip sections taken from sprouting adult bulbs. The diploid number, 2n = 24, confirms Newton's data. The chromosomes can be divided by the length ratios of the arms into four groups (fig.11i, 17g).

Group I: Absent.

Group II, one pair: The long arms 2.5μ, the short arms 1.0μ (described by Newton as 'very short chromosomes').

Group III: Absent.

Group IV, one pair: The long arms 6.5μ, the short arms 1.5μ.

Group V, one pair: The long arms 5.62μ, the short arms 1.0μ.

Group IX, nine pairs of 'capitate' chromosomes: In this group the dimensions of the spherical heads or arms vary; the 1st pair with the long arms 8.5μ, the short arms 0.5μ; the 2nd pair 7.4 and 0.75μ; the 3rd pair 6.0 and 0.5μ; the 4th pair 5.75 and 0.5μ; the 5th pair 5.75 and 0.4μ; the 6th pair 5.75 and 0.25μ; the 7th pair 5.5 and 0.25μ; the 8th pair 5.4 and 0.5μ; the 9th pair 5.13 and 0.5μ resp.

The total length of the haploid complement is 77.45μ; of the long arms 69.8μ and of the short arms 7.65μ. Total index: 9.1.

We found two pairs of chromosomes with short arms instead of spherical heads. Newton (1927, in fig.1 of plate 7) gives a metaphase plate of *T.linifolia.* In the lower lefthand corner a chromosome can be seen of which the short arm can hardly be described as spherical since it is twice as long as broad; it is an arm. Of the remaining chromosomes the short arms are slightly raised, but owing to their very small dimensions, this is barely noticeable.

T.batalinii is in its karyotype very similar to *T.linifolia* (Newton, 1927, plate 7, fig.2). There can be seen one pair of very small chromosomes. The other chromosomes are difficult to disentangle, but the size of the heads varies greatly, as in *T.linifolia.* The dimensions of the chromosomes are in both species the same.

Tulipa wilsoniana Hoog. The individuals studied came from Turkmenia (coll. A.I. Vvedenskiĭ) and from the foothills near the Prokhladnoye settlement (coll. A.F. Gnezdilo).

The chromosomes were examined only in root-tip sections taken from sprouting adult bulbs. The diploid number is 2n=24 (fig.11j) and the complement can be divided into four groups (fig.11b, 17h).

Group I: Absent.

Group II, one pair: Small chromosomes (as found by Newton in *T.linifolia* and *T.batalinii);* the long arms 2.25, the short arms 1.0μ.

Group III, three pairs: The 1st pair with the long arms 5.0μ, the short arms 1.5μ; the 2nd pair 3.5 and 1.0μ; the 3rd pair 3.0 and 1.0μ; besides in the short arms of the 1st pair an akinetic constriction approximately halfway and in the long arms of the 3rd pair an a.con. at ca. 1.25μ from the centromere.

Group IV, three pairs: The 1st pair with the long arms 5.6μ, the short arms 1.25μ; the 2nd pair 4.5 and 1.0μ; the 3rd pair 4.0 and 1.0μ resp.; besides in the long arms of the 3rd pair an a.con at ca. 1.75μ from the centromere; the distal end of these arms is not rounded as in the other chromosomes but tapers to a thread-like appendage; no satellites were found.

Group IX, five pairs, 'capitate' chromosomes: The spherical segments in all chromosomes ca. 0.75μ long; the long arms of the 1st pair 6.25μ, of the 2nd pair 5.75μ; of the 3rd pair 5.6μ; of the 4th pair 5.13μ of the 5th pair 4.4μ; besides on the 2nd, 3rd and 4th pairs with an a.con. in the long arms at 1.0, 1.75 and 1.5μ resp. from the centromere.

The total length of the haploid complement is 67.47μ; of the long arms 54.97μ and of the short arms 12.5μ. Total index: 4.4.

Although *T.linifolia* and *T.maximowiczii* differ but little in their external morphological characters there is a striking difference in karyotypes. This seems to be a case of a simple rearrangement of the chromatin substance, but it remains possible that these two species differ also in a number of essential characters not yet recorded by taxonomists. This is all the more probable since Vvedenskiĭ in 'Flora of the USSR', Vol. IV, notes: '*T.maximowiczii* and *T.linifolia* are very closely related and supplementary observations on live material are needed for definite distinction'. The discrepancy with Newton's karyotype studies is probably due to this external similarity and the consequent error in the identification of the species. There is no way of checking this, because geographical or taxonomic information about the species studied was not given. The same applies to *T.batalinii* when we examine the metaphase plates of figure 7 (Newton, 1927). The karyotype of *T.batalinii* is identical with that of *T.linifolia.* Vvedenskiĭ (Flora of the USSR, Vol. IV) mentions *T.batalinii* as a synonym of *T.maximowiczii.* Apparently the taxa studied were not accurately identified.

T.wilsoniana is linked with *T.linifolia* by one pair of small chromosomes in group II, one pair in group IV and a number of 'capitate' chromosomes. As the shorter arms are in general longer than in the other species described, one could say that there are in all seven pairs of chromosomes with short arms instead of 'capitate' chromosomes. Further there are a number of typical features in its karyotype, not found in the other species, as for instance the distinct secondary constrictions in the chromosomes. *T.wilsoniana* also differs in its external morphological characters from the other species in this group.

Leiostemones, fifth group

The fifth group of Middle Asian *Leiostemones* includes four taxa: *T.behmiana* Regel, the typical *T.lehmanniana* Merckl., the *T.lehmanniana* from Kushka and *T. borszczowii* Regel. We did not find any cytological information about these species in the literature.

Tulipa lehmanniana Merckl. The individuals studied were collected in Uzbekistan, the environs of Malik railway station (by us in 1931 and by T.A. Romanovich in 1933).

The chromosomes were studied in root-tip sections taken from adult bulbs and in pollen tubes in vitro during spermatogenesis. The 24 chromosomes fall into five main groups (fig.15g, 16f).

Group I, one pair: The long arms 7.75, the short arms 4.0μ; in the short arms the usual constriction at ¼ of the arm's length from the centromere.

Group II, two pairs: The 1st pair with the long arms 8.13μ, the short arms 3.0μ; the 2nd pair 6.25 and 2.25μ resp.; besides at the distal end of the long arms an a.con. below a spherical segment, which in some plates looks like a satellite.

Group III, four pairs: The 1st pair with the long arms 8.0μ, the short arms 2.25μ; the 2nd pair 6.25 and 2.0μ; the 3rd pair 6.0 and 2.0μ; the 4th pair 6.0 and 1.75μ resp.; besides in the short arms of the 4th pair an a.con. in the middle.

Group IV, three pairs: The 1st pair with the long arms 9.6μ, the short arms 2.0μ; the 2nd pair 7.25 and 1.75μ; the 3rd pair 6.5 and 1.5μ resp.; besides, the 3rd pair with on the distal end of the long arms an a.con. below a spherical segment, also unstable, cf. group II, the 2nd pair.

Group V, two pairs: The 1st pair with the long arms 11.75μ, the short arms 2.3μ; the 2nd pair 10.0 and 1.75μ resp.

The total length of the haploid complement is 120.13μ; of the long arms 93.47μ and of the short arms 26.66μ. Total index: 3.51.

Tulipa lehmanniana Merckl. from Kushka (Turkmenistan). Vvedenskiĭ makes a distinction between this taxon and the typical *T.lehmanniana,* but apparently because of insufficiently clear characters, he does not regard it as an independent species.

We studied root-tip sections taken from adult bulbs. The number of chromosomes is the same as that of the typical *T.lehmanniana,* 2n=24, but the distribution of the chromosomes over the five main groups is somewhat different (fig.16j, 18a).

Group I, one pair: The long arms 5.75, the short arms 3,5μ; in the long arms an akinetic constriction approximately halfway, in the short arms an a.con. below a ca. 1.0μ segment next to the centromere.

Group II, two pairs: The 1st pair with the long arms 6.5μ, the short arms 2.25μ; the 2nd pair 5.13 and 2.0μ resp.

Group III, five pairs: The 1st pair with the long arms 6.5μ, the short arms 1.9μ; the 2nd pair 5.25 and 1.75μ; the 3rd pair 5.25 and 1.75μ; the 4th pair 5.0 and 1.6μ; the 5th pair 5.0 and 1.5μ resp.; besides, in the long arms of the 3rd pair an a.con. approximately halfway; in the long arms of the 4th pair two a.cons. at 0.75μ from the distal end and at 1.5μ from the centromere; the last two pairs with a satellite at the distal end of the long arms.

Group IV, three pairs: The 1st pair with the long arms 8.5μ, the short arms 2.0μ; the 2nd pair 7.0 and 1.5μ; the 3rd pair 7.0 and 1.5μ resp.; besides in the long arms of the 1st pair an a.con. at ca. 3.25μ from the distal end.

Group V, one pair: The long arms 9.25μ, the short 1.75μ.

The total length of the haploid complement is 99.13μ; of the long arms 76.13μ and of the short arms 23.0μ. Total index: 3.31.

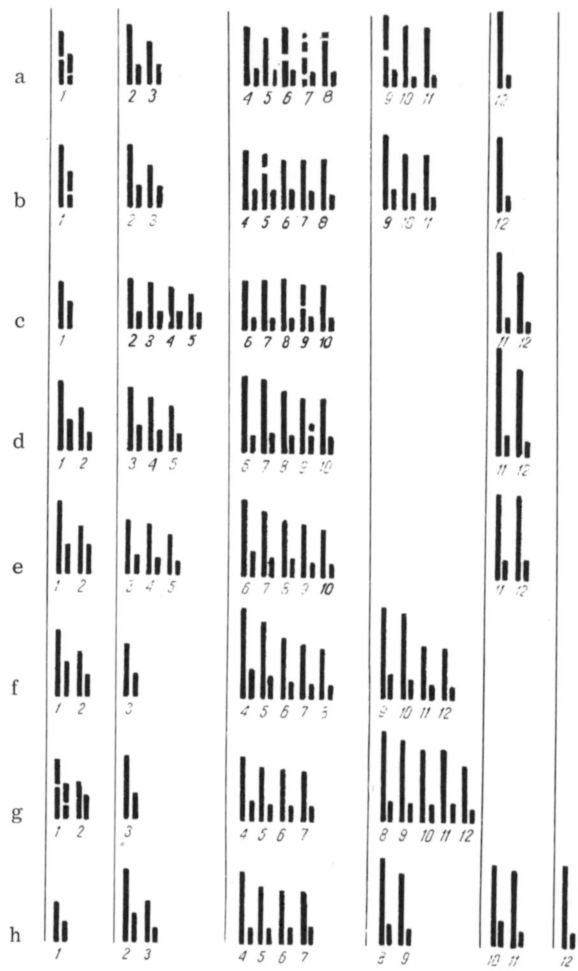

Figure 18. Idiograms of measured plates (enlarged 600 times) a. *T. lehmanniana* Merckl.; b. *T. borszczowii* Regel; c. *T. nitida* Hoog; d. *T. korolkowii* Regel; e. *T. rosea* Vved.; f. *T. schrenkii* Regel; g. *T. kolpakowskiana* Regel; h. *T. ferganica* Vved. (Judging by the dimensions and figures 18d and e, chromosome no 1 should fall into group II, in which case *T. korolkowii* and *T. rosea*, like *T. nitida*, have one pair of chromosomes in group I and four pairs in group II.) (editor)

Tulipa behmiana Regel. The individuals studied were collected in Kazakhstan, the environs of Iliisk (coll. Granitova, 1931).

We examined only root-tip sections taken from adult bulbs. The diploid complement of 24 chromosomes falls into five main groups (fig.15h, 16i).

Group I, one pair: The long arms 6.25μ, the short arms 3.25μ; we did not find any akinetic constriction in the short arms.

Group II, two pairs: The 1st pair with the long arms 6.75μ, the short arms 2.75μ; the 2nd pair 5.5 and 2.25μ resp.; besides in the short arms of the 1st pair a large a.con. at ca. 1.25μ from the centromere; on the long arms of the 2nd pair an a.con. below an oblong segment at ca. 1.5μ from the distal end.

Group III, five pairs: The 1st pair with the long arms 6.4μ, the short arms 1.75μ; the 2nd pair 5.5 and 1.75μ; the 3rd pair 5.5 and 1.72μ; the 4th pair 5.5 and 1.75μ; the 5th pair 5.25 and 1.75μ resp.; besides in the long arms of the 4th pair an a.con. approximately halfway and in the long arms of the 5th pair two a.cons. dividing them into three parts, the central section slightly longer than the two almost equally long other sections.

Group IV, three pairs: The 1st pair with the long arms 8.0μ, the short arms 2.0μ; the 2nd pair 6.6 and 1.5μ; the 3rd pair 6.0 and 1.5μ resp.; besides in the long arm of the 2nd pair an a.con. at ca. 3.0μ from the distal end and in the long arms of the 3rd pair an a.con. below a small, ca. 0.5μ long segment at the distal end.

Group V, one pair: The long arms 10.25μ, the short arms 2.0μ; in the long arms an a.con. below a capitate segment, ca. 0.5μ long.

The total length of the haploid complement is 101.5μ; of the long arms 77.5μ and of the short arms 24.0μ. Total index: 2.23.

Tulipa borszczowii Regel. The individuals studied were collected near Tyuratam railway station in Kazakhstan (coll. T.S. Aganesova, 1931).

We examined root-tip sections taken from adult bulbs. The diploid number is 24 and the chromosomes can be divided into five main groups (fig.16c, 18b).

Group I, one pair: The long arms 7.0μ, the short arms 4.0μ and an akinetic constriction in the lower part of the short arms.

Group II, two pairs: The 1st pair with the long arms 7.0μ, the short arms 2.5μ; the 2nd pair 4.75 and 1.9μ resp.

Group III, five pairs: The 1st pair with the long arms 6.5μ, the short arms 2.0μ; the 2nd pair 6.0 and 2.0μ; the 3rd pair 5.5 and 1.75μ; the 4th pair 5.5 and 1.75μ; the 5th pair 5.5 and 1.5μ resp.; besides in the long arms of the 2nd pair an a.con. below an oblong segment, ca. 1.0μ long.

Group IV, three pairs: The 1st pair with the long arms 8.25μ, the short arms 2.0μ; the 2nd pair 6.0 and 1.5μ; the 3rd pair 6.0 and 1.25μ resp.

Group V, one pair; the long arms 8.25μ, the short arms 1.5μ.

The total length of the haploid complement is 99.15μ; of the long arms 75.25μ and of the short arms 23.9μ. Total index: 3.15.

Summarizing the results of the karyotypic studies in this group, it appears that all species have in common: the one pair of chromosomes in group I, the two pairs in group II, four pairs in group III, three pairs in group IV and one pair in group V.

The Turkmenian *T.lehmanniana* from Kushka shows more similarity in karyotype to *T.behmiana* than to the typical *T.lehmanniana*.

In karyotype *T.borszczowii* is closer to *T. behmiana* than to *T.lehmanniana,* which has longer chromosomes. In the chromosomes of *T.borszczowii* we found fewer secondary constrictions than in any other species of this group.

Leiostemones, sixth group.

This group includes three species: *T.korolkowii* Regel, *T.nitida* Hoog and *T.rosea* Vved. The plants of the taxa we investigated rather varied. *T.nitida* has a stem up to 30 cm long, leaves which do not reach the flower and bright red, glossy tepals, whereas the stem of *T.korolkowii* is short, the leaves extend beyond the flower and the tepals are red, but of a deeper shade than in *T.nitida.* Vvedenskiĭ states: 'The characters distinguishing this species from *T.korolkowii* are not clear at present and specimens from the northern parts of the Pamir-Alay can be referred with equal justification to either species'. The unusual bright pink colour of the flowers of *T.nitida* distinguishes it from the two other species.

Tulipa nitida Hoog. The individuals studied were collected in the Denauskiy district of the Uzbekskaya SSR.

We examined root-tip sections taken from adult bulbs. The diploid chromosome number is 24 and the chromosomes can be divided into four groups (fig.16d 18c).

Group I, one pair: The long arms 5.5, the short arms 2.75μ; we did not find any akinetic constriction in the short arms.

Group II, four pairs: The 1st pair with the long arms 5.6μ, the short arms 2.0μ; the 2nd pair 5.3 and 2.0μ; the 3rd pair 4.75 and 2.0μ; the 4th pair 4.25 and 1.9μ resp.

Group III, five pairs: The 1st pair with the long arms 5.75μ, the short arms 1.5μ; the 2nd pair 5.75 and 1.5μ; the 3rd pair 5.75 and 1.5μ; the 4th pair 5.25 and 1.75μ; the 5th pair 5.25 and 1.5μ resp.; besides the 4th pair has an a.con. in the long arm at ca. 3.0μ from the centromere.

Group IV: Absent.

Group V, two pairs: The first pair with the long arms 9.0μ, the short arms 1.75μ; the 2nd pair 7.0 and 1.25μ resp.

The total length of the haploid complement is 90.5μ; of the long arms 69.1μ and of the short arms 21.4μ. Total index: 3.23.

Tulipa korolkowii Regel. The plants studied were collected in the environs of Darbaza station on the Middle Asian railway.

We examined root-tip sections taken from adult bulbs. The diploid chromosome number is 24. They fall into four groups (fig.16g, 18d).

Group I, two pairs: The 1st pair with the long arms 7.75μ, the short arms 3.75μ; the 2nd pair 4.5 and 2.25μ resp.; it should be noted that here also there is no secondary constriction in the short arms of the 1st pair (cf. *T.lehmanniana, T.borszczowii).*

Group II, three pairs: The 1st pair with the long arms 7.25μ, the short arms 3.0μ; the 2nd pair 6.25 and 2.25μ; the 3rd pair 5.2 and 1.9μ resp.

Group III, five pairs: The 1st pair with the long arms 8.5μ, the short arms 2.2μ; the 2nd pair 8.5 and 2.25μ; the 3rd pair 7.0 and 2.0μ; the 4th pair 6.25 and 2.0μ; the 5th pair 6.25 and 1.9μ resp.

Group IV: Absent.

Group V, two pairs: The 1st pair with the long arms 12.0, the short arms 2.25μ; the 2nd pair 9.75 and 1.7μ resp.

The total length of the haploid complement is 113.2μ; of the long arms 85.2μ and of the short arms 28.0μ. Total index: 3.04.

Tulipa rosea Vved. The individuals studied were collected in the environs of Namangan in the Pamir-Alay.

The chromosomes were examined during spermatogenesis in pollen tubes in vitro. The diploid number is 24 and they fall into four main groups (fig.18e).

Group I, two pairs: The 1st pair with the long arms 8.25μ, the short arms 3.13μ; the 2nd pair 5.5 and 3.13μ resp.

Group II, three pairs: The 1st pair with the long arms 6.25μ, the short arms 2.25μ; the 2nd pair 5.5 and 2.0μ resp.

Group III, five pairs: The 1st pair with the long arms 8.5μ, the short arms 2.5μ; the 2nd pair 7.25 and 2.25μ; the 3rd pair 6.0 and 1.75μ; the 4th pair 5.75 and 1.5μ; the 5th pair 5.5 and 1.5μ resp.

Group IV: Absent.

Group V, two pairs: The 1st pair with the long arms 9.5μ, the short arms 1.85μ; the 2nd pair 9.25 and 1.75μ resp.

Secondary segmentation was not observed in this species. The total length of the haploid complement is 106.9μ; of the long arms 81.5μ and of the short arms 25.4μ. Total index: 3.2.

The species *T.korolkowii, T.nitida* and *T.rosea* are similar both in external morphology and in karyotype; the only difference is that *T.korolkowii* and *T.rosea* have an extra pair of small chromosomes more in group I, but a comparison of this pair with the 4th pair in group II of *T.nitida* reveals their close affinity. The chromosomes of *T.korolkowii* and *T.rosea* are somewhat larger than those of *T.nitida.* The karyotype of *T.rosea* is nearly a replica of that of *T.korolkowii.*

Leiostemones, seventh group.

This group comprises five species: *T.schrenkii* Regel, *T.kolpakowskiana* Regel, *T.ferganica* Vved., *T.zenaidae* Vved. and *T.ostrowskiana* Regel. Except for *T.schrenkii,* the species of this group are closely related. For example *T.kolpakowskiana* of the valleys differs from *T.ostrowskiana* of the mountain shrublands, but where they share areas of distribution, they are practically indistinguishable. Vvedenskiĭ notes in 'Flora of the USSR' Vol. IV: 'Numerous hybrids between *T.kolpakowskiana* and *T.ostrowskiana* are known, for instance from the Alma Ata area, displaying various combinations in flower colour and habit and thus completely obliterate the distribution between the species, making their identification difficult'. Newton investigated only one species of this group, *T.kolpakowskiana,* for which he found 2n = 24. His successors studied garden forms of *T.ostrowskiana.*

Tulipa schrenkii Regel. The individuals studied were collected by an expedition of our botanical garden near Berchogur in Kazakhstan.

The chromosomes were studied in root-tip sections taken from adult bulbs. The diploid chromosome number is 24. They can be divided into four main groups (fig.16h, 18f).

Group I, two pairs: The 1st pair with the long arms 7.5μ, the short arms 3.75μ; the 2nd pair 4.6 and 2.6μ resp.

Group II, one pair: The long arms 6.0μ, the short arms 2.5μ.

Group III, five pairs: The 1st pair with the long arms 10.25μ, the short arms 3.0μ; the 2nd pair 8.5 and 2.25μ; the 3rd pair 6.75 and 1.75μ; the 4th pair 6.0 and 1.75μ; the 5th pair 5.25 and 1.6μ resp.

Group IV, four pairs: The 1st pair with the long arms 10.5μ, the short arms 2.6μ; the 2nd pair 9.5 and 2.25μ; the 3rd pair 6.0 and 1.5μ; the 4th pair 5.75 and 1.25μ resp.

Secondary segmentations were not found. The total length of the haploid complement is 113.15μ; of the long arms 86.6μ and of the short arms 26.55μ. Total index: 3.22.

Tulipa kolpakowskiana Regel. The individuals studied originated from the Sary-Bastau in the Chu-Iliyskiye mountains, environs of Androkai station on the Turkestan-Siberian railway (coll. Z.P. Botschantzeva, 1932); from the Zailiyskiye Alatau; the Talgar river valley near Talgar station and Razdol'ye (coll. Z.P. Botschantzeva); from the Malyy Kemin river basin near the upper reaches of the Karta-Bulga river (coll. Z.P. Botschantzeva).

We examined root-tip sections taken from adult bulbs and found 2n=24, confirming the number found by Newton (1927), who draws attention to the size of the chromosomes of *T.kolpakowskiana* holding an intermediate position between those of *T.linifolia* and those of the normal type for the genus. He mentions that the complement of *T.kolpakowskiana* does not include chromosomes with a submedian attachment of the spindle threads and that there is no pair of exceptionally small chromosomes typical of *T.linifolia.*

An examination of the chromosome morphology reveals two karyotypes. The first was found in *T.kolpakowskiana* from the Chu-Iliyskiye mountains and the Zailiyskiye Alatau (fig.11d). The 24 chromosomes fall into five main groups (fig.17b).

Group I, one pair: The long arms 4.5, the short arms 2.25μ.

Group II, three pairs: The 1st pair with the long arms 6.25μ, the short arms 3.0μ; the 2nd pair 4.5 and 1.75μ; the 3rd pair 4.5 and 1.6μ resp.; besides in the long arms of the 1st pair an akinetic constriction at 2.0μ from the centromere.

Group III, two pairs: The 1st pair with the long arms 6.75μ, the short arms 2.0μ; the 2nd pair 6.0 and 2.0μ resp.

Group IV, four pairs: The 1st pair with the long arms 8.0μ, the short arms 2.0μ; the 2nd pair 5.0 and 1.25μ; the 3rd pair 4.75 and 1.0μ; the 4th pair 4.5 and 1.0μ resp.; besides in the long arms of the 2nd pair an a.con. below a spherical segment, ca. 0.5μ long, at the distal end.

Group V, two pairs: The 1st pair with the long arms 8.0μ, the short arms 1.5μ; the 2nd pair 7.75 and 1.5μ resp.; besides in the long arms of the 2nd pair an a.con. at the distal end separating an oblong segment just over 1.0μ long.

The total length of the haploid complement is 91.35μ; of the long arms 70.5μ and of the short arms 20.85μ. Total index 3.38.

The second karyotype found in *T.kolpakowskiana* from the Malyy Kemin river basin also has 24 chromosomes in somatic cells, but their distribution over the length-ratio groups is different. There are four main groups (fig.18g).

Group I, two pairs: The 1st pair with the long arms 6.5, the short arms 3.7μ; the 2nd pair 4.0 and 2.6μ resp.; the 1st pair has two a.cons., one in the long arms at ca. 2.5μ from the distal end and the other in the short arms at ca. 1.5μ from the centromere.

Group II, one pair: The long arms 7.0μ. the short arms 2.75μ.

Group III, four pairs: The 1st pair with the long arms 7.0μ, the short arms 2.0μ; the 2nd pair 5.75 and 1.9μ; the 3rd pair 5.75 and 1.5μ; the 4th pair 5.5 and 1.5μ resp.

Group IV, five pairs: The 1st pair with the long arms 10.5μ, the short arms 2.4μ; the 2nd pair 9.35 and 2.0μ; the 3rd pair 8.0 and 2.0μ; the 4th pair 7.9 and 1.5μ; the 5th pair 6.25 and 1.25μ resp.

The total length of the haploid complement is 107.6μ; of the long arms 83.5μ and of the short arms 24.1μ. Total index 3.46.

Thus *T.kolpakowskiana* has two different karyotypes. The one found in the plants from Malyy Kemin is similar to that of *T.schrenkii* and *T.korolkowii,* while the karyotype of the plants from the Alma Ata and the Chu-Iliyskiye mountains comes close to that of *T.ostrowskiana* and *T.zenaidae.*

Tulipa ferganica Vved. The individuals studied were collected in the Ferganskiy range at various locations along the rivers Kugart and Yassy (coll. Z.P. Botschantzeva, 1933).

We examined the chromosomes in root-tip sections taken from adult bulbs and in pollen tubes, grown in vitro, during division of the generative nucleus. The diploid number is 24; they fall into six main groups (fig.11f, 18h).

Group I, one pair: Small chromosomes with the long arms 4.5μ, the short arms 2.25μ.

Group II, two pairs: The 1st pair with the long arms 8.25μ, the short arms 3.5μ; the 2nd pair 4.5 and 1.6μ resp.

Group III, four pairs: The 1st pair with the long arms 7.0μ, the short arms 1.9μ; the 2nd pair 6.5 and 1.75μ; the 3rd pair 6.0 and 1.9μ; the 4th pair 5.75 and 1.75μ resp.

Group IV, two pairs: The 1st pair with the long arms 9.75μ the short arms 2.25μ; the 2nd pair 7.75 and 1.6μ resp.

Group V, two pairs: The 1st pair with the long arms 9.0μ, the short arms 1.5μ; the 2nd pair 8.5 and 1.5μ resp.

Group VI, one pair: The long arms 9.0μ, the short arms 1.4μ.

No secondary segmentations were observed in this species. The total length of the haploid complement is 141.1μ; of the long arms 90.5μ and of the short arms 23.6μ. Total index: 3.83.

Tulipa zenaidae Vved. The individuals studied were brought from Mount Shekule in the Aleksandrovskiy range (coll. Z.P. Botschantzeva, 1932).

We examined root-tip sections taken from adult bulbs. The diploid chromosome number is 24; they can be divided into six main groups (fig.11k, 17a).

Group I, one pair: The long arms 4.0μ, the short arms 2.25μ.

Group II, two pairs: The 1st pair with the long arms 7.0μ, the short arms 3.0μ; the 2nd pair 6.5 and 2.25μ resp.

Group III, two pairs: The 1st pair with the long arms 6.0μ, the short arms 1.6μ; the 2nd pair 5.6 and 1.5μ resp.

Group IV, four pairs: The 1st pair with the long arms 7.75μ, the short arms 1.75μ; the 2nd pair 7.0 and 1.4μ; the 3rd pair 6.0 and 1.25μ; the 4th pair 6.0 and 1.25μ resp.

Group V, two pairs: The longest of the complement; the 1st pair with the long arms 10.25μ, the short arms 2.0μ; the 2nd pair 9.25 and 1.6μ resp.

Group VI, one pair: The long arms 7.75μ, the short arms 1.0μ.

Secondary constrictions were not observed. The total length of the haploid complement is 103.95μ; of the long arms 83.1μ and of the short arms 20.85μ. Total index: 3.99.

Tulipa ostrowskiana Regel. The individuals studied were obtained from the Zailiyskiye Alatau, the Baibulak ravine in the environs of Talgar, on clayey and shingly slopes (coll. Z.P. Botschantzeva, 1933, 1934 and 1936).

We examined chromosomes in root-tip sections and in germinating pollen during spermatogenesis. The chromosome number, $4x = 2n = 48$, is constant for all colour forms of this species, from pale yellow, orange, orange-red to bright dark red. It was found in the sixteen different colour forms which we investigated. The chromosomes can be divided into five main groups (fig.16k, 17c).

Group I, one pair: The long arms 6.5μ, the short arms 3.75μ; in the short arms an akinetic constriction at ca. 1.0μ from the centromere.

Group II, five pairs: The 1st pair with the long arms 6.9μ the short arms 2.5μ; the 2nd pair 6.75 and 2.5μ; the 3rd pair 6.0 and 2.25μ; the 4th pair 5.0 and 2.0μ, the 5th pair 4.5 and 1.8μ resp.

Group III, nine pairs: The 1st pair with the long arms 9.4μ, the short arms 2.5μ; the 2nd pair 8.5 and 2.25μ; the 3rd pair 7.0 and 2.0μ; the 4th pair 6.25 and 2.0μ; the 5th pair 5.5 and 1.8μ; the 6th pair 5.25 and 1.75μ; the 7th pair 5.25 and 1.5μ; the 8th pair 5.25 and 1.5μ; the 9th pair 4.4 and 1.3μ resp.

Group IV, seven pairs: The 1st pair with the long arms 9.5μ, the short arms 2.25μ; the 2nd pair 8.25 and 1.9μ; the 3rd pair 8.0 and 1.75μ; the 4th pair 7.0 and 1.4μ; the 5th pair 6.5 and 1.25μ; the 6th pair 6.5 and 1.25μ; the 7th pair 5.0 and 1.25μ resp.

Group V, two pairs: The 1st pair with the long arms 8.75μ, the short arms 1.5μ; the 2nd pair 8.25 and 1.4μ resp.

The total length of the haploid complement is 205.9μ; of the long arms 159.7μ and of the short arms 46.2μ. Total index: 3.46.

A close examination of the karyotypes of the natural group: *T.schrenkii, T.kolpakowskiana, T.ferganica, T.zenaidae* and *T.ostrowskiana,* shows that *T.schrenkii* directly links this group with the karyotype of *T.korolkowii.* In the distributional pattern of the chromosomes over the length ratio groups the karyotype of *T.kolpakowskiana* from the Malyy Kemin is almost an exact replica of that of *T.schrenkii.*

T.ferganica and *T.zenaidae* have the same number of chromosomes in groups I, II, V and VI. The former has four pairs of chromosomes in group III and two pairs in group IV, whereas the corresponding groups of *T.zenaidae* have two and four pairs respectively.

It can be concluded from a comparison of the karyotype of *T.ostrowskiana* with those of the closely related species, that *T.ostrowskiana* is not an autotetraploid but an amphidiploid with one of its components the 'typical' *T.kolpakowskiana.* The latter is what we use to call '*T.kolpakowskiana* from the plains'; its karyotype is represented by the samples from Razdol'ye (Talgar, Zailiyskiye Alatau). The shortened chromosomes in groups II and IV are very characteristic and undoubtedly also present in the karyotype of *T.ostrowskiana.* The other component species is not known. Apparently this group is now in the process of species formation. The study of a large number of species and infraspecific taxa and the information on the ecological conditions of their habitats should throw more light on this process in nature. With due consideration to our still insufficient knowledge of the karyotypes in this group, we could give an illustration of the relationships between its members by the following sequence: *T.schrenkii* – *T.kolpakowskiana* (Malyy Kemin) – *T.ferganica* – *T.zenaidae* – *T. kolpakowskiana* (Alma Ata) – *T.ostrowskiana*.

Leiostemones, eighth group

Tulipa tetraphylla Regel. The individuals studied were collected in the environs of Talgar in the Zailyskiye Alatau (coll. Z.P. Botschantzeva, 1933-1936).

The chromosome complement of this species is no less distinct than its external morphological characters (fig.11g). Although the chromosome number is as a rule 2n = 24, tetraploid forms occur frequently. The distribution of the chromosomes over the various groups is quite unusual for species of the section *Leiostemones*. There are five main groups (fig.17d).

Group I, three pairs: The 1st pair with the long arms 8.5, the short arms 4.5μ; the 2nd pair 7.5 and 4.25μ; the 3rd pair 7.0 and 4.1μ resp.; besides in the middle of the short arms of the 1st pair an akinetic constriction and in the short arms of the 2nd pair an a.con. at ca. 1.0μ from the centromere.

Group II, three pairs: The 1st pair with the long arms 8.75μ, the short arms 3.25μ; the 2nd pair 6.5 and 3.0μ; the 3rd pair 6.25 and 2.5μ resp.; besides in the short arms of the 3rd pair an a.con. in the middle.

Group III, three pairs: The 1st pair with the long arms 7.4μ, the short arms 2.0μ; the 2nd pair 6.25 and 1.75μ; the 3rd pair 5.75 and 1.5μ resp.

Group IV, two pairs: The 1st pair with the long arms 9.0μ, the short arms 2.25μ; the 2nd pair 7.0 and 1.75μ resp.

Group V, one pair: The long arms 7.0μ , the short arms 1.4μ.

The total length of the haploid complement is 119.15μ; of the long arms 86.9μ and of the short arms 32.25μ. Total index: 2.69.

The karyotype of *T. tetraphylla* appears to be unique among the species of this section that have been studied by us. The three pairs of chromosomes, with the arms-ratio >2, link it more closely to certain varieties of *Eriostemones* species than to *Leiostemones*. As far as the latter group is concerned the karyotype of *T. tetraphylla* comes nearest to that of *T. schrenkii* or *T. korolkowii*, but the chromosomes show clearly secondary constrictions. Polyploid forms are frequent.

It remains possible that the study of *T. iliensis* Regel and *T. dykesiana* Vved. would class *T. tetraphylla* with *T. kolpakowskiana* and the species grouped around it, because both in nature and in cultivation we found a whole series of transitional forms which are undoubtedly hybrids between these species.

SECTION *SPIRANTHERA* VVED.

In "Flora of the USSR", Vol. IV, this section includes only one species, *T. kaufmanniana*. However, considering the mode of dehiscence of the anthers, it should include also *T. dubia* (section *Leiostemones* in "Flora of the USSR") and the two new species *T. anadroma* from Lake Sarychelek and *T. tschimganica* from the Bostandykskiy district.

T. kaufmanniana Regel. The individuals studied were taken from the collection in the botanical garden of the Tashkent State University. The parent material had been collected in the Western Tien Shan, the Pskem and Ugam river valleys, and in Kazakhstan, near Tyul'kubas station and on the second right-hand tributary of the Tavdy-Bulak gorge (coll. V.P. Bochantsev and A. Ya. Butkov).

The diploid chromosome number, 2n=24, first found by Newton (1927), was confirmed by our counts on root-tip sections and on young pollen (n=12). The chromosomes fall into six main groups according to the length ratio of the arms (fig.11a, 17e).

Group I, one pair: The long arms 8.5, the short arms 4.37μ; besides in the short arms an akinetic constriction as in all chromosomes of this type, but closer to the centromere, at ¼ of the length of the arms from it; in the long arms an a.con. below an oblong segment, 1.75μ long.

Group II, one pair: The long arms 7.25μ, the short arms 2.75μ and no secondary segmentations.

Group III, six pairs: Differing but little in the length of the arms; the long arms of the 1st pair 8.5μ, the short arms 2.75μ and in the short arms an a.con. almost in the middle; the 2nd pair 7.7 and 2.4μ; the 3rd pair 7.25 and 2.0μ; the 4th pair 7.0 and 2.0μ with an a.con. in the long arms at ca. 2.0μ from the centromere; the 5th pair, with the arms of the same length as the 4th pair, has a small spherical satellite at the distal end of the long arms, which seems rather unstable since it could not be observed in all plates; the 6th pair of the same size as the two preceding pairs, the long arms only one tenth of a micron shorter; in the short arms a tiny, spherical satellite which is unstable.

Group IV, two pairs: The 1st pair with the long arms 7.25, the short arms 1.75μ and an a.con. in the long arms at ca. 1.0μ from the centromere; the 2nd pair with the long arms 7.0μ and the short arms 1.75μ.

Group V, one pair: The long arms 7.25μ, the short arms 1.25μ; in the long arms an a.con. nearly in the middle.

Group VI, one pair: The long arms 9.5μ, the short arms 1.5μ; in the long arms an a.con. at ca. 3.0μ from the distal end.

The total length of the haploid complement is 117.75μ; of the long arms 91.25μ and of the short arms 26.5μ. Total index: 3.44.

The karyotype of the only representative of this section that we examined shows more similarity to the karyotypes of the third group of *Leiostemones* than to that of any species we have studied in other groups and sections.

SECTION *LOPHOPHYLLON* VVED.

This is a monotypic section consisting of only one species *T.regelii* Krassn. The literature does not give any indication that the species has been cytologically studied.

Tulipa regelii Krassn. The individuals studied were

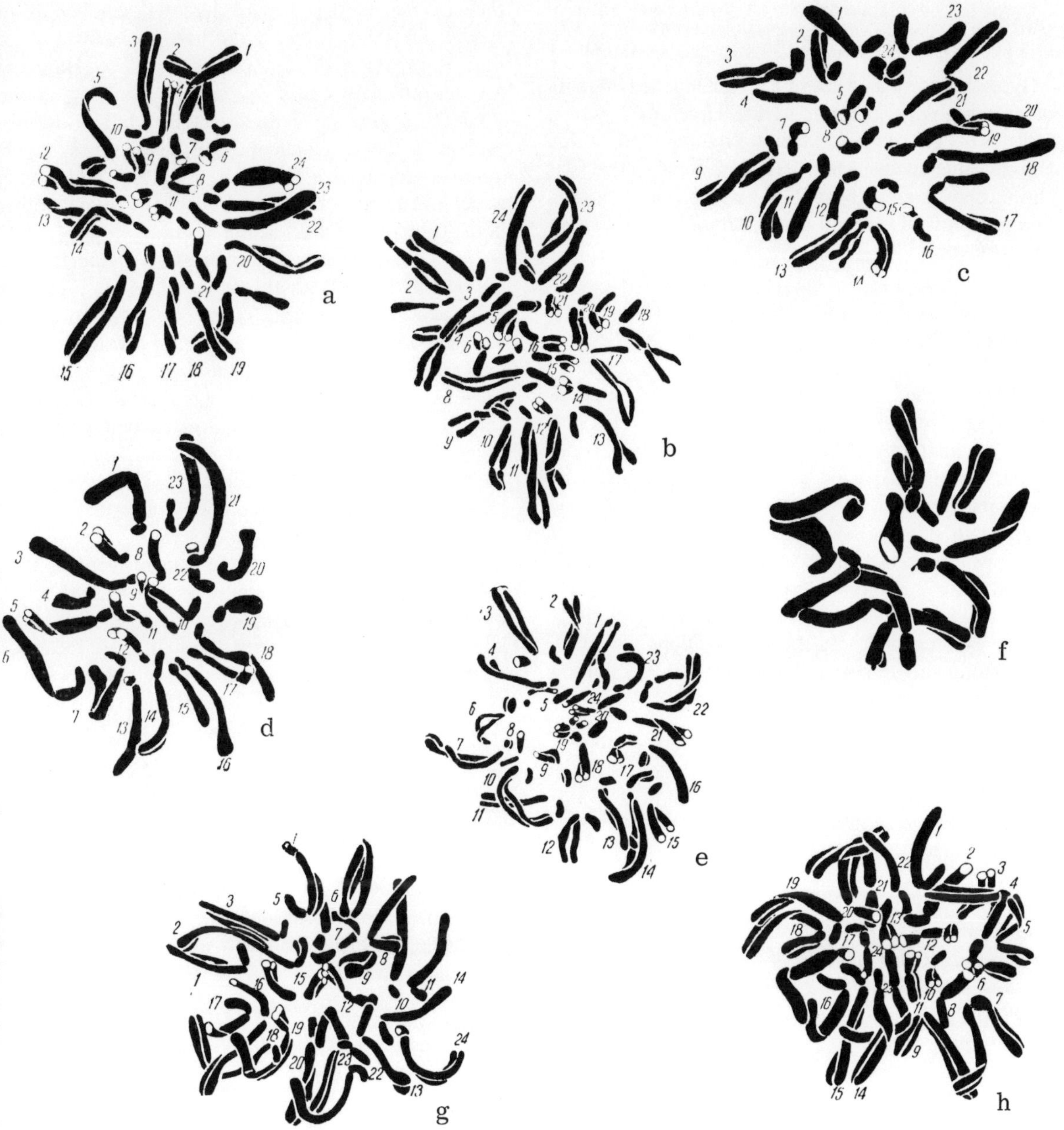

Figure 19. Metaphase plates
a. *T. biebersteiniana* Schult.; b. *T. patens* Agardh. ex Schult.; c. *T. biflora* Pall.; d *T. sogdiana* Bunge; e. *T. androssowii* Litw.; f. *T. turcomanica* B. Fedtsch.; g. *T. regelii* Krassn.; h. *T. dasystemonoides* Vved.

raised from seed of herbarium specimens which are preserved at the Kazakhstan branch of the Academy of Sciences of the USSR. They had been collected at Andrakai by Dmitriev in 1935 and other specimens had been brought from the Chokpar railway station (coll. Bochantsev, 1947).

The diploid chromosome number is 24; they can be divided into five main groups (fig.19g, 20g).

Group I, two pairs: The 1st pair with the long arms 7.0μ and the short arms 5.8μ; in the short arms an akinetic constriction at ca. 1.5μ from the centromere; the 2nd pair 5.0 and 2.63μ resp.

Group II, two pairs: The 1st pair with the long arms 7.0μ, the short arms 2.5μ; the 2nd pair 5.75 and 2.25μ resp.

Group III, four pairs: The 1st pair with the long arms 7.0μ, the short arms 2.13μ; the 2nd pair 6.5 and 2.0μ; the 3rd pair 6.25 and 1.75μ; the 4th pair 6.0 and 2.0μ resp.

Group IV, three pairs: The 1st pair with the long arms 10.0μ, the short arms 2.25μ and in the short arms an a.con. at ca. 0.75μ from the centromere; the 2nd pair 9.25 and 2.25μ; the 3rd pair 6.63 and 1.5μ resp.

Group V, one pair: The long arms 7.4, the short arms 1.13μ.
The total length of the haploid complement is 111.65μ; of the long arms 83.78μ and of the short arms 27.87μ. Total index: 3.01.

Typical of the karyotype of *T.regelii* are the two pairs of chromosomes in group I, the larger pair with a secondary constriction, which is characteristic of this chromosome in the genus *Tulipa*. Likewise the first chromosome pair of group IV has a secondary constriction, which is also found in the karyotypes of certain *Eriostemones*.

SECTION *ERIOSTEMONES* BOISS.

In "Flora of the USSR", Vol. IV, Vvedenskiĭ gives the following diagnosis of this section: 'Bulb tunics glabrous or hirsute or arachnoid-lanate on the inside; flowers in most species small, inconspicuous; filaments pubescent, in one species (*T.sogdiana*) glabrous; ovary with subsessile stigma or with a more or less developed style, but this never longer than the ovary'.

Hall (1940) mentions for the whole section 21[3] species, which he believes to be sufficiently distinct, and he divides them into three subsections.

The following 18 species have distributional areas in the Soviet Union:

T.violacea Boiss. et Buhse
T.caucasica Lipsky
T.sylvestris L.
T.biebersteiniana Schult.*
T.patens Agardh. ex Schult.*
T.sogdiana Bunge*
T.buhseana Boiss.*
T.turkestanica Regel*
T.turcomanica B. Fedtsch.*
T.bifloriformis Vved.*
T.biflora Pall.*
T.polychroma Stapf
T.koktebelica Junge[4]
T.callieri Halácsy et Levier[4]
T.orythyioides Vved.*
T.dasystemonoides Vved.*
T.dasystemon (Regel) Regel*
T.tarda Stapf*

We have studied the species marked with an asterisk including a number of taxa that are mentioned in "Flora of the USSR". Vol. IV, either as a synonym or in explanatory notes: *T.androssowii* Litw., *T.neustruevae* Pob.,[5] *T.binutans* Vved., *T.talievii* Vved. and *T.orthopoda* Vved.[6] This brings the total number to 16 species of which several from various habitats.

A comparison of the species listed by Hall and Vvedenskiĭ shows that the species growing in the Soviet Union belong predominantly to Hall's third subsection, *Biflores,* which includes only three, in his opinion 'good species', but he expresses his doubts about the status of nine more species which other botanists have referred to this section.

Hall's first subsection, *Australes,* includes three species with areas in the Soviet Union: *T.biebersteiniana*, *T.patens* and *T.sylvestris*. Of these we studied only the first two. *T.biebersteiniana* is brought under the synonymy of *T.australis,* which seems doubtful because the areas of these taxa are disjunct. The solution of this problem requires further study, the more so as *Eriostemones* possess a high plasticity in their morphological characters (including the karyotype), which do not show clear-cut discontinuities between species. In describing them it will therefore be necessary to give precise geographical data on the origin of the specimens studied. The tulips of this section have surprisingly variable karyotypes, which even in one species vary with different locations in its area of distribution. Of the second subsection only *T.violacea* grows in the Soviet Union; according to "Flora of the USSR", it occurs in the Talyshkiye mountains.

Tulipa biebersteiniana Schult. The individuals studied were collected near Berchogur railway station by Ponomareva in 1932.

We examined root-tip sections taken from adult bulbs. The diploid chromosome number is 24. The chromosomes can be divided into five main groups (fig.19a, 20a).

Group I, one pair: The long arms 6.75μ, the short arms 4.25μ; in the long arms an akinetic constriction at ca. 3.0μ from the centromere; in the short arms an a.con. at ca. 1.0μ from the centromere.

Group II, two pairs: The 1st pair with the long arms 7.5μ, the short arms 2.75μ; the 2nd pair 7.0 and 2.5μ resp.; in the long arms of the 1st pair an a.con. at ca. 3.0μ from the centromere; in the short arms of the 2nd pair an a.con. at ca. 1.0μ from the centromere.

Group III, four pairs: The 1st pair with the long arms 8.0μ, the short arms 2.5μ; the 2nd pair 6.0 and 2.0μ; the 3rd pair 6.0 and 1.74μ; the 4th pair 6.0 and 1.75μ resp.; besides in the long arms of the 1st pair an a.con. at ca. 2.0μ from the distal end.

Group IV, three pairs: The 1st pair with the long arms 8.25μ, the short arms 2.0μ; the 2nd pair 7.0 and 1.75μ; the 3rd pair 6.0 and 1.25μ resp.; besides in the long arms of the 2nd pair an a.con. at 2.25μ from the centromere.

Group V, two pairs: The 1st pair with the long arms 7.5μ, the short arms 1.5μ; the 2nd pair 7.25 and 1.25μ resp.; besides in the long arms of the 2nd pair three a.cons. at 0.75μ and 2.6μ from the distal end and at ca. 0.75μ from the centromere resp..

The total length of the haploid complement is 108.5μ; of the long arms 83.25μ and of the short arms 25.25μ. Total index: 3.30. Index chromosome A1: 1.59.

Tulipa patens Agardh. ex Schult. We studied root-tip

3. Hall (1940) describes 22 species, although he does not commit himself as to their discontinuities in nature and regards some of them as belonging to one good species by indicating them as 'subsidiary species'. (editor)

4. See footnote on p. 78.

5. *T. neustruevae* Pob. sp. nov. in: Not. Syst. Herb. Inst. Bot. Acad. Sci. USSR 9: 62 (1949). (editor)

6. See footnote on p. 68.

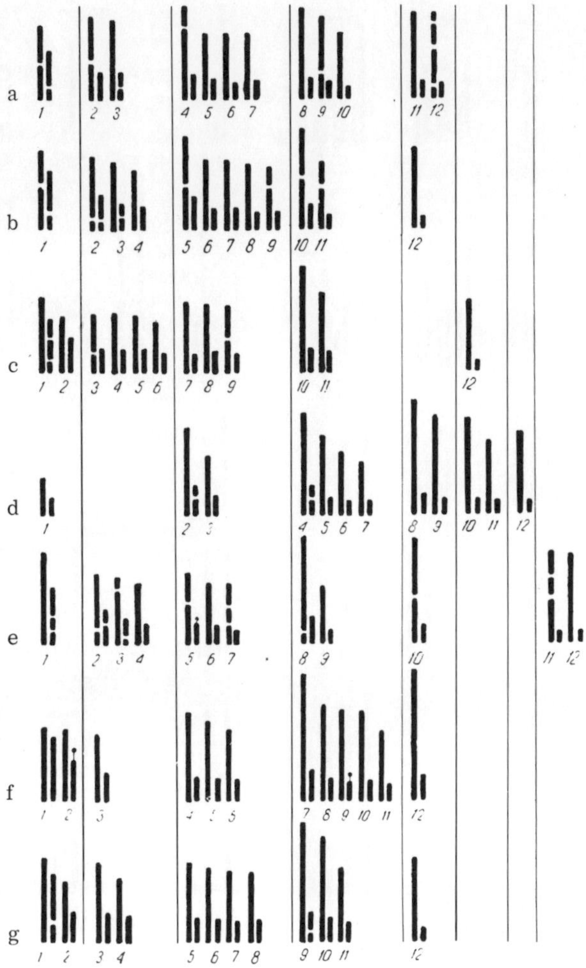

Figure 20. Idiograms of measured plates (enlarged 600 times) a. *T. biebersteiniana* Schult.; b. *T. patens* Agardh. ex Schult.; c. *T. biflora* Pall.; d. *T. sogdiana* Bunge; e. *T. androsowii* Litw.; f. *T. turcomanica* Regel; g. *T. regelii* Krasn. (According to the dimensions given chromosome no 12 of fig. 20g should fall into group VI.) (editor)

sections from adult bulbs, which had been collected in the Semipalatinsk region, environs of Bobrovka (coll. Ermakov in 1933). The diploid chromosome number is 24. They can be divided into five main groups (fig.19b, 20b).

Group I, one pair: The long arms 6.75μ, the short arms 5.0μ; in the long arms an akinetic constriction at 3.0μ from the distal end; in the short arms an a.con. at 1.0μ from the centromere.

Group II, three pairs: The 1st pair with the long arms 6.25μ, the short arms 2.75μ; the 2nd pair 6.0 and 2.25μ; the 3rd pair 5.5 and 2.25μ resp.; besides in the long arms of the 1st pair an a.con. at 0.75μ from the centromere and in the short arms one at 5.0μ; in the short arms of the 2nd pair an a.con. at 1.0μ from the centromere.

Group III, five pairs: The 1st pair with the long arms 8.25μ, the short arms 2.9μ; the 2nd pair 6.25 and 2.0μ; the 3rd pair 6.25 and 2.0μ; the 4th pair 5.75 and 1.75μ; the 5th pair 5.5 and 1.75μ resp.; besides an a.con. in the long arms of the 1st pair at ca. 3.75μ from the centromere and an a.con. in the long arms of the 5th pair at 1.75μ from the distal end.

Group IV, two pairs: The 1st pair with the long arms 8.25μ and the short arms 2.0μ; the 2nd pair 5.75 and 1.25μ resp.; besides in the long arms of the 1st pair an a.con. at 4.0μ from the centromere and in the long arms of the 2nd pair an a.con. at 2.0μ from the distal end.

Group V, one pair: The long arms 6.75μ, the short arms 1.25μ.

The total length of the haploid complement is 104.4μ; of the long arms 77.25, and of the short arms 27.15μ. Total index: 2.85. Index chromosome A1: 1.35.

Tulipa biflora Pall. The individuals studied came from the Chërnyy Irtysh (coll. F.H. Rusanov, 1932).

We examined root-tip sections taken from adult bulbs. The chromosome number is 2n = 24; they fall into five main groups (fig.19c, 20c).

Group I, two pairs: The 1st pair with the long arms 6.75 and the short arms 5.0μ; the 2nd pair 5.25 and 3.0μ resp.; besides in the short arms of the 1st pair two akinetic constrictions at 1.0 and 2.75μ from the centromere resp.

Group II, four pairs: The 1st pair with the long arms 5.5μ and the short arms 2.5μ; the 2nd pair 5.5 and 2.25μ; the 3rd pair 5.25 and 2.0μ; the 4th pair 4.5 and 1.75μ resp.; besides in the long arms of the 1st pair an a.con. at 1.75μ from the centromere.

Group III, three pairs: The 1st pair with the long arms 6.75μ and the short arms 1.75μ; the 2nd pair 6.0 and 2.0μ; the 3rd pair 6.0 and 1.75μ resp.; besides in the long arms of the 3rd pair an a.con. at 2.75μ from the centromere.

Group IV, two pairs: The 1st pair with the long arms 9.25μ and the short arms 2.25μ; the 2nd pair 7.0 and 1.75μ resp.

Group V: Absent.

Group VI, one pair: The long arms 6.0μ and the short arms 1.0μ.

The total length of the haploid complement is 100.75μ; of the long arms 73.75μ and of the short arms 27.0μ. Total index: 2.73. Index chrom. A1: 1.35. Index chrom. A2: 1.75.

Tulipa sogdiana Bunge. The individuals studied were received from N.A. Androsov in 1931 together with a note stating 'at the beginning of the sands near Ashkhabad'.

The root-tip sections were taken from sprouting adult bulbs. The 24 chromosomes can be divided into six main groups (fig.20d).

Group I: Absent.

Group II, one pair: The long arms 3.0μ, the short arms 1.5μ.

Group III, two pairs: The first pair with the long arms 7.5μ, the short arms 2.5μ; the 2nd pair 5.25 and 1.75μ resp.; besides in the short arms of the 1st pair an akinetic constriction at 1.0μ from the centromere.

Group IV, four pairs: The 1st pair with the long arms 9.0μ, the short arms 2.25μ; the 2nd pair 7.0 and 1.5μ; the 3rd pair 5.5 and 1.25μ; the 4th pair 4.5 and 1.0μ resp.; besides in the short arms of the 1st pair an a.con. at 1.0μ from the distal end.

Group V, two pairs: The 1st pair with the long arms 10.0μ, the short arms 1.75μ; the 2nd pair 8.75 and 1.5μ resp.

Group VI, two pairs: The 1st pair with the long arms 8.25μ, the short arms 1.25μ; the 2nd pair 6.5 and 1.0μ resp.

Group VII, one pair: The long arms 7.0μ, the short arms 1.0μ.

The total length of the haploid complement is 100.5μ; of the long arms 82.25μ and of the short arms 18.25μ. Total index: 4.5. Index chrom. A.: 2.

These plants from Ashkhabad have chromosomes showing higher arms ratios than the preceding *Eriostemones* species.

Tulipa androssowii Litw. The material which we received under this name was sent by N.V. Androsov from Serakhs (Turkmenia). It had been collected c' to the river Tedgasan on a clayey soil.

The chromosomes (2n = 24) fall into six main groups (fig.19e, 20e).

Group I, one pair: The long arms 7.5, the short arms 4.25μ; besides in the short arms an akinetic constriction, typical of the chromosomes in this group, at ca. 1.0μ from the centromere.

Group II, three pairs: The 1st pair with the long arms 6.0μ, the short arms 2.75μ and in both the long and the short arms an a.con. at 1.0μ from the centromere; the 2nd pair with the long arms 5.5μ, the short arms 2.0μ, besides an a.con. in the long arms at 0.75μ from the centromere and another in the short arms at 0.5μ from the centromere; the 3rd pair with the long arms 5.0μ, the short arms 2.0μ.

Group III, three pairs: The 1st pair with the long arms 5.75μ, the short arms 1.75μ and an a.con. in the long arms at 2.5μ from the distal end; on the short arms a spherical satellite connected by a thin thread; the 2nd pair with the long arms 5.25μ, the short arms 1.5μ; the 3rd pair 5.0 and 1.5μ resp. and two a.cons. in the long arms at ca. 1.75μ and 2.75μ resp. from the centromere.

Group IV, two pairs: The 1st pair with the long arms 9.0μ, the short arms 2.25μ and in the long arms an a.con. at 0.75μ from the centromere; the 2nd pair with the long arms 5.0μ, the short arms 1.25μ.

Group V, one pair: The long arms 8.75μ, the short arms 1.5μ and in the long arms an a.con. at 3.75μ from the centromere.

Groups VI, VII, VIII, IX: Absent.

Group X, two pairs: Capitate, equally long pairs of chromosomes; the long arms 7.5μ, the short arms 0.75μ; in the long arms of the 1st pair two a.cons. at 3.75 and 5.25μ resp. from the centromere.

The total length of the haploid complement is 100.25μ; of the long arms 77.75μ and of the short arms 22.5μ. Total index: 3.46. Index chrom. A: 1.76.

Tulipa turcomanica B. Fedtsch. The individuals studied were collected at different localities in the western Kopet-Dag (coll. A.F. Gnezdilo). This species has very distinct morphological characters.

We examined chromosomes in root-tip sections taken from adult bulbs, in smears from young anthers and in pollen tubes at the stage of spermatogenesis. The diploid chromosome number is 24; they fall into five main groups (fig. 19f, 20f).

Group I, two pairs: The 1st pair with the long arms 6.5, the short arms 5.75μ; the 2nd pair 6.25 and 3.75μ resp. and with a satellite on the short arms.

Group II, one pair: The long arms 5.75μ, the short arms 2.5μ.

Group III, three pairs: The 1st pair with the long arms 7.75μ, the short arms 2.1μ; the 2nd pair 7.0 and 2.0μ; the 3rd pair 6.25 and 2.0μ resp.

Group IV, five pairs: The 1st pair with the long arms 11.0μ, the short arms 2.75μ; the 2nd pair 8.5 and 2.0μ; the 3rd pair 7.75 and 1.75μ with a satellite on the short arms; the 4th pair 7.75 and 1.75μ; the 5th pair 6.25 and 1.65μ resp.

Group V, one pair: The long arms 11.25μ, the short arms 2.25μ.

The total length of the haploid complement is 121.25μ; of the long arms 92.0μ and of the short arms 29.25μ. Total index: 3.15. Index chrom.A1: 1.13. Index chrom. A2: 1.66.

Tulipa dasystemonoides Vved. The individuals studied were brought from the environs of the Pervomayskiy settlement (coll. V.P. Bochantsev, 1935).

We examined root-tip sections taken from adult bulbs. The diploid number of chromosomes is 24; they can be divided into six main groups (fig.19h, 21a).

Group I, one pair: The long arms 7.75μ, the short arms 6.0μ with the akinetic constriction, typical of *Tulipa* chromosomes in this group, at 1.75μ from the centromere.

Group II, three pairs: The 1st pair with the long arms 7.75μ, the short arms 3.75μ; the 2nd pair 7.25 and 3.25μ; the 3rd pair 6.75 and 2.5μ resp.; besides in the short arms of the 2nd pair an a.con. at ca. 1.0μ from the distal end and in the short arms of the 3rd pair an a.con. at 1.0μ from the centromere.

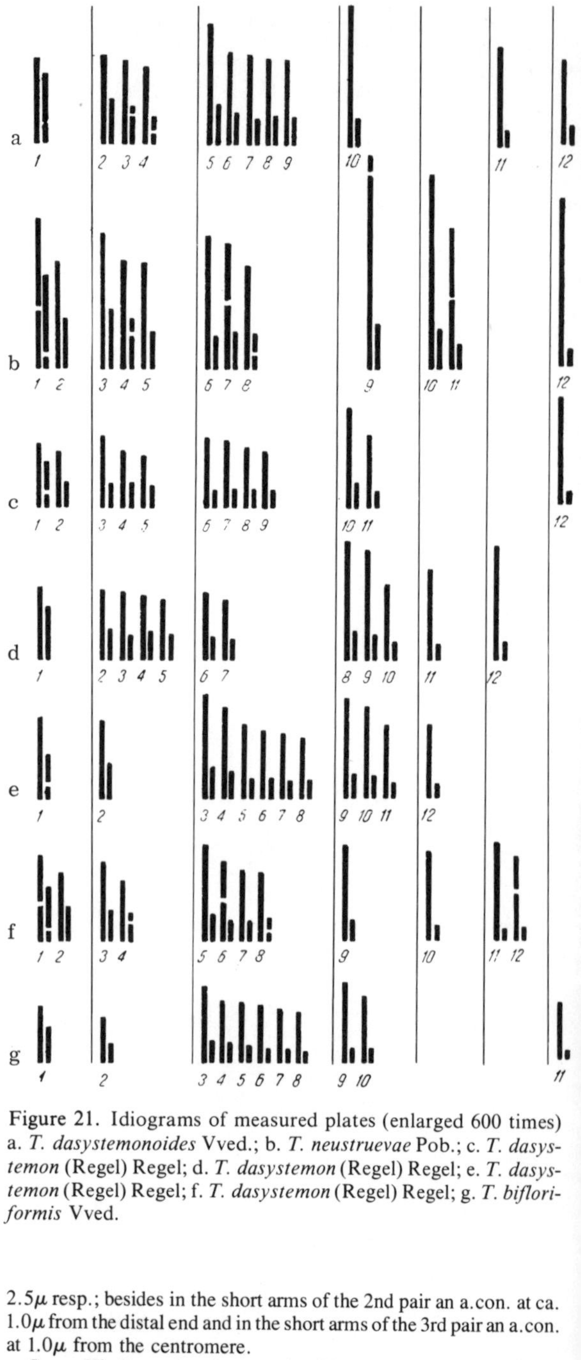

Figure 21. Idiograms of measured plates (enlarged 600 times) a. *T. dasystemonoides* Vved.; b. *T. neustruevae* Pob.; c. *T. dasystemon* (Regel) Regel; d. *T. dasystemon* (Regel) Regel; e. *T. dasystemon* (Regel) Regel; f. *T. dasystemon* (Regel) Regel; g. *T. bifloriformis* Vved.

Group III, five pairs: the 1st pair with the long arms 10.68μ, the short arms 3.55μ; the 2nd pair 8.25 and 2.75μ; the 3rd pair 8.05 and 2.5μ; the 4th pair 7.75 and 2.6μ; the 5th pair 7.6 and 2.25μ resp.

Group IV, one pair: The long arms 12.5μ, the short arms 2.75μ.

Group V: Absent.

Group VI, one pair: The long arms 9.0μ, the short arms 1.6μ.

Group VII, one pair: The long arms 10.62μ, the short arms 1.75μ.

The total length of the haploid complement is 140.15μ; of the long arms 104.95μ and of the short arms 35.2μ. Total index: 2.98. Index chrom. A: 1.29.

Tulipa sp. The individuals studied were brought from mount Kyzykurt (coll. V.P. Bochantsev and A. Ya.Butkov, 1933).

We examined root-tip sections taken from adult bulbs. The diploid number of chromosomes is 24; they fall into seven main groups (fig. 22h, 23a)

Group I, one pair: The long arms 6.5μ, the short arms 4.25μ.
Group II, two pairs: The 1st pair with the long arms 7.0μ, the short arms 2.73μ; the 2nd pair 5.0 and 1.75μ resp.
Group III, three pairs: The 1st pair with the long arms 6.75μ, the short arms 1.75μ; the 2nd pair 6.62 and 2.06μ; the 3rd pair 6.33 and 1.69μ resp.
Group IV, two pairs: The 1st pair with the long arms 7.45μ, the short arms 1.57μ; the 2nd pair 6.53 and 1.62μ resp.
Group V, one pair: the long arms 7.7μ, the short arms 1.5μ; in the distal end of the long arms an akinetic constriction below a segment, long 1.25μ.
Group VI, two pairs: The 1st pair with the long arms 11.44μ, the short arms 1.86μ; the 2nd pair 9.18 and 1.52μ resp.
Group VII, one pair: The long arms 8.75μ, the short arms 0.78μ.
The total length of the haploid complement is 111.25μ; of the long arms 88.25μ and of the short arms 23.0μ. Total index: 3.85. Index chrom. A1: 1.53.

Tulipa neustruevae Pob. The individuals studied were collected on the left bank of the river Kugart opposite Dmitrievskiy railway station, the upper terrace, in shingle and fine earth (Dzhalal-abad district).

We examined chromosomes in root-tip sections and in pollen tubes in vitro during division of the generative nucleus. The diploid number of chromosomes is 24; they can be divided into six main groups (fig.21b)

Group I, two pairs: The 1st pair with the long arms 13.13, the short arms 8.13μ; in the long arms an akinetic constriction at 5.25μ from the centromere; in the short arms an a.con. at 1.25μ from the centromere; the 2nd pair with the long arms 9.5μ, the short arms 4.4μ.
Group II, three pairs: The 1st pair with the long arms 12.0μ, the short arms 5.4μ; the 2nd pair 9.5 and 4.25μ resp. and in the short arms an a.con. at 1.5μ from the distal end; the 3rd pair 9.5 and 3.5μ resp.
Group III, three pairs: The 1st pair with the long arms 12.0μ, the short arms 3.13μ; the 2nd pair 11.25 and 3.5μ resp. and an a.con. in the long arms at 5.75μ from the centromere; the 3rd pair 9.5 and 2.75μ resp. and in the short arms an a.con. at 1.25μ from the centromere.
Group IV, one pair: The long arms 19.8μ, the short arms 4.1μ.
Group V, two pairs: The 1st pair with the long arms 17.5μ, the short arms 3.6μ; the 2nd pair 11.25 and 2.25μ resp.
Group VI: Absent.
Group VII, one pair: The long arms 18.28μ, the short arms 1.5μ.
The total length of the haploid complement is 195.38μ; of the long arms 148.88μ and of the short arms 46.5μ. Total index: 3.20. Index chrom. A1: 1.61. Index chrom. A2: 2.16.

Tulipa dasystemon (Regel) Regel. This species, which we studied from various parts of its area, appears to have a highly variable karyotype. The present individuals had been brought from the Kugart Pass in the Ferganskiy range, where they were found growing in a matted, grey soil.

The chromosome numbers in individuals from this habitat were 2n = 24 and 2n = 48, i.e. both diploid and polyploid forms were found. The chromosomes of the diploid form fall into six main groups (fig. 21f):

Group I, two pairs: The 1st pair with the long arms 7.5μ, the short arms 4.75μ; in the long arms an akinetic constriction at 3.0μ from the centromere and in the short arms an a.con. at 0.75μ from the centromere; the 2nd pair with the long arms 6.25μ, the short arms 3.25μ.
Group II, two pairs: The 1st pair with the long arms 7.0μ, the short arms 3.0μ; the 2nd pair 5.2 and 2.0μ resp.; in the short arms of this pair an a.con at 0.5μ from the distal end.
Group III, four pairs: The 1st pair with the long arms 8.75μ, the short arms 2.5μ; the 2nd pair 7.0 and 2.0μ resp., besides in the long arms of this pair an a.con at 3.5μ from the centromere; the 3rd pair with the long arms 6.25μ, the short arms 2.0μ; the 4th pair 6.25 and 1.9μ resp., besides in the short arms of this pair an a.con. at 0.5μ from the centromere.
Group IV, one pair: The long arms 8.75μ, the short arms 2.0μ.
Group V, one pair: The long arms 8.25μ, the short arms 1.5μ.
Group VI, two pairs: The 1st pair with the long arms 9.25μ, the short arms 1.25μ; the 2nd pair 7.5 and 1.25μ resp., in the long arms of this pair an a.con. at 3.25μ from the centromere.
The total length of the haploid complement is 116.35μ; of the long arms 87.95μ and of the short arms 28.4μ. Total index: 3.1. Index chrom. A1: 1.58. Index chrom. A2: 1.92.

Tulipa dasystemon (Regel) Regel. The individuals studied were collected in the eastern part of the Alay valley area (coll. M. Velichko, 1935). We believe these plants to be *T.dasystemon*. However, Vvedenskiĭ, describing *T.turkestanica* in "Flora of the USSR", remarks: 'Plants from the Alay valley need further study. According to Velichko they had pale yellow flowers'.

We counted 24 chromosomes in root-tip sections. They fall into five main groups (fig.21c, 22b).

Group I, two pairs: The 1st pair with the long arms 6.0μ, the short arms 4.0μ and an akinetic constriction at 1.75μ from the centromere; the 2nd pair 5.25 and 2.75μ resp.
Group II, three pairs: The 1st pair with the long arms 6.75μ, the short arms 2.5μ; the 2nd pair 5.25 and 2.25μ; the 3rd pair 4.74 and 2.25μ resp.
Group III, four pairs: The 1st pair with the long arms 6.5μ, the short arms 1.63μ; the 2nd pair 6.25 and 1.75μ; the 3rd pair 5.25 and 1.75μ; the 4th pair 5.0 and 1.63μ resp.
Group IV, two pairs: The 1st pair with the long arms 9.15μ, the short arms 2.25μ; the 2nd pair 6.5 and 1.5μ resp.
Groups V – VIII: Absent.
Group IX, one pair: The long arms 9.5μ, the short arms 1.0μ.
The total length of the haploid complement is 99.63μ; of the long arms 74.0μ and of the short arms 25.63μ. Total index: 2.88. Index chrom. A1: 1.5. Index chrom. A2: 1.9.

Tulipa dasystemon (Regel) Regel. In Kazakhstan, the Zailiyskiy Alatau near Lake Issyk-Kul', two types occur: the diploid with 2n = 24 and the tetraploid with 2n = 48. The chromosomes of the first type can be divided into six main groups (fig. 21d, 22c).

Group I, one pair: The long arms 6.25μ, the short arms 4.75μ.
Group II, four pairs: The 1st pair with the long arms 6.5, the short arms 2.6μ; the 2nd pair 6.5 and 2.25μ; the 3rd pair 5.75 and 2.6μ; the 4th pair 5.4 and 2.4μ resp.
Group III, two pairs: The 1st pair with the long arms 6.0μ, the short arms 2.0μ; the 2nd pair 5.4 and 1.75μ resp.
Group IV, three pairs: The 1st pair with the long arms 10.75μ, the short arms 2.6μ; the 2nd pair 9.9 and 2.25μ; the 3rd pair 6.75 and 1.5μ resp.
Group V, one pair: The long arms 8.25μ, the short arms 1.6μ.
Group VI, one pair: The long arms 10.0μ, the short arms 1.5μ.
The total length of the haploid complement is 125.26μ; of the long arms 99.63μ and of the short arms 25.63μ. Total index: 3.89. Index chrom. A: 1.32.

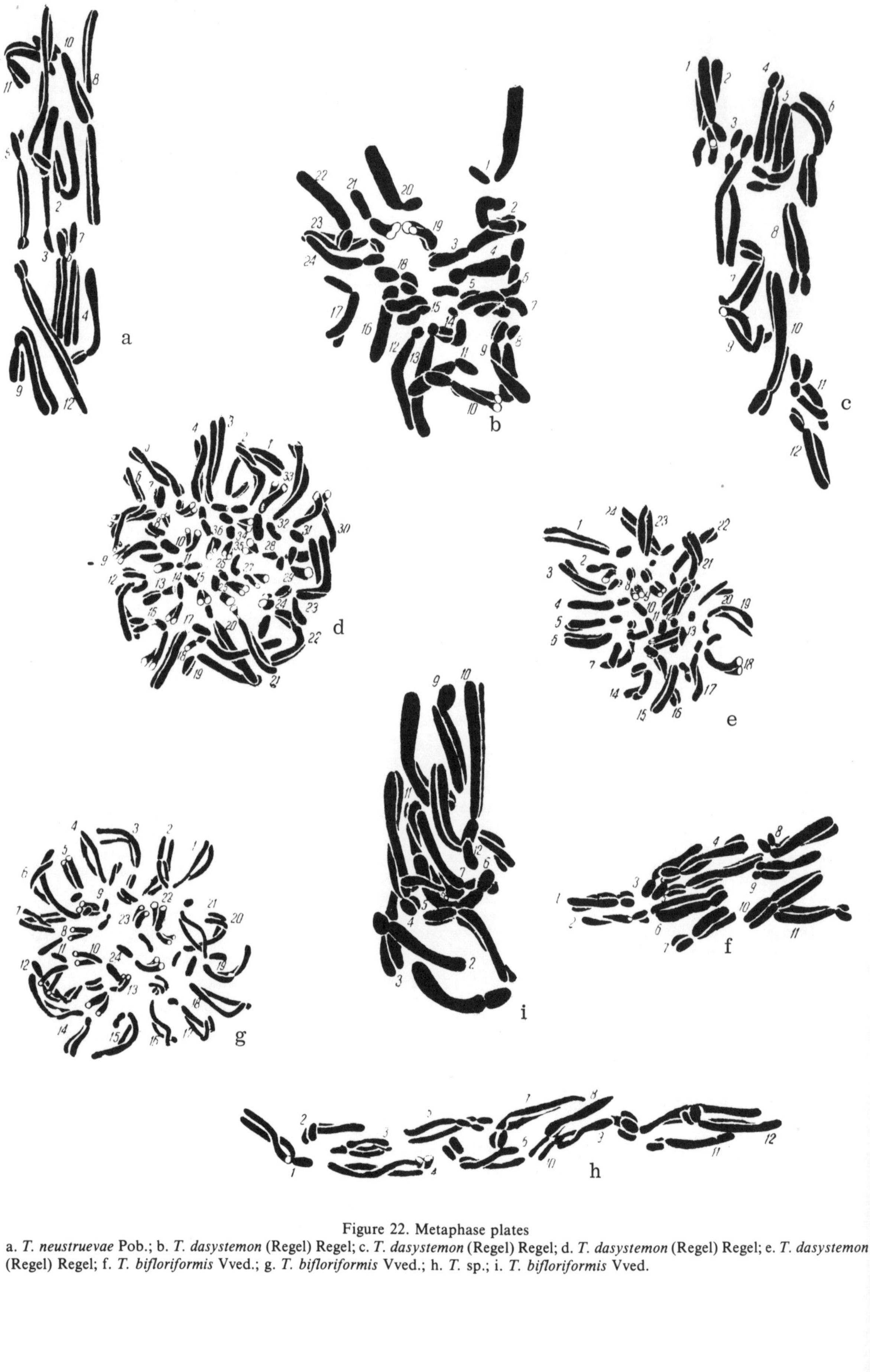

Figure 22. Metaphase plates
a. *T. neustruevae* Pob.; b. *T. dasystemon* (Regel) Regel; c. *T. dasystemon* (Regel) Regel; d. *T. dasystemon* (Regel) Regel; e. *T. dasystemon* (Regel) Regel; f. *T. bifloriformis* Vved.; g. *T. bifloriformis* Vved.; h. *T.* sp.; i. *T. bifloriformis* Vved.

Tulipa dasystemon (Regel) Regel. This material was collected in the Terskey-Alatau, Dzhetyoguz (coll. A.I. Vvedenskiĭ, 1931).

The sample contained two forms: the tetraploid with 2n=48 (fig.22d) and the diploid with 2n=24 (fig. 21e, 22e). The chromosomes of the diploid form fall into five main groups.

Group I, one pair: The long arms 7.5μ, the short arms 4.9μ; in the short arms and akinetic constriction at 1.5μ from the centromere.
Group II, one pair: The long arms 7.0μ, the short arms 3.25μ.
Group III, six pairs: The 1st pair with the long arms 9.25μ, the short arms 2.75μ; the 2nd pair 8.25 and 2.5μ; the 3rd pair 6.75 and 1.75μ; the 4th pair 6.25 and 1.75μ; the 5th pair 6.1 and 1.75μ; the 6th pair 5.75 and 1.75μ resp.
Group IV, three pairs: The 1st pair with the long arms 9.25μ, the short arms 2.25μ; the 2nd pair 8.5 and 2.0μ; the 3rd pair 6.75 and 1.5μ resp.
Group V, one pair: The long arms 6.75μ, the short arms 1.25μ.

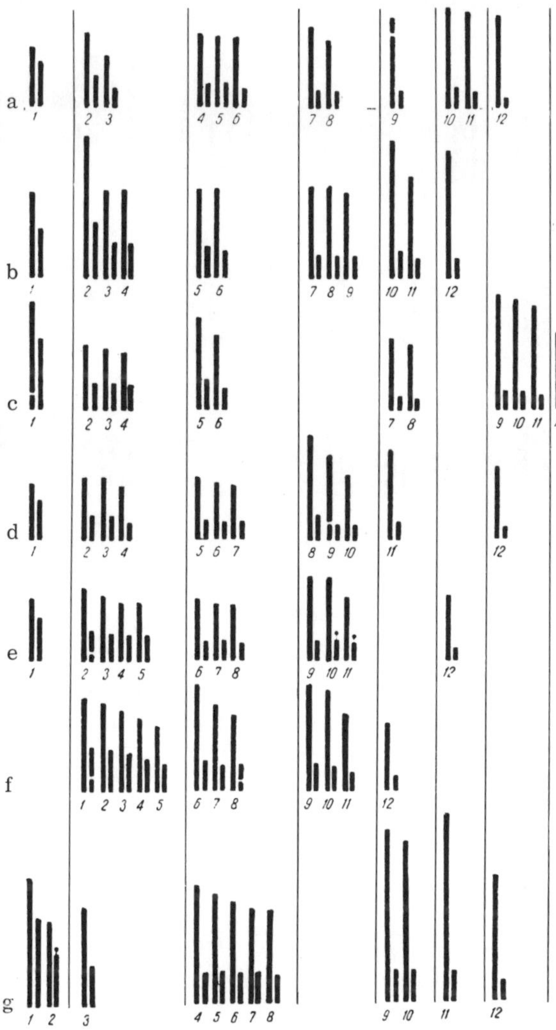

Figure 23. Idiograms of measured plates (enlarged 600 times) a. *T.* sp.; b. *T. bifloriformis* Vved.; c. *T. bifloriformis* Vved.; d. *T. bifloriformis* Vved.; e. *T. binutans* Vved.; f. *T. buhseana* Boiss.; g. *T. turkestanica* Regel.

The total length of the haploid complement is 124.85μ; of the long arms 97.05μ and of the short arms 27.8μ. Total index: 3.49. Index chrom. Al: 1.88.

Tulipa bifloriformis Vved. This species shows entirely different karyotypes for populations from different geographical areas. The individuals studied originated from the Talasskiy Alatau, where the population was found growing on a scree in a gully close to the village of Novonikolayevka (coll. A.Ya. Butkov and V.P. Bochantsev, 1930).

We examined chromosomes in root-tip sections and in pollen tubes during spermatogenesis. The diploid chromosome number is 24. The chromosomes can be divided into six main groups (fig.22i, 23b).

Group I, one pair: The long arms 8.7, the short arms 4.5μ.
Group II, three pairs: The 1st pair with the long arms 13.62μ, the short arms 5.0μ; the 2nd pair 8.5 and 3.25μ; the 3rd pair 8.5 and 3.25μ resp.
Group III, two pairs: The 1st pair with the long arms 8.25μ, the short arms 2.75μ; the 2nd pair 8.15 and 2.32μ resp.
Group IV, three pairs: The 1st pair with the long arms 8.75μ, the short arms 1.95μ; the 2nd pair 8.62 and 1.87μ; the 3rd pair 8.0 and 1.87μ resp.
Group V, two pairs: The 1st pair with the long arms 13.8μ, the short arms 2.70μ; the 2nd pair 9.75 and 1.87μ resp.
Group VI, one pair; The long arms 12.0μ, the short arms 1.75μ.
The total length of the haploid complement is 147.27μ; of the long arms 115.39μ and of the short arms 31.88μ. Total index: 3.62. Index chrom. Al: 1.93.

Tulipa bifloriformis Vved. This population from the Ferganskiy range was growing on the left bank of the Kugart river near the village of Dmitrievskiy, on a slope above a mill, in the scrub zone (coll. Z.P. Botschantzeva, 1933).

The chromosomes were examined in pollen tubes in vitro during spermatogenesis. The chromosomes in somatic cells number 24 and fall into six main groups (fig.21g, 22f).

Group I, one pair: The long arms 5.25μ, the short arms 3.38μ.
Group II, one pair: The long arms 4.25μ, the short arms 1.86μ.
Group III, six pairs: The 1st pair with the long arms 6.75μ, the short arms 1.94μ; the 2nd pair 5.5 and 1.75μ; the 3rd pair 5.5 and 1.7μ; the 4th pair 5.35 and 1.63μ; the 5th pair 5.0 and 1.5μ; the 6th pair 4.3 and 1.35μ resp.
Group IV, three pairs: The 1st pair with the long arms 7.5μ, the short arms 1.8μ; the 2nd pair 6.78 and 1.63μ; the 3rd pair 5.0 and 1.25μ resp.
Groups V – VI: Absent.
Group VII, one pair: The long arms 5.35μ, the short arms 0.69μ.
The total length of the haploid complement is 85.85μ; of the long arms 66.09μ and of the short arms 19.76μ. Total index: 3.34. Index chrom. A: 1.55.

Tulipa bifloriformis Vved. The individuals studied were taken from the collection of the Institute for the Dry Subtropics, in 1935. They originated from the environs of Namangan.

We examined chromosomes in pollen tubes in vitro during spermatogenesis. The chromosome number found was 2n = 24. The chromosomes fall into six main groups (fig. 12a, 23c).

Group I, one pair: The long arms 10.0, the short arms 5.5μ; in the long arm an akinetic constriction at 1.25μ from the centromere.

Group II, three pairs: The 1st pair with the long arms 6.13μ, the short arms 2.4μ; the 2nd pair 6.0 and 2.4μ; the 3d pair 5.13 and 2.35μ resp.

Group III, two pairs: The 1st pair with the long arms 8.75μ, the short arms 2.75μ; the 2nd pair 7.0 and 2.0μ resp.

Group IV: Absent.

Group V, two pairs: The 1st pair with the long arms 6.63μ, the short arms 1.25μ; the 2nd pair 6.25 and 1.13μ resp.

Group VI: Absent.

Group VII, three pairs: The 1st pair with the long arms 11.0μ, the short arms 1.5μ; the 2nd pair 11.55 and 1.75μ; the 3rd pair 9.9 and 1.35μ resp.

Group VIII, one pair: The long arms 7.55μ, the short arms 0.84μ.

The total length of the haploid complement is 120.38μ; of the long arms 94.88μ; of the short arms 25.5μ. Total index: 3.72. Index chrom. A1: 1.82.

Tulipa bifloriformis Vved. These plants also came from a collection of the Institute for the Dry Subtropics and were grown on a suburban plot. There was no label indicating their origin, but an expedition of the institute had previously been collecting in the environs of Samarkand, Denau and Dushanbe.

We examined chromosomes in pollen tubes in vitro during spermatogenesis. The haploid number of chromosomes was n = 12. The chromosomes fall into six main groups (fig.12b, 23d).

Group I, one pair: The long arms 5.25μ, the short arms 4.13μ.

Group II, three pairs: The 1st pair with the long arms 5.75μ, the short arms 2.5μ; the 2nd pair 5.75 and 2.15μ; the 3rd pair 4.95 and 1.75μ resp.

Group III, three pairs: The 1st pair with the long arms 5.75μ, the short arms 1.8μ; the 2nd pair 5.25 and 1.5μ; the 3rd pair 5.13 and 1.63μ resp.

Group IV, three pairs: The 1st pair with the long arms 9.75μ, the short arms 2.25μ; the 2nd pair 7.75 and 1.58μ; the 3rd pair 5.75 and 1.4μ resp.; in the long arms of the 2nd pair an akinetic constriction at 1.55μ from the centromere.

Group V, one pair: The long arms 8.25μ, the short arms 1.5μ.

Group VI: Absent.

Group VII, one pair: The long arms 6.88μ, the short arms 0.95μ.

The total length of the haploid complement is 99.03μ; of the long arms 76.2μ and of the short arms 22.83μ. Total index: 3.34. Index chrom. A1: 1.27.

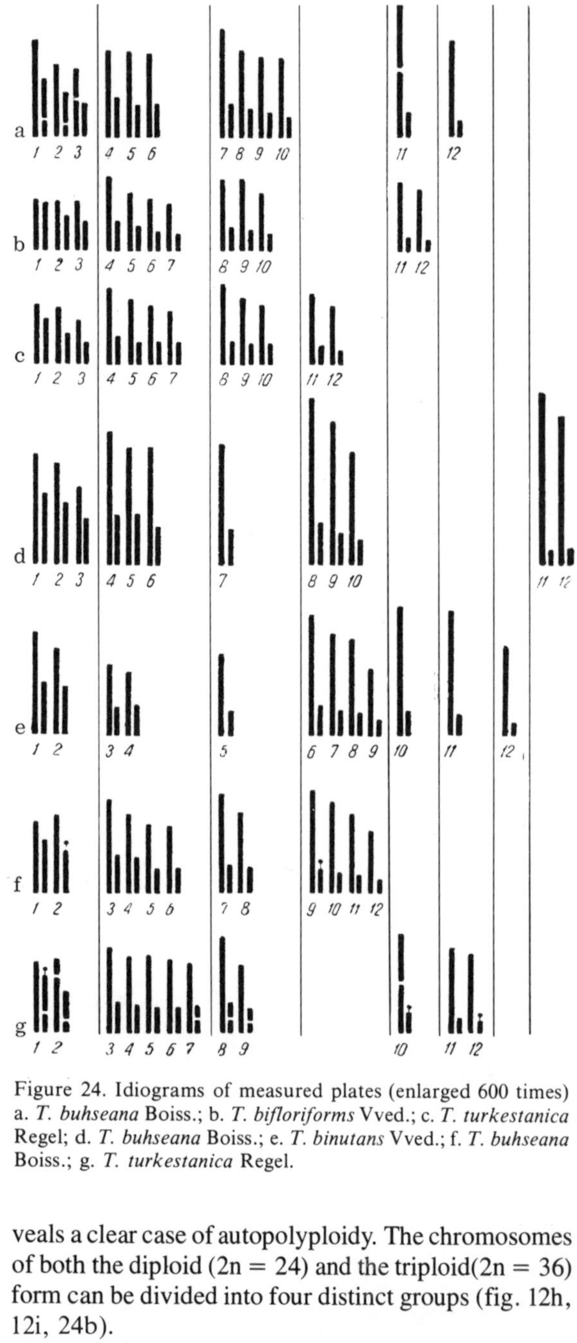

Figure 24. Idiograms of measured plates (enlarged 600 times) a. *T. buhseana* Boiss.; b. *T. bifloriforms* Vved.; c. *T. turkestanica* Regel; d. *T. buhseana* Boiss.; e. *T. binutans* Vved.; f. *T. buhseana* Boiss.; g. *T. turkestanica* Regel.

Tulipa bifloriformis Vved. The individuals studied were taken from plants growing in the Botanical Garden under trees and on the hills towards Ankhor. It has not been possible to trace their origin. They are very similar to samples of this species from Aktash and Khumsan.

We examined chromosomes in root-tip sections taken from adult bulbs, in young pollen grains during division of the primary nucleus and in pollen tubes during spermatogenesis.

The plants were predominantly triploid and only in a few rare cases did we find a diploid. The triploid is a free-flowering form and produces large flowers but no seeds. It shows vigorous vegetative reproduction by forming large clonal populations. A study of the karyotype of the triploid form from this location reveals a clear case of autopolyploidy. The chromosomes of both the diploid (2n = 24) and the triploid(2n = 36) form can be divided into four distinct groups (fig. 12h, 12i, 24b).

Group I, three pairs: The 1st pair with the long arms 4.5, the short arms 4.0μ; the 2nd pair 4.25 and 2.75μ; the 3rd pair 4.0 and 2.25μ resp.

Group II, four pairs: The 1st pair with the long arms 6.25μ, the short arms 2.75μ; the 2nd pair 4.75 and 2.0μ; the 3rd pair 4.5 and 1.75μ; the 4th pair 4.0 and 1.5μ resp.

Group III, three pairs: The 1st pair with the long arms 6.25μ, the short arms 2.0μ; the 2nd pair 6.25 and 1.75μ; the 3rd pair 5.0 and 1.5μ resp.

Group IV: Absent.

Group V, two pairs: The 1st pair with the long arms 6.0μ the short arms 1.1μ; the 2nd pair 5.25 and 1.0μ resp.

The total length of the haploid complement is 85.35μ; of the long arms 61.0μ and of the short arms 24.35μ. Total index: 2.5. Index chrom. A1: 1.13. Index chrom. A2: 1.55. Index chrom. A3: 1.78.

Tulipa buhseana Boiss. The material was received from V.P. Bochantsev and A.Ya. Butkov, in 1933. The label was inscribed: 'plain near Chardara with slight salinization'.

We examined chromosomes in root-tip sections from adult bulbs and in pollen tubes in vitro during spermatogenesis. The diploid chromosome number is 24 and the chromosomes can be divided into five main groups (fig.12g, 24a).

Group I, three pairs: The 1st pair with the long arms 8.5μ, the short arms 5.0μ and in the short arms an akinetic constriction at 1.75μ from the centromere; the 2nd pair 6.25 and 4.0μ resp. and in the short arms an a.con. at 1.3μ from the centromere; the 3rd pair 5.75 and 3.0μ resp.

Group II, three pairs: The 1st pair with the long arms 7.5μ, the short arms 3.25μ; the 2nd pair 7.25 and 3.0μ; the 3rd pair 7.5 and 2.7μ resp.

Group III, four pairs: The 1st pair with the long arms 9.25μ, the short arms 3.0; the 2nd pair 7.5 and 2.5μ; the 3rd pair 6.75 and 2,25μ; the 4th pair 6.75 and 1.75μ resp.

Group IV: Absent.

Group V, one pair: The long arms 11.25μ, the short arms 2.25μ; in the long arms an a.con. in the middle.

Group VI, one pair: The long arms 8.5μ, the short arms 1.25μ.

The total length of the haploid complement is 125.7μ; of the long arms 92.75μ and of the short arms 32.95μ. Total index: 2.81. Index chrom. A1: 1.7. Index chrom. A2: 1.56. Index chrom. A3: 1.92.

Tulipa buhseana Boiss. The plants were brought to us in 1935 from shunting-station 146 on the Tashkent railway. The habitat is a plain with clear signs of salinization and in some places transitions to marshy solonchaks with tamarisks.

We examined chromosomes in root-tip sections from adult bulbs and in pollen tubes. This form is diploid, 2n = 24. The chromosomes fall into six main groups (fig.12k, 24d).

Group I, three pairs: The 1st pair with the long arms 9.25, the short arms 6.0μ; the 2nd pair 8.87 and 5.25μ; the 3rd pair 6.75 and 4.0μ resp.

Group II, three pairs: The 1st pair with the long arms 11.65μ, the short arms 4.5μ; the 2nd pair 10.0 and 4.25μ; the 3rd pair 9.25 and 3.87μ resp.

Group III, one pair: The long arms 12.5μ, the short arms 3.5μ.

Group IV, three pairs: The 1st pair with the long arms 14.5μ, the short arms 3.55μ; the 2nd pair 12.5 and 2.75μ; the 3rd pair 9.75 and 2.25μ resp.

Group V – VII: Absent.

Group VIII, one pair: The long arms 12.87μ, the short arms 1.4μ.

Group IX – XII: Absent.

Group XIII, one pair: The long arms 15.12μ, the short arms 1.12μ.

The total length of the haploid complement is 175.33μ; of the long arms 133.01μ and of the short arms 42.32μ. Total index: 3.14. Index chrom. A1: 1.54. Index chrom. A2: 1.69. Index chrom. A3: 1.69.

Tulipa buhseana Boiss. (*T.talievii* Vved.). The individuals studied were collected in stony soil near Dzhambul, Mount Tekturmas (coll. A. Ponomareva, 1934).

We examined the chromosomes in root-tip sections from adult bulbs and in young pollen grains during division of the primary nucleus. The diploid chromosome number is 24. The chromosomes fall into four main groups (fig.24f, 25a).

Group I, two pairs: The 1st pair with the long arms 6.25μ, the short arms 4.5μ; the 2nd pair 7.0 and 3.75μ resp; on the short arms of the 2nd pair a small, spherical satellite.

Group II, four pairs: The 1st pair with the long arms 8.0μ, the short arms 3.0μ; the 2nd pair 7.0 and 3.25μ; the 3rd pair 6.25 and 2.25μ; the 4th pair 6.0 and 2.25μ resp.

Group III, two pairs: The 1st pair with the long arms 8.75μ, the short arms 2.5μ; the 2nd pair 7.0 and 2.25μ resp.

Group IV, four pairs: The 1st pair with the long arms 9.0μ; the short arms 2.0μ; the 2nd pair 8.0 and 2.0μ; the 3rd pair 7.0 and 1.75μ; the 4th pair 5.5 and 1.25μ resp.

The total length of the haploid complement is 117.5μ; of the long arms 85.75μ and of the short arms 31.75μ. Total index: 2.70. Index chrom. A1: 1.39. Index chrom. A2: 1.87.

Tulipa buhseana Boiss. The individuals studied were collected in the environs of Kushka (coll. A. Ponomareva, 1935).

We examined chromosomes in root-tip sections taken from sprouting adult bulbs. The diploid number is 24. The chromosomes can be divided into four main groups (fig.12e, 23f).

Group I: Absent.

Group II, five pairs: The 1st pair with the long arms 8.5, the short arms 3.75μ and an akinetic constriction at 1.0μ from the centromere (this pair of chromosomes is typical of group I, but falls in this case in group II, which is probably due to an increase in the length of the longer arms since the akinetic constrictions in the shorter arms divide these into segments which have retained the characteristic ratio for this pair); the 2nd pair with the long arms 8.25μ, the short arms 3.75μ; the 3rd pair 8.0 and 3.45μ; the 4th pair 7.0 and 2.75μ; the 5th pair 6.25 and 2.75μ resp.

Group III, three pairs: The 1st pair with the long arms 10.0μ, the short arms 2.75μ; the 2nd pair 8.0 and 2.25μ; the 3rd pair 7.0 and 2.25μ resp.; in the short arms of the 3rd pair an a.con. at 0.75μ from the centromere.

Group IV, three pairs: The 1st pair with the long arms 10.0μ, the short arms 2.5μ; the 2nd pair 9.35 and 2.25μ; the 3rd pair 7.4 and 1.75μ resp.

Group V, one pair: The long arms 6.4μ, the short arms 1.25μ.

The total length of the haploid complement is 126.86μ; of the long arms 95.65μ and of the short arms 31.21μ. Total index: 3.06.

Tulipa binutans Vved. The individuals studied were collected in Kirgiziya, about 7 km above the confluence of the rivers Alamedin and Chonkurchak at the foot of the lower ridge of rocks.

We examined chromosomes in root-tip sections taken from adult bulbs and in young anthers during division of the primary nucleus (fig.12d). The diploid number of chromosomes is 24. They can be divided into five main groups (fig.12c, 12d, 23e).

Group I, one pair: The long arms 5.75μ, the short arms 4.0μ.

Group II, four pairs: The 1st pair with the long arms 6.75μ, the short arms 2.75μ and an akinetic constriction at 0.75μ form the centromere; the 2nd pair 6.25 and 2.5μ; the 3rd pair 5.5 and 2.25μ; the 4th pair 5.5 and 2.0μ resp.

Group III, three pairs: The 1st pair with the long arms 5.4μ, the short arms 1.75μ; the 2nd pair 5.25 and 1.75μ; the 3rd pair 5.25 and 1.5μ resp.

Group IV, three pairs: The 1st pair with the long arms 8.0μ, the short arms 2.0μ; the 2nd pair 8.0 and 1.75μ; the 3rd pair 6.0 and 1.5μ resp.; the short arms of the 2nd and 3rd pairs have at the distal ends small spherical satellites attached by thin threads.

Group V: Absent.

Group VI, one pair: The long arms 6.0μ, the short arms 1.0μ.

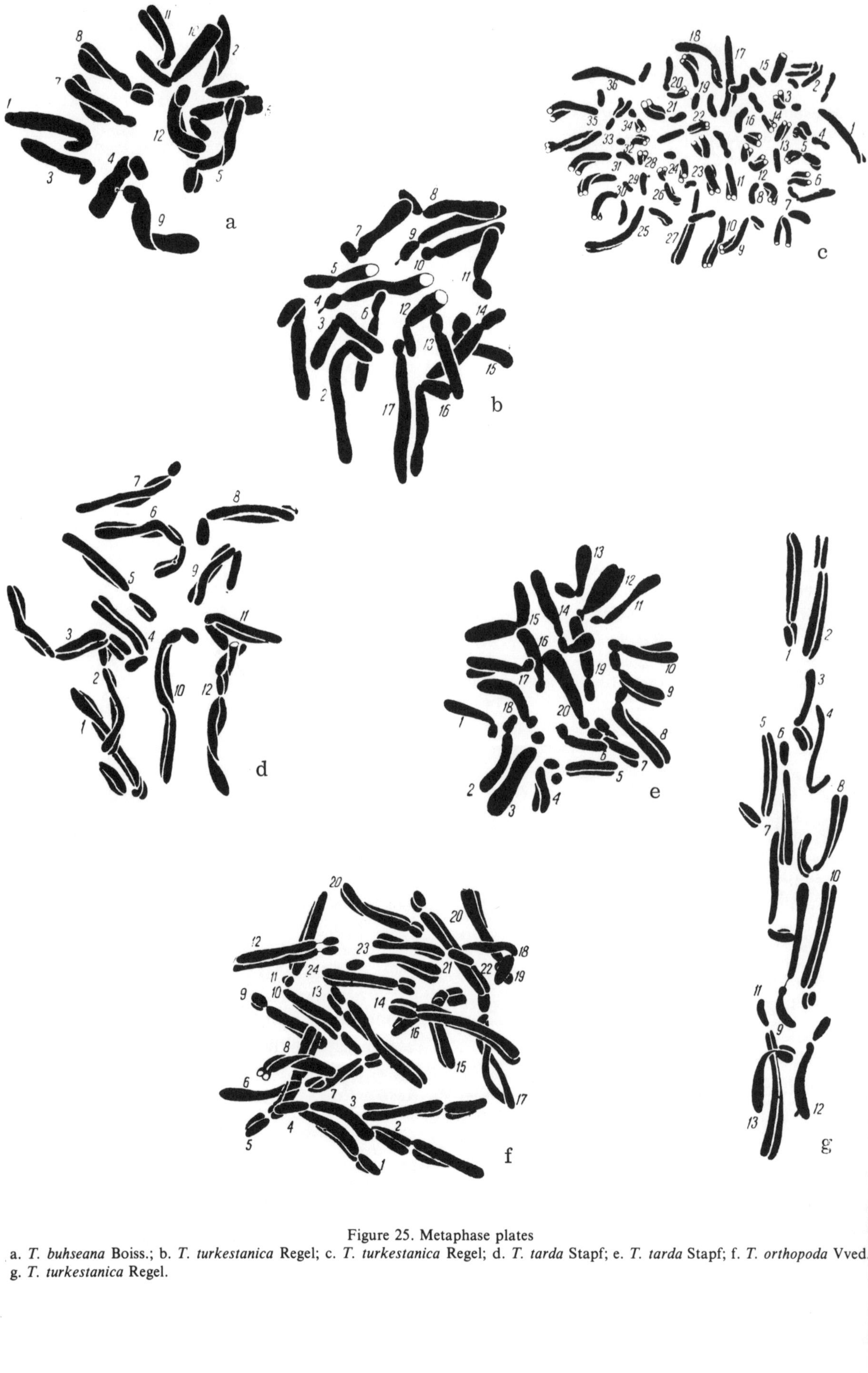

Figure 25. Metaphase plates

a. *T. buhseana* Boiss.; b. *T. turkestanica* Regel; c. *T. turkestanica* Regel; d. *T. tarda* Stapf; e. *T. tarda* Stapf; f. *T. orthopoda* Vved. g. *T. turkestanica* Regel.

The total length of the haploid complement is 98.4μ; of the long arms 73.65μ and of the short arms 24.75μ. Total index: 2.99. Index chrom. A1: 1.44.

Tulipa binutans Vved. The individuals studied were collected in the Chulak-Kaindy gorge of the Kirgizskiy Alatau, northern Kirgiziya (coll. Krasovskiĭ, 1934).

We examined chromosomes in pollen tubes during spermatogenesis. The chromosome number was 2n = 24. They can be divided into seven main groups (fig.12i, 24e).

Group I, two pairs: The 1st pair with the long arms 8.75, the short arms 4.5μ; the 2nd pair 7.5 and 4.12μ resp.

Group II, two pairs: The 1st pair with the long arms 6.25μ, the short arms 2.25μ; the 2nd pair 5.75 and 2.7μ resp.

Group III, one pair: The long arms 7.12μ, the short arms 2.12μ.

Group IV, four pairs: The 1st pair with the long arms 10.75μ, the short arms 2.6μ; the 2nd pair 9.0 and 2.2μ; the 3rd pair 8.75 and 2.0μ; the 4th pair 6.0 and 1.5μ resp.

Group V, one pair: The long arms 11.5μ, the short arms 2.25μ.

Group VI, one pair: The long arms 11.07μ, the short arms 1.8μ.

Group VII, one pair: The long arms 8.0μ, the short arms 1.13μ.

The total length of the haploid complement is 130.84μ; of the long arms 100.19μ and of the short arms 30.65μ. Total index: 3.27. Index chrom. A1: 1.94. Index chrom. A2: 1.82.

Tulipa turkestanica Regel. The individuals studied originated from the Nuratau mountains, the Koytash range, where they were growing in a meadow in the upper gully close to the village of Karakon (coll. V.P. Bochantsev, 1934).

We examined chromosomes in root-tip sections and in pollen tubes. The diploid number is 24. The chromosomes fall into four main groups (fig.12j, 24c).

Group I, three pairs: The 1st pair with the long arms 5.0μ, the short arms 3.75μ; the 2nd pair 4.75 and 3.0μ; the 3rd pair 3.75 and 2.0μ resp.

Group II, four pairs: The 1st pair with the long arms 6.75μ, the short arms 2.25μ; the 2nd pair 5.75 and 2.1μ; the 3rd pair 5.0 and 2.0μ; the 4th pair 4.5 and 1.9μ resp.

Group III, three pairs: The 1st pair with the long arms 7.0μ, the short arms 2.0μ; the 2nd pair 5.75 and 1.75μ; the 3rd pair 5.25 and 1.5μ resp.

Group IV, two pairs: The 1st pair with the long arms 6.6μ, the short arms 1.5μ; the 2nd pair 5.5 and 1.25μ resp.

The total length of the haploid complement is 89.85μ; of the long arms 65.6μ and of the short arms 24.25μ. Total index: 2.71. Index chrom. A1: 1.33. Index chrom. A2: 1.58. Index chrom. A3: 1.87.

Tulipa turkestanica Regel. The individuals studied were brought from Mount Kungus-Buchak, the northern slope towards Tashbuch spring (coll. A.F. Gnezdilo, 1935).

We examined chromosomes in pollen tubes during spermatogenesis in vitro and found 2n = 24, n = 12. The chromosomes can be divided into six main groups (fig.12f, 23g).

Group I, two pairs: The long arms of the 1st pair 12.0, the short arms 8.0μ; the 2nd pair 8.0 and 4.87μ resp.; the short arms of the 2nd pair carry a spherical satellite attached by a thin thread.

Group II, one pair: The long arms 9.2μ, the short arms 3.62μ.

Group III, five pairs: The 1st pair with the long arms 11.12μ, the short arms 2.87μ; the 2nd pair 10.5 and 3.0μ; the 3rd pair 9.5 and 3.0μ; the 4th pair 9.0 and 3.0μ; the 5th pair 9.0 and 2.7μ resp.

Group IV: Absent.

Group V, two pairs: The 1st pair with the long arms 16.5μ, the short arms 3.0μ; the 2nd pair 15.37 and 3.0μ resp.

Group VI, one pair: The long arms 18.0μ, the short arms 3.0μ.

Group VII, one pair: The long arms 11.75μ, the short arms 1.65μ.

The total length of the haploid complement is 181.55μ; of the long arms 139.74μ and of the short arms 41.81μ. Total index: 3.34. Index chrom. A1: 1.5. Index chrom. A2: 1.64.

Tulipa turkestanica Regel. The individuals studied came from the Kugitang range and were growing near the Maidan winter camp in an *Acantophyllum* thicket (coll. A.F. Gnezdilo, 1932).

We examined chromosomes in root-tip sections from adult bulbs, in young pollen during division of the primary nucleus and in pollen tubes during spermatogenesis. The chromosome number for somatic cells was 2n = 24, but some forms were triploid, 2n = 36. A study of the haploid complements showed n = 12, n = 17 and n = 18. The first and last numbers were more frequent; n = 17 was recorded only once. The chromosomes of both the diploid and the triploid forms, which were found to be autopolyploids, can be divided into five main groups (fig.24g, 25b,c).

Group I, two pairs: The 1st pair with the long arms 6.0μ, the short arms 4.75μ and an akinetic constriction at 1.5μ from the centromere including a small spherical satellite attached by a thin thread at the distal end; the 2nd pair with the long arms 6.25μ, the short arms 3.25μ; on the long arms of this pair an a.con at 1.75μ from the distal end and on the short arms an a.con. at 0.75μ from the centromere.

Group II, five pairs: The 1st pair with the long arms 7.5μ, the short arms 2.9μ; the 2nd pair 6.75 and 2.75μ; the 3rd pair 7.0 and 2.5μ; the 4th pair 6.5 and 2.5μ; the 5th pair 6.25 and 2.25μ resp.; in the middle of the short arms of the 5th pair an a.con.

Group III, two pairs: The 1st pair with the long arms 8.75, the short arms 2.5μ with in the middle an a.con.; the 2nd pair with the long arms 6.95μ, the short arms 2.25μ with in the middle also an a.con.

Group IV: Absent.

Group V, one pair: The long arms 8.75μ with an a.con. in the middle, the short arms 1.75μ with a small satellite at the distal end.

Group VI, two pairs: The 1st pair with the long arms 7.25μ, the short arms 1.25μ; the 2nd pair 7.0 and 1.0μ resp. and on the short arms a satellite.

The total length of the haploid complement is 123.3μ; of the long arms 84.25μ and of the short arms 39.05μ. Total index: 2.16. Index chrom. A1: 1.26. Index chrom. A2: 1.92.

Tulipa turkestanica Regel. The individuals studied originated from the Mal'guzar mountains. They were growing between stones on a mountain top in the eastern part of the range (coll. P.A. Gomolitskiĭ, 1934).

We examined chromosomes in root-tip sections from adult bulbs and in pollen tubes during spermatogenesis in vitro. The somatic cells have 2n = 24 and the generative cells n = 12. The chromosomes can be divided into five main groups (fig.25g, 26a).

Group I, two pairs: The 1st pair with the long arms 7.5μ, the short arms 4.25μ and an akinetic constriction at 0.75μ from the centromere; the 2nd pair 7.5 and 4.12μ resp.

Group II, four pairs: The 1st pair with the long arms 8.25μ, the short arms 2.87μ; the 2nd pair 7.87 and 2.7μ; the 3rd pair 7.0 and 2.5μ; the 4th pair 6.6 and 2.3μ.

Group III, three pairs: The 1st pair with the long arms 8.65μ, the short arms 2.52μ; the 2nd pair 8.48 and 2.37μ; the 3rd pair 8.4 and 2.23μ resp.

Group IV, two pairs: The 1st pair with the long arms 11.32μ, the short arms 2.27μ; the 2nd pair 8.7 and 2.0μ resp.

Groups V and VI: Absent.

Group VII, one pair: The long arms 9.75μ, the short arms 1.27μ.

The total length of the haploid complement is 131.22μ; of the long arms 100.02μ and of the short arms 31.2μ. Total index: 3.20. Index chrom. A1: 1.76. Index chrom. A2: 1.82.

Tulipa tarda Stapf. The individuals studied were brought from the Kastek river valley, 10 km from Mikhalovskoye, where they were growing in the scrub zone with *Prunus prostrata* between numerous outcrops of bedrock; also from the Malyy Kemin river area, south of the Dzhol-Bulak Pass, growing on the southern slope of a ridge formed by large stones covered with a thin layer of very porous soil (coll. Z.P. Botschantzeva, 1932). The plants show a strong tendency to vegetative reproduction.

We examined the chromosomes in root-tip sections from adult bulbs and in young pollen at the stage of division of the primary nucleus. We found both diploid and tetraploid forms growing together with $2n = 2x = 24$ and $2n = 4x = 48$. The chromosomes fall into six main groups (fig.25d, 25e, 26b).

Group I, two pairs: The 1st pair with the long arms 9.25, the short arms 6.75μ and the characteristic akinetic constriction at 2.25μ from the centromere; the 2nd pair 6.5 and 3.75μ resp.

Group II, five pairs: The 1st pair with the long arms 8.75μ, the short arms 4.13μ; the 2nd pair 8.5 and 3.95μ; the 3rd pair 7.79 and 3.13μ; the 4th pair 7.5 and 3.0μ; the 5th pair 6.75 and 2.4μ resp.

Group III, two pairs: The 1st pair with the long arms 12.0μ, the short arms 3.75μ; the 2nd pair 8.25 and 2.4μ resp.

Group IV: Absent.

Group V, one pair: The long arms 12.5μ and an a.con. at 5.75μ from the centromere, the short arms 2.5μ.

Group VI, one pair: The long arms 9.13μ, the short arms 1.5μ.

Group VII – VIII: Absent.

Group IX, one pair: The long arms 15.25μ with an a.con. at 1.5μ from the centromere, the short arms 1.6μ.

The total length of the haploid complement is 150.70μ; of the long arms 112.65μ and of the short arms 38.05μ. Total index: 2.96. Index chrom. A1: 1.37. Index chrom. A2: 1.73.

Tulipa orthopoda Vved. The individuals studied were brought from the Karatau range; they were growing in gullies, the Dzhinagan-Atal, and others (coll. P.A. Gomolitskiĭ, 1930). The northern form of *T.buhseana* also grows in the Karatau.

We examined the chromosomes of *T.orthopoda* in root-tip sections and in young pollen during spermatogenesis and found the chromosome numbers $2n = 2x = 24$ and $2n = 4x = 48$. The chromosomes fall into six main groups (fig.25f, 26c).

Group I, two pairs: The 1st pair with the long arms 7.0μ, the short arms 3.88μ; the 2nd pair 6.63 and 3.35μ resp.

Group II, two pairs: The 1st pair with the long arms 7.25μ, the short arms 2.44μ; the 2nd pair 6.75 and 2.63μ resp.

Group III, four pairs: The 1st pair with the long arms 10.0μ, the short arms 2.69μ; the 2nd pair 6.75 and 1.75μ; the 3rd pair 6.55 and 1.63μ; the 4th pair 5.75 and 1.75μ resp.

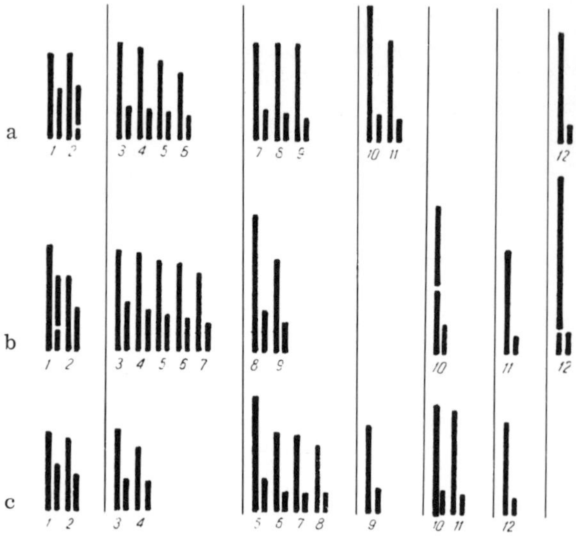

Figure 26. Idiograms of measured plates (enlarged 600 times) a. *T. turkestanica* Regel; b. *T. tarda* Stapf; c. *T. orthopoda* Vved.

Group IV, one pair: The long arms 7.75μ, the short arms 1.9μ.

Group V, two pairs: The 1st pair with the long arms 9.7μ, the short arms 1.94μ; the 2nd pair 9.25 and 1.63μ resp.

Group VI, one pair: The long arms 7.94μ, the short arms 1.14μ.

The total length of the haploid complement is 116.57μ; of the long arms 90.86μ and of the short arms 25.71μ. Total index: 3.53. Index chrom. A1: 1.80. Index chrom. A2: 1.98.

Although a large number of *Eriostemones* have been the subject of cytological studies it remains difficult to draw any definite conclusions about evolutionary trends in their karyotypes and about the phylogeny of this group. We studied 14 of the 18 taxa mentioned in "Flora of the USSR":

T.biebersteiniana (2n = 24), *T.patens* (2n = 24), *T.biflora* (2n = 24), *T. turcomanica* (2n = 24) and *T.orthopoda* (2n = 24, 36, 48) each from one location. *T.sogdiana* (2n = 24), *T.tarda* (2n = 24, 36, 48) and *T.binutans* (2n = 24), each from two locations. *T.turkestanica* (2n = 24, 36) from three locations. *T.dasystemon* (2n = 24, 36, 48) from four locations. *T.buhseana* (2n = 24) and *T.bifloriformis* (2n = 24, 36, 48) from five locations.

The basic chromosome number is $x = 12$ for all the locations, yet each species and polyploid shows a new karyotype, which is difficult to relate to either other populations belonging to the same taxon but occurring in another area, or to other taxa from adjacent areas. For instance *T.buhseana* from Chardara, *T.bifloriformis* from Tashkent and *T.turkestanica* from the Koytash range all have in group I three pairs of chromosomes with the arms ratio < 2 (fig.24a, b, c), whereas in other locations these species have only one chromosome or none at all in group I, e.g. *T.buhseana* (fig.23f) and *T.bifloriformis* from Namangan (fig.23c).

There was found a comparatively high degree of secondary segmentation in *Eriostemones* with northern areas of distribution, like *T.biebersteiniana*, *T.patens* and *T.biflora*. The karyotype of *T.sogdiana* from the environs of Ashkhabad shows a clear tendency towards a further shortening of the short arms. There are also karyotypic races of this species and one of them from Serak has a distinct secondary segmentation of the chromosomes. *T.dasystemon* from the Tien Shan (fig.21f) differs from the Pamir-Alay form (fig.21e) by the longer chromosomes and a stronger tendency towards secondary constrictions.

To obtain a more or less correct image of the phylogeny within the *Eriostemones*, it will be necessary to investigate a large number of species from the areas which were not dealt with in this study. No doubt the evolution of the karyotypes in sections *Lophophyllon* and *Eriostemones* took a different course from that in the sections *Tulipanum*, *Leiostemones* and *Spirantha*. Their roads must have parted in the very early stages of phylogenesis.

CHAPTER SIX

Polyploidy in tulips

Without karyological data it is difficult to decide whether a given species, a section or a genus are primitive or advanced. The basic haploid chromosome number in the genus *Tulipa* is 12. This number is found in most representatives of the Liliaceae family and is apparently the primary basic number and an ancient family character. In the group *Tulipeae* of this family it is the basic haploid number for the genera *Lilium, Fritillaria* and *Gagea* (Darlington, 1956). According to Tischler, this number was found by Saki in 1935 for one species of the genus *Lloydia,* viz. *Lloydia serotina.* These data are important because Buxbaum (1937) holds the view that the genus *Lloydia* occupies a key position among the genera *Erythronium, Tulipa* s.1., *Fritillaria* s.1. and *Lilium* s.1.

Referring to Oikawa (1937), Tischler (1938) gives 12 as the haploid chromose number for *Cardiocrinum cordatum* and *C.glehnii.* The same number was found by Darlington (1936) for *Nomocharis saluensis.*

A deviation towards multiples of the basic number is a secondary development. The resulting polyploidy or aneuploidy are an indication that the given taxon is comparatively more recent than the diploid. Also in the case of a reduction in the basic number the new species is derived.

In the genus *Tulipa* two tendencies are in evidence: an increase and a reduction in the basic number. The latter was found in *T.maximowiczii* from the Pamir-Alay, growing on a red sandstone near the village of Sangtud. The diploid complement is 2n = 22. In the reduction division 11 bivalents are observed, confirming n = 11. The literature mentions other chromosome numbers for *T.maximowiczii,* but the results of repeated investigations on plants, raised from bulbs collected in the natural habitat, contradict the data first published by Newton, which were based on plants from botanical gardens and afterwards cited by other botanists. Romanov (1944), who studied the embryo sac of this species, confirmed our results, i.e. 2n = 22 chromosomes. Besides, it can be seen from the description of the karyotype of *T.maximowiczii* in the preceding chapter, that the smaller arms of the chromosomes are very short. In this case the reduction in the number of chromosomes is accompanied by a reduction in size of the shorter arms.

We also found a reduction of the basic chromosome number in pollen tubes of *T.lanata,* when studying its spermatogenesis (Botschantzeva, 1937). The bulbs had been collected in the Pamir-Alay, near the village of Sina. In some pollen tubes eleven chromosomes were found in the metaphases of the generative nucleus during the division into two sperm nuclei. Other pollen tubes showed metaphases with 12 chromosomes, but we observed none with 13 chromosomes. Apparently in *T.lanata* pollen grains with 13 chromosomes do not survive. The plants which we examined set seed with difficulty but they easily multiplied vegetatively. Both examples indicate that *Tulipa* species from the Pamir-Alay tend to reduce their basic chromosome number. A reduction in basic number is an exception in the genus *Fritillaria.* In *F.ruthenica* 2n = 18, but in *F.pudica* 2n = 26; in the latter case an increase of the basic number. According to Cooper *Erythronium albidum* has 2n = 11.

In view of the general, although weakly pronounced, tendency of some representatives of the tribus *Tulipeae* towards a reduction in the basic chromosome number, we feel that it is not justified to exclude the genus *Calochortus* from the *Tulipeae.* The haploid numbers counted in this genus are 6, 7, 8, 9, 10 and multiples thereof. It is possible that, contrary to the views of Newton (1937) and Beal (1939), the basic and primary chromosome number of *Calochortus* may have been x = 12 and underwent reductions later on. These changes are of course reflected in the external morphology of the plants. In *Calochortus* species chromosomes with a nearly terminal attachment of the spindle threads are more frequent than in other representatives of the tribus. In *T.maximowiczii* a reduction in the basic number led to morphological changes in the karyotype or vice versa, morphological changes in the chromosomes resulted in a reduced basic number. Besides *T.maximowiczii* several other tulips from the Pamir-Alay have chromosomes with a nearly terminal centromere: *T.linifolia, T.wilsoniana* and others.

The genus *Calochortus* may be regarded as further

advanced than any other representative of the tribus *Tulipeae* and therefore spread to new areas and evolved new morphological characters. This is even more likely because a similar development, although not affecting the basic chromosome number, is observed in the east of the distributional areas of the genus *Tulipa*. In China and Japan occur *T.edulis* and *T.graminifolia* with linear leaves more or less in pairs, drooping flowers and rounded seeds. The first species is a polyploid. Both are morphologically so much different from other tulips that Hall (1937) dealt with them in a separate genus *Amana*. In the Tien Shan, also in the east, but closer to the main centres of diversity of the genus, grows *T.heterophylla* from section *Orithyia*. Because of its angular and sharp-edged seeds Popov (1936) placed it in a separate genus, *Eduardoregelia*, although its karyotype differs but little from that of other tulips and it is not a polyploid.

The shape of the seeds and the number and morphology of the chromosomes of *T. uniflora* and *T.heteropetala* need studying, as it may be possible that the whole section *Orithyia* has to be included in a separate genus, although in karyotype *Eduardoregelia* does not differ much from other *Tulipa* species.

In the genus *Gagea*, which Buxbaum regarded as being closest to the archetype of the tribus *Tulipeae*, both rounded and flat seeds occur. The shape of the seeds is a character used in subdividing this genus, e.g.: Subgenus *Eugagea* with globose seeds; subgenus *Platyspermum* with flat seeds; section *Foliatae* with cylindric seeds; section *Euchloranthae* with flat seeds.

An increase in the basic chromosome number of Tulipa was only observed in rare and not very convincing cases. We found this in one tulip received from Tehran, a tetraploid with four extra chromosomes (T.sp., 4n = 52). Newton counted extra chromosomes in somatic cells of *T.galatica* with 2n = 32, which makes four pairs of chromosomes extra.

In pollen tubes of *T.lanata* we found gametes with 11 and 12 chromosomes, but none with 13. However, in some years the gametes may show more vitality and then certain forms of this species may have 26 chromosomes in the diploid complement. Thus we find both reduced and increased basic numbers in the genus, which could be expected since an increase in number was also found in *Fritillaria pudica*.

Of particular interest are the chromosome numbers in the genus *Gagea*. They are known for the three species: *G.spathacea*, n = 51; *G.lutea*, n = 36; *G.minima*, n = 12. If 12 is taken as the basic number in this genus, then the polyploidy of the first two species is too high and raises serious doubts about the key position accorded by Buxbaum (1937) to the genus *Gagea*, because such high polyploid numbers are not typical of the subfamily and do not agree with the general trends of its evolution. Moreover polyploidy is a secondary development and it is therefore hardly justified to place the genus *Gagea* close to the prototype of the *Tulipeae*.

Polyploid forms of tulips have long been known from the literature. De Mol (1925, 1927) first discovered polyploidy of garden tulips in which he found 36 chromosomes. He explained this by the failure of the reduction division of the chromosomes. In his first cytological study of the genus Newton (1927) found two tetraploids: *T.sylvestris* and *T.whittallii*. The former is regarded as derived from *T.austrális*. However, *T.australis* comes from habitats in the Apennines, Southern France, Spain and Portugal, while *T.sylvestris* is spread all over the Mediterranean area as a weed in vineyards. It occurs also in England and in Iran near Täbriz. *T.sylvestris* is believed to be the result of autopolyploidy as is also *T.whittallii*. Hall (1940) surmises that the latter is a descendant of the highly variable Greek species *T.orphanidea*.[1] Newton (1927) found another tetraploid, *T.turkestanica*, but gave no precise information about its origin.

Tulipa saxatilis, a triploid, has been in cultivation for a long time and *T.cretica* is a tetraploid from Crete (Hall, 1940).[2]

An interesting species is *T.clusiana* with a large area of distribution and a wide range of polyploidy. Newton (1927) determined the pentaploid character, 2n = 60, of the Mediterranean form of this species and found a tetraploid form, 2n = 48, from Chitral. Among the collection of a horticultural institute[3] in England a diploid form (2n = 24) was found, but all information about the provenance of this individual was lost. In *T.chrysantha*,[4] which differs from *T.stellata* only by the yellow-coloured tepals, 48 chromosomes were counted. Morphologically close to this tetraploid is a small yellow tulip[5] from Kashmir which proved to be diploid.

A tetraploid form of *T.korolkowii*, viz. *T.korolkowii concolor*, is well known abroad. Among the great number of plants, which we collected in the natural habitats of this species, we did not discover any tetraploids. This gives us reason to believe that, like many other polyploid forms of Middle Asian species cited in the literature, the polyploid *T.korolkowii* evolved in cultivation.

1. Hall (1940) mentions under *T. whittallii* (p. 64): 'It can only be surmised to have originated from *T. hageri*, which has been reported to grow in the Smyrna region' and further on p. 62: 'Cytological evidence suggests *T. hageri* as the older form, *T. orphanidea* as possibly originating from it by hybridisation with *T. australis*'. (editor).

2. According to Hall (p. 71) *T. cretica* is a diploid species. (editor).

3. The John Innes Institute at Norwich. (editor).

4. The *T. chrysantha* mentioned here is *T. clusiana* DC. var. *chrysantha* (A.D. Hall) Sealy. See Sealy, J.R. in: Curtis's Bot. Mag. n. ser. 165, t 13, (1948). (editor).

5. A yellow form of *T. aitchisonii* A.D. Hall. (editor).

The tetraploid, scarlet, *T.kolpakowskiana* which comes from the Soviet Union is undoubtedly a natural hybrid of *T.kolpakowskiana* and *T.ostrowskiana*. It was found by an expedition of our botanical garden in the Zailiyskiy Alatau, environs of Talgar, and later exported.

Besides the polyploids cited above, Hall mentions *T.praecox* as a triploid along with two more closely related polyploids from Asia Minor and Palestine that formed part of the collection of the mentioned horticultural institute.

T. oculus solis which, as Hall points out, is known in the trade as *T.borszczowii*, has several fragments in its complement. Apparently this *'T.borszczowii'* must be closely related to *T.oculus-solis*, because the Middle Asian *T.borszczowii* we have studied is a diploid that shows the usual reduction division without any fragments. Hall is of the opinion that polyploids possess the stoloniferous habit which facilitates vegetative propagation, because their sexual reproduction is very much suppressed or nonexistent.

There have appeared numerous studies on the adaptation of tetraploids to certain ecological conditions. The prevailing view in literature is that, because of the changed physiological homeostasis, polyploids often have other habitats than the diploids. The polyploids can grow in different, but not necessarily worse, conditions than the diploids. The changes resulting from polyploidy permit these plants to grow under conditions entirely different from those under which the initial species and forms grow and flourish. This is shown by *T.australis – T.sylvestris, T.whittalii – T.orphaninidea* and the forms of *T.clusiana*, which in their different areas have different numbers of chromosomes. It can therefore be concluded that the Mediterranean area s.str. may be seen as a secondary centre of the genus, considering the wide-spread multiples of the basic chromosome complement, occurring more often along its present shores than in any other region. There are also many polyploids among the cultivated tulips which found a second home in Western Europe.

Of the Middle Asian tulips studied the largest number of polyploids was found among the *Eriostemones*. In this section polyploids occur in *T.bifloriformis* from the Tashkent area and from the Ugam river valley in the western Tien Shan. In these areas both diploid (2n = 24) and tetraploid (2n = 48) forms are found. An analysis of their karyotypes showed the latter to be autotetraploids.

In *T.turkestanica* from Kugitang we found diploid (2n = 24) and triploid (2n = 36) forms. As a rule the gametes produced by triploids have 18 chromosomes, but in some cases the number was 17 or 19. The chromosomes of this species were counted in cells from root-tips and in pollen tubes during spermatogenesis.

Two *Eriostemones, T.buhseana* (2n = 24) and the tetraploid *T.orthopoda* (2n = 48), grow in the Karatau mountains. *T.tarda* from the Kastek pass has both diploid and tetraploid forms. The tetraploids produce gametes with 14, 20 and 24 chromosomes. *T.dasystemon*, usually its diploid form, is widespread in the high mountain areas of Middle Asia, but forms from the Kugart pass, the Terskey Alatau and the Zailiyskiy Alatau appear to be tetraploids. Of this species diploids, tetraploids as well as triploids grow in the Terskey Alatau and they are all autopolyploids.

We see that with the exception of *T.bifloriformis* of the Tashkent area, the polyploid *Eriostemones* occur in the Pamir-Alay and the Tien Shan and that there are no polyploids among the lowland tulips of this section. The polyploids of *T.dasystemon* and *T.tarda* are more or less high mountain forms, while *T.turkestanica* and *T.orthopoda* belong to the middle mountain belt.

Among the members from the other sections we have studied is a rather interesting *Leiostemones*, viz. *T.ostrowskiana*. It has 48 chromosomes and appears to be amphidiploid. The idiogram combines one complete set of chromosomes of *T.kolpakowskiana* and a set of some other species, unknown to us (fig.17c). No diploid forms have been found among plants of *T.ostrowskiana* from the Zailiyskiy Alatau. However, in the same mountain range both diploid and tetraploid forms of *T.tetraphylla* occur. A more detailed study of the karyotype of this species showed that the form with 48 chromosomes is an autotetraploid.

Of the seven species in which we found polyploidy, six grow in the Tien Shan and one in the Pamir-Alay mountain system. The polyploid species from the Tien Shan are distributed as follows: one species in the Karatau range, one in the Western Tien Shan and four in the Zailiyskiy Alatau. In the last area one species, *T.tarda*, inhabits the mountain slopes which face the valley of the river Chu. The other three were found in the environs of Alma Ata and Talgar.

Of particular interest is the altitudinal distribution of the species over the desert zone, the foothills, the mountain scrub belt, the spruce belt and the alpine meadows. In the deserts reaching to the Chu-Iliyskiye mountains grow *T.greigii, T.kolpakowskiana, T.albertii, T.regelii, T.buhseana, T.talievii*, all of them diploid with 24 chromosomes. On the sands, where the river Talgar discharges into the Chu, occurs *T.behmiana* (2n = 24). In the upper reaches of the river Talgar we find *T.kolpakowskiana*, 2n = 24, growing along its banks in a wide valley between low hills covered by grassland communities. *T.kolpakowskiana* climbs a little higher up these hills to where it meets with *T.ostrowskiana*, which has 48 chromosomes of which one complement is identical with that of *T.kolpakowskiana*. These two species produce numerous natural hybrids with 24, 36 and 48 chromosomes. The authentic *T.ostrowskiana*

(2n = 48) grows higher in the mountain scrub belt e.g. in the Akbulak gorge. Going still higher up the gorge through the mixed forest belt, right up to the upper boundary of the spruce belt, one finds *T.tetraphylla* with 2n = 24 and 2n = 48. The upper spruce belt and the dry alpine meadows are the habitats of *T.dasystemon* also with 24 and 48 chromosomes. Still higher grows *Eduardoregelia* with 24 chromosomes.

It appears that there are no polyploids in the deserts and the lower mountain belt. They are all concentrated in the middle and upper mountain belts of the Zailiyskiy Alatau, which form the typical altitudinal and eastern boundaries of the habitats of the polychrome representatives of the genus. Thus the largest numbers of polyploids are to be found along the western and close to the eastern boundaries of the area of distribution of the genus, whereas hardly any occur in its centre, with the exception of *T.turkestanica* from Kugitang. However, this exception merely confirms the rule since the *Eriostemones* are in the structure of the embryo-sac, in polyploidy and in their external morphological characters basically different from the tulips of the sections *Tulipanum* and *Leiostemones.*

CHAPTER SEVEN

The seed and its germination

The Seed

The flat tulip seeds which develop from the anatropous ovules are arranged in six piles in a trilocular capsule; two piles in each locule (fig.27b). They are of different shades of brown and have a rounded triangular shape resembling a distorted section of a circle between two unequal radii and a segment of its circumference, equalling one sixth of its length (fig.27a). The vascular bundle, entering the ovule from the placenta through the funicle, runs along its narrow edge, turns inwards into the ovarian locule and ends in the chalaza next to the nucellus. With the ripening of the seed this bundle is obliterated, but the tissues of the ovule lying partly inside and partly outside the place where the vascular bundle ends (in the ovarian locule) have stored chloroplasts and remain alive for a long time after. In dry seeds of many species and varieties they shine through as a dark green, nearly black, crescent-shaped blotch. The brown colour of the seeds is caused by the walls of the seed coats. A dried membrane, up to 1 mm wide and slightly turned downward, forms a thin, brown, transparent margin around the seed (fig.27b). It aids seed dispersal from the capsule. The epidermis cells of the seed coat above the endosperm are drawn out across the seed while those along the margin are drawn out parallel to the edge of seed.

On ripening the walls of the epidermal cell layer of the integuments, which form the seed coat, undergo hardly any changes. Hence the seed coat is very thin and in mature seeds consists of two cell layers, the inner and outer epidermis of an integument (fig.28b). The cells of the other layers of the integuments are obliterated or flattened and form a thin membrane in which almost no trace of them is left. The cells along the edge of the seed coat which form the margin decompose long before full maturation, which is when the endosperm can still be easily pierced with a needle before it becomes horny.

In normal seeds the endosperm usually does not completely fill the seed coat. The milky white, semi-transparent, cellular endosperm of tulip seeds has a horny consistency and in cross section shows seven or eight rows of thick-walled cells. A microscopic examination of the dry endosperm reveals the layered structure of its cell walls, while the cell cavities appear as minute light slits of an irregular shape. Their protoplasts form amorphous lumps which are pressed against the walls of the dry cells (fig.28). It follows that in tulip seeds the reserve materials are laid down in the thickened cell walls of the endosperm. These walls cannot be stained by chlorine-zinc-iodine, but if after treatment with this reagent, glycerine is dropped on the endosperm sections, they will swell and look like mucous formations which are much larger than the cavities they are lying in. The stored material in the endosperm of tulip seeds consists of mucilage in the layered cell walls. The embryo and the protoplasts regulate its dissolution and transformation into easily soluble and transportable sugars. Without this regulation system, the seed may rot and fail to germinate under conditions of excessive humidity and at certain temperatures. This frequently happened when we were trying to germinate seeds in which only a defective endosperm had developed and the embryo was often lacking. The reserve materials stored in the cell walls of the endosperm are absorbed by the growing embryo through the protoplasts. As the cell walls become quite thin the embryo stimulates by its excretions the dissolution of cell walls by the protoplasts, finally destroying the protoplasts themselves and absorbing them. It is therefore only after the stored materials are dissolved by the action of the revitalised protoplasts that these materials become available to the embryo, which cannot use them directly.

The straight, or comma-shaped, embryo lying in a small hollow on the surface of the dry seed can be easily seen through the endosperm and the thin seed coat on the flat upperside of the seed and in many species it is just noticeable also on the underside. When seeds are soaked for germination the swollen embryo slightly raises the endosperm and seed coat and shows as a low ridge on the seed surface. In a dry seed it lies at the acute end and reaches approximately halfway up the endosperm (fig.27a). It consists in this state of two unequal parts (fig. 27d,e). The upper part, being 6 to 12 times as large as the lower part, is formed by the haustorium of the only cotyledon. The lower part of the cotyledon forms a sheath connected with the outer

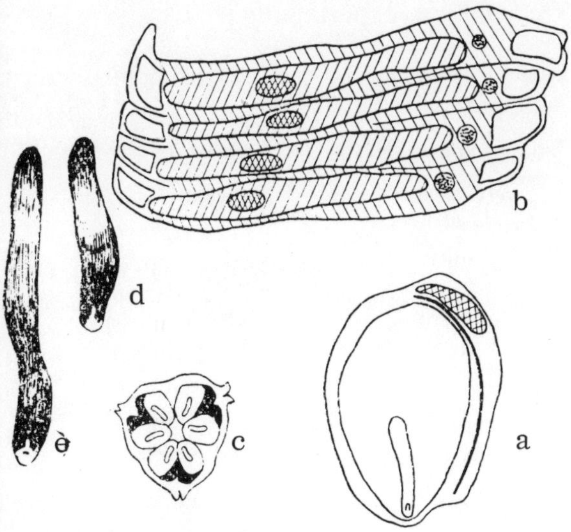

Figure 27. Tulip seed
a. the anatropous ovule before maturation; b. cross section of the micropylar part of young seeds; c. cross section of the ovary of *T. kaufmanniana*; d. dissected embryo of a maturing seed of *T. fosteriana*; e. dissected embroy of a maturing seed of *T. kaufmanniana*.

medium by a narrow channel and tightly envelops a budlet. This budlet, the primordial stem and root, forms the lower part of the embryo. The stem part, the future basal plate, only a few microns large, lies directly under the budlet and gradually merges into a tiny root primordium. Although the embryo is rather large, it is not possible to distinguish its component parts

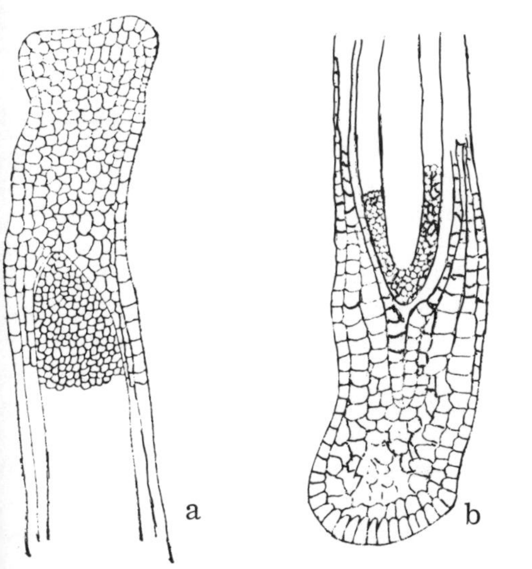

Figure 28. Longitudinal section of the seed of *T. subpraestans*.
a. the part where the chalaza ends; b. micropylar part.

without a powerful magnifying glass and special staining.

The embryo of the dry seed lies rather loosely in the cavity of the endosperm. The better the conditions under which the plants have been growing, the smaller will be the number of seeds without embryos. Many tulip seeds are empty, in particular those towards the bottom and the top of the capsule, where they always become smaller and loose their characteristic shape. The empty seeds look like light brown or whitish, transparent membranes and contain neither endosperm nor embryo. Moreover, one finds near the top and the bottom of the capsule often underdeveloped hard seeds of the same brown colour as the normal ones. These contain an underdeveloped endosperm and no embryo. Their reserve materials are not stored in the layers of the thickened cell walls of the endosperm, but have remained in the cell cavities in the form of mucilage and readily soluble sugars. During germination, even at low temperatures, such seeds initially absorb large quantities of water and swell because of the high concentration of the sugar solution, but they soon turn mouldy.

The reserve material is accumulated in the endosperm when the temperatures during the active cycle of the adult plant are highest (for the tulip May, June), but its dissolution and transformation into a form which can be used by the embryo is only possible at low temperatures during the first active period in the life of the new plant, from December to February (Botschantzeva, 1951).

Germination

Tulip seeds germinate only at rather low temperatures (from 0° to 10°C) and high humidity in the autumn and winter (Botschantzeva, 1951). However, even under favourable conditions, the embryo development inside the seed continues for a long period. The length of this period depends on the conditions under which the inherited properties of the given species have evolved.

The germination period is long and extends over the autumn, winter and early spring. During this process the embryo gradually fills the whole cavity in the endosperm and presses itself against its cells. When the protoplasts of the endosperm cells become active again, their plasma swells and fills the enlarging cavities of the cells. The nuclei of the endosperm cells grow and nucleoli and lumps of chromatine show up clearly. The observed processes in the nucleus are morphologically similar to those of the telophase. The interconnection between the growing embryo and the protoplasts of the endosperm cells is quite evident at this stage. The closer the endosperm cells in the germinating seed are to the embryo, the larger are their protoplasts. While the growing embryo stimulates the development of the protoplasts in the endosperm cells, the protoplasts dissolve the thickened cell walls in which the reserve material is stored.

At the first stages of embryo development inside the seed, the upper part of the cotyledon, the haustorium functioning as sucking organ, grows quickly. It does not contain chloroplasts and for some time grows towards the centre of the seed, which is in particular noticeable in the large seeds of cultivated tulips. However, at approximately one third from the end of the endosperm in the direction of the top of the seed, it bends slightly and continues to grow sideways. It is only after the haustorium has stopped growing that the lower part of the cotyledon starts growth by pushing apart and breaking through the endosperm cells and emerges from the seed with the apical bud and the root primordium.

For a long time the embryo lives on the reserve material in the endosperm. The cotyledon, the rootlet, the young bud and the bag-shaped outgrowth, the stolon, which carries the budlet far away from the place of its initiation, all develop at the expense of this reserve material. The empty seed coat falls off the haustorium afterwards, but the cotyledonous leaf has become green long before, while its haustorial part remains white and dries up soon after the seed coat has been shed.

We found that the vernalization period of young tulip shoots lasts throughout embryo development in the seed and, after emergence from the seed, till the appearance of the green colour and subsequent straightening; more or less in the same way as in germinating stone fruits (Rodionov, 1950). After the shoots have become green and have straightened they require different conditions, viz. moderate temperatures and light.

When the seeds are sown in autumn they usually sprout in spring, but particular during warm winters, the seeds of many montane species and of some cultivated varieties do not receive enough cold for germination. In these cases, after having dried in spring and summer and become wet again in the autumn, they will get the additional number of cold days in the following winter and sprout next spring.

A study of seed germination clearly shows the great influence of the environment on the hereditary constitution of the plants. Information on this subject is very scarce as far as tulip seeds are concerned. Proskoryakov (1926) was the only one who, when studying temperature conditions on the germination of spring perennials in 1922-'23, also mentioned three tulip species: *T.biebersteiniana, T.biflora* and *T.schrenkii.* Later the same author (Proskoryakov, 1952) examined a large number of seeds of representatives of various families including the genus *Tulipa.*

Germination experiments

In 1949 we germinated seeds of the following species and cultivars:

Section *Tulipanum*

T. tubergeniana
T. kuschkensis
T. hoogiana

Section *Leiostemones*

T. ingens
T. fosteriana
T. praestans
T. subpraestans
T. albertii
T. greigii
T. micheliana
T. maximowiczii
T. wilsoniana
T. korolkowii
T. kolpakowskiana
T. vvedenskyi
T. hissarica

Section *Spiranthera*

T. dubia
T. kaufmanniana

Section *Lophopyllon*

T. regelii

Section *Eriostemones*

T. turkestanica
T. bifloriformis

Darwin tulips *(T. gesneriana)*

'Elephant'
'Jubilee'
'Kathleen Parlow'
'King George V'

Cultivars formerly classed by gardeners with *T. suaveolens*

'Duc van Tol Yellow maxima'
'Van der Eerden'
'Thomas Moore'

Cottage tulips

'Mother's Day'

Origin of material

The seed of the cultivars was collected from plants grown in the Botanical Garden of the Academy of Sciences of the Uzbek SSR. The bulbs were received from Holland in 1932 and from the Central Botanical Garden of the Academy of Sciences of the USSR.

T. tubergeniana. The seed came from plants which had been grown in the botanical gardens since 1933, the year in which the bulbs were collected on loess slopes of mount Sebistan in the Pamir-Alay (coll. V.P. Bochantsev). They had been growing without transplantation till 1940 when they were dug up from a depth of 60 cm and replanted 15-20 cm deep.

T. kuschkensis. The seed came from plants which had been growing on a hill in the botanical garden of Tashkent State University since 1931, in which year the bulbs were collected on loess hills in the environs of Morgunovskiy village in the Kushka district (coll. A.A. Mali). In 1949 they were transplanted on to fertilized beds and afterwards transplanted every other year.

T.hoogiana. The bulbs were collected on screes in the western Kopet-Dag (coll. A.F. Gnezdilo, 1935). They were growing on a hill in the botanical garden from 1935 till 1949 and were transplanted every other year after 1949.

T. fosteriana. We collected this species in the Zeravshanskiy range, in the vicinity of the Aman-Kutan forestry, at altitudes of ca. 1 500 m and higher. They grow mainly on screes between large blocks of marble in a marble/limestone substrate, in rock crevices filled with fine earth and sometimes on slopes covered with fine earth and broken stones and then usually on hill crests. The bulbs lie at a depth of 35 cm.

T.subpraestans. The seeds came from plants which had been grown in the botanical garden since 1933. These bulbs were collected on clayey slopes in the western spurs of the Pamir-Alay near the village of Tutkaul (coll. V.P. Bochantsev). They easily multiply vegetatively, forming clumps at a depth of 20 cm. This species is often multiflorous.

T.greigii. We collected the seeds near Darbaza station on the Tashkent railway in 1949. The plants grow on clayey hill slopes and in the valleys in between. They stand very close in the valleys and cover them almost completely whenever the soil contains much gypsum. In these places the bulbs lie at a depth of 35 cm.

T.korolkowii. The seed was also collected in 1949 near the Darbaza station, in the Kyzkuduk valley. These tulips grow on sandstone bânks covered with a thin layer of sandy soil. Where the soil layer becomes deeper they disappear, giving way to *T.greigii.* The bulbs do not lie deeper than 25 cm.

T.kaufmanniana. We collected the seeds in the western Tien Shan, the Sukok river valley in the middle mountain belt; on rock benches in rifts filled with fine earth. The bulbs do not lie deeper than 25 cm.

T.turkestanica. The seeds were collected in the Zeravshanskiy range in the vicinity of the Aman-Kutan forestry on the northern, western and eastern stony slopes just above the valleys. As a rule this species grows lower than *T.fosteriana,* but the upper boundary of its habitat merges with the lower boundary of *T.fosteriana.* The bulbs do not lie deeper than 10 cm.

T.bifloriformis. We collected seeds near Darbaza station on clayey slopes and in the desert mountains of the Alymtau, northwest of Tashkent. The soil is a red clay with a high gypsum content. The bulbs lie at a depth of 10-16 cm.

Methods, results and discussion

In practice seeds with hard seed coats and firm endosperms are often scalded with boiling water to speed up germination. In September 1948, we tried to germinate seeds of two species, *T.bifloriformis* and *T.greigii.* The seeds came from two lots collected in 1941 and 1948.

On September 2 dry seeds were scalded with boiling water. On September 3 the water was poured off, half of the seeds were again scalded with boiling water and the other half soaked in water at room temperature. On September 4 half of the seeds scalded September 3 were once more immersed in boiling water. Thus we obtained seeds that were scalded once, twice and three times. All seeds were kept in gauze bags to be sown in boxes on September 7. Seeds which had been soaked in water at room temperature served as a control.

The sown seeds were watered daily but did not germinate during the autumn. Although the temperatures did not drop below zero in September and October, they became low enough during the winter months so that the earth in the boxes froze. After thawing in February both species started to germinate. Fifty percent of the *T.greigii* and *T.bifloriformis* seeds

Table 3

Germination of seeds (%); sown at temperatures of —2 to +3°, on 25/1.

Treatment	21/2	23/2	25/2	27/2	1/3	3/3	6/3	8/3	10/3	13/3	Total
					T. korolkowii						
1	2	0	0	0	6	0	4	0	2	4	18
2	6	12	32	26	4	4	14	2	0	0	100
3	4	32	16	10	14	14	0	0	6	4	100
4	22	16	28	4	0	8	2	4	0	0	84
5	20	8	24	26	4	8	8	2	0	0	100
Control	2	20	8	8	22	0	0	0	0	4	62
					T. greigii						
2	—	12	6	30	2	10	14	4	4	2	84
3	—	12	12	12	12	4	6	18	2	8	86
4	—	14	14	16	6	4	10	0	0	2	66
5	—	4	6	20	8	14	8	2	8	18	88
Control	—	12	18	0	14	6	6	10	2	0	68

collected in 1941 and 1949, and scalded only once, germinated. The seeds of both species that had been scalded twice and thrice did not germinate. The control gave 100% germination.

Apart from these sowings of scalded seeds in the warm season we sowed seeds of wild species and cultivars in a frame, a glasshouse and in a room. Despite careful tending they failed to sprout. We also sowed seeds of *T.korolkowii, T.greigii, T.bifloriformis* and *T.fosteriana* which had been immersed in biostimulators and were kept in a thermostatically controlled chamber at 22°C. After two months at this temperature even the embryos of the seeds had failed to swell.

Seeds sown in the open in August or September do not sprout until the next spring. Some workers hold the view that seeds in general contain special inhibitors delaying germination (Isip, 1939; Yarkova, 1939). Isip mentions that when the seeds are soaked in large quantities of water the 'protective substances' are dissolved and can be partially removed with the water. By drying the soaked seeds, in particular at a high temperature, it is possible to destroy the 'protective substance' even further. He recommends for some species to soak and dry twice.

Following Nikolaeva (1948), who found that water at room temperature has a delaying effect on the germination of *Ferula karatavica,* a species growing under conditions similar to those of tulips, we soaked tulip seeds in water at 0° and 18°C.

Seeds of *T.korolkowii, T.greigii, T.kaufmanniana* and the cultivars 'Jubilee', 'Kathleen Parlow' and 'Duc van Tol Yellow Maxima' were treated as follows (100 seeds per treatment):

1. carefully ground in a porcelain mortar together with washed river sand to remove seed coats, then thrice soaked in water at 0°C and dried after each soaking;
2. once soaked in water at 0°C and then dried;
3. twice soaked in water at 0°C and dried each time;
4. three times soaked in water at 0°C and dried each time;
5. three times soaked in water at 18°C and dried each time;

Each soaking and drying lasted for 24 hours. The controls did not receive any treatment.

Before sowing on January 25th each batch was divided in half and sown in Koch dishes on filter paper lying on lignin moistened with water. Half of the seeds were placed in a room at 16–18°C and the other half was left to germinate at low temperatures ranging from −2° to +3°C.

T.korolkowii seeds began to sprout on the 25th day in all treatments carried out at low temperatures (table 3).

It is difficult to conclude from the data in table 3 whether or not tulip seeds contain protective substances, since the seeds of the controls and of all the different treatments started to germinate simultaneously. The data are however interesting in view of the ecological aspects of seed germination of plants adapted to the climatic conditions in Middle Asia.

All tulip species require a rather long cold period and considerable humidity for germination or else they do not germinate. In the environs of Tashkent, the origin of the *T.korolkowii* plants from which the seeds were collected, humid periods often alternate in the autumn and winter with very dry ones and low temperatures with longer periods of high temperatures. Although the periods of drying and soaking were too short in this experiment, the results have thrown some light on the germination process under these radically changing conditions. It appears that preliminary soaking and drying does not advance germination, but in treatments 2, 4 and 5, it slightly reduces the length of the germination period and somewhat changes its course. The rates of sprouting are not uniform. With treatments 4 and 5, sprouting starts very rapidly and then gradually proceeds at the same rate as in treatment 2. With treatments 2 and 3 the percentages of germinating seeds over the first ten-day period are not much different. The longer we soaked the *T.korolkowii* seeds in water at low or even raised temperatures and dried them afterwards, the earlier they started sprouting, but in treatments 2 and 3 germination proceeded more evenly than in treatments 4 and 5 and in the control. In treatment 1, where the seed coats had been removed, the germination percentage was very low, probably because of mechanical damage to the seeds and the embryos.

The seeds of *T.greigii* started to germinate on the 27th day in all treatments, whereas seeds from which the seed coats had been removed failed to sprout altogether. In this case it was also difficult to conclude whether or not germination was accelerated in comparison with that of the seeds which were sown in pots and sprouted in the early spring of 1949. Soaking and drying, only once, appeared to be sufficient to affect the rates of germination. Treatment 3 gave a maximum rate of germination on March 18 and treatment 5 on March 13. With treatment 4, soaking thrice at 0°C with subsequent drying, germination was very poor in both *T. greigii* and *T.korolkowii.* In general repeated soaking and drying of the seeds considerably affected the rates of germination.

The seeds of all species and cultivars sown at 16-18°C or in a thermostatically controlled chamber at 22-25°C, whether or not subjected to repeated soaking and drying, became dark brown after ten days and mouldy after about 25 days. This was particularly noticeable in treatment 1 where the seed coats had been removed. 'Duc van Tol Yellow maxima' and to a lower degree also 'Jubilee' appeared to be very susceptible to moulding. We could prevent moulding by dusting the seeds with charcoal powder but they did not germinate. The embryos, although increasing in size when moistened, stopped growing.

It is clear that the method proposed by Isip does not accelerate germination of this type of seed. An influence of the protective substance in tulip seeds, which could have slowed down or arrested germination, could not be noticed either. Apparently there are other physiological factors governing the germination period, but they are more complicated than suggested by T.D. Lysenko (1943) when he discusses the seed of certain weedy grasses: 'As a rule not all seeds of wild oat sprout without having wintered; roughly only 10-30% do. The rest do not germinate in summer and seem to be dead, but they are actually alive, waiting for the next spring to germinate. Until the seed coats are frozen the embryo is hardly accessible to the air, hence the seeds are said to be dormant'. Repeated soaking and drying of the seed should have a strong deleterious effect on the seed coats and ensure a better access of the air to the endosperm. However, the interrelations between the environment and the seed are not so simple and it is not very well possible to explain an accelerated or delayed germination merely by the destructive action of one agent on the seed coat.

The experimental results show qualitative differences between the seeds of the various species. In nature some of them germinate rather quickly after alternating periods of thawing, drought and frost, while in others germination is delayed and protracted over a longer period with the possibility that part of the seeds will only germinate the following year. Warm autumn rains alternating with dry periods can change the rate of germination. The adaptibility of the seed to varying conditions of germination over years with different temperature regimes ensures the survival of the species. In years with alternating periods of rains and droughts the development of some seeds is retarded and they do not germinate within a year of being shed from the capsule. After having become dry again and given favourable conditions, they germinate the next year. The results of these experiments again show the close connection between the plant and its environment.

The rate and percentage of germination of spring-sown seeds depend on the wheather conditions throughout the year.

We sowed seeds of *T. bifloriformis, T. greigii, T. fosteriana* and *T. kaufmanniana* in pots, late in February 1949, without any preliminary treatment, placed them out of doors and watered when necessary. The air temperature in that period dropped to —1°C by night and rose to +12.5°C by day. Under these conditions the seeds of *T. bifloriformis* started to sprout after 22 days and those of *T. greigii* after 25 days, which is nearly as early as in the preceeding experiment. After March 20 the temperatures began to rise and the seeds of *T. kaufmanniana* and *T. fosteriana,* which need a longer cold period to germinate, only sprouted in the spring of 1950 together with the lots sown in the autumn of 1949.

Cutting the endosperm

Nikolayeva (1948) claims that clipping the endosperm covering the root-tip of the embryo accelerates germination of *Ferula karatavica.* At temperatures, ranging between 0° and 10°C, germination took 45 days. We therefore decided to rupture the seed coat and to clip off part of the endosperm next to the root-tip of the embryo. The clipped seeds were left to germinate on moistened filter paper at temperatures of 18-20°C in a humidity chamber. In these conditions the damaged seeds easily start rotting. However, part of the clipped seeds with undamaged embryos did not become infected by mould fungi and after five days showed signs of sprouting, but actually the embryo was being pushed out by the swelling endosperm.

A young tulip plant lives for a long time on the reserve material in the endosperm. It reaches the young shoot through the upper part of the cotyledon, which functions as haustorium. Normally the young plant remains connected with the seed for a long time. If after clipping a thin layer of endosperm is left over the growing root-tip, the seed will readily germinate at room temperature (18-20°C). Because of the rather hard endosperm and the thin layer covering the root-tip, it is very difficult to leave this layer undamaged in clipping. However, when this succeeds, the method results in rapid germination. In a matter of 35 days we obtained several young growing shoots of *T. fosteriana*, looking very much like those produced under normal conditions.

In the autumn of 1949 we experimented on a larger scale to determine the optimal conditions for germination. The results of all our previous experiments in the autumn of 1948 and spring 1949 had pointed to low temperature requirements.

Proskoryakov (1926) studied germination of three species: *T. biflora,* a characteristic component of the *Artemisia* association, growing in complex semideserts in southeast Middle Asia; *T. schrenkii* growing in heavy clayey soils in vast areas of the Transvolga steppes; *T. biebersteiniana* with a still wider geographical distribution and often found in steppes and forests. He germinated tulip seeds at temperatures ranging from 0°-6°C. In his experiments germination continued at —0.5°C. Seeds of *T. biflora* began to sprout 18 days after sowing and 100% germination took 31 days (from March 20 to April 20). Only 47% of the *T. schrenkii* seeds germinated in the course of 150 days. Of *T. biebersteiniana* 96% germinated in a period of 176 days. It appeared from his experiments that light impedes germination. At high temperatures the seeds did not germinate at all, neither in darkness nor in the light.

For our experiments we selected from each species and variety 500 well-formed seeds with fully developed embryos, each species or variety to be sown in one container (Koch dishes or clay pots). Sowing was done at three different dates: early in August, in sand (Koch dishes); on August 29, in small pots, which were plunged in sand; and on October 1st, on filter paper placed on lignin and the seeds being covered with filter paper with lignin on top. The seeds were covered with either sand or lignin to shield them from light which, according to Proskoryakov, has a retarding influence on germination. All dishes and pots were put in a refrigerator, EKD-120, and watered when necessary.

During the first days, from August 4 to August 29, we switched the refrigerator on at 8.30 hrs and off at 17 hrs. During this period the temperature on the upper shelf next to the freezer dropped to –5° and on the middle shelf to +2°C. By the morning it rose to 18°C. The temperatures in the room with the refrigrator reached 30°C and sometimes even higher. The high temperatures favoured the appearance of fungi and bluegreen algae on the pots and the dishes. From August 29 we prolonged the refrigeration period with 7 hrs. to 22-23 hrs. The lowest temperature on the upper shelf was −8° and in the morning before switching off the highest temperature was +7°C. These temperatures proved to be more favourable for germination and delayed the appearance of bluegreen algae.

In the experiments, of which the results are given in table 4, we tried to create the natural growing conditions, taking into account the special environmental requirements for tulip seeds. We therefore used a wide

Table 4

Germination of tulip seed in the refrigerator

Species or cultivar	Date of sowing	Germination started	No. of days from sowing to start of germination	End of germination period	No. of days from sowing to end of germination	Duration of germination period	Remarks
			Sown early in August				
T. bifloriformis	5/8	14/9	40	19/10	76	36	
T. korolkowii	4/8	15/9	42	19/10	76	34	
T. turkestanica	6/8	29/9	54	2/12	118	64	
T. kaufmanniana	—	—	—	—	—	—	germination still
T. fosteriana	—	—	—	—	—	—	continuing 18/1; only
T. tubergeniana	—	—	—	—	—	—	50% germinated
'Jubilee'	4/8	27/10	84	11/11	99	15	
'Elephant'	5/8	29/10	55	10/11	97	42	
'Kathleen Parlow'	5/8	10/10	66	4/11	91	25	
'Duc van Tol Yellow maxima'	5/8	10/10	66	4/11	91	25	
			Sown late in August				
T. bifloriformis	28/8	8/10	40	26/10	58	18	
T. korolkowii	28/8	8/10	40	25/10	57	17	by March 1st, 80% had germinated
T. greigii	28/8	8/10	40	25/10	57	17	
*T. fosteriana**	29/8	10/1	134	—	—	—	
'Jubilee'*	29/8	10/1	134	—	—	—	
*T. kaufmanniana**	29/8	10/1	134	—	—	—	
			Sown early in October				
T. buhseana from Alymtau	1/10	24/10	24	10/11	41	17	
T. korolkowii	1/10	24/10	24	10/11	41	17	
T. greigii	1/10	14/11	43	—	—	—	
T. kaufmanniana	1/10	20/12	80	—	—	—	
T. subpraestans	1/10	1/3†	151	—	—	—	
T. fosteriana	1/10	—	—	17/2	140	—	
T. tubergeniana	1/10	—	—	—	—	—	

* Germination not completed by November 14; were frozen and thawed under snow January 1 and subsequently sprouted by January 10.

† Were brought outside and covered with snow; germinated by March 1, after thawing. These data need further verification.

range of temperatures. Under natural conditions in Middle Asia day and night temperatures fluctuate sharply, particularly during late autumn, winter and early spring when tulip seeds germinate and the living plants (bulbs) are protected.

It appears that a nearly equal number of cold days is necessary for germination of the seeds of *T. bifloriformis*, *T. korolkowii* and *T. greigii*, which grow under similar climatic conditions in the clayey semideserts. It should be noted that *T. bifloriformis* belongs to a different section of the genus and that all three species are also found on different soil types near Darbaza railway station. In the first two treatments *T. korolkowii* and *T. bifloriformis* needed 40 days to germinate, in the third treatment, *T. korolkowii* and *T. buhseana*, 24 days (table 4). Apparently the environment in which they germinated had some effect.

The seeds of *T. greigii* in the first treatment proved to be of bad quality. They became black and failed to sprout. The second and third treatments gave very similar results, 40 and 43 days resp. to first germination.

T. turkestanica, *T. fosteriana* and *T. kaufmanniana* were brought from the mountains where the last two occur in ecologically very similar habitats. Their seeds require the same cold period of germination. *T. turkestanica*, which grows much lower in the mountain belt, needs a shorter cold period for germination (table 4). *T. tubergeniana* from loess slopes, not high in the mountains, requires fewer cold days than *T. fosteriana* and *T. kaufmanniana*, which originate from the high mountain belt where they are growing between rocks in crevices filled with fine earth.

The cultivars germinate more rapidly at higher

Table 5

Rate of germination (%)

Species		July						August						September						October					
	30	5	10	15	20	25	30	5	10	15	20	25	30	5	10	15	20	25	30	5	10	15	20	25	30
T. regelii	3.1	—	3.1	6.2	9.3	15.6	28.0	15.6	9.3	—	3.1	3.1	—	—	3.1	—	—	—	—	—	—	—	—	—	—
T. albertii	—	2.2	4.4	8.8	13.6	13.6	15.8	31.2	6.6	6.6	6.6	—	—	—	—	—	—	—	—	—	—	—	—	—	—
T. tubergeniana	—	2.2	15.0	17.0	24.0	16.5	11.6	3.3	2.5	1.5	1.3	1.4	—	0.8	0.7	—	—	1.3	1.3	—	—	—	—	—	—
T. ingens	—	2.8	—	2.8	2.8	—	2.8	8.3	16.6	—	8.3	16.6	—	—	—	14.0	—	2.8	—	2.8	—	—	—	—	5.6
T. wilsoniana	—	0.5	3.3	2.3	1.9	3.8	4.7	3.8	3.8	19.7	15.5	13.1	3.3	15.5	5.6	6.1	2.5	—	0.9	—	0.5	—	0.9	—	—
T. kuschkensis	—	—	7.1	13.6	22.4	14.4	8.2	8.0	7.2	10.4	—	4.8	—	—	3.2	—	—	—	—	—	—	—	—	—	—
T. micheliana	—	—	0.7	1.4	3.5	4.9	11.9	4.9	7.0	8.7	7.7	7.0	2.8	1.4	4.2	7.0	1.7	4.2	6.3	6.8	1.4	—	—	—	1.4
T. greigii	—	—	—	—	—	—	1.7	8.6	6.9	6.9	12.0	3.4	5.2	6.9	8.6	6.9	5.2	1.7	1.7	3.4	3.4	5.2	6.9	3.4	1.7
T. hoogiana	2	2	2	2	2	—	—	0.9	2.6	7.1	15.1	2.5	5.0	7.1	18.7	10.7	2.6	1.8	3.6	—	0.9	—	—	—	—
T. vvedenskyi	—	—	—	—	—	—	—	0.5	1.5	7.8	9.9	4.0	4.6	7.8	9.3	3.6	1.0	4.1	3.1	1.5	—	—	0.5	—	—
T. fosteriana	—	—	—	—	—	—	—	—	—	—	1.5	1.5	2.2	4.3	1.5	3.7	2.2	1.5	1.5	2.2	2.9	0.7	0.8	2.2	4.3
T. kaufmanniana	—	—	—	—	—	—	—	—	—	—	—	—	1.0	1.0	2.0	8.0	10.0	7.0	3.0	—	1.0	3.0	—	—	1.0
T. subpraestans	—	—	—	—	—	—	—	—	—	—	—	—	1.2	—	—	—	1.2	6.2	6.2	9.9	13.6	12.3	6.2	2.4	1.2
T. maximowiczii	—	—	—	—	—	—	—	—	—	—	—	—	—	—	—	—	—	2.0	—	—	2.0	—	6.0	—	3.0

Table 5 continued

Species	November						December						January						February			
	5	10	15	20	25	30	5	10	15	20	25	30	5	10	15	20	25	30	5	10	15	20
T. regelii	—	—	—	—	—	—	—	—	—	—	—	—	—	—	—	—	—	—	—	—	—	—
T. albertii	—	—	—	—	—	—	—	—	—	—	—	—	—	—	—	—	—	—	—	—	—	—
T. tubergeniana	—	—	—	—	—	—	—	—	—	—	—	—	—	—	—	—	—	—	—	—	—	—
T. ingens	—	—	—	5.6	—	—	5.6	—	—	—	—	—	—	—	—	—	—	—	—	—	—	—
T. wilsoniana	—	—	—	—	—	—	—	—	—	—	—	—	—	—	—	—	—	—	—	—	—	—
T. kuschkensis	—	—	—	—	—	—	—	—	—	—	—	—	—	—	—	—	—	—	—	—	—	—
T. micheliana	1.4	1.4	—	1.4	—	1.4	—	0.7	—	—	—	—	—	—	—	—	—	—	—	—	—	—
T. greigii	—	—	—	—	—	—	—	—	—	—	—	—	—	—	—	—	—	—	—	—	—	—
T. hoogiana	—	—	—	—	—	—	—	—	—	—	—	—	—	—	—	—	—	—	—	—	—	—
T. vvedenskyi	—	—	1.0	—	—	1.0	10.2	10.2	20.0	—	—	—	—	—	—	—	—	—	—	—	—	—
T. fosteriana	8.6	13.2	5.1	2.9	0.7	0.8	1.5	1.5	2.2	4.3	6.6	8.6	4.3	2.9	1.5	1.5	—	—	—	—	—	—
T. kaufmanniana	2.0	6.0	11.0	3.0	11.0	—	1.0	3.0	6.0	12.0	9.0	1.0	—	—	—	—	—	—	—	—	—	—
T. subpraestans	1.2	2.4	3.7	6.2	3.7	24.0	1.2	—	1.2	1.2	1.2	2.4	1.2	1.2	1.2	1.2	1.2	1.2	1.2	1.2	2.4	2.4
T. maximowiczii	—	3.0	3.0	8.0	9.0	12.0	20.0	14.0	7.0	6.0	5.0	4.0	3.0	2.0	4.0	5.0	2.0	—	—	—	—	—

temperatures. 'Kathleen Parlow', which in our conditions flowers later than any other tulip, and 'Duc van Tol Yellow Maxima' need the same number of cold days for germination. This stresses the importance of the environment in which the tulip cultivars evolved. Both varieties were bred in Holland and imported as adult bulbs. The seeds of Dutch tulips require about the same number of cold days as the wild tulips from montane areas. With lower temperatures the period between sowing and first germination for cultivars becomes longer and approaches that of montane tulips. The number of cold days required for germination does not depend on the species, but on the environment in which they have evolved.

In June 1953 the experiments were continued and thirteen more species were included with the eight already studied.

Of the section *Tulipanum* we took, besides *T.tubergeniana,* already studied in 1949, two other species: *T.hoogiana* and *T.kuschkensis.*

Of the *Leiostemones* we again used *T.fosteriana, T.greigii* and *T.subpraestans* for comparison with related species and added *T.ingens, T.praestans, T.albertii, T.micheliana, T.maximowiczii, T.wilsoniana, T.kolpakowskiana, T.dubia, T.vvedenskyi* and *T.hissarica.*

We again used *T.kaufmanniana* of section *Spiranthera* and included for the first time *T.regelii* of the monotypic section *Lophophyllon.*

Besides these 18 species we selected the four cultivars 'Van der Eerden', 'King George V', 'Mother's Day' and 'Thomas Moore', which represent three different groups.

The seeds were placed on wet filter paper in Petri dishes covered with paper. The dishes were put in a refrigerator at temperatures ranging from —8° to +7°C. Every eight days the seeds which had germinated were counted and transplanted into pots, while those which had begun to rot, or appeared to be damaged, were removed from the dishes. The paper was changed when necessary.

Table 5 gives the results for the species in order of their time of germination. The percentages of germinated seeds were calculated for successive periods of five days, taking the total number of germinated seeds per species as 100% and not the number of seeds sown. While giving a true picture of the rate of germination of a species, the absolute values do not permit a comparison of the different species. However, the arrangement of the columns, running from the upper left to the lower right corner of the table, gives interesting information. The desert and foothill species *T.albertii, T.kuschkensis, T.regelii* and *T.tubergeniana* fall into the upper left corner, while species from the higher montane altitudes, such as *T.fosteriana, T.kaufmanniana, T.subpraestans* and *T.dubia* are found in the lower right corner. Thus the species from the deserts and foothills

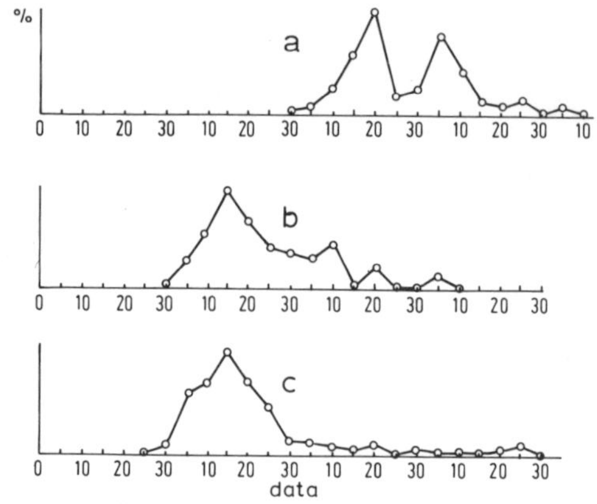

Figure 29. Germination of species from the section *Tulipanum.* a *T. hoogiana*; b. *T. kuschkensis*; c. *T. tubergeniana.*

start germination earlier after sowing and germination is completed within a shorter period, *T.ingens* being the only exception. In *T.fosteriana, T.kaufmanniana, T.kolpakowskiana, T.dubia, T.subpraestans* and *T.hissarica* germination is slow and takes a considerably longer period (figs.29,30,31).

The data in table 6 were used to compare species which are partially sympatric and for which the ecological conditions differ but little. The germination behaviour of such species is similar although taxonomically they may be very different; cf. *T.albertii* of section *Leiostemones* and *T.regelii* of the monotypic section *Lophophyllon.* The bulbs of both species came from the environs of Chokpar station on the Turkestan-Siberian railway, where *T.albertii* grows on screes and *T.regelii* on stoney slopes, covered with a layer of fine soil. They were planted on neighbouring beds in the old part of the botanical garden and the seeds we used had been collected from these plants.

The interval between sowing and first germination was 21 days for *T.regelii* and 30 days for *T.albertii;* the duration of the germination was 72 and 82 days resp. *T.regelii* seeds start to germinate and complete growth earlier than all the others. By its morphological and biological characters this species is closer to species of section *Eriostemones* than to those of the sections *Tulipanum* and *Leiostemones.*

T. greigii and *T. korolkowii* are very much alike in their germination. Although belonging to the same section they are widely different in morphological characters. *T. greigii* grows near Darbaza on the Tashkent railway in deep grey soils, often containing a high percentage of gypsum, while *T. korolkowii* occurs mostly on sandstone ridges covered with sand and on sandy slopes. The seeds of *T. greigii* (Darbaza) required 40 cold days from sowing to first germination, those of *T. korolkowii* 24-42 days (tables 4 and 6). The length of the germination periods was 17 and 17-34 days resp.,

Table 6
Germination of tulip seeds

Species	Date of sowing	Germination started	No. of days from sowing to start of germination	End of germination period	No. of days from sowing to end of germination	Duration of germination period
Section *Tulipanum:*						
T. tubergeniana	8/6	7/7	30	7/9	95	65
T. kuschkensis	8/6	7/7	30	10/9	94	64
T. hoogiana	8/6	5/8	58	9/10	123	65
Section *Leiostemones:*						
T. ingens	8/6	3/7	26	3/12	178	152
T. fosteriana	30/6	19/9	81	13/2	229	148
T. praestans	8/6	28/10	142	13/2	250	108
T. subpraestans	8/6	29/8	82	13/2	250	168
T. albertii	8/6	7/7	30	28/9	112	82
T. greigii (Karankul'saí)	30/6	5/8	36	23/1	207	171
T. micheliana	8/6	7/7	30	9/7	183	153
T. wilsoniana	30/6	21/7	22	6/12	159	137
T. maximowiczii	30/6	25/9	87	13/2	228	141
T. korolkowii	—	—	42	—	76	34
T. kolpakowskiana	8/7	19/9	103	13/2	250	147
T. vvedenskyi	8/6	5/8	58	12/1	218	160
T. dubia	30/6	28/9	90	13/2	228	138
T. hissarica	30/6	13/2	229	—	—	—
Section *Spiranthera:*						
T. kaufmanniana	30/6	14/9	76	13/2	228	152
Section *Lophophyllon:*						
T. regelii	8/6	29/6	21	9/9	93	72
Section *Eriostemones:*						
T. turkestanica	—	—	54	—	118	64
T. bifloriformis	—	—	40	—	76	36

which was for those two species the shortest of the section, but it should be kept in mind that we studied only a comparatively small number of species. Thus from sowing to the end of the germination period *T. greigii* (Darbaza) needed 57 days and *T. korolkowii* 41 days and in another case 76 days.

T. greigii from the slopes of Karankulsay, in the Chirchik valley, needed 36 days to start germinating and 207 days for completion of germination, instead of 57 days, meaning that germination continued for 171 days (cf. 17 days for the Darbaza provenance). The bulbs of all plants from which we had collected these seeds had been replanted on beds in the Tashkent Botanical Garden. The lasting effects of the environments in which these plants evolved is evident. *T. greigii* from Darbaza inhabits clayey semideserts with an annual precipitation of 360 mm, whereas the Karankulsai provenance grows in the lower mountain belt (800-1 200m) with an annual precipitation of up to 600 mm. The course of the temperatures over the year is also different in these areas.

The adaptable characters of plants maintaining themselves in different environments persisted in their varying germination patterns even after they had been grown for several years under the changed conditions. During the development of numerous generations the environment had a lasting effect on the heredity of the organisms with the result that within the boundaries of a species different types with a different germination behaviour could develop. Yet the germination curve for *T. greigii* from Karankulsai is similar to that of *T. micheliana* of Aman-Kutan in the Samarkand mountains (fig.30). Since both species grow under the same conditions, it follows that close taxonomic affinity together with similar ecological conditions results in a comparable germination behaviour. This is also shown by *T. tubergeniana*, T. kuschkensis and *T. hoogiana* of the *Tulipanum* section, which, although found in different areas of Middle Asia, occupy rather similar habitats. The similarity of the curves and the almost equally long periods of germination point not only to the corresponding ecological conditions of their habitats, but also to phylogenetic affinity: *T. kuschkensis* is closer to *T. tubergeniana* than to *T. hoogiana* (table 5, 6, 7; fig.29).

In certain features of its juvenile development and in the pubescence of the tunic *T. ingens* represents a link between section *Tulipanum* and the *Leiostemones* grouped around *T. fosteriana*. Seed of *T. ingens* starts germinating soon after sowing, even earlier than *Tulipanum* species, but germination is slow and protracted in the same manner as in *T. fosteriana* and its close relations (fig.30).

In 1952 we studied germination in *T. kolpakowskiana* from Talgar, which had been growing in the Tashkent Botanical Garden for several decades. In nature it is

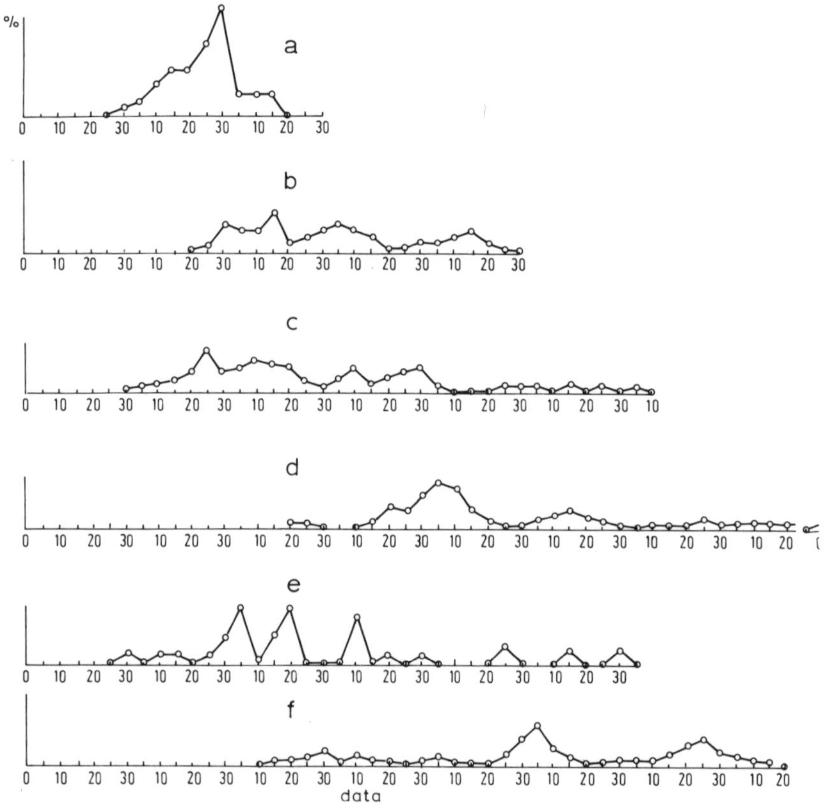

Figure 30. Germination of species from the section *Leiostemones* (refrigerator).
a. *T. albertii*; b. *T. greigii*; c. *T. micheliana*; d. *T. subpraestans*; e. *T. ingens*; f. *T. fosteriana*.

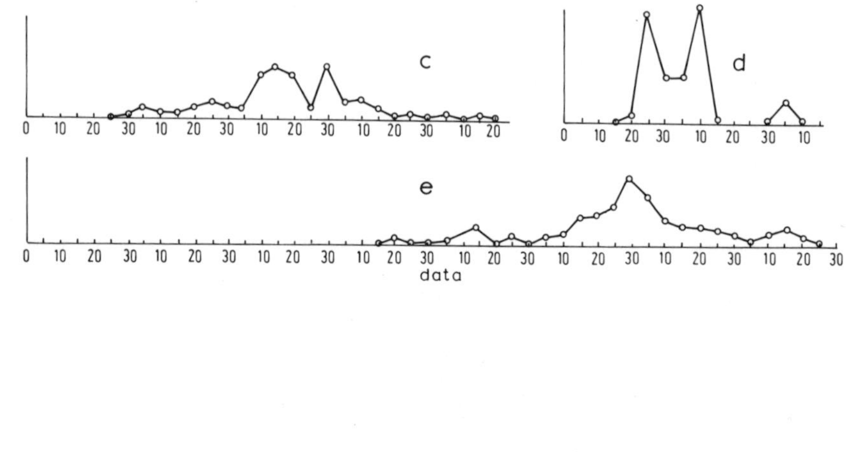

Figure 31. Germination of species from the sections *Leiostemones* and *Spiranthera* (refrigerator).
a. *T. kaufmanniana*; b. *T. wilsoniana*; c. *T. vvedenskyi*; d. *T. korolkowii*; e. *T. maximowiczii*.

Table 7

Length of the germination period of tulip seeds in days

Species	From sowing to first germination	From first germination to the end of germination	Duration of the germination period
T. albertii	30	112	82
T. dubia	90	228	138
T. fosteriana	81	228	147
T. greigii	36	207	171
T. hissarica	228	—	—
T. hoogiana	58	123	65
T. ingens	26	178	152
T. kaufmanniana	76	228	152
T. kolpakowskiana	103	250	147
T. kuschkensis	30	94	64
T. maximowiczii	87	228	141
T. micheliana	30	183	153
T. praestans	142	250	108
T. regelii	21	93	72
T. subpraestans	82	250	168
T. tubergeniana	30	95	65
T. vvedenskyi	58	218	160
T. wilsoniana	22	159	137

found in the foothills, but its germination behaviour is more like that of a montane species. However, the temperatures are lower and the precipitation is higher in the foothills of the Zailiyskiy Alatau than in the foothills of the western Tien Shan and the Chirchik river basin. For these reasons the ecological conditions for the Zailiyskiy Alatau provenance are more comparable to those of the higher regions of the western Tien Shan. The germination behaviour of this species fully corresponds to its ecology.

In these experiments the two species *T.wilsoniana* and *T.maximowiczii* behaved differently. *T.wilsoniana* was the first to germinate, 22 days after sowing in the refrigerator. Germination continued for 137 days. *T.maximowiczii* started germinating after 87 days and the whole process took 141 days (tables 6 and 7). There is but little similarity in germination behaviour between the two species (fig.31) and the curve for *T.wilsoniana* differs to the same extent from that of *T.korolkowii;* the latter being more contracted. With the exception of *T.linifolia,* which we did not include in these experiments, there do not occur related species of *T.wilsoniana* or *T.maximowiczii* in Middle Asia. The other *Clusianae* are found in Iran and Afghanistan. The deviating germination behaviour of the two species may be explained by their genetical constitution, which is different from that of most other species we have investigated.

The example of *T.korolkowii* illustrates the importance of the influence of the ecological conditions on germination behaviour.

In table 6 the species are arranged in the order in which they appear in the "Flora of the USSR". However, this arangement does not give a complete image of the phylogenetic and ecological relations, probably because the number of species studied was too small. In table 7 the species are arranged according to the beginning, the end and the length of the germination period. Each column in this table confirms in some measure the interspecific relations described above.

In 1953, we germinated also seeds of five cultivars belonging to four different groups: 'Van der Eerden' (Mendel Tulips); 'Thomas Moore' (Single Early Tulips); 'Mother's Day' (Cottage Tulips); 'King George V' and 'Elephant' (Darwin Tulips).

The seeds sown at low temperatures in a refrigerator started to germinate after periods of 42-58 days. The seeds of 'Van der Eerden' germinated before those of 'Thomas Moore'. The cultivars 'Mother's Day', 'King George V' and 'Thomas Moore' completed germination in this order.

On the whole the germination periods for all cultivars were rather similar and not very long. It was only

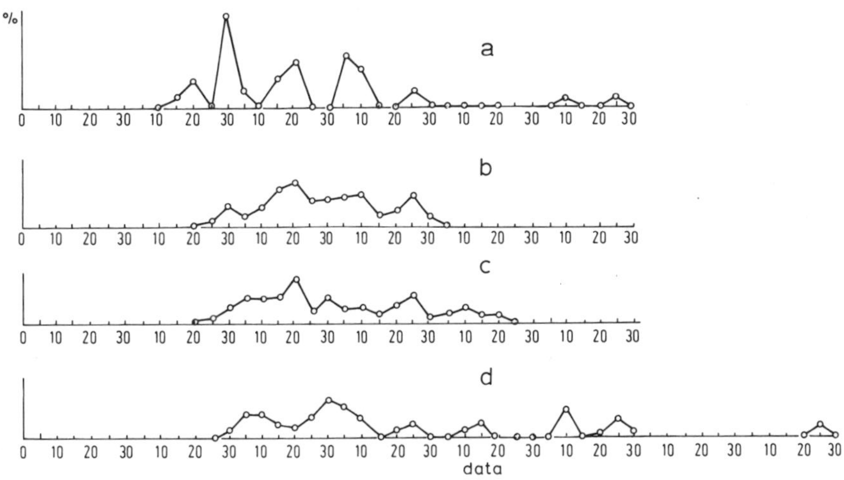

Figure 32. Germination of tulip cultivars.
a. 'Van der Eerden'; b. 'Mother's Day'; c. 'King George'; d. 'Thomas Moore'.

in the earlier flowering cultivars 'Van der Eerden' and 'Thomas Moore' that a small portion of the seeds germinated later than the bulk. In the first case often after two months and in the second after three and a half months. They show a similar behaviour to that of the wild species among their progenitors. The rates of germination for 'Van der Eerden' and 'Thomas Moore' are almost identical (fig. 32), which is probably because both cultivars belong to early flowering groups, derived from *T.suavolens.* The germination of 'Mother's Day' and 'King George V' is more alike than that of 'Van der Eerden' and 'Thomas Moore'. There is some relationship between 'Mother's Day' and 'King George V' since the origin of both species is connected with *T.gesneriana.*

Germination of immature seed

In 1953 we experimented with immature seeds of which the endosperm had not yet hardened and become horny. We used the following material:

Among the *T. greigii* plants from Karankulsai, growing in the Tashkent Botanical Garden, we selected a green capsule containing seeds that were just beginning to turn yellow. They measured 9.2 x 7.7 mm and were still green at the micropyle and all along the vascular bundle up to the chalaza.

Other green seeds were collected from *T. greigii,* growing on the top of a ridge, close to the Tashkent reservation, near the village of Sanginak (Parkent district). Despite their size, 8.9 x 6.9 mm, they were still quite soft and could easily be pierced with a pair of sharp pincers.

Seeds of the cultivars 'Kathleen Parlow', 3.5 x 7.0 mm, and 'Elephant', 12.9 x 8.6 mm, were taken from green capsules of plants growing in the botanical garden. From 'Elephant' only the seeds from one half of the valves were used. The seeds in the other halves were left to dry in the valves on a table in the laboratory before sowing on August 3. Dry seeds of 'Elephant', collected in the same place in 1951, served as control.

The seeds of *T. greigii* from Sanginak, sown in a refrigerator, remained green for a long time and then began to turn yellow. Some white crystals appeared on the seeds, a phenomenon which we had never observed before in any other tulip species when sowing dry and mature seeds. By August 10 the seeds had turned a dark colour and in some of them the large embryo, acquiring gradually the typically curved shape, became visible. Later, however, dark spots appeared on the seeds (disease?) and by August 28 all seeds were rotten.

The *T. greigii* seeds, which had begun to turn yellow before sowing, but were still green around the micropyle and the chalaza, were put in the refrigerator on June 13 and began to germinate on August 5, i.e. 53 days after sowing and 17 days later than dry and mature seeds from the same origin. Apparently these 17 days had been necessary for after-ripening. It shows that tulip seeds can complete maturation even at low temperatures. However, before germination started half of the seeds had rotted and by January 25, at the end of the germination period, we had to discard a considerable number of bad seeds.

The seeds of 'Kathleen Parlow' and 'Elephant', which had been collected ca. 54 days after flowering, were sown green on July 13. Of the 'Kathleen Parlow' seeds only one embryo became visible on August 10, after which all the seeds developed black spots and were decayed by August 29. Of 'Elephant' not a single embryo started to develop and by the end of August all seeds were rotten.

On August 3 we sowed the 'Elephant' seeds from the other half of the valves which had become quite dry after having been left from July 13 to August 3 on a laboratory bench. They began to germinate on September 14, 30 days after sowing.

Sowing immature seed after a drying period shows the variability of its biological properties. Seeds of some plants may need only 30 cold days to germinate, while other seeds require no less than 160 days. The seeds of a plant which experiences a drought before the end of the vegetation period are able to extend their germination over a period of several years. The conditions prevailing during the ripening period have therefore undoubtedly an effect on germination. In tulips, and perhaps also in other plants, this ability is a very useful property.

Germination after supersonic treatment

The capacity of seed to germinate under natural conditions only at low temperatures and the unusually long germination period is an effective adaptation to climates in which the precipitation period coincides with the cold season. The tulips acquired this capacity during their evolution. However, when raising tulips from seeds which in our conditions do not experience sufficient cold days and therefore germinate only after two and sometimes even three winters, it may be necessary to break this dormancy and make the seeds germinate more readily.

In 1952 N.N. Dolgopolov, head of the laboratory of the Mendeleev Institute in Moscow, suggested to speed up germination of the two species, *T.fosteriana* and *T.kaufmanniana,* by giving supersonic treatment.

The seeds were put in beakers, filled with water at room temperature, and exposed to supersonic waves with a frequency of 30.000 kilohertz for 90, 60 and 30 seconds. *T. fosteriana* received the first two treatments and *T.kaufmanniana* all three. Untreated seeds collected in the same year from the same plants served as controls.

Table 8

Germination of tulip seeds after supersonic treatment at a frequency of 30 000 kHz

Species	Duration of treatment (seconds)	Period of sowing to beginning of germination (days)	Period of sowing to end of germination (days)	Length of germination period (days)	Germination (%)
T. fosteriana	90	76	158	82	81.4
T. fosteriana	60	90	129	40	89.0
T. fosteriana, control	0	81	144	63	87.0
T. kaufmanniana	90	41	168	127	36.7
T. kaufmanniana	60	54	164	110	68.8
T. kaufmanniana	30	87	141	54	90.4
T. kaufmanniana, control	0	87	141	54	98.8

When we received the treated seeds, in 1952, it was not possible to sow them in a refrigerator and therefore we placed them on moist filter paper in petri dishes for germination. Under the conditions of a Tashkent summer this experiment did not provide reliable data. However, on July 30, 1953, we could place dishes with treated seeds and controls in a refrigerator.

The longer the seeds of both species had undergone supersonic treatment the earlier they started to germinate (table 8). *T.kaufmanniana,* treated for 90 sec., after 41 days; treated for 60 sec., after 54 days and treated for 30 sec., after 87 days, the same as the control. An exposure for 30 seconds had no effect in this experiment.

While a longer exposure resulted in an earlier germination, it prolonged the germination period, showing the possible variation in their germination pattern as was the case in the experiments with seeds collected some two weeks before full maturation and then dried. A longer exposure reduces the germination percentage (table 8).

The results for *T.fosteriana* are very similar, the only difference being that the treatment for 60 sec. resulted in later germination (9 days) and reduced the length of the germination period by 23 days, as compared with the control; the germination percentage increased from 87 to 89%.

The results of our studies and of the experimental work of Ruban and Komarov, who treated seed of *Lonicera* and *Ligustrum* with supersonic waves, lead to the conclusion that this treatment may be used within certain limits to accelerate germination of seeds of widely different species. The practical application of this method should be worked out for single species or groups of closely related taxa. Moreover, it will be important to follow closely the development of the plants obtained after this treatment of the seeds, since disturbances in their growth pattern and the occurrence of various malformations, particularly when high frequencies and longer exposures are applied, may be possible.

Poptsov (1954) assumes that germination already begins during stratification and that the seeds are then not in the after-ripening stage.

All seeds needing stratification belong to a type which requires a low temperature optimum for germination. It should be investigated whether the embryo is dormant or if the delay in germination is due to some other cause. It is thought that the seed coats, the endosperm and the perisperm have a great influence on the germination process. Many workers have succeeded in germinating excised embryos of non-stratified seeds and concluded that the embryos of these seeds were not dormant and did not require after-ripening. Others have pointed out that by no means all excised embryos will germinate.

Nikolaeva (1950) observed earlier germination when the seed coat and the endosperm next to the root of the embryo were broken, but does not give any information about further development of the young plant. She found that the difficulty with which water is transported in the embryos is apparently due to insufficient permeability of their tissues. Sometimes the embryo is not capable of rapid germination, and when it germinates slowly the plants grow badly and enter the fruit-bearing phase very late (Rodionov, 1950).

The seedcoat, endosperm and perisperm, which tightly envelop the embryo, form barriers to germination. Poptsov (1954) arrived at the well-founded conclusion that these barriers are not purely mechanical, but physiological, since they impede the respiratory gas exchange. An embryo with insufficiently permeable tissues experiences the same difficulties.

As a rule excised embryos from unstratified seeds also sprout at higher temperatures and the low temperature optima are not the best for the germination of excised embryos.

Poptsov cites the following essential points which may contribute to understanding why some types of seed can germinate only at a certain low temperature:

1. the oxygen needed for respiration can enter the seed only in an aqueous solution;

2. the solubility of oxygen in water increases at lower temperatures;

3. the rate of respiration and consequently the quantity of oxygen needed by the embryo is considerably reduced at low temperatures (the respiratory quotient ranging from 2 - 3);

4. the permeability of the tissues also decreases, although to a lesser extent, because this property is connected with processes of diffusion and osmosis (temperature coefficient 1.1-1.3);

Thus at lower temperatures the oxygen requirements of the embryo decrease much more than the permeability of the tissues. Under these conditions the embryo can therefore obtain sufficient oxygen to start its life functions.

Proskoryakov (1952) used Raunkiaer's system for the classification of life forms. Agreeing with the criticism levelled at this system and admitting that it is rather artificial, Kul'tiasov (1950) states that there is at present no other workable system of life forms, and that the Raunkiaer system, if correctly applied, gives good results. Using this system he classifies the tulips under the 'geoxerophytic ephemeroids' and characterizes them as follows: 'The geoxerophytic ephemeroids represent a biologically well-defined form of life of the Mediterranean region, not dependent on taxonomic position and morphological habit. The seeds have a germination optimum at low temperatures. By this adaptation to the specific climate the plant can make a more efficient use of the low precipitation in this region.'

Summary

We have seen that in nature some seeds accelerate their germination under conditions of repeated thawing, droughts and frost, whereas others delay germination and protract this period with the possibility that part of the seeds defer germination till the next year.

The adaptibility of the seeds to variable conditions of germination over years with a different seasonal weather pattern has survival value.

The rates of germination and the percentage of seeds that will germinate in spring are dependent on the weather conditions prevailing throughout the year.

The experiments with immature seeds have shown that green seeds with a still soft endosperm, collected some two weeks before ripening, do not germinate. Of the seeds which were beginning to turn yellow, 50% germinated. The seeds which were harvested green and were dried afterwards germinated more readily, but had a much longer germination period.

Seeds of plants that have experienced dry conditions at the end of their vegetation period are capable of extending germination over several years, which clearly shows their variability.

The desert and foothill species germinate more readily, have a higher rate of germination and consequently a shorter germination period (e.g. *T.regelii, T.albertii, T.kuschkensis* and *T.tubergeniana).* Germination of the later starting species is slow and extends over a much longer period (e.g. *T.fosteriana, T.kaufmanniana, T.kolpakowskiana, T.dubia, T.praestans* and *T.hissarica).*

Sympatric or partially sympatric species growing under comparable ecological conditions show a similar germination behaviour (e.g. *T.regelii* and *T.albertii; T.korolkowii* and *T.greigii).*

Seeds of one species collected from different populations, occupying ecologically different habitats, show a deviating germination behaviour. This property is hereditary (e.g. *T.greigii* from Karankulsai and Darbaza).

Phylogenetic affinity may also be expressed in the germination pattern of related species, for example by those of the section *Tulipanum* we studied. The experiments have shown that they have a similar germination behaviour and it may be concluded that *T.kuschkensis* is closer to *T.tubergeniana* than to *T.hoogiana* and that *T.ingens* is intermediate between the species of the section *Tulipanum* and the *Leiostemones* grouped around *T.fosteriana.*

CHAPTER EIGHT

Growth and development of the seedling

The young shoot

The seedlings of the seventeen species and the many varieties we studied formed in the first year only one primary root, which died towards the end of the vegetation period. It was only in a single case that we found a first-year seedling of *T.bifloriformis*, with a branched primary root.

In 1953 seeds were germinated under humid conditions in a refrigerator. Three weeks after sprouting up to 3 mm long root hairs were observed on the primary roots of *T.dubia, T.ingens, T.maximowiczii* and *T.wilsoniana* (fig. 33). We never found root hairs in other species and the older plants do not form root hairs either. In the first year at the end of the dormant period adventitious roots are initiated on the basal plate to replace the dead primary root and in the second year all species and varieties have formed varying numbers of unbranched adventitious roots.

Under Tashkent conditions tulips start sprouting in spring as soon as the midday temperatures rise above zero, approximately from the middle of February to mid March. When the night temperatures fall rather low, after an early sprouting, the shoots may take a long time to become green. At first they are lying on the ground, but later they become erect and are brightly coloured by anthocyanin. Regardless of the time the young plants have started sprouting, they will as a rule have completed their vegetative growth with the arrival of higher temperatures near the end of April. Of the seedlings, which were grown in the botanical garden, *T.subpraestans* had the longest vegetation period.

A first-year seedling produces only one cotyledon. After germination the upper colourless part of the cotyledon, which functions as haustorium, remains covered by the seed coat for quite a long time. This part is flattened, has an obtuse top, and its vascular bundles are not arranged in a circle as in the assimilative part, but spread out flatly.

The green part in the middle of the cotyledon, referred to by Vasil'chenko (1945) as 'the cotyledon connective', is nearly cylindrical, broadest in the middle and tapers gradually to both ends. This photosynthetically active part is covered by an epidermis consisting of rectangular cells, which are drawn out longitudinally (*T.greigii* and *T.kaufmanniana,* fig.34). The outer walls of the cells are thickened and covered with a cuticle. The rather small stomata have bean-shaped guard cells and the stomatal apertures are also stretched longitudinally.

A cross section of the tissue under the epidermis shows a large-celled, rather loose and still undifferentiated parenchymatous cell tissue with well-developed air ducts (fig.35a). The vascular bundles are deeply embedded in the parenchyma in a more or less circular arrangement (fig.35b). *T.kaufmanniana* shows as a rule five or seven vascular bundles, three large and two or four small ones (fig.36). *T.greigii* has five and *T.korolkowii* has two vascular bundles. The distributional pattern of the vascular bundles in the cotyledon has one axis of symmetry. In *T.kaufmanniana* the three largest bundles are found in one half of the cotyledon and the two small ones in the opposite half, while there are often two still smaller bundles between these two groups. As far as can be concluded from its anatomy the rounded shape of the cotyledon resulted from reduction of the upper plane of the cotyledonary leaf as is confirmed by the arrangement of the xylem and phloem in the vascular bundles (fig.36a). The adaxial xylem consists only of a few annular vessels, the abaxial phloem is better developed. The phloem and xylem are separated by small and rounded, thin-walled cells and the vascular bundles are surrounded by two or three rows of parenchymatous cells, which are smaller than the other cells of the mesophyll. The cotyledon does not contain sclerenchyma or single sclerenchyma cells (fig.36b). In *T.kaufmanniana* the 8-10 rows of epidermal cells on the side of the two smaller vascular bundles form a shallow groove, which apparently is a remnant of the upper plane of the cotyledonary leaf (fig.36c).

Before fixation in the soil the cotyledon shrinks considerably because the air ducts in the parenchyma disappear. Below ground the cotyledon retains its rounded shape and does not contain chlorophyll. Although it could be called a petiole, it is more correct to refer to this part as a sheath.

The initial sinking of the cotyledon proceeds, as is

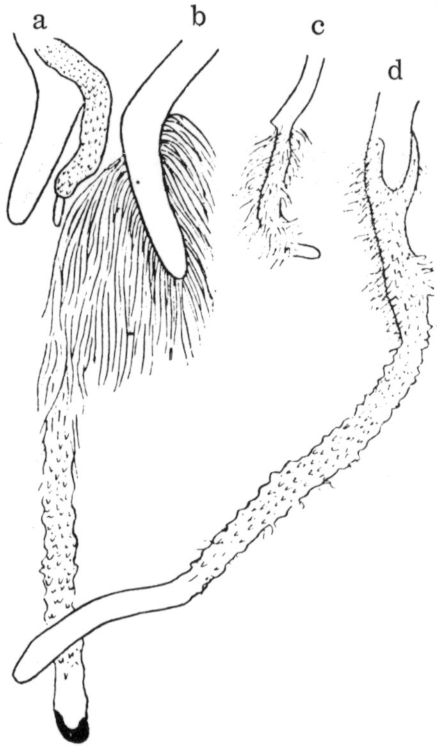

Figure 33. Root hairs of the primary root of the seedling.
a. *T. ingens;* b. *T. wilsoniana;* c. *T. maximowiczii;* d. *T. dubia.*

usual in Liliaceae, by positive geotropical growth of its sheath, which carries the bud of the young shoot several centimeters below ground level. In the first year the young shoot of most species we have studied sinks deeper than the stolons which develop later on.

In the sheath, next to where the cotyledon is attached to the basal plate and opposite the primary root, is the aperture of a cavity which extends down to the sunken bud. In *T.kaufmanniana* this cavity is located in the parenchymatous tissue between the two small vascular bundles, i.e. near the reduced, flattened upper side of the cotyledon. We have sometimes noticed in *T.kaufmanniana* and *T.fosteriana* above the aperture a small ligulate outgrowth, consisting of 4 to 6 cell layers, which resembles the epiblast of grasses. In *T.greigii* and most other species the aperture is formed by the appearance of a groove in the lower part of the cotyledonal sheath on the side of the reduced upper plane. This groove deepens gradually and forms the cavity of the sheath by drawing its edges closer together.

The stolon

On the side of the aperture in the cotyledonal sheath, where the basal plate should develop, appears a rounded hollow outgrowth, which is several times thicker than the primary root, but not twisted. Its development can be followed with the naked eye as soon as the cotyledon and primary root start growth and are not yet functioning as organs. In the literature this outgrowth, which has been described by many authors, has no generally accepted name. Irmisch (1850) and still earlier Germain de Saint Pierre and Fabre called it

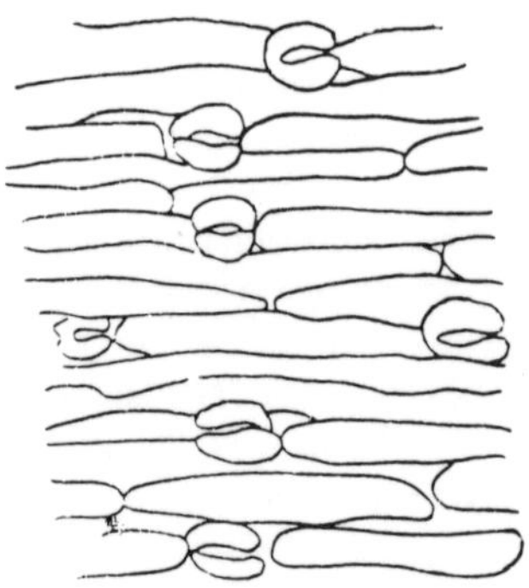

Figure 34. Epidermis of the assimilatory part of the cotyledon of *T. greigii.*

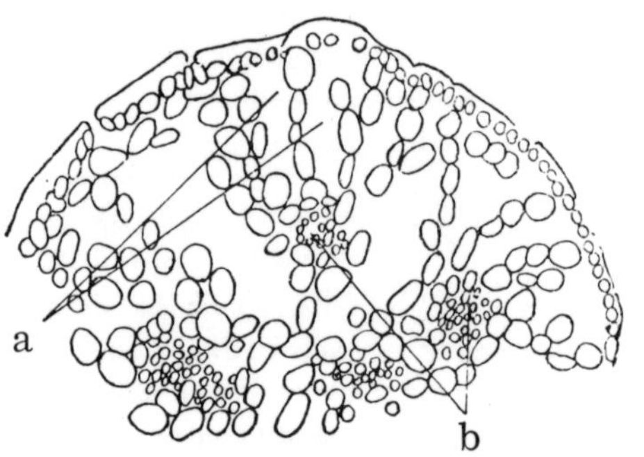

Figure 35. Cross section of the assimilatory part of the cotyledon of *T. greigii.*
a. air ducts; b. vascular bundles.

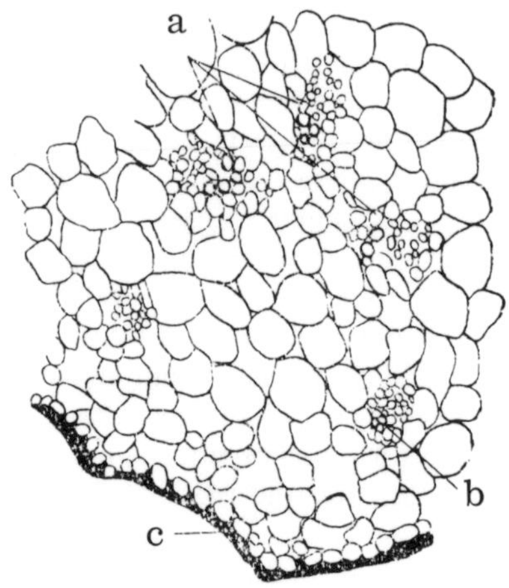

Figure 36. Cross section of the lower part of the cotyledon of *T. kaufmanniana.*
a. vascular bundles; b. smaller vascular bundles close to the periphery; c. reduced upper plane of the cotyledon.

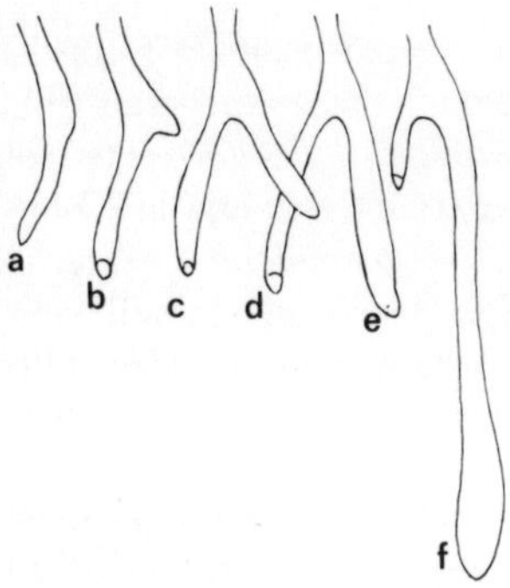

Figure 37. Initial growth stages of the stolon of a young tulip seedling.
a. on emergence from the seed; b, c, d and e. consecutive stages of development; f. stolon after completion of its extension growth.

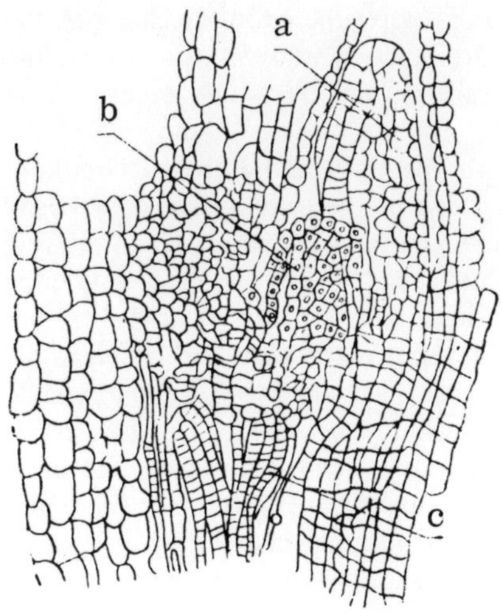

Figure 38. Apical bud of a young shoot of *T. kaufmanniana* after displacement by stolon formation (longitudinal section).
a. dorsal part of the fleshy scale primordium; b. median vein; c. vascular bundles of the basal plate.

'Schporen'. Hall (1929) calls it 'dropper' when it grows vertically downward and 'stolon' when it grows horizontally. Yanishevskiĭ (1934) used the terms 'proliferous outgrowth' and 'rhizomorphous outgrowth'. Serebryakov (1952) refers to it as 'bulb stolon', 'typical stoloniferous organ' and 'stoloniferous dropper'. In the "Flora of the USSR", Vol. IV, it is called 'underground rhizome'. Following the experts we feel fully justified to maintain the term 'stolon'.

Morphologically and also in its organogeneses it differs from the stolons of the other higher plants but resembles them in its functions of vegetative reproduction, removal of the daughter bulb from the parent plant and the transport of nutrients from parent plant to daughter bulb.

The stolon is initiated a few days after the seedling emerges from the seed coat. At first, as a result of unilateral development of the first node of the young shoot and the swollen part of the cotyledonal sheath, the apical bud begins to move towards the aperture of the cotyledonal sheath. This process can be easily followed in young tulip shoots during sprouting of the seeds even before the cotyledons become green. The first stage is a slight swelling on one side of the shoot at the base of the cotyledon (fig.37a). Then, because of the intercalary growth in the stem and the cotyledonal sheath, a bag-shaped, gradually elongating, outgrowth

Table 9

Average size of first-year tulip seedlings (cm)

Species	Length of the haustorium	Cotyledonal blade		Depth of initial sinking of the apical bud by the cotyledonal sheath	Diameter of the young shoot	Depth of sinking of the apical bud by the stolon	Diameter of the stolon	Total depth of sinking in the first year
		Length	Diameter					
T. kaufmanniana	0.7	8.5	0.30	4.3	0.10	2.5	0.10	6.8
T. greigii	0.7	8.0	0.18	3.0	0.10	5.0	0.12	8.0
T. albertii	0.5	4.7	0.10	4.0	0.05	4.0	0.07	8.0
T. micheliana	0.7	6.0	0.16	6.0	0.12	1.5	0.18	7.5
T. subpraestans	0.7	10.5	0.15	4.0	0.11	4.5	0.14	8.5
T. praestans	0.6	6.5	0.13	3.6	0.05	2.6	0.05	6.2
T. ostrowskiana	0.4	7.0	0.15	4.0	0.08	3.7	0.16	7.7
T. kolpakowskiana	0.35	6.0	0.12	5.0	0.08	3.5	0.10	8.5
T. schrenkii	0.6	5.0	0.10	4.5	0.03	1.5	0.10	6.0
T. korolkowii	0.5	6.5	0.12	5.0	0.06	3.0	0.16	8.0
T. bifloriformis	0.3	4.5	0.08	5.0	0.05	3.0	0.07	8.0
T. maximowiczii	0.4	4.5	0.08	2.5	0.05	1.5	0.07	4.0
T. hoogiana	0.7	6.0	0.15	3.5	0.10	1.5	0.10	5.0
T. ingens	0.6	6.5	0.15	4.0	0.08	4.0	0.08	8.0
T. fosteriana	0.5	7.0	0.30	3.0	0.10	3.0	0.10	6.0

is formed which develops like a root by changing the negative geotropic reaction into a positive one; the stem growing at the top and the cotyledon at its base (fig.37 b,c,d,e).

The apical bud, which has been shifted to the stolon, consists at this stage of the primordium of a fleshy scale. It has a cylindrical shape and is more strongly developed on the abaxial side of the cotyledon (fig.38a). This primordium is the sheath of the first leaf which has only a rudimentary blade and becomes the first fleshy bulb scale. Its median vein becomes visible somewhat later opposite the side of the cotyledon, showing that the young shoot has already the distichous leaf arrangement typical of Liliaceae. There is at this stage no meristematic zone in the centre of the bud which has hardly increased in size during the extension growth of the stolon.

The stolon of all first-year seedlings grows as a sinker and is an important organ in adjusting bulb depth. As the light and flat seeds cannot penetrate into the soil, they usually germinate on the surface or in crevices where they have been blown by the wind. However, the primary sinking of the apical bud is caused by the positive geotropic reaction of the cotyledonal sheath and only later by that of the stolon (table 9).

The process of stolon formation in a dropper by concrescence of the tissues of two organs, the stem and the cotyledonal sheath, has been described by Yanishevskiĭ (1934) for adult plants of *T.schrenkii* and *T.biebersteiniana* and by Botschantzeva (1956) for many other species of various ages.

The number of vascular bundles running from the stem of the young shoot through the stolon varies from two to seven: in *T.korolkowii* and *T.bifloriformis* two or three, in *T.greigii* and *T.fosteriana* five and in *T.kaufmanniana* five or seven. In *T.kaufmanniana* the three or five bundles of the cotyledon (3 large and 2 small ones, fig. 39a) unite in the basal plate with the bundles of the root (fig. 39b) and then continue in the stolon (fig. 39c). The two marginal bundles (fig. 39d) run directly from the cotyledonal sheath to the stolon. It is the node and not, as in dicotyledonous plants, the internode which takes part in stolon formation, because part of the vascular bundles of the cotyledonal sheath enter the stem at the basal plate while the other bundles join them later at the base of the bud. In juvenile and rejuvenated individuals the typical transverse growth of the nodes was found to have attained a length of 10-30cm (50cm in daughter bulbs of *T.fosteriana),* without any extension of the internodes. The mentioned ligulate outgrowth above the aperture in the cotyledonal sheath, similar to the epiblast in grasses, very much resembles a rudimentary second cotyledon (fig. 40a). However, when accepting this view, the position of the dorsal side of the first leaf next to this 'rudimentary cotyledon' does not agree with the dis-

Figure 39. Vascular bundles in a young shoot of *T. kaufmanniana* extending from the leaf and the root into the stolon.
a. Large vascular bundles of the cotyledon; b. vascular bundles of the root; c. vascular bundles extending from the basal plate to the stolon; d. marginal small bundles of the cotyledon directly extending into the stolon.

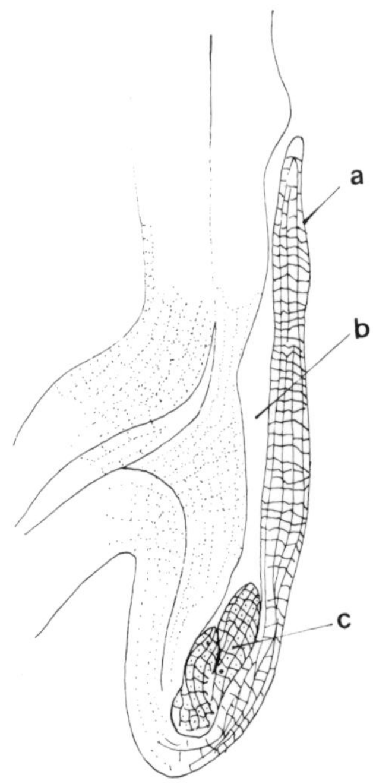

Figure 40. Dropper stolon of *T. fosteriana* with ligulate outgrowth.
a. ligulate outgrowth on the cotyledonal sheath; b. air channel; c. dorsal part of the first leaf (fleshy scale).

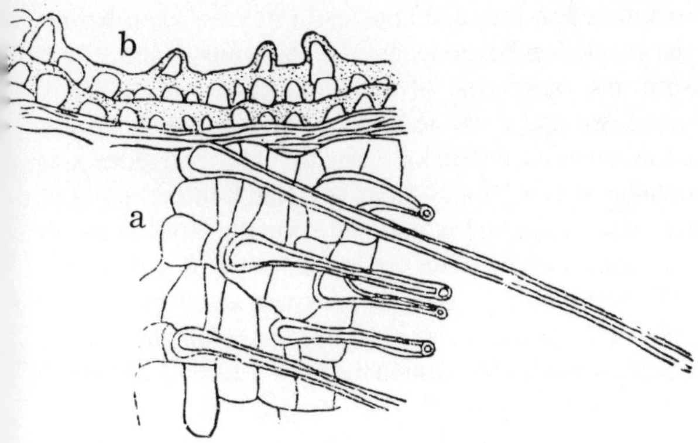

Figure 41. Epidermal tissue of stolon channel of *T. greigii*. a. erect hairs in stolon channel; b. cells of outer epidermis with thickened walls at the top of the thickening stolon.

tichous leaf arrangement of Liliaceae (fig.40b). The ligulate outgrowth consists of six cell layers including the inner and the outer epidermis. It has no vascular bundles, nor are there any in the wall of the stolon below it.

The apical bud is found at the bottom of the air cavity running from the aperture in the sheath through the centre of the stolon (fig.40c). It was brought there by the unilateral growth of the node and the base of the cotyledonal sheath. The cavity connects the bud with the air outside and is covered on the inside, almost from the basal plate to the bud, with long, erect, unbranched hairs (fig.41a).

The growing stolon is covered on the outside with a smooth epidermis. However, when it is lifted from humid soil before it begins to thicken, the whole stolon

Table 10

Growth of stolons during the first year, under field conditions

Species	Appearance of the stolon	End of extension growth and beginning of swelling	End of vegetation period	Period of growth (days)		Duration of the vegetation period (days)
				Elongation	Swelling	
T. bifloriformis	28/2	10/3	15/4	10	36	46
T. fosteriana	28/2	23/3	18/4	23	26	49
T. kaufmanniana	28/2	20/3	20/4	20	31	51
T. greigii	28/2	26/3	19/4	26	24	50

Table 11

Relation between the growth of the apical bud and the daughter bulb of a sinker

Species	Second year*					Third year					Fourth year*					Fifth year				
			Sinking (cm)					Sinking (cm)					Sinking (cm)					Sinking (cm)		
	Stolon	Daughter bulb	Apical bud	Daughter bulb bud	Depth of bulb (cm)	Stolon	Daughter bulb	Apical bud	Daughter bulb bud	Depth of bulb (cm)	Stolon	Daughter bulb	Apical bud	Daughter bulb bud	Depth of bulb (cm)	Stolon	Daughter bulb	Apical bud	Daughter bulb bud	Depth of bulb (cm)
T. fosteriana	+	+	3.0	0	9.0	+	+	15.0	12.0	23.5	+	+	9.0	8.0	25.0	+	+	16.0	8.0	21.0
T. carinata	+	+	3.5	1.5	10.5	+	+	3.7	3.3	13.0	+	+	3.5	4.0	12.5	+	+	29.0	20.0	33.0
T. kaufmanniana	+	+	5.0	0	10.5	+	+	6.5	0	14.0	+	+	8.0	0	16.5	—		—	—	—
T. greigii	+	+	4.0	0	14.0	+	+	7.5	0	16.0	+	+	10.0	0	28.0	—		—	—	—
T. subpraestans	+	+	12.7	4.0	17.0	+	+	7.7	5.4	13.0	+	+	13.0	10.0	26.0	—		—	—	—
T. praestans	+	+	5.5	0	16.0	+	+	11.5	0	18.0	+	+	7.0	5.0	15.5	—		—	—	—
T. korolkowii	+	—	3.0	0	12.0	+	+	4.5	1.8	14.5	—	—	0	0	16.5	—		—	—	—
T. kolpakowskiana	+	+	9.0	0	13.0	+	+	16.0	2.0	20.0	—	—	0	0	20.0	—		—	—	—
T. hoogiana	+	+	3.0	0	11.0	+	+	11.0	0	17.0	—	—	—	—	—	—		—	—	—
T. bifloriformis	+	+	2.0	0	9.0	+	+	3.4	1.0	11.0	+	+	3.0	3.0	15.0	—		—	—	—
T. ingens	+	—	4.0	0	12.0	+	+	11.4	0	20.0	—	—	—	—	—	—		—	—	—
T. ostrowskiana	+	+	6.0	0	10.5	+	+	11.0	0	16.5	—	—	—	—	—	—		—	—	—
T. schrenkii	+	—	4.0	0	15.0	+	+	14.5	0	17.5	—	—	—	—	—	—		—	—	—
T. maximowiczii	+	—	3.3	0	15.0	+	+	11.0	0	17.0	—	—	—	—	—	—		—	—	—
T. micheliana	+	+	10.0	0	20.0	+	+	—	—	—	—	—	—	—	—	—		—	—	—

* After the second and the fourth growing season the bulbs were transplanted at a depth of 7–10 cm, depending on their size.

is coated with small soil particles, only the growing tip being free of them. After extension growth has stopped, the top of the stolon starts swelling and enters the next stage of development when the apical bud of the young seedling begins to grow (table 10). During this period the bud grows by enlargement of the organs already initiated. The epidermal tissue of the swollen part of the stolon changes. Some of the cells acquire a bottle-shape and clearly protrude above the adjacent cells, giving the impression of thick-walled root hairs (fig.41b). Below these is another layer of cells with thickened radial and tangential walls. Many botanists, who found these elongated bottle-shaped cells at the tip of a stolon, assumed them to be root hairs. Actually these solid cells are formed when the stolon starts swelling. They help to fix the young bulb in the soil when the stolon withers away and protect it from being squashed by the soil.

Thus, the development of the stolon passes through three stages. During the first stage the bud of the seedling is displaced sideways; during the second stage the stolon elongates; in the third stage it thickens when the displaced apical bud develops into a bulb (tables 10, 11).

Already in the first stages of development the stolon has stored food reserves. The starch which had accumulated in the seed appears in its cells even before the cotyledon becomes green. The walls of the stolon form the outer coat of the bud, which only slightly elongated and remained very narrow when the stolon completed its extension growth. The bud does not contain starch. Nor is there any starch in the tissue at the base of the bud which will form the basal plate, in the inner and the outer epidermis of the stolon, in the cells surrounding the vascular system and in the oblong cells of the provascular bundles. The remaining stolon tissue, in particular around the bud, is filled with starch which will be gradually translocated to new cells with further thickening of the stolon.

The first-year-bulb

During thickening of the stolen the seedling does not initiate new organs but the single basal leaf, the fleshy scale, grows vigorously. It is the storage organ of the first-year-bulb. Under our conditions the rootlet of the seedling dies and the cotyledon withers in the second half of April. The starch accumulated in the bulb tunic, formed by the basal part of the stolon, is translocated to the fleshy scale and the tunic itself dries up. This is the beginning of the dormant period.

Table 12

Bulbs from seedlings

Species or cultivar	First year					Second year				
	Weight (mg)			Sinkers (%)	Daughter bulbs (%)	Weight (mg)			Sinkers (%)	Daughter bulbs (%)
	Large	Medium	Small			Large	Medium	Small		
T. tubergeniana	130	73	30	100	0	782	612	550	100	5
T. kuschkensis	125	90	32	100	0	717	580	440	100	6
T. hoogiana	140	75	35	100	0	985	651	450	100	30
T. ingens	100	60	25	100	0	625	512	420	99	0
T. fosteriana	120	77	30	100	0	550	412	250	100	50
T. praestans	155	100	50	100	0	795	595	230	100	100
T. subpraestans	160	122	60	100	0	1000	325	80	100	90
T. albertii	80	66	25	100	0	1220	726	390	94	0
T. greigii	145	80	24	100	0	1050	220	31	100	16
T. micheliana	140	100	40	100	0	725	526	360	100	2
T. eichleri	—	—	—	100	0	460	332	265	73	0
T. maximowiczii	30	20	10	100	0	200	125	80	100	2
T. wilsoniana	45	33	15	100	0	370	293	200	10	8
T. korolkowii	60	43	20	100	0	370	280	160	31	0
T. schrenkii	80	62	25	100	0	—	—	—	—	0
T. ostrowskiana	130	72	20	100	0	770	466	290	98	13
T. kolpakowskiana	135	90	25	100	0	450	329	106	92	5
T. dubia	120	94	27	100	0	970	636	380	100	0
T. vvedenskyi	150	100	40	100	0	1350	984	635	100	50
T. kaufmanniana	180	104	37	100	0	1900	1160	65	100	20
T. regelii	54	28	16	100	0	474	134	43	40	32
T. bifloriformis	27	18	10	100	0	77	43	18	—	—
'D. v. Tol Yellow'	80	58	20	100	0	1400	97	35	60	10
'Jubilee'	77	50	28	100	0	372	167	72	20	27
'Elephant'	80	55	25	100	0	327	133	15	20	5
'Kathleen Parlow'	138	104	57	100	0	372	162	59	30	14

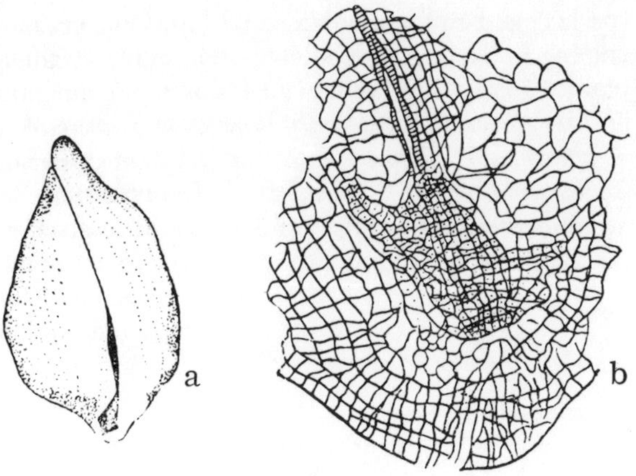

Figure 42. Longitudinal section of a first-year bulb in mid-May. a. general view; b. the basal plate at greater magnification, showing the living cells of the inner epidermis.

Although the seedling has in its first season only one green organ, the small cotyledon, which varies in size from one species to another (table 9), and only one rootlet, it manages to store food reserves and water in the young bulb. These reserves are sufficient for the bulb to survive the hot and dry dormant period, to initiate the assimilatory organs for the next season and to recommence growth towards the end of the dormant period. The first-year bulbs of the larger species do not weigh more than 180 mg, those of the smaller species may be only 10 mg (table 12).

Around the middle of May a first-year *T. greigii* bulb has only one fleshy scale containing much starch and is covered by a brown tunic (fig.42a). The initiation of other organs sets in later. In the fleshy scale two tissue layers of nearly equal thickness can be distinguished, an outer layer containing much starch and an inner layer of more loosely arranged cells without starch (fig.42a). In the centre of the scale runs a narrow slit lined with a monolayered, thin-walled epidermis, also without starch. When the first green leaf is being initiated the epidermis does not rupture, because its cells can elongate and their walls can continue growing. Judging by their appearance the cells of the inner epidermal layer are even capable of division because there are no signs of senescence (fig.42b). When examining these cells in May, it appears that they have not become vacuolated and that their nuclei have not accumulated reserve or waste products. In the structure of their protoplasts they resemble the group of meristematic cells at the base of the bulb (fig.42b), where the first adventitious roots are initiated during the second half of the dormant period near the end of July. The primordia of the first green leaf and the second fleshy scale will appear at the base of the narrow slit in the first fleshy scale.

The leaf

The tulip has leaves which are arranged in distichous rosettes. The midrib of the first green leaf is situated at the side where the cotyledon has been, because the only leaf left of the first year, which functions as a storage organ, stands between them. The second fleshy scale, appearing the next season, has its midrib on the side of the first scale from which it is separated by an internode formed by the elongated stolon. Thus, during the years preceding the flowering phase, the fleshy scales develop on one side of the bulb until a larger bulb is formed consisting of several fleshy scales.

The first green leaf, which required some time to develop inside the bulb, emerges from the bulb late in autumn and appears above ground the next spring. It will be the only photosynthetically active organ during the second year of the bulb's life. It has a green blade and underground a white petiole and sheath. The dimensions of the blade are very variable (table 13), but the leaf of a two-year-old seedling already shows the characteristic length to width ratio of the leaf of a juvenile plant. Although rather small, it has the structure, shape and colour which is typical of the given species. For example in *T. micheliana* the anthocyanin stripes, in *T. greigii* the anthocyanin spots, in *T. regelii* the ribbed upper surface, in *T. fosteriana* the horny and often red-edged margins, in *T. korolkowii* the undulate margins, in *T. kaufmanniana* the bright green longitudinal veins and the reticulate lateral pattern shining through the whitish epidermis.

The blade, which has numerous stomata, serves as respiration and transpiration organ and collects the light needed for photosynthesis. The petiole is only transport organ and carries the blade above ground. Its length depends on the position of the bulb, the deeper it lies the longer is the petiole. The sheath of the green leaf of juvenile individuals envelops the bud from the early stages of its initiation. Together with the stem part of the plant it forms the outer scale of the growing bulb and its tunic during the next dormant period. When the apical bud sinks, the sheath of the first green leaf fuses with part of the stem and forms a stolon in the same way as the cotyledon. The dual nature of the stolon is even more pronounced in stolons of bulbs of the second and following years than in those of a first-year-seedling.

Stolons of juvenile plants

The cross section of the stolon of the two-year-old *T. greigii* plant in fig. 43 shows the distribution of its vascular bundles. In the vascular bundle, extending from the stem, the adaxial xylem and the abaxial phloem are clearly visible. On both sides are vascular bundles extending directly from the leaf to the stolon.

In all species studied we observed the same pattern and it follows that, also in these cases, it is the node and not the internode which takes part in the development of a stolon (figs.43,44).

In some of the 22 species studied we found individuals that did not make sinkers at the end of the first year (*T. ingens, T. albertii, T. ostrowskiana, T. kolpakowskiana, T. dubia, T. regelii* and *T. bifloriformis*). The species can be roughly divided into four groups, according to the degree of stolonisation in the juvenile plants. In the first group all bulbs make sinkers (*T. tubergeniana, T. kuschkensis, T. hoogiana, T. fosteriana, T. praestans, T. subpraestans, T. greigii, T. micheliana, T. maximowiczii, T. vvedenskyi, T. dubia* and *T. kaufmanniana*). In the second group at least 90% of the juvenile plants make sinkers (*T. ingens,*

Table 13
Linear dimensions of leaves of non-flowering tulip seedlings (cm)

	Second year						Third year					
	Length			Width			Length			Width		
Species	Maximum	Average	Minimum	Maximum	Average	Minimum	Maximum	Average	Minimum	Maximum	Average	Minimum
T. hoogiana	14.0	9.0	6.0	0.65	0.44	0.15	15.0	11.0	8.7	2.2	1.5	0.5
T. ingens	16.0	12.7	10.0	1.0	0.8	0.4	15.0	12.2	8.8	3.9	2.6	0.6
T. fosteriana	12.0	9.5	6.0	2.0	1.0	0.5	15.4	12.3	6.0	6.5	4.0	1.4
T. carinata	9.0	8.0	7.0	0.6	0.5	0.4	11.0	9.4	9.0	1.0	0.7	0.6
T. praestans	13.2	7.0	6.0	1.5	1.0	0.6	14.0	11.6	6.0	1.9	1.3	0.9
T. subpraestans	16.2	12.7	11.6	1.2	0.7	0.4	21.0	19.4	18.0	2.5	1.8	0.5
T. greigii	10.0	8.2	6.0	1.4	1.1	1.0	15.0	10.0	9.2	4.3	2.8	1.7
T. micheliana	10.5	8.0	7.0	4.1	0.9	0.7	—	—	—	—	—	—
T. maximowiczii	16.0	14.2	13.0	0.3	0.2	0.12	18.0	16.0	14.0	0.5	0.4	0.3
T. korolkowii	9.0	7.0	5.5	0.3	0.5	0.15	13.0	12.0	9.0	0.9	0.4	0.3
T. schrenkii	9.2	6.0	4.5	0.8	0.6	0.3	17.0	11.0	7.0	1.9	1.2	0.6
T. ostrowskiana	15.0	13.0	9.0	0.7	0.43	0.35	17.0	14.0	11.9	2.0	1.4	0.9
T. kolpakowskiana	7.5	4.5	3.0	1.75	0.7	0.15	19.0	13.0	10.0	2.5	1.0	0.5
T. kaufmanniana	17.1	9.0	6.0	2.2	1.8	1.3	16.4	10.9	5.8	7.8	4.0	2.1
T. bifloriformis	9.0	8.0	6.0	0.4	0.3	0.15	13.0	10.0	7.0	0.8	0.5	0.3

Table 13 continued

	Fourth year						Fifth year					
	Length			Width			Length			Width		
Species	Maximum	Average	Minimum	Maximum	Average	Minimum	Maximum	Average	Minimum	Maximum	Average	Minimum
T. hoogiana	—	—	—	—	—	—	—	—	—	—	—	—
T. ingens	—	—	—	—	—	—	—	—	—	—	—	—
T. fosteriana	19.0	15.0	13.5	6.2	3.8	2.7	19.0	16.0	10.0	16.0	10.0	5.0
T. carinata	25.0	22.3	19.0	6.0	3.5	1.5	19.5	15.0	8.0	5.3	4.9	1.3
T. praestans	16.0	12.5	12.0	3.8	2.8	1.4	—	—	—	—	—	—
T. subpraestans	25.0	17.0	7.0	4.8	3.2	1.2	—	—	—	—	—	—
T. greigii	25.0	14.8	12.0	8.0	3.1	all flowering						
T. micheliana	—	—	—	—	—	—	—	—	—	—	—	—
T. maximowiczii	partly flowering											
T. korolkowii	21.0	15.2	12.0	1.6	1.2	0.7	—	—	—	—	—	—
T. schrenkii	—	—	—	—	—	—	—	—	—	—	—	—
T. ostrowskiana	flowering											
T. kolpakowskiana	flowering											
T. kaufmanniana	18.0	16.4	14.0	8.5	6.3	4.0	all in fruit-bearing phase					
T. bifloriformis	17.0	15.0	13.0	1.2	1.0	0.7	all in fruit-bearing phase					

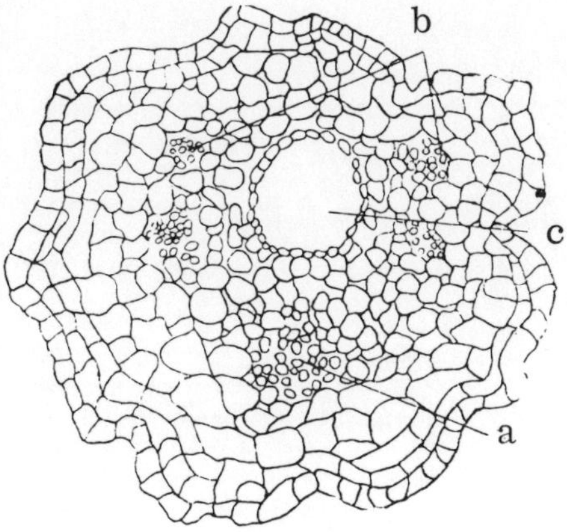

Figure 43. Cross section of a *T. greigii* stolon in the second year. a. vascular bundle extending from the stem to the stolon; b. vascular bundles from the leaf; c. central cavity of the stolon.

T. albertii, T. ostrowskiana and *T. kolpakowskiana).* In the third group 40-75% of the second-year bulbs make sinkers *(T. eichleri* and *T. regelii).* Of the fourth group, which includes *T. wilsoniana* and *T. korolkowii,* less than 40% of the second-year bulbs make sinkers (table 12).

The adult plants of the first group have large bulbs with the exception of *T. maximowiczii,* which holds a place of its own in the genus. The second, third and fourth groups include species with comparatively small bulbs and a number of cultivars (tables 12, 14).

Under Tashkent conditions sinkers are formed towards the end of the dormant period in the first and following years, as a rule until the adult phase is reached. With the exception of first-year-seedlings, the stolons begin to develop in late September or early October and in lifted bulbs even before replanting. At this stage the stolon primordium does not show on the outside of the bulb, as it consists in the beginning only of the displaced apical bud, which is growing in the direction of the concave side of the bulb (fig.44a). At the same time there appears in the wall of the primordial leaf sheath, covering the bud, an outgrowth which can be seen as a small projection at the base of the apical bud. In the thickened half of the bulb scale, where it makes contact with this projection, the cell tissue loosens and the accumulated starch is dissolved and disappears (fig.44b). The displaced part of the node expands because of active cell division and pushes the apical bud towards the loosened cell tissue of the bulb scale.

The basal plate of the bulb is a condensed shoot of which the one node next to the fleshy scale and the true leaf has retained in juvenile individuals and their daughter bulbs, its potential for lateral growth, which attains in some species more than 10 cm (table 14). The difference between a stolon and the shoot of an adult individual is that the node of the apical bud of a juvenile individual has a positively geotropic growth response and expands laterally, whereas the node of an adult individual does not grow, but the internodes elongate by a negatively geotropic growth response when developing the shoot.

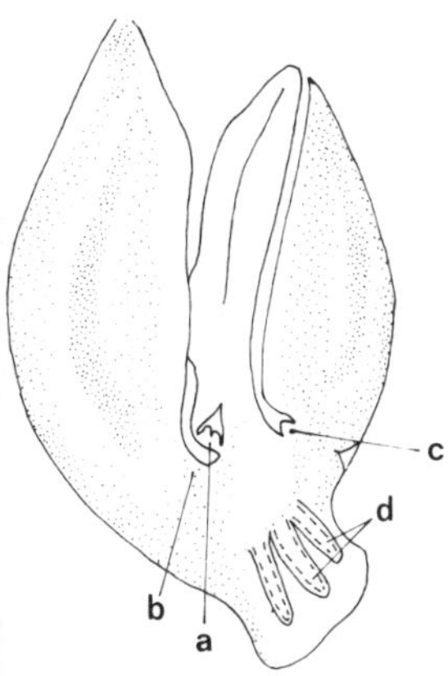

Figure 44. Displacement of the apical bud of a one-year-old bulb of *T. kaufmanniana* towards the convex side of the bulb, in October.
a. displaced apical bud; b. loosening of scale tissue in front of of the shifting apical bud; c. initiation of daughter bulb bud; d. root primordia.

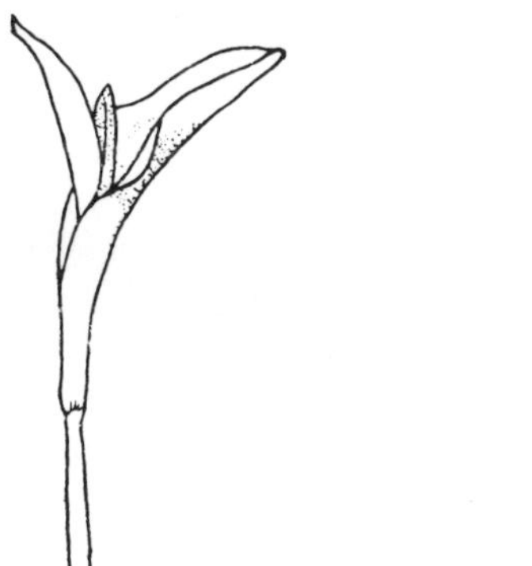

Figure 45. Rupture of fleshy scale tissue by growing stolon of *T. kaufmanniana* (February).
a. stolon breaks through bulb scale; b. attachment of the stolon to the basal plate (scales removed).

Table 14
Length of the stolons and total depth of the bulbs developed from the apical bud (cm)

Species or cultivar	First year		Second year				Third year				Fourth year			
			Length of the stolon				Length of the stolon				Length of the stolon			
	Length of the stolon	Total depth	Maximum	Average	Minimum	Total depth	Maximum	Average	Minimum	Total depth	Maximum	Average	Minimum	Total depth
T. tubergeniana	8.0	22.0	7.0	4.0	2.0	29	29.0	9.2	4.0	40	—	—	—	—
T. kuschkensis	4.0	16.0	3.5	2.7	2.0	20	8.0	4.6	2.0	20	—	—	—	—
T. hoogiana	1.6	5.0	7.0	4.2	2.0	21	7.0	6.1	3.0	18	—	—	—	—
T. ingens	4.0	8.0	7.0	4.0	2.0	20	—	—	—	—	—	—	—	—
T. fosteriana	3.0	6.0	5.0	3.0	2.0	16	—	—	—	—	—	—	—	—
T. praestans	3.6	6.2	5.0	3.0	1.0	15	8.0	5.4	3.0	15	—	—	—	—
T. subpraestans	4.0	8.5	8.0	5.4	3.0	20	—	—	—	—	9.0	5.0	4.0	24
T. albertii	4.0	8.0	10.0	8.0	5.0	10	—	—	—	—	—	—	—	—
T. greigii	3.0	8.0	7.0	4.3	2.0	20	7.0	5.7	5.0	17	15.0	6.7	4.0	37
T. micheliana	1.7	7.5	6.5	2.8	2.0	27	—	—	—	—	—	—	—	—
T. eichleri	3.0	13.0	4.0	2.0	0	17	—	—	—	—	—	—	—	—
T. maximowiczii	1.5	4.0	2.5	1.5	0	17	—	—	—	—	—	—	—	—
T. wilsoniana	1.5	8.0	5.0	0	0	13	—	—	—	—	—	—	—	—
T. korolkowii	3.0	8.0	4.5	4.0	3.0	11	7.0	0	0	10	5.0	2.0	0	17
T. schrenkii	1.5	6.0	—	—	—	—	11.0	7.0	4.0	16	—	—	—	—
T. ostrowskiana	3.7	7.7	8.0	5.4	4.0	21	4.0	3.6	3.0	22	—	—	—	—
T. kolpakowskiana	3.5	8.5	5.0	3.5	3.0	20	—	—	—	—	—	—	—	—
T. dubia	2.5	4.0	7.0	5.5	4.0	10	—	—	—	—	—	—	—	—
T. vvedenskyi	5.0	13.0	7.0	4.3	2.0	20	—	—	—	—	—	—	—	—
T. kaufmanniana	2.5	6.8	7.0	4.6	3.0	15	5.0	4.1	3.0	23	12.0	6.4	4.0	34
T. regelii	9.5	18.5	5.5	1.9	0.7	—	—	—	—	—	—	—	—	—
'D. v. Tol Yellow'	4.0	8.0	5.0	4.0	3.0	13	—	—	—	—	7.0	5.0	3.0	19
'Jubilee'	2.0	6.0	6.0	4.2	3.0	10	11.0	9.0	6.7	18	6.0	4.3	3.0	19
'Kathleen Parlow'	—	—	10.0	6.0	3.0	16	9.0	8.0	5.9	23	8.0	7.0	5.0	17

The positive geotropism of the node of juvenile individuals persists in most species and varieties until the adult phase. In some of them this node may after two or three years lose its capacity for further intercalary growth and does not regain it until reaching the adult phase. This can occur in wild species with shallow-lying bulbs (table 14). The earlier the plant reaches the flowering phase, the sooner the apical bud of the young bulb will produce idividuals that do not sink (tables 12, 14).

The displaced apical bud, covered by the coalescent tissues of stem and leaf, looks like a rather thick rootlet when it emerges from the longitudinal slit in the fleshy scale or scales (fig.45).

There are specific differences in stolon formation. For example the fragile filamentous stolons of juvenile *T.praestans* plants and their globose bulbs are strikingly similar to the stolons and bulbs of *T.fosteriana.* But in *T.praestans* the apical bud, or all axillary buds, sink, leaving only the dead scales of the mother bulb, whereas in *T.fosteriana* there often remain daughter bulbs around the remnants of the mother bulb (figs.46: 1 and 2). Of certain adult individuals of *T.praestans* the secondary bulb and the other daughter bulbs will sink somewhat sideways, a habit which is retained in subsequent seasons. This comparatively rare habit has also been observed in *T.kaufmanniana* and 'King George V', a variety of the Darwin group, but the stolons of these plants are morphologically different from those of *T.praestans.* The main daughter bulb of an adult *T.fosteriana* plant does not sink and remains between the scales of the old bulb, while the sinkers have developed from the apical buds of the other daughter bulbs. They are as a rule considerably rejuvenated and after separation from the mother plant propagate vegetatively, requiring many years to reach the flowering phase.

T.subpraestans, which in flower type and other characters comes closest to *T.praestans,* has very different stolons and bulbs in comparison with *T.fosteriana* and *T.praestans.* They are much thicker and stronger and grow nearly vertically downwards instead of sideways (cf. *T.praestans*). *T.subpraestans* will often form a cluster of non-descending bulbs. The bulbs which have reached the adult phase are radish-shaped and usually do not sink.

The long, thick and soft stolons of *T.ingens,* broadest in the middle, terminate in a narrow, long bulb. In these characters *T.ingens* is quite different from *T.fosteriana* and closely resembles such species as *T.tubergeniana*

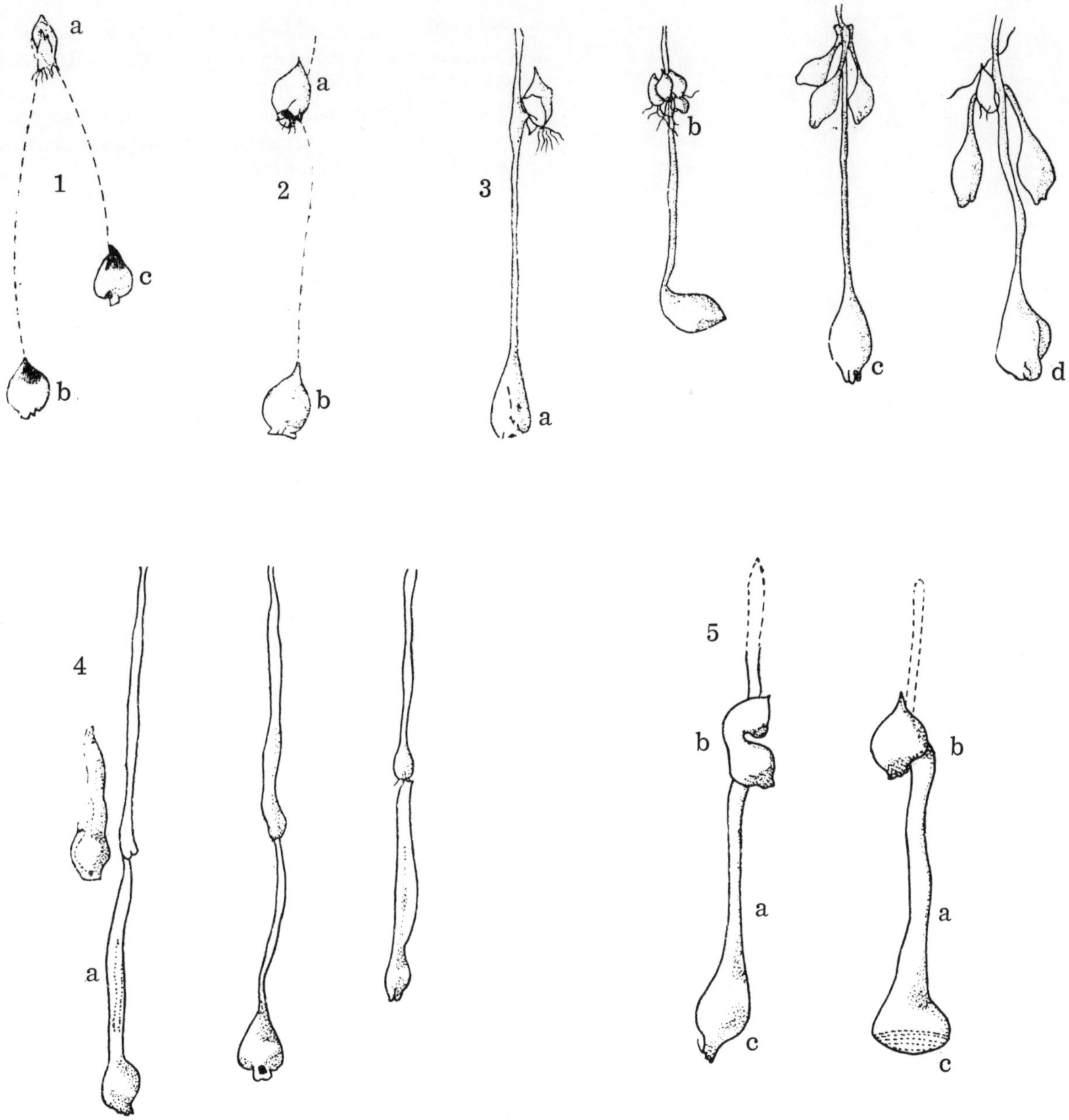

Figure 46. Sinkers
1. *T. praestans*—a. previous year's bulb, b. main daughter bulb, c. other daughter bulb;
2. *T. fosteriana* (two-year-old)—a. main daughter bulb, b. other daughter bulb;
3. *T. subpraestans* (two-year-old)—a. the main daughter bulb has descended and the other daughter bulb remained between the old scales, b. clump of collateral daughter bulbs which have not descended, c. all bulbs have descended, d. the main daughter bulb and two daughter bulbs have descended and one daughter bulb remained in place;
4. *T. ingens* (two-year-old)—a. broad and soft stolon;
5. *T. kuschkensis* (three-year-old)—a. broad and soft stolon, b. daughter bulbs, c. main daughter bulb.

and *T.kuschkensis* of the section *Tulipanum* (fig.46: 4 and 5).

Beginning with the first year every tulip bulb has at least two buds, an apical bud and an axillary bud (figs.47, 48), which is always initiated in the axil of a fleshy scale, but quite often it does not develop beyond a meristematic boss in which terminates a large vascular bundle. In many species the development of the axillary buds depends on the nutrition of the bulb during the previous season. However, in some species axillary buds as a rule do not develop into daughter bulbs even when the conditions for bulb growth are

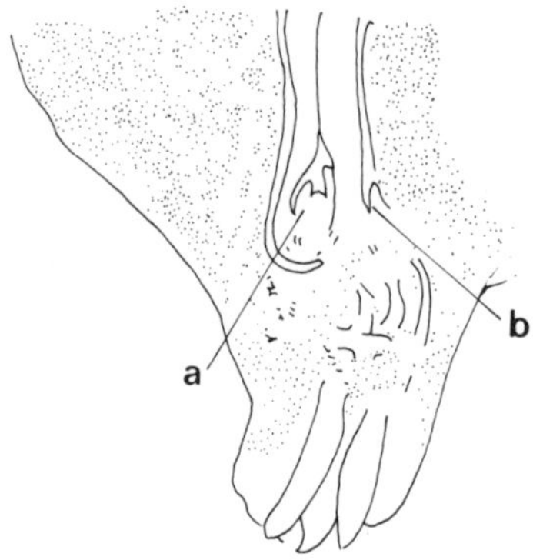

Figure 47. Initiation of daughter bulb in axil of fleshy scale of a one-year-old *T. greigii* bulb.
a. apical bud being displaced with the stolon; b. axillary bud.

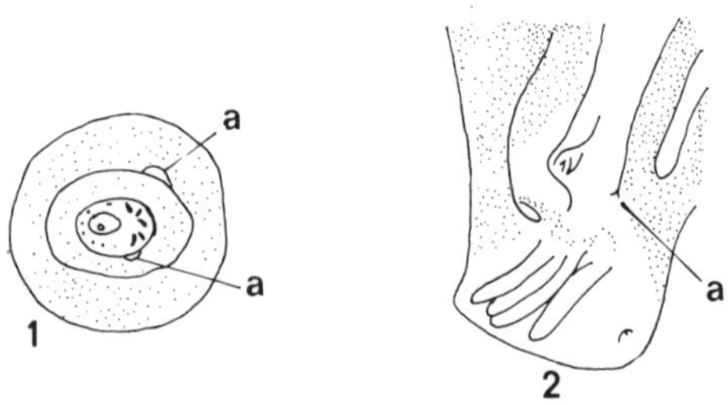

Figure 48. Initiation of daughter bulbs (a) in axil of the fleshy scales of a four-year-old *T. fosteriana* bulb.
1. Cross section: a. daughter bulb bud.
2. Longitudinal section: a. daughter bulb bud.

favourable (e.g. *T.korolkowii, T.micheliana, T.dubia, T.maximowiczii* and certain formae of *T.schrenkii*, (tables 12, 15).

When juvenile individuals have initiated buds the apical bud usually forms a sinker, but the other daughter bulbs remain between the scales of the mother bulb. The tunic of such daughter bulbs consists only of leaf sheath tissue without any coalescent stem parts as in the sinker. It is glossy, thinner and of a brighter colour than the tunic of a sinker.

The habit to form sinkers may vary in different individuals of one species. For example in *T.subpraestans*, we found collaterally initiated daughter bulbs, which remained where the mother bulb had been, while in other individuals they all descended or some daughter bulbs sank and some did not (fig.46; 3a, b, c, d). The vegetative reproduction of this species often exceeds our conception of this capacity where it is based on observations of wild species and even garden varieties. Judging by its vigorous vegetative reproduction *T.subpraestans* was expected to reach the adult phase quite late, but in fact 25.3% of the plants sown in 1949 began to flower in 1953 i.e. after three years.

Early flowering

Hall (1929) found that tulips raised from seed will start flowering in their fifth, sixth or seventh year. However, there are species which, having made only a small number of bulb scales, can flower in the third or fourth

Table 15
Weight (mg) and number of bulbs from seedlings

	Third year					Fourth year				
Species or variety	Large	Average from 100	Small	Sinkers (%)	Daughter bulbs (%)	Large	Average from 100	Small	Sinkers (%)	Daughter bulbs (%)
T. tubergeniana	3100	1000	412	100	26	—	—	—	—	—
T. kuschkensis	2150	1372	900	100	47	—	—	—	—	—
T. hoogiana	4080	2275	1100	100	47	—	—	—	—	—
T. fosteriana	1050	674	400	100	—	4500	3606	800	55	—
T. ingens	—	—	—	—	—	—	—	—	—	—
T. praestans	2000	1050	370	100	100	—	—	—	—	—
T. subpraestans	870	730	50	100	100	10440	2826	50	100	90
T. subpraestans, other form	—	—	—	—	—	15900	10450	4737	4	—
T. greigii	650	348	32	100	17	26500	16500	9400	100	20
T. korolkowii	1100	400	100	17	5	1700	720	280	10	0
T. schrenkii	6400	2765	800	100	100	—	—	—	—	—
T. ostrowskiana	5500	2301	500	100	100	—	—	—	—	—
T. ostrowskiana, other form	—	—	—	—	—	8260	6036	4290	100	—
T. kaufmanniana	1760	1311	600	100	24	11220	5500	3073	100	—
T. bifloriformis	205	63	19	100	76	5020	2464	860	—	—
'Jubilee'	1630	800	250	—	—	—	—	—	—	—
'Elephant'	1920	910	500	25	25	—	—	—	—	—
'Kathleen Parlow'	1250	820	500	—	—	—	—	—	—	—
T. tarda	2250	539	320	100	100	5600	400	100	37	26

Table 16

Flowering of tulip seedlings in the third year

Species or variety	Total number of plants	Flowering	
		Number	%
	Sown 1950		
T. kaufmanniana	33	3	9.0
T. kaufmanniana	40	8	20.0
T. kaufmanniana	466	6	1.3
T. ostrowskiana	22	4	18.2
T. tarda	96	2	2.1
'Orange Delight'	85	2	2.4
	Sown 1952		
T. kolpakowskiana	364	16	4.4
T. tetraphylla	4	2	50.0
T. vvedenskyi	175	3	1.7
T. subpraestans	120	1	0.8

year (table 16). Among the plants which flowered in the third year were species which produced large bulbs with a comparatively large number of fleshy scales, such as *T.kaufmanniana* and the variety 'Orange Delight', and also species with medium-sized and small bulbs, e.g. *T.ostrowskiana* and *T.tarda.* All species and varieties produce at their first flowering not more than 2 or 3 fleshy scales and it is only in subsequent years that this number increases (table 17). In *Leiostemones* the development of two fleshy scales is an indication that the bulb will flower next season. An exception is *T.fosteriana,* which often does not flower although having made four fleshy scales.

Of the species we had sown in the autumn of 1949, a much larger number flowered in 1953, their fourth year, as compared with the number of species flowering in their third year (table 16, 18). Besides *T.kaufmanniana, T.ostrowskiana, T.tarda* and the variety 'Orange Delight', also *T.greigii, T.korolkowii, T.subpraestans, T.bifloriformis,* and several varieties and hybrids between varieties and wild species began to flower in that year. This group includes species which differ in the size of leaves and bulbs and their capacity for vegetative reproduction. Despite considerable differences in bulb morphology and development such species as *T.korolkowii, T.greigii, T. bifloriformis* and *T.subpraestans* started flowering simultaneously (table 13, 16, 18, 19).

In a population raised from seed not all individuals reach the adult phase at the same time, because there are both early and late-flowering individuals. Among three species, sown in 1948, a large number of plants did not flower even in their fifth year, whereas some individuals had already produced flowers for the second time (table 19). However, if calculated from the number of bulbs planted in 1952, instead of 1953, the percentage of plants flowering in 1953 is much higher,

Table 17

Number of scales in tulip seedlings of different ages

Species	First year		Second year		Third year		Fourth year	
	Tunic	Fleshy scales	Tunic	Fleshy scales	Tunic	Fleshy scales	Tunic	Fleshy scales
T. greigii	1	1	1	1	1	2	1	3
			1	2				
T. kaufmanniana	1	1	1	1	1	2	1	3
			1	2				
T. fosteriana	1	1	1	1	1	1	1	2
T. praestans	1	1	1	1	1	1	1	1
T. korolkowii	1	1	1	1	1	1	1	2
T. bifloriformis	1	1	1	1	1	1	1	1
					1	2	1	2
T. ingens	1	1	1	1	1	1		
T. schrenkii	1	1	1	1	1	1		
T. kolpakowskiana	1	1	1	1	1	1		
			1		1	2		
T. ostrowskiana	1	1	1	1	1	1		
				2	1	2		
T. maximowiczii	1	1	1	1	1	1		
T. hoogiana	1	1	1	1	1	1		
					1	2		
T. micheliana	1	1	1	1				

Table 18
First flowering of tulip seedlings in the fourth year

Species or variety	Total number of plants	Number of plants flowering	% of plants flowering
	Sown 1949		
T. korolkowii	351	1	0.3
T. subpraestans	207	7	3.4
T. subpraestans	75	19	25.3
T. kaufmanniana	238	34	14.2
T. greigii	77	37	48.0
T. greigii	106	30	28.3
T. binutans	39	1	2.6
T. sp.	210	10	4.8
T. sp. (*Eriostemones*)	72	7	9.7
T. bifloriformis	202	2	1.0
T. tarda	16	3	18.8
T. tarda	39	12	30.8
'Proserpine'	122	2	1.6
'Duc v. Tol Yellow'	702	2	0.3
'Keizerskroon'	51	1	2.0
'Kathleen Parlow'	1005	5	0.5
'Elephant'	248	8	3.2
'Jubilee'	900	10	1.1
	Sown 1951		
T. tubergeniana	580	8	1.4
T. kuschkensis	940	3	0.3
T. hoogiana	679	28	4.1
T. julia	67	1	1.5
T. ingens	800	40	5.0
T. praestans	400	35	8.8
T. albertii	131	11	8.4
T. micheliana	482	54	11.2
T. maximowiczii	300	10	3.3
T. wilsoniana	640	4	0.6

because that number also includes new daughter bulbs. In *T. kaufmanniana* it is over 100 and in *T. greigii* close to 100% (table 18).

The prolonged juvenile period of sown populations occurs under Tashkent conditions not only in the wild species but also in cultivars. Without proper selection for many years it is therefore difficult to obtain a decorative effect from seedling populations. This can be explained by the adaptation of the plants to the adverse conditions of growth and development which may occur in certain years. The long germination period and the varying periods required to reach the adult phase have evolved in the past as an adaptive response of the genus.

In the juvenile, as well as in the adult phase, all species and varieties initiate in the axils of the fleshy scales buds which can develop into daughter bulbs. As organs of vegetative reproduction these daughter bulbs are to some extent rejuvenated as compared with the apical bud of a juvenile plant and the main daughter bulb of an adult one. The degree of rejuvenation varies in different species and cultivars. After separation from the mother bulb the daughter bulbs of *T. fosteriana* continue to multiply vegetatively for several years by forming sinkers from apical buds and axillary buds and, although frequently having formed several large fleshy scales, do not flower. Daughter bulbs of *T. kaufmanniana* or *T. greigii*, of at least 1.5 cm diameter, start to flower under Tashkent conditions in the next season. From seedling populations of wild species forms can be selected with a high capacity for vegetative reproduction and through cultivation and subsequent selection it is possible to obtain forms similar to garden varieties.

The daughter bulbs are the organs of vegetative reproduction which has replaced in the many garden varieties sexual reproduction. The well-known cultivar 'Keizerskroon' of the Single Early Tulips has existed for over 200 years. It hardly produces any seeds and in spite of its being reproduced through daughter bulbs it will, if well tended, bear large flowers. The variety has been maintained without showing any signs of degen-

Table 19
First and second flowering of tulip seedlings, sown 1948

Species	1952			1953			
		Plants flowering			Main daughter bulbs flowering		
	Total number of plants	Number	%	Total number of plants	Number	% of 1953 plants	% of 1953 plants
T. kaufmanniana	273	68	24.9	211*	128	60.7	—
T. kaufmanniana	667	127	19.0	760	599	78.8	89.9
T. kaufmanniana	40	21	52.5	46	41	89.1	102.5
T. kaufmanniana	60	20	33.3	68	58	85.3	96.7
T. greigii	15	5	33.3	15	12	80.0	80.0
T. greigii	16	1	6.3	18	15	83.3	93.8
T. bifloriformis	20	3	15.0	21	15	71.3	75.0
T. fosteriana	100	0	0	200	0	0	0

* Part of these bulbs were in 1953 destroyed by mole rats.

Figure 49. First flowering of *T. greigii* in the fourth year after sowing (photo taken at the Institute for Horticultural Plant Breeding, Wageningen).

eration because the daughter bulbs are partly rejuvenated. Avakyan (1948) mentions, that the termination of the developmental phases of the onion is accompanied by the initiation of reproductive organs, whether sexual or not sexual. However, in the vegetative reproduction of tulips, the bulbs are only partly rejuvenated, the degree depending on the species. Daughter bulbs of *T.kaufmanniana* and *T.greigii* are but little rejuvenated, those of *T.fosteriana* considerably.

During the cycle of growth and development of a tulip plant reproductive growth changes from monopodial to sympodial. From the first years of ontogenesis until the adult phase all species grow monopodially from the displaced growing point. On reaching sexual maturity the flowering shoot is always formed by the apical bud and in multiflorous plants of 'the third type' (Botschantzeva, 1953, 1954) also by the apical bud of a daughter bulb. The individual continues life by development of the axillary bud of the innermost fleshy scale, i.e. sympodially.

Influence of the environment

Tulip seedlings take several years to reach the flowering phase. In nature growth recommences at low temperatures and low humidity in the autumn and continues underground during winter. With the advent of the first warm days at the end of winter or early in spring a bud appears which, depending on the phase of development, produces a single leaf or a rosette of leaves. Only in the latter case will the plant come into flower. Plants raised from seed, which flower for the first time, are smaller than those produced in following years. The specific characters of seedlings from free-pollinated species are very distinct and as a rule similar to those of the mother plants. Exceptions are the seedlings of *T.albertii*, *T.tubergeniana* and *T.vvedenskyi.* The inherited characters in the wild species are very stable compared with those of seedlings from free-pollinated cultivars, which show much variability and the progeny is entirely different from the mother plants.

Table 20

The effect of daylength on average bulb weight of 100 two-year old bulbs (mg)

Species	Daylength		
	Normal	8 hours	4 hours
T. korolkowii	280	280	—
T. vvedenskyi	984	736	—
T. hoogiana	651	487	—
T. fosteriana	412	217	180
T. kaufmanniana	1160	633	—

Table 21

The effect of daylength on average bulb weight of 100 bulbs (mg) and the percentage of flowering plants in the fourth year

Species	Normal days		8 hour days	
	Average bulb weight	% flowering plants	Average bulb weight	% flowering plants
T. korolkowii	1700	0.3	561	0
T. fosteriana	2605	0	1253	0
T. kaufmanniana	9740	17.1	7000	18.0
T. greigii	12650	48.0	7950	25.0

We found that under normal growth conditions there exists an interrelation between daylength and the temperatures prevailing during the vegetation period. When, in 1952 and 1953, we studied the effect of shortened days on a number of species of various ages, it appeared that shortened days affected bulb weight (table 20, 21). Two-year-old bulbs of *T.fosteriana, T.hoogiana, T.kaufmanniana and T.vvedenskyi,* which had received eight hours of daylight during the vegetation period, showed lower average bulb weights than under the normal Tashkent conditions. Only the *T.korolkowii* bulbs did not show a decrease in weight. Average bulb weight for *T.kaufmanniana* was nearly half the normal weight. Also the four-year-old bulbs when grown under an 8-hour day yielded considerably lower average bulb weights than under normal conditions.

Among the four-year-old plants three species began to flower (table 21). Of *T.korolkowii* only 0,3% flowered under normal conditions and not a single flower was produced when given 8 hours of daylight. Of *T. greigii* 48% flowered under normal conditions and 25% when daylight was reduced to 8 hours during two seasons. The 8-hour daylight treatment reduced in this case the percentage of flowering plants nearly by half.

As flower formation starts towards the end of the dormant period in the year preceding flowering, the reduction of the photoperiod to eight hours in 1953 had no direct influence on flower initiation. However the same reduction in 1952 would have affected flowering in 1953. Yet the shortened days had no effect on the percentage of flowering plants of *T.kaufmanniana,* because we observed a slight increase in the number of flowering individuals under short days.

These experiments produced nevertheless some unexpected results, because the shortened days greatly reduced the resistance of the plants to low night temperatures, which dropped to −6° in the beginning of April. When grown under normal conditions the seedlings of various ages and the adult plants were not affected, but when grown under shortened days, from 30 to 80% of the foliage was frozen although they had been covered with cardboard boxes. The plants which had been given only 4 hours of daylight were most damaged. Of the two- and four-year-old plants *T.greigii, T.kaufmanniana, T.korolkowii* and *T.fosteriana* were damaged by frost and *T.vvedenskyi* was the only species that did not suffer. The reduced resistance to low temperatures must be due to the fact that by the time the temperatures had dropped to their lowest level at dawn, the products of photosynthesis had been translocated from the leaves to the bulb. The concentrations of cell solutes and leaf plasma were by that time probably much lower than in the leaves of the plants grown under normal daylight, which made them sensitive to frost. The concentration of cell solutes drops most radically in the plants with the smallest quantity of stored assimilates, i.e. in the plants subjected to a 4-hour day and then in those grown under 8-hour day conditions. In these plants the interval between the termination of the photosynthetic active period and the onset of the falling night temperatures was longer than in the plants grown under normal conditions.

Moskov (1937) carried out experiments in which shortened day treatments led to better wood maturation towards the end of the vegetation period and resulted in a higher frost resistance. However, his results do not contradict ours, which are of interest in solving problems connected with low temperature resistance in tulips and their propagation in higher latitudes.

Cultivated tulips grow and multiply very well both in Tashkent and in Leningrad. During growth and development in Leningrad they have to withstand night frosts much more frequently and yet they do not suffer any damage because the days there are much longer than in Tashkent during that period.

The experiments with different periods of daylight have shown that there is a direct relation between daylength and temperature. The plants which did not sustain any frost damage at low temperatures under long days were killed by the same low temperatures under short days.

CHAPTER NINE

Growth of the juvenile leaves and the reproductive shoots after emergence

The aboveground part of a juvenile tulip plant is always a single leaf, while the formation of a rosette marks the transition from the juvenile to the adult phase. The ephemeral generative shoot of the adult plant dies rather soon after emergence from the bulb. The bulbous habit of the genus is well-adapted to the very short vegetation period as initiation, development and primary growth of the bud are all accomplished underground during the autumn and winter. After emergence of the bud the organs mainly grow in size and the most radical changes take place in the reproductive organs. These processes include coloration of the floral parts, opening of the flower buds, ripening of the anthers, pollen and embryo sacs, fertilization, growth and development of the embryos and finally development and ripening of the seeds. The study of the growth rates of the aboveground organs provided interesting data on the influence of the environment.

In 1954 we started with the phenological observations on the aboveground parts of 2-, 3- and 4-year-old juvenile individuals and on 4-year-old adult individuals of *T.greigii* and *T.kaufmanniana.* They were made from the moment the small leaf or bud emerged from the soil, at the stage when these were still brightly coloured by anthocyanin. All the juvenile plants produced one green leaf and the adult plants a rosette of leaves and a flower. Of each group a sample of 10 individuals was measured for the length and the width of the leaf (the juvenile plants) and the length and width of the rosette leaves, the bud, the flower, the ovary and the pedicel (the adult plants). The measurements were taken at five-day intervals, simultaneously for all samples.

When the first measurements were taken, on February 20, glass labels were placed with the plants under observation. Towards the evening rain and afterwards snow began to fall which started a period of cold weather lasting until March. In the beginning the temperatures dropped to −23°C, but on March 15 the conditions changed; the snow melted and we resumed our observations. The temperatures rose gradually and on April 25 a spell of hot weather set in during which several plants became scorched. The sharp fluctuations in the temperature had a marked effect on the growth and development of our plants.

Growth of the leaves

As a result of the low-temperature period of 25 days growth was arrested and the leaves of the 4-year-old juvenile plants of *T.kaufmanniana* had only attained twice their length while growth in width was not much more. The 2- and 3-year old plants had grown even less (fig.50,51). The long cold period which considerably reduced extension growth of both species hardly affected growth in width (the 2- and 3-year-old *T.greigii* plants).

The older plants of both species showed higher growth rates of the leaves, because they were larger and suffered less from heat. After April 25 their average growth rate increased sharply despite the loss of some individuals from scorching. Growth in width of the juvenile plants of both species stopped five days earlier than their extension growth (fig.50,51).

A young tulip plant continues growth until the top of its leaf begins to turn yellow. When hot weather sets in the older, upper part, of the leaf may become yellow and dry up while the lower part remains fresh and green.

The 4-year-old adult plants of *T.greigii* and *T.kaufmanniana* flowered for the first time and the growth pattern of the leaves of both species was similar to that of the juvenile plants (fig.50,51,52,53,54,55). During bud development and flowering and for 10 days after that period the rosette leaves of *T.kaufmanniana* grew rather uniformly. During flowering (March 20-31) extension growth of the top leaf declined but was restored in the second half of that period. It stopped however five days earlier than the extension growth of the other leaves (fig.52). During the first five days after mass flowering (March 31) the growth rate of the second leaf decreased but was restored in the next 5-day period, after which it diminished again and stopped altogether. Extension growth of the basal leaf was hardly affected by the phenophases. The growth in width proceeded

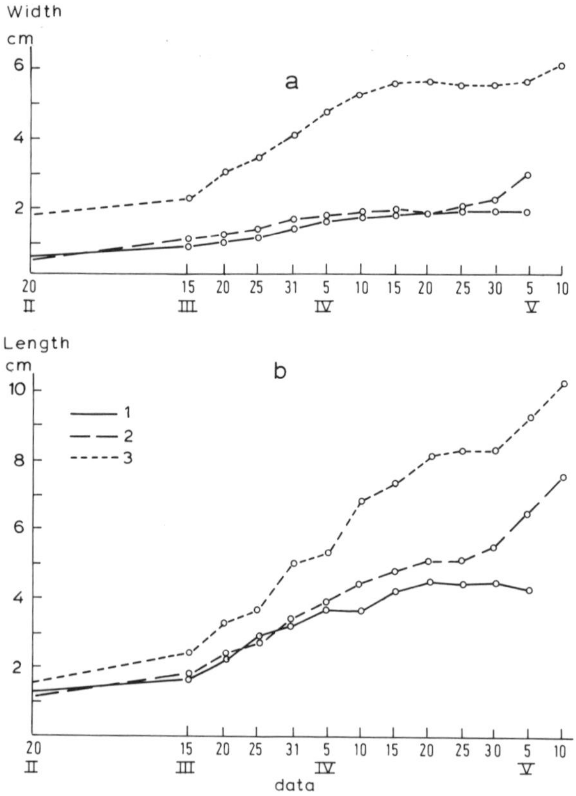

Figure 50. Growth of leaves of juvenile plants of *T. kaufmanniana*. a. width; b. length; 1. two years old; 2. three years old; 3. four years old.

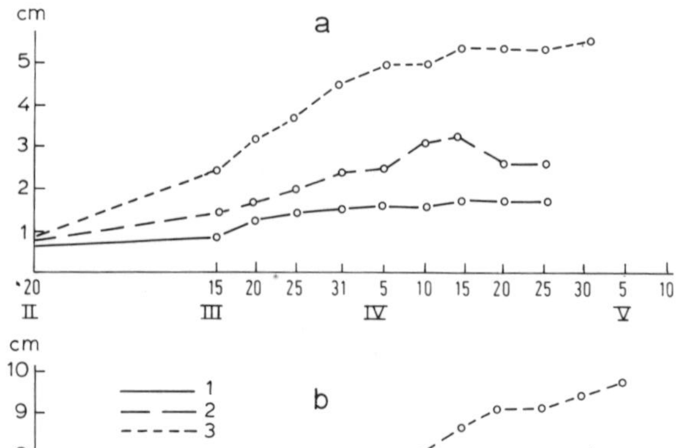

Figure 51. Growth of leaves of juvenile plants of *T. greigii*. a. width; b. length; 1. two years old; 2. three years old; 3. four years old.

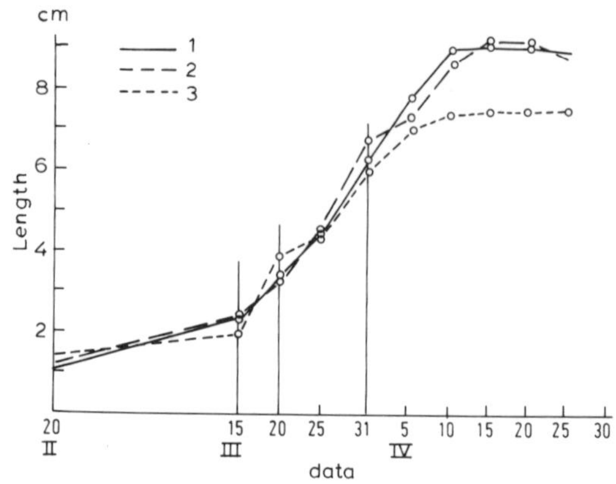

Figure 52. Extension growth of the rosette leaves of *T. kaufmanniana* at first flowering (the scale is reduced by half). 1. the basal leaf; 2. the second leaf; 3. the top leaf.

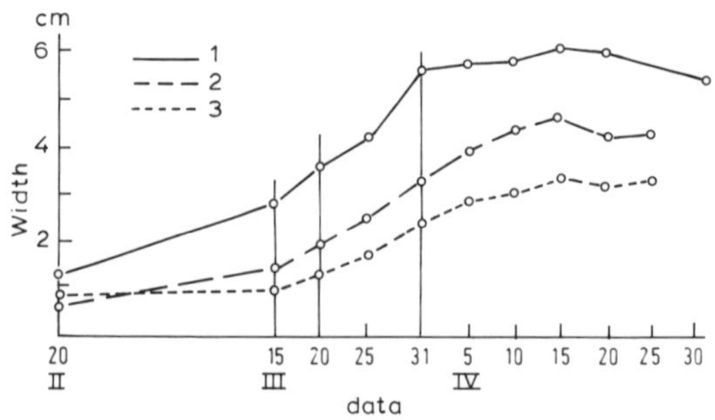

Figure 53. Growth in width of the rosette leaves of *T. kaufmanniana* at first flowering.
1. the basal leaf; 2. the second leaf; 3. the top leaf.

uniformly in all the rosette leaves during bud development and flowering, but after flowering (March 31) growth of the basal leaf stopped while that of the second leaf and the top leaf continued for another 15 days (fig.53).

Extension growth of the rosette leaves of the adult *T.greigii* plants continued in most cases till April 25, the basal leaf and the second leaf terminating growth somewhat later than the two upper leaves (*T.greigii* made four leaves). During the first five days following the thaw on March 15, marking the beginning of a warm spell, all leaves grew rapidly both in length and in width (fig.54,55). The next 5 days growth in both dimensions somewhat fell off, but the two lower leaves kept growing more rapidly than the top leaves. After March 25 the growth rates increased again up to the beginning of flowering (April 6). During the first five days of flowering growth of all leaves accelerated and the two lower leaves continued extension growth at this rate; their growth in width decreased a little. The rate of extension growth of the two upper leaves fell sharply

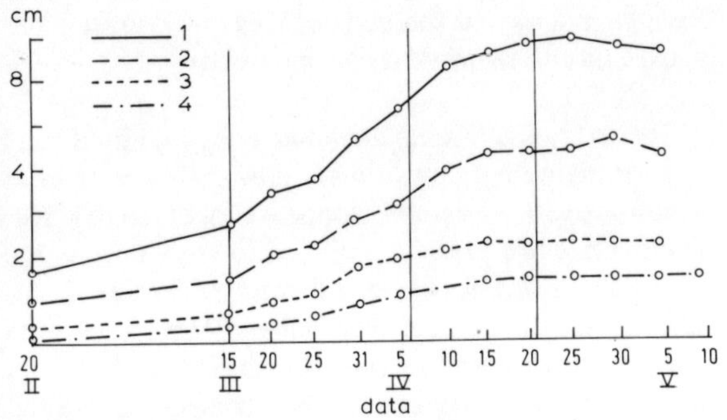

Figure 54. Growth in width of the rosette leaves of *T. greigii* at first flowering.
1. the basal leaf; 2. the second leaf; 3. the third leaf; 4. the top leaf.

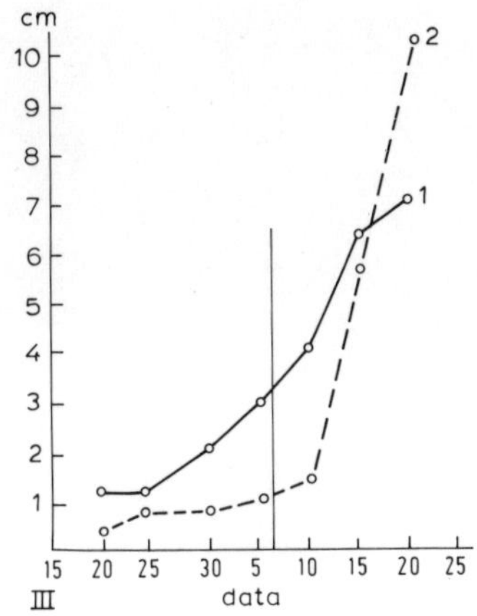

Figure 56. Growth of the bud and flower of *T. greigii.*
1. extension growth; 2. growth in width.

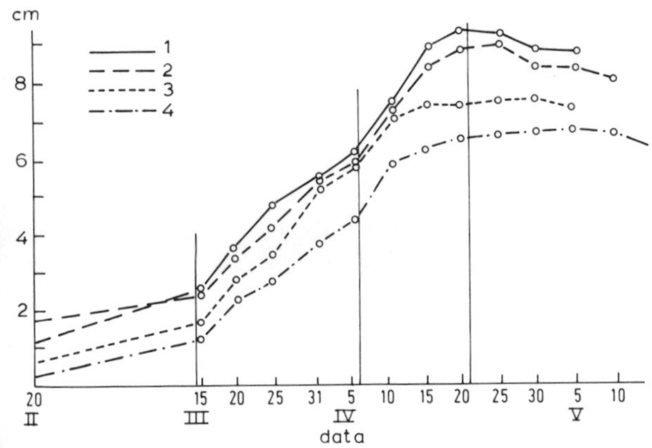

Figure 55. Extension growth of the rosette leaves of *T. greigii* at first flowering.
1. the basal leaf; 2. the second leaf; 3. the third leaf; 4. the top leaf.

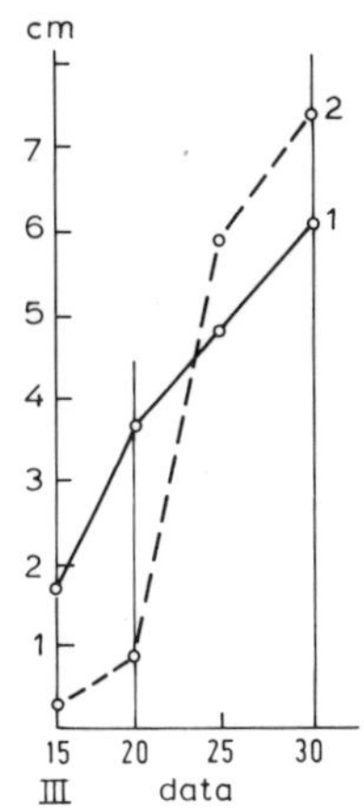

Figure 57. Growth of the bud and flower of *T. kaufmanniana.*
1. extension growth; 2. growth in width.

during the second 5 days after the beginning of flowering, the top leaf continuing growth for another five days after the third leaf had stopped it (fig.55). The latter can be explained by the position of the top leaf of *T.greigii,* which is usually set rather high on the stem and probably intercepts part of the nutrients that are being translocated to the fast growing pedicel.'Growth in width of the two upper leaves during flowering proceeded uniformly but at a lower rate. All leaves of the rosettes, however, stopped growing in both dimensions before the end of this period.

Growth of the flower

On emergence the bud of an adult individual is already of considerable size. Before the flower opens it grows much more in length than in width. As a rule the bud of *T.kaufmanniana* becomes coloured and opens rather soon after appearing aboveground. When opening it increases much in width, and extension growth somewhat decreases (fig.57).

In particular the early-flowering forms open their flowers as soon as the bud appears in the rosette, showing already the coloured tepals. In some cases during the first two days only half of the opened flower with dehisced anthers protruded from the soil. After flowering the pedicel of such forms grows rapidly and lifts the ovary 15-20 cm aboveground. There are other forms that carry a, still green, bud on a long stem.

Owing to the unfavourable weather conditions in the winter period of 1954, many tulips did not set seed and the ovaries of the *T.kaufmanniana* plants were much smaller than usual. During bud development and the flowering period (March 15-April 1) they grew vigorously after which the growth rates fell off markedly. The pedicels, however, continued to grow during and after the flowering period (fig.58). Despite a small decrease in growth rate the increment in length of the pedicel during the 25 days after flowering was only 0.5 cm less than the length attained during the 19 days of bud development and flowering, because growth had continued at a very uniform rate till April 25.

In the different forms of *T.kaufmanniana* there is a

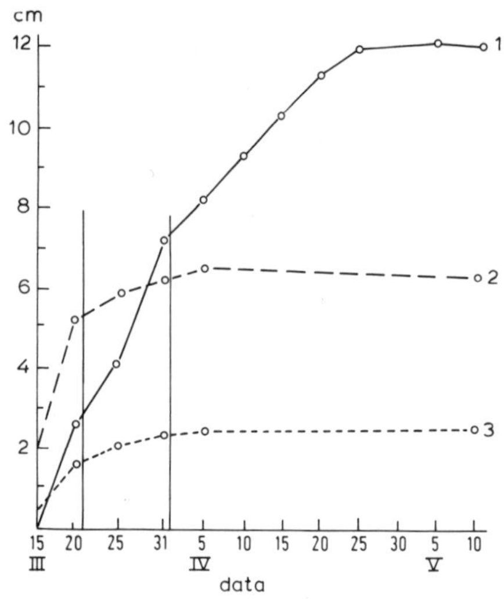

Figure 58. Growth of the pedicel and the ovary of *T. kaufmanniana*.
1. length of the pedicel; 2. length of the ovary; 3. width of the ovary.

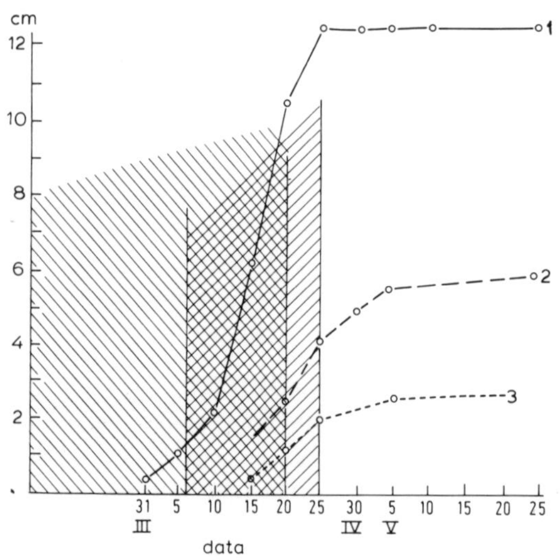

Figure 59. Growth of the pedicel and ovary of *T. greigii*.
1. length of the pedicel; 2. length of the ovary; 3. width of the ovary.

relation between leaf shape and the growth rate and length of the pedicel. The forms with broad and rounded leaves have as a rule short pedicels, whereas forms with long and narrow leaves produce buds and flowers on long pedicels.

The bud of *T.greigii* starts to grow very slowly, after which growth suddenly increases and equals that of the flowers (fig.56). The sharp increase in width of the buds is in both species due to their opening. However, opening of the buds and flowering proceed differently. The *T.kaufmanniana* plants needed a shorter period between emergence of the bud and opening than the *T.greigii* plants and the buds of *T.kaufmanniana* elongated more rapidly than the buds of *T.greigii* (fig.56,57).

The ovaries of *T.kaufmanniana* grew rapidly during bud formation but more slowly during flowering after which growth very soon stopped altogether (fig.58). The ovaries of *T.greigii* continued to grow during and after the flowering period, till May 5 (fig.59). It appeared that the pedicel of *T.kaufmanniana* continues growth for a long time after flowering but the ovary stops quite soon. By contrast the pedicel of *T.greigii* stops growing at the end of the flowering period while the ovary continues to grow 15 days longer than the ovary of *T.kaufmanniana*.

The transition from the juvenile to the adult phase

As already mentioned in the previous chapter, there occurs during growth and development of the tulip plant a transition from monopodial to sympodial growth. From the first year of ontogenesis until the adult phase the bulb grows from the apical growing point, i.e. monopodially. The changes in this growing point, which occur gradually over several years, bring the plant to sexual maturity. When the adult phase is reached the flowering shoot is always initiated by an apical bud and in the case of multiflorous plants by the apical buds of daughter bulbs. The adult plant as a rule continues life sympodially by developing the axillary bud of the innermost fleshy scale. Thus, the monocarpic shoot, which develops monopodially for several years, forms the basic structural element of the tulip. It is composed of alternating long and short stem segments.

In the first year the axis of the first node grows rapidly and fuses with the ventral side of the cotyledonal sheath. Because of its positive geotropic growth response it carries the apical bud several centimeters below the place of initiation on the shoot. The prolonged coalescence of the cotyledonal sheath with the axial part of the stolon leads to the conclusion that the formative organ is not an elongating internode but the unilaterally developed nodal part of the shoot. In its anatomical structure this part resembles more an internode than a node. However, the lateral vascular bundles of, for example, the leaf of a *T.kaufmanniana* seedling runs freely along the stolon to the bud without fusing with vascular bundles of the stem, showing that we are dealing with a node and not with an internode.

The positive geotropism and lateral growth of the node towards the convex half of the basal plate are features of seedlings which have not yet reached the adult phase and are also retained by the vegetatively multiplying daughter bulbs.

The first rosette of the tulip is composed of the cotyledon and the single basal leaf, a fleshy scale. Towards the end of the dormant period the next node begins to develop laterally and after emerging from the bulb it shows a positive geotropic growth response. As in the first year, it fuses with the sheath of the first green leaf and forms the second rosette consisting of the first true green leaf and the second basal leaf, again a fleshy scale. Every year the one green leaf of the rosette appears above ground.

The initiation of a bud consisting of one green leaf and two or more fleshy scale primordia marks the transition from the juvenile to the flowering phase. In most species stoloniferous growth stops with the approaching formation of the generative shoot. Exceptions are *T.kaufmanniana, T.praestans* and the cultivar 'King George V'.

With the arrival of the next dormant period the bulb completes the cycle by developing the generative shoot when the apical bud is stimulated into a negative geotropic growth response, forming long internodes, sessile, sheathless leaves in the centre of the shoot and a flower. This ends the monopodial growth of the polycyclic shoot and life of the individual is carried on by the bud in the axil of the innermost fleshy scale of the bulb.

CHAPTER TEN

The annual cycle of the adult bulb

INTRODUCTION

Considering the brevity of the active period of the tulip's aerial parts, the study of the annual development cycle of the adult bulb is essential for understanding the formative processes of the tulip bulb. In recent years Russian researchers have paid much attention to the vegetative reproduction of cultivars and the annual cycle of the bulb. Among them are Silina (1952, 1953, 1955), Migushova (1954), Danilevskaya (1955), Timofeeva (1956) and Zaitseva (1958).

Silina (1953) and Migushova (1954) studied the annual cycle of the bulbs of imported cultivars, grown in the Leningrad area, and described the development of the bulb and its vegetative progeny in general. Silina investigated in 1949, 1950 and 1951 the cultivars 'Bin I', 'Turner', 'William Copland' (Darwin tulips), 'Pink Ball' (Cottage tulips) and 'Prince of Austria' (Single Early Tulips). However, all characters in the cultivar 'Bin I' are typical of the tulips that belong to the Cottage tulip group and the cultivar that is called here 'Pink Ball' is really the cultivar 'Margaret' (synonym 'Gretchen') which is one of the oldest Darwin tulips. The bulbs of 'William Copland', a typical Darwin tulip, are described in detail and in some cases compared with those of 'Bin I'. Bulb development of the other varieties is dealt with briefly in the beginning of the chapters. It seems that there is not much difference in the annual cycle between the different groups of cultivars under Leningrad conditions. According to the data available 'Turner' and 'Prince of Austria' differ in development of the aerial parts, the latter starting the flowering period some days earlier, between May 23 and 26. It should be mentioned that under Tashkent conditions in 1954 and following years the difference between the beginning of the flowering periods was much more pronounced. 'Turner' started to flower April 21 and 'Prince of Austria' opened the first flower on April 12. 'Prince of Austria' appears to be one of the latest of the early flowering tulips in Tashkent.

Migushova (1954) mentions that in Leningrad initiation of daughter bulbs is later in the early flowering cultivars than in the late flowering ones. For instance in 'Zwanenburg' (Darwin tulips) the main daughter bulb was initiated in May, and in 'Maréchal Niel' (Double Early Tulips) one month later. This author studied the morphogenesis of 14 cultivars, but they were not treated as groups. Besides the two cultivars mentioned she also included the Triumph tulip 'Kansas'.

As these investigations were carried out in the northern part of the USSR, the resulting data should be used with some reservations when comparing them with the growth and development of wild species. Moreover the annual cycle of wild species has not yet been studied by other workers. Of particular interest will be a comparison of the formation of daughter bulbs of cultivars and wild species, growing under the severe conditions of continental Middle Asia since the cultivated tulips have already existed for centuries under the much milder conditions created by man, who consciously has selected those forms which lend themselves to vegetative propagation. The important organ, the bulb, represents the whole plant during its summer dormant period, whereas in the autumn, winter and spring it forms only part of the plant. For these reasons we decided to treat in our investigations two cultivars on a comparable basis with the wild Middle Asian species and to include in the observations also the aerial parts of the plant.

MATERIALS AND METHODS

During the period 1950-1955 growth and development of five wild species and two cultivars were studied. In 1950 we planted 150 first quality bulbs each of *T.fosteriana* and the variety 'Jubilee' (Darwin group) on a special plot. In the course of a year at least two bulbs of each lot were lifted every two weeks for examination. Because many of the *T.fosteriana* bulbs appeared to be still in the juvenile phase the study of this species was continued in following years. The lifted plants were described and all parts measured from the aboveground organs down to the roots. As a rule the bulb scales were removed one by one, measured, and drawings were made of them. The exposed axillary buds were placed on moist filter paper in the order of their initiation, then dissected and examined. With the aid of a dilute iodine solution (I-KI) the growing meristematic bulb tissues could be distinguished from the storage tissues. All organs of the central bud were consecutively measured, while development of the generative organs was studied by using a more rapid method with acetocarmine. Phloroglucinol and hydrochloric acid were used to trace the vascular bundles in the basal plate and in the organs already lignified. Because of the absence of starch in the parenchyma surrounding the vascular bundles a dilute iodine solution (I-KI) was used to locate the vascular system in the scales and buds of young bulbs, while in old bulbs the vascular system was made visible the other way round by the cells containing starch, which as a rule sheath the vascular bundles. Sugars were determined with a Fehling solution. Cellulose membranes were stained with chlorzinciodine and fats with a Sudan III solution.

In the following years we included other tulip species in our studies and to some extent changed the selection criterion for the plants to be analyzed.

All bulbs of cultivated varieties, e.g. 'Jubilee' and 'Duc van Tol Yellow', are vegetatively reproduced progenies of a single or several, but not many, mother bulbs. They are clones and for this reason the study of several bulbs of the same variety and size, grown under uniform conditions, will produce comparable data. However vegetative reproduction of wild species in their natural habitats is not frequent and every bulb has grown from a seed into an organism with a specific heredity and a development pattern determined by the influence of a different environment. They belong also to different age classes.

In the first few years we were unable to obtain by propagating vegetatively progenies of a single mother bulb all the stages to be studied. We therefore studied the annual cycle of *T.kaufmanniana* on bulbs grown from the seeds of one capsule, which were sown in 1948. They began to flower for the first time in 1952 and in 1954 we selected 150 high quality bulbs to be planted on special beds. Of *T.korolkowii* we selected bulbs of flowering size, grown from seed sown in 1950, which flowered for the first time in 1954. Of *T.subpraestans* we used a first selection obtained through vegetative propagation.

The bulbs were examined every month on the 2nd and 16th. When the examinations could not be carried out in due time, the bulbs and the plants were measured and preserved in alcohol for a further analysis later on.

In Middle Asia the vegetation period of tulips begins every year at different times, but always much earlier than in the more northern regions of the Sovjet Union. In Tashkent the rosettes appear in February and sometimes even late in January. The early tulips begin to flower in the first ten days of February or March, sometimes early in April. The *Eriostemones* are the first to flower, followed by species from sections *Spiranthera* and *Lophophyllon*, while *Leistemones* and *Tulipanum* species start almost simultaneously. The periods in which tulips of all sections complete vegetation and disperse their seeds lie close together.

There are three important stages of development: the appearance of roots, the emergence of the bud above the soil and the beginning of summer dormancy. At the first and the second stage we find that the organs of both vegetative and generative reproduction have already been initiated and made some growth. Summer dormancy begins towards the end of the vegetation period when the leaf tips become dry. During this period the flattened part of stem next to the bulb becomes limp and soft while the remaining stem part, the leaves and the capsule still retain their green colour. The scales of the mother bulb are already completely depleted and the tunics of the daughter bulbs have become thin and transparent and are beginning to dry. The latter, however, are not yet separated from the mother bulb and bud development continues inside them. In Tashkent summer dormancy starts very early, soon after flowering, e.g. with *T.fosteriana* early in May. This period beginning with the onset of wilting of the green organs, the leaves and the stem, and underground the roots, may start at different times depending on the age of the plant.

The termination of the vegetation period also depends on the group or species to which the individual belongs and on plant condition. In Tashkent *Eriostemones* complete this period earlier than the species of other sections. All wild species of Middle Asian origin, with the exception of *T.subpraestans,* mature earlier than cultivated varieties. Individuals that have not reached the adult phase, or plants from which the ovary has been removed, complete their vegetation period earlier than sexually mature individuals of the same population.

The annual development cycles of the different organs are so closely integrated that it is not possible to treat them separately without detriment to the true image of the plant's life processes. Inside the still vigorous mother bulb develop new bulbs which will replace the old one in the next year. As a rule even before the organs have stopped growth and the green shoot begins to dry up, the first leaf of the next year's shoot is initiated in the main daughter bulb as a tiny protuberance at the base of the old shoot. It is from this stage that we shall begin to describe the annual cycle of development.

It took us some time to find an adequate way of representing graphically the annual cycle of growth and development, because it was necessary to show the interrelations of the growing organs with respect to succession or simultaneity of their development, i.e. the latent growth of the organs inside the bulb and the visible growth of the aboveground shoot. We therefore used circular diagrams which made it possible to represent the growth of the principal organs with respect to time and correlation (figs.60,64,70,75,79,83 and 94). The somewhat abstract and sketchy character of the diagrams is compensated by their clear representation of the whole organism. The diagrams are supplemented by rectilinear graphs representing growth of the separate organs.

The circle is divided into 12 sectors, designated by Roman numerals, which correspond to the months of the year. The spiral windings and circles correspond to the various organs shown in the diagram. Functioning organs are indicated by solid lines and organs still in the initial stages of development and not yet functioning by dotted lines.

The first winding next to the centre of the diagram (1) includes the processes from the initiation of the bulb and its life as an independent organism till senescence. The next circle (2) represents growth and development of the roots; winding (3) shows the development of the shoot including the stem, the flower and the leaves; (4) the basal leaf; (5) the second leaf; (6) the third leaf. Some species produce an additional fourth and fifth leaf in which case their initiation and subsequent growth are shown in the corresponding places near the periphery of the diagram. However, as a rule windings (7) and (8) represent growth and development of the stem and flower and (9) that of the ovary, by which the cycle is completed.

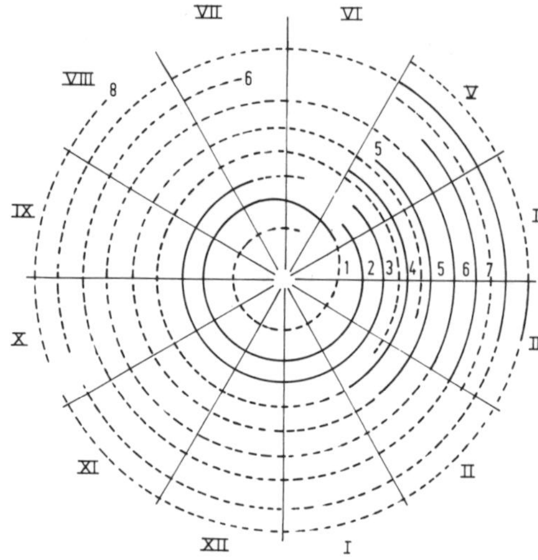

Figure 60. *T. kaufmanniana.* Annual cycle of development. I–XII. months; 1. bulb; 2. roots; 3. shoot; 4. basal leaf; 5. second leaf; 6. upper leaf; 7.stem; 8. flower.

TULIPA KAUFMANNIANA REGEL

The main daughter bulb was initiated in mid June (fig.60,1). It only developed inside the mother bulb until the third week in May of the next year. After the scales of the mother bulb had dried up it became separated and continued life as an independent organism until the second half of May in the third year of its life. Its life span from initiation till senescence is 23 months.

The roots

Root initiation on the basal plate of the mother bulb started almost simultaneously with the initiation of the main daughter bulb in the fourth week of June (fig.60,2). The roots developed slowly and towards October reached the epidermis of the basal plate. At low temperatures and increased soil humidity they grow very quickly, break through the epidermis and begin to function immediately by carrying water and nutrients from the soil to the bulb. The bulb and its roots die in the latter half of May of the next year.

The meristem of the basal plate does not last long. The first roots were initiated in the central part of the crescent, formed by the basal plate, from where they spread towards its periphery. The sturdier the bulb the larger the number of roots initiated.

During sprouting of the bulbs the roots grew in the order in which they were initiated. Those in the central part of the crescent were the first to break through the epidermis; the peripheral roots following one or two weeks later. If some roots are cut off in the autumn during sprouting, part of the basal plate remains bare because the meristem in that place has been used up. The active period of the meristem is very short and there is no regeneration. For these reasons cultural methods, such as root pruning, are not possible and growing plants, of which only a part of the root system is damaged by transplantation, are set back for many years. In *T.kaufmanniana* the period of root initiation and growth on the basal plate was three and a half months. The active period, during which the roots function as organs, was seven months and some 10-20 days. In a dry autumn root growth starts later and the latent growth period is longer. From initiation to senescence the roots lived for ca. 11 months. For one month of the annual cycle the plant has neither old roots nor root primordia. Then, during a period of 4½ to 5 months after the roots of the old bulbs have died and while the roots of the new plants are not active, there is no root contact with the soil. This is the dormancy period, the most favourable period for severing the plant's contact with the environment while keeping it alive.

The shoot

The apical bud (fig.60,3) was initiated in the period when the main daughter bulb sits inside the scales of the mother bulb and the shoot of the latter had just begun to unfold its first leaf. This occurs late in spring even before the shoot of the mother bulb emerges above ground. In our plants of 1954 the first meristematic bulges of the next year's shoots in the main daughter bulbs appeared during the last days of February or early in March. Initiation and growth inside the bulb took rather more than 11 months. The active growing phase above ground till complete withering was nearly four months, which made the life span of the shoot from initiation to senescence 15 months.

The leaves

The basal leaf (fig.60, 4) appeared as a barely noticeable bulge on the shoot meristem two weeks after shoot initiation, usually in mid March. We regard the period between initiation and unfolding of the leaf as the latent period, which was 10 months in *T.kaufmanniana.* The active period from emergence and unfolding of the basal leaf in mid February till senescence towards the end of May was three months and some days.

The second leaf (fig.60, 5) was initiated much later than the basal leaf in most forms of this species. Its primordium appeared in the last 10 days of May when the second leaf on the shoot of the mother plant died. As a rule it emerges in the beginning of March of the next year and soon unfolds. It withers away a few days earlier than the basal leaf and its life span is shorter. The latent growth period was nearly 9 months and 10 days and the active period approximately 2 months and 20 days. Its duration of life was 12 months.

The top leaf (fig.60, 6) is initiated much later than the first two leaves. The meristematic protuberance could be observed in mid July, 4 months later than that of the first leaf and nearly 2 months later than that of the middle leaf, but it emerged almost simultaneously with the second leaf and grew for almost the same number of days. Latent growth from initiation till emergence was 7½ months and as the active period is just under 2½ months, its life span of 10 months was 2 months less than the corresponding period of the middle leaf and four months less than that of the basal leaf. Its active period, however, is a little shorter, because development and growth are accelerated owing to its position on the shoot. The higher up the shoot a leaf is initiated the shorter its period of latent growth and development and the sooner it matures.

The stem

The stem (fig.60, 7) started to grow much later than the leaves. The whole of the shoot and stem was initiated early in March, but the latter, although functioning as an organ, did not make any extension growth during the period when the leaves were initiated. The part of the stem between the bulb and the basal leaf was the first to show extension growth, which could be observed inside the bulb in the beginning of September. The next spring, in the first decade of March, the stem lifted the flower bud aboveground and continued extension growth till the end of May.

Growth and development of the stem can be divided into three periods: organ formation from the beginning of March till the end of September (ca. 195 days); latent underground growth from mid September till the beginning of March (ca. 180 days); growth from emergence of the bud in the first decade of March till maturation of the capsule and senescence (ca. 75 days). The entire life span of the stem was 450 days.

Comparing the extension growth of the leaves and the stem we found that intensive leaf growth of this species started long before the plant had rooted. During the winter months, from December to the beginning of February both stem and leaf growth stopped or were very much slowed down. After winter dormancy the stem resumed growth earlier than the leaves.

The lower part of the stem can be divided into two sections. The first from the basal plate to soil surface (6.0 cm) and the second from soil surface to the first leaf (2.3 cm). Hall (1940) described species of which the lower part of the stem lifts the leaves more or less high aboveground. Of the Middle Asian species *T.hoogiana* shows this habit. Hall mentioned in all his diagnoses the height of the first leaf above the ground.

The flower

The flower primordium, initiated after the leaves, (fig.60,8) showed around mid August the three protuberances which develop into the outer tepals. Flower formation was completed by the beginning of September, while the reduction division in the anthers and mother cells occurred in late October and the pollen hibernated as mononucleate pollen grains.

The flowers of our plants appeared aboveground in mid March and had wilted by the first of April; the ovary reached maturation late in May. The period of development of the reproductive organs, from initiation of the first tepals to emergence, was 7 months; the flowering period 15 days, after which ripening of the ovary took another 2 months.

The aboveground organs of *T.kaufmanniana* did not stop growth after shedding the flower. The peduncle continued to grow while the developing ovary stimulated growth of the peduncle and the leaves. During and after flowering it grew vigorously. The ovary of the

Table 22
Growth of the reproductive shoot of *T. kaufmanniana* (cm)

	Leaves						Stem (lower part)		Internodes				Peduncle		Ovary		Stigma	
	Basal		Middle		Upper				Lower		Upper							
	Length	Width	Length	Width	Length	Width	Length	Width	Length	Width	Length	Width	Length	Width	Length	Width	Length	Width
16/2	5.7	2.0	5.0	1.3	4.4	0.9	6.0	0.5	0.5	0.4	0.4	0.4	1.0	0.3	0.1	0.1	0.1	0.2
1/3	8.5	4.0	8.5	2.5	7.5	2.0	6.5	0.5	0.6	0.5	0.5	0.5	1.3	0.4	1.4	0.3	0.4	0.4
16/3	10.4	5.8	10.5	3.4	9.8	1.8	8.5	0.7	1.2	0.6	0.9	0.5	3.0	0.5	1.9	0.4	0.2	0.6
Flowering period																		
1/4	11.3	8.4	11.5	5.9	11.0	4.8	8.9	0.7	1.3	0.7	1.5	0.5	8.9	0.5	3.5	0.7	0.5	0.7
18/4	12.5	9.0	12.0	6.0	12.0	4.5	10.5	0.7	2.1	0.6	2.5	0.5	12.6	0.5	5.3	1.5	0.2	0.6
Starting to turn yellow																		
3/5	11.5	7.0	11.2	4.0	11.0	3.0	8.0	0.8	2.3	0.6	2.5	0.6	11.3	0.4	6.5	2.6	0.8	0.7
16/5	became dry and broke						11.0	0.7	2.0	0.4	1.9	0.4	10.0	0.4	6.3	2.4	—	—
2/6	the whole plant dry, except the seed																	

plants which started to flower in mid March was 2.0 cm long and 0.4 cm wide, but on April 1, one week after the tepals had fallen, it was 3.5 cm long and 0.7 cm wide. These were low-growing, sturdy plants with leaves of the following dimensions: basal leaf 11.3 cm long, 8.4 cm wide; second leaf 11.5 cm long, 5.9 cm wide; top leaf 11.0 cm long, 4.8 cm wide.

Table 22 gives the chronological data for the development of the reproductive shoot of *T.kaufmanniana* from its initiation till senescence of the leaves. It shows that the aboveground organs of this species do not function for very long. On February 16 the reddish green apical bud was 3 cm aboveground. On March 1 the tip of the basal leaf began to unfold and the second leaf, still rolled into a tube, became visible. On March 16 the plant had a rosette of three leaves and the flower bud started to open. By April 1 it had finished flowering. The plant from which the ovary had not been removed was still green on April 18, but by May 3 the upper third of the leaves had turned yellow and dry while the capsule had retained its light green colour. By June 2 all aboveground organs had dried off but the seeds were still damp. The aboveground parts of *T.kaufmanniana* plants lived in 1955 for 107 days after the emergence of the apical bud.

In the first two weeks after emergence of the bud the lower part of the stem grew very slowly, the internodes were active, but the peduncle hardly grew at all. In the following two weeks, till the beginning of flowering, growth of the lower stem part sharply increased, the internodes grew evenly and the peduncle actively. During the flowering period growth of the lower part of the stem slowed down a little, but the peduncle enlarged more than the other green organs; the internodes continued to grow slowly. Unlike the other parts of the axis, they were not affected by flowering and it was only during the first week after this period that their growth showed a sharp increase, after which it stopped altogether.

For two weeks after completion of flowering the peduncle made still extension growth; the lower part of the stem grew little and after this period all parts of the axis stopped growth abruptly.

During the first two weeks after emergence the basal leaf made vigorous extension growth which somewhat diminished in the following two weeks preceding flowering. There was some increase during and after flowering, but after a month growth stopped altogether. This stage is marked by the yellow discoloration at the tips of the leaves. It is quite possible that the base of the leaf continues to grow while the upper part already begins to wither.

The ovary grew intensively during the first two weeks after emergence of the flower bud, but nearly stopped growing two weeks before flowering. In the period between opening and withering of the flower growth again speeded up and in the following weeks continued steadily. After flowering all other organs, the underground part of the stem, the internodes, the peduncle and the leaves stopped growth completely. It appears that the ovary of *T.kaufmanniana* continues to grow two weeks longer than the other aerial organs.

Initiation and growth of the daughter bulbs fig. 61

Weather conditions during the spring and autumn of 1954 in Tashkent were not propitious for studying vegetative reproduction of wild growing species. How-

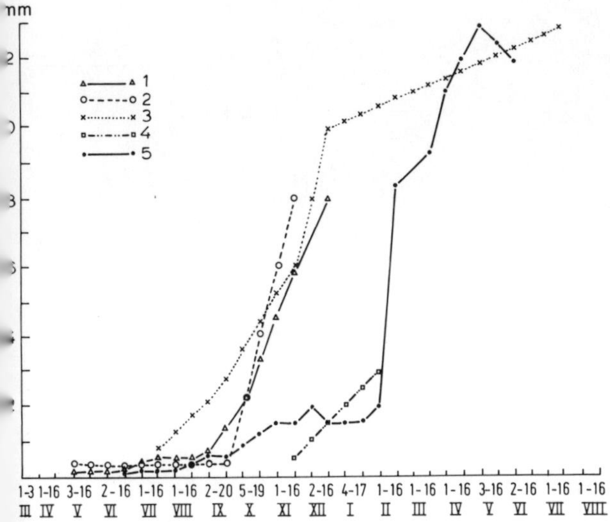

Figure 61. *T. kaufmanniana.* Initiation and growth of the daughter bulbs.
1. daughter bulb in the axil of the first fleshy scale; 2. daughter bulb in the axil of the 2nd fleshy scale; 3. daughter bulb in the axil of the 3rd fleshy scale; 4. daughter bulb in the axil of the 4th fleshy scale; 5. main daughter bulb.

ever, in the preceding growing season conditions had been favourable for the initiation of daughter bulbs in the large bulbs and therefore the initial growing stages could be observed.

As a rule *T.kaufmanniana* has no daughter bulbs in the axil of the tunic. The daughter bulb in the axil of the first fleshy scale was initiated in the first days of May, when the main bulb became separated from its mother bulb and for about six weeks after initiation it did not increase in size. From the beginning of June till mid July, it grew 0.5 mm. Around the end of July growth stopped. It was resumed in the latter half of August and continued throughout September. In October growth accelerated and it continued to mid November after which it slightly slowed down and finally stopped altogether towards December 15, when the young bulb entered the winter dormant period. Although this daughter bulb had not grown sufficiently by that time, it did not grow any more after reaching a length of 1 cm.

Almost simultaneously with the daughter bulb just described another one was initiated in the axil of the second fleshy scale. Till mid September growth was very slow, but then it grew very intensively for a period of two months and by mid November it had attained the size of the first daughter bulb after which development stopped completely.

The daughter bulb in the axil of the third fleshy scale was initiated in mid July. Unlike the first two daughter bulbs it did not show such a long period of latent growth because the situation of the nutrient reserves in the scales was now quite different. After initiation this daughter bulb immediately began to grow actively. From mid September growth increased and it reached a maximum between mid November and mid December. Then after a sharp drop it continued steadily till the end of the vegetation period.

Sometimes, in vigorous bulbs, a daughter bulb was initiated in the axil of the fourth fleshy scale in mid November. Till mid January it made rapid and uniform growth after which a sudden increase in extension growth occurred. We were unable to follow further development of these daughter bulbs because they were found only in a limited number of plants.

Growth of the main daughter bulb
fig. 62

The main daughter bulb was initiated in mid June at the

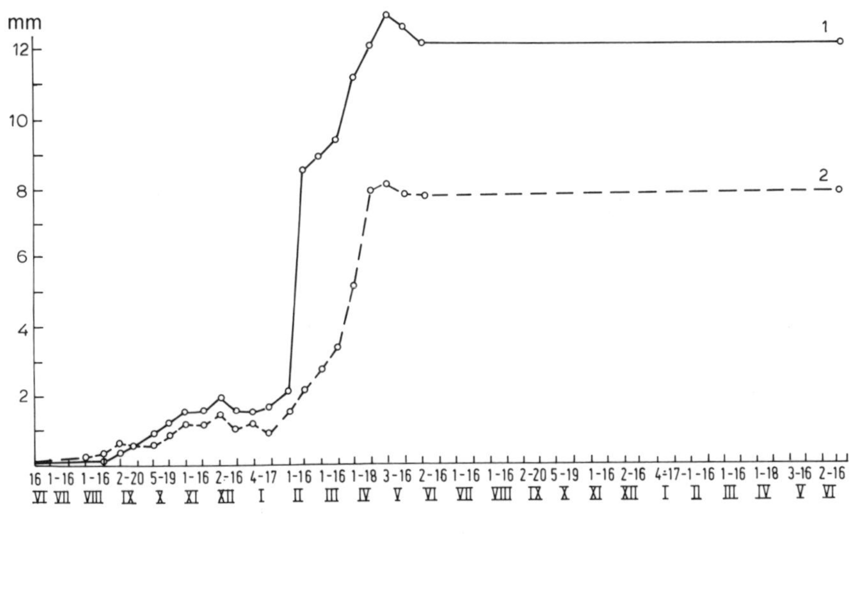

Figure 62. *T. kaufmanniana.* Growth of the main daughter bulb.
1. length; 2. width.

base of the reproductive shoot, earlier than the daughter bulbs in the axils of the third and fourth fleshy scales. Till August it made hardly any growth, after which it grew slowly till the beginning of September. In the course of the following two weeks growth had slowed down again, but beginning in the last decade of September the bulb grew steadily till winter dormancy, lasting from November till February. There could be observed a short burst of growth around the middle of this period probably because of thawing of the soil and resumption of root acitivity. The spring phase of growth began in the latter half of January and was quite pronounced between February 1 and 16. Growth slowed down in the following month, but increased again during the latter half of March. The bulb continued growing steadily until the first days of May after which it seemed to diminish in size owing to shrinkage of the tunic.

So far we have described the outward growth of the main daughter bulb without paying attention to the formation of its organs. The bud was initiated in mid June as a small bulge, but the ridge of the primordial tunic did not become visible till mid August. When making a cross section of the bud early in September a second ridge, the primordium of the first fleshy scale, could already be observed inside the tunic primordium. Immediately afterwards the ridges of the second and third fleshy scale primordia appeared. The fourth fleshy scale, initiated somewhat later, may be lacking in many adult individuals of this species.

We see that the scales in the main daughter bulb of *T.kaufmanniana* were initiated from 2 to 3 months after the appearance of its bud primordium, but before the period of active growth in the autumn. By winter all the organs of the main daughter bulb are in their initial stages.

Initiation and growth of the flower
fig. 63

Initiation of the flower organs proceeds in an upward direction. The three crescent-shaped bulges of the outer tepals appeared first, early in August, and the primordial bulges of the inner tepals soon followed. The stamens were initiated in the same order, the primordia of the outer set standing out clearly while those of the inner set were barely visible. The carpels appeared in the first days of September as three separate bulges. Their tops remained separate while their lower parts grew together somewhat later on.

All flower organs showed about the same growth pattern (fig.63), as they all began to grow intensively right after initiation and continued it till October. During the first half of that month the growth rates fell off, but apparently because of commencing root activity, they accelerated again afterwards. During this period the stamens grew more actively than the other flowering parts. By the first November development of the organs came to a halt, but the ovary which did not grow as rapidly as the stamens, continued to grow till December. Growth of the tepals, which considerably slowed down in the beginning of October, had continued at this low rate till December when the winter dormant period started. The anthers had stopped growing two weeks earlier.

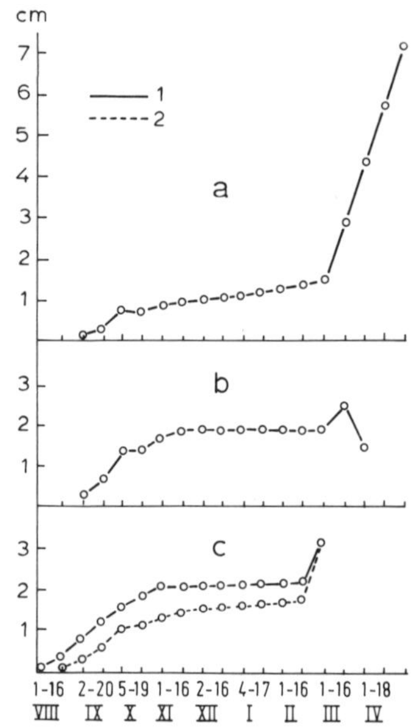

Figure 63. *T. kaufmanniana.* Initiation and growth of the flowering parts.
a. ovary; b. anthers; c. flower bud and flower; 1. length; 2. width.

The pistil was the first to break winter dormancy by resuming growth in the beginning of January. Growth continued at a slow rate till early March and then showed a sharp increase. The tepals started again six weeks later, in mid February, and grew very actively till the end of flowering. The stamens, which had almost completed their development in the autumn, finished this process with the extension growth of the filaments early in March.

TULIPA FOSTERIANA HOOG EX IRV.

Initiation and growth of the main daughter bulb
fig. 64

In the latter half of July the main daughter bulb was initiated at the base of the reproductive shoot and for the next ten months, until mid May of the following

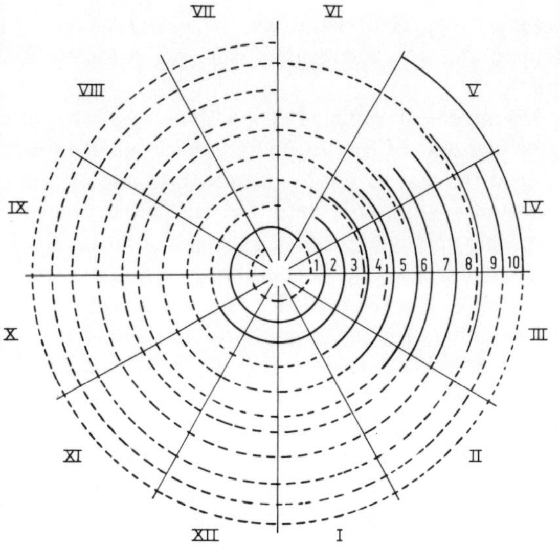

Figure 64. *T. fosteriana.* Annual cycle of development. I–XII. months; 1. bulb; 2. roots; 3. shoot; 4. basal leaf; 5. 2nd leaf; 6. 3rd leaf; 7. upper leaf; 8. stem; 9. flower; 10. ovary.

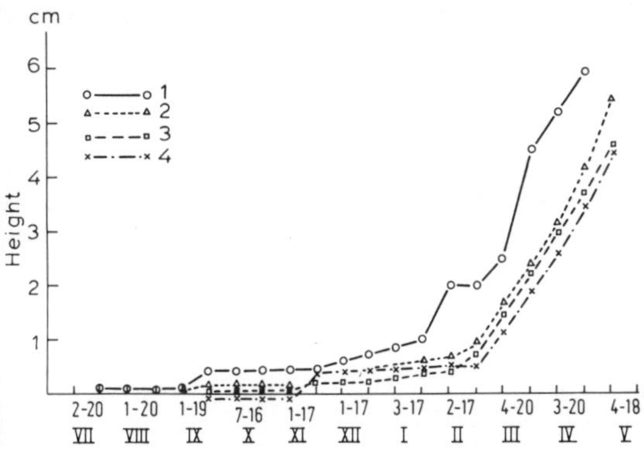

Figure 65. *T. fosteriana.* Growth of the main daughter bulb. 1. tunic; 2. 1st fleshy scale; 3. 2nd fleshy scale; 4. 3rd fleshy scale.

year, it grew inside the mother bulb. After separation from the tunic of the mother bulb it lived for another year as an independent organism. The life span of *T.fosteriana* bulbs from initiation till senescence was 22 months.

The meristematic bulge of the bud at the base of the reproductive shoot became visible after July 20 (fig.65). It grew but slowly till the beginning of September and during that period no initiation of scales was observed. The primordia of the tunic and the first and second fleshy scales appeared early in September; the first two primordia were quite pronounced, the third was barely noticeable. Two weeks later, in mid September, two more scale primordia were initiated. Until the mother bulb took root in the autumn they hardly grew at all, with the exception of the tunic, which during the two weeks after initiation enlarged considerably, but it showed no further growth between mid September and November, during the rainy period when the mother bulbs formed roots.

In the first half of November the first and the third fleshy scale grew but little. After November 15 the tunic began to grow slowly and continued growth at a constant rate till mid January. During the latter half of that month growth increased but it slowed down again in the beginning of February. From mid February till mid April the tunic again grew vigorously. Its extension growth stopped two weeks earlier than that of the fleshy scales, which during the period between their initiation and mid February, had hardly shown any growth at all. In that period the first fleshy scale grew to 0.9 cm high and the third to 0.5 cm. Between mid February and the beginning of May all fleshy scales grew rapidly and uniformly. Typically, the main daughter bulb of *T.fosteriana* chiefly grows in spring before the summer dormant period.

The roots
fig. 64:2

In the first days of July the beginning of root formation was clearly visible in the meristematic zone of the basal plate; it was completed by October. Under humid soil conditions the roots may start to grow, otherwise growth may be delayed till November or later. They die off in May, but in juvenile individuals already towards the end of April.

The shoot
fig. 64:3

In mid March the reproductive shoot was initiated in the centre of the main daughter bulb before growth of the latter was completed and at the stage when it was still closely attached to the mother bulb. At that time the future shoot is only discernible as a very small meristematic protuberance. Soon the basal leaf primordium was initiated and towards the beginning of April also the primordium of the second leaf. Further initiation of leaves was delayed till the mother bulb entered the period of summer dormancy. The third leaf was initiated in mid June and a fourth early in July. Adult plants of this species usually have three leaves. The time interval between initiation of the first and the second leaf was shorter than that between the second and the third leaf and as the interval between the initiation of the third and a fourth leaf was also comparatively shorter, it looks as if the leaves are initiated in pairs on the shoot.

Flower initiation started in July and the shoot

emerged from the bulb in October. It appeared above-ground between mid February and the beginning of March and died around the middle of May. The capsule matured much later.

Growth of the leaves

figs.64:4-5-6, 66

The basal leaf and the second leaf of the shoot, initiated in the main daughter bulb before it becomes an independent organism, hardly grew at all for a period of 2½ months. Early in June, after the main daughter bulb had finished growing and was separated from the mother bulb, both leaves suddenly showed a short increase in growth. However, by mid June growth stopped and was not resumed for approximately 2½ months.

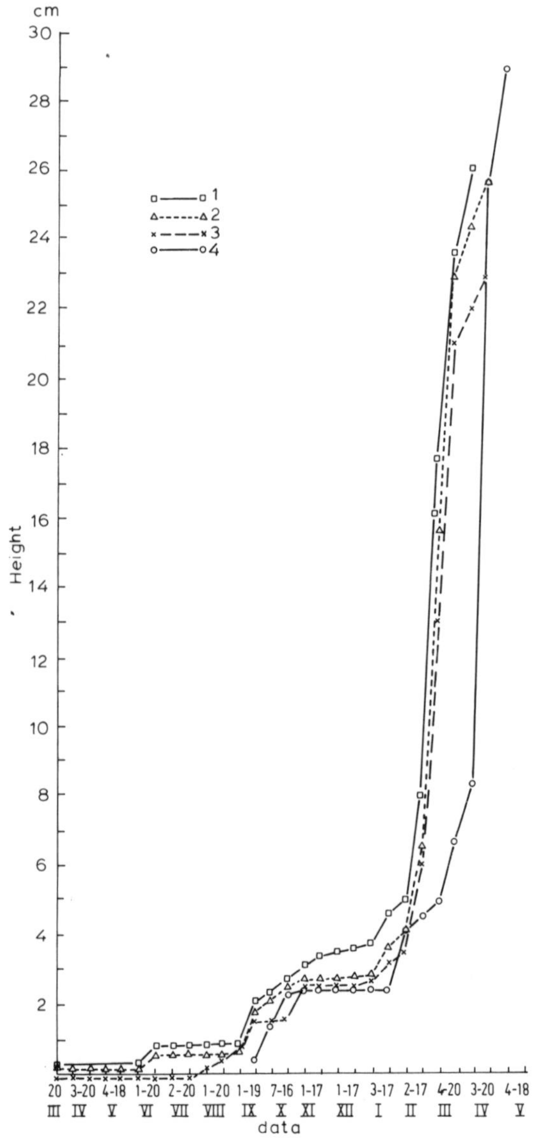

Figure 66. *T. fosteriana.* Growth of the leaves and the stem. 1. basal leaf; 2. 2nd leaf; 3. upper leaf; 4. stem.

The upper leaf, initiated early in June, began to grow in the first days of August. In September it had already attained the size of the two lower ones and all three leaves grew vigorously until mid September. Because there is no precipitation during that period in Tashkent, the planted bulbs could not take root and further growth of the leaves was retarded. The upper leaf was the first to stop growth for one month, until mid October. The two lower leaves continued to grow, but less rapidly. In the first half of October the second leaf also stopped growth while the basal leaf continued to increase in size. From the latter half of October till the beginning of November all three leaves resumed active growth. Then, because of the winter dormant period, growth of the two upper leaves stopped till January and they made but little extension growth in the latter half of this period. The basal leaf continued growth at a slow but steady rate.

From the beginning of January all three leaves began to grow again actively. They slowed down somewhat in the latter half of the month owing to the vigorously growing stem. During the latter half of February and early in March the leaves grew rapidly while stem growth slowed down considerably. The 'spring phase' of growth of the stem started early in April when leaf growth again diminished and stopped altogether. Thus we see, that the more rapidly the leaves grow, the slower the growth of the stem and vice versa.

The flower

figs. 64:9, 67

Flower differentiation started early in June when the root meristem and upper leaf were initiated. The primordia of the tepals and the stamens appeared almost at the same time, but till September they grew very little. The active growing period of the shoot and flower organs, the tepals and the stamens, began in the first days of September when the carpels were initiated and started to grow. This first autumnal growing phase of the stamens and the carpels is very short and was interrupted for two weeks in mid September, after which the stamens resumed growth at the same rate, but the carpels at a much slower rate. The converse is true in the second half of October when the carpels grew more intensively than the stamens. The tepals had continued to grow actively and by November at a slightly reduced rate.

From November till early January the flower organs stopped growing because of the winter dormancy. Some growth was observed early in January, apparently due to thawing of the soil and increased root

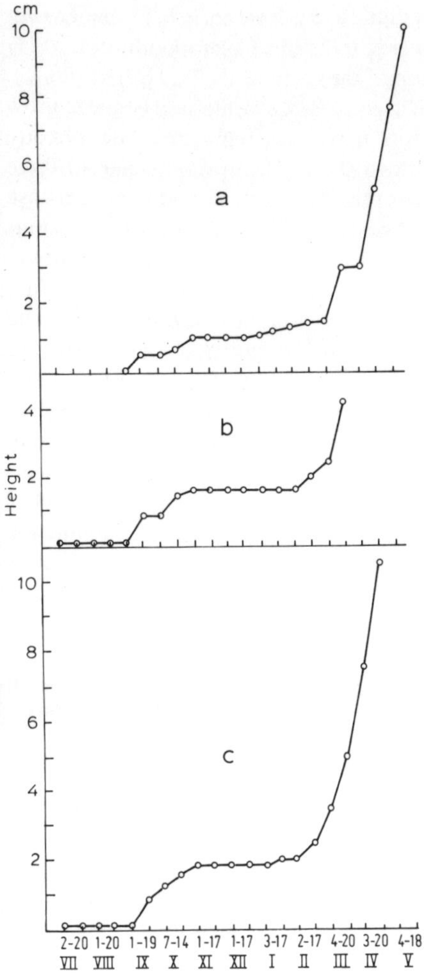

Figure 67. *T. fosteriana.* Initiation and growth of the flower organs.
a. ovary; b. stamens; c. flower bud and developing flower.

activity, but it stopped again in the second half of the month. When winter dormancy of the flower organs was broken early in February, they started growing at an increasing rate till the end of the flowering period in the second half of April.

The stamens grew steadily but slowly throughout February and increased their growth sharply in the first half of March. Towards flowering, early in April, they became somewhat larger by elongation of the filaments. After the flowers had unfolded the anthers dehisced and shrunk.

The winter dormant period of the carpels was shorter than that of the peripheral organs. The carpels and the tepals terminated this period in mid January and showed some microscopical growth till the beginning of March. Before emergence from the soil, in the first half of March, their growth was retarded, but in the latter half of that month it sharply accelerated to drop again towards the end of March, before unfolding of the flowers. After that the ovary began to grow vigorously and for a longer period than the other flower organs, till the second half of May.

Initiation and growth of the daughter bulbs
fig. 68

Only a very small number of our *T.fosteriana* bulbs initiated a daughter bulb in the axil of the tunic and therefore we could not follow the development of such daughter bulbs throughout the year.

When the plants entered the dormant period and the main daughter bulbs had become separated, early in June, a meristematic protuberance was observed on the median vein in the axil of the first fleshy scale. It appeared in the wedge-shaped meristematic tissue between the scales which was at that time in active growth (fig.69). This primordium started to grow slowly and after a month it had grown only a few millimeters. Two weeks after the appearance of the first primordium another daughter bulb was initiated in the axil of the second scale, but in its first stages this primordium grew much slower than the first one. It took another six weeks before the next daughter bulb was initiated. During this interval the plant initiated a large number of other organs: the roots, the upper leaves and the different flower parts. In the first days of August a daughter bulb was initiated in the axil of the third fleshy scale, after the initiation of the bud of the main daughter bulb (cf. fig.64).

In August the daughter bulb primordia grew a little, but soon stopped. Similar insignificant growth was observed when the bulbs were taking root, from the second half of October till mid November. The daughter bulb bud of the third scale, the last to be initiated, was the first to start development in mid December and in one month attained twice its size. Then growth gradually slowed down and although it continued till

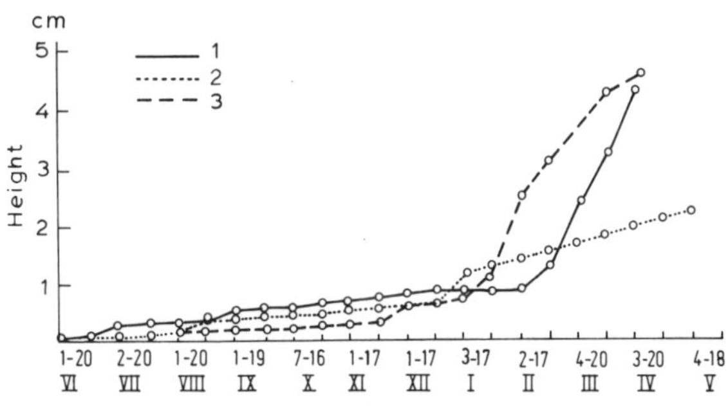

Figure 68. *T. fosteriana.* Initiation and growth of the daughter bulbs.
1. daughter bulb in the axil of the 1st fleshy scale; 2. daughter bulb in the axil of the 2nd fleshy scale; 3. daughter bulb in the axil of the 3rd fleshy scale.

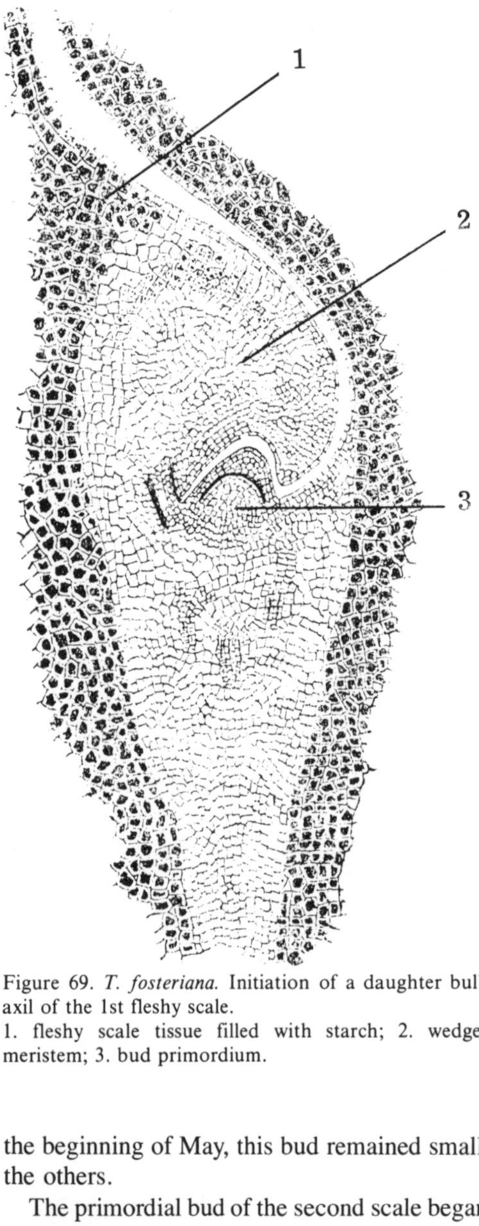

Figure 69. *T. fosteriana.* Initiation of a daughter bulb in the axil of the 1st fleshy scale.
1. fleshy scale tissue filled with starch; 2. wedge-shaped meristem; 3. bud primordium.

the beginning of May, this bud remained smaller than the others.

The primordial bud of the second scale began active growth in the second half of January and continued at a high growth rate till mid April. The primordium in the axil of the first scale was the last to start active growth. It broke winter dormancy early in February, making rapid extension growth from the second half of February until the beginning of April, i.e. till the flowering period of the plant.

It appears that the daughter bulb buds complete their winter dormancy one by one, centrifugally, but later on the outer buds on the contrary stop their growth earlier than the central ones.

The initiation and development of the scales of the daughter bulbs proceeded in the following order:

The tunic of the bud primordium in the axil of the first fleshy scale swells up early in July; its first fleshy scale late in October; its second fleshy scale early in November; its third fleshy scale early in December and the fourth towards the end of that month.

The initiation of the tunic of the bud primordium in the axil of the second fleshy scale was observed also early in July, but it was not before the beginning of November that two other scale primordia became visible, a large one of the first scale and a barely discernible one of the second scale. The third scale primordium appeared early in December and the fourth a month later.

The tunic of the bud primordium in the axil of the third fleshy scale, the last initiated scale, appeared towards the first of September; its first fleshy scale primordium on November 1 and the second and third fleshy scale primordia early in December, which completed scale initiation in this daughter bulb.

Collateral daughter bulbs were rarely observed in *T.fosteriana,* but serial daughter bulbs may be initiated in every bud and in the axil of any scale. They were generally initiated on the vascular bundle at the base of the daughter bulb primordium after the initiation of its tunic, but prior to the initiation of its first fleshy scale. We found these buds also in the main daughter bulb of *T.fosteriana.*

Initiation of the organs

May and February are the only months no organs are initiated. The months in which the various parts of the plants were initiated are given in table 23. The shoot and two leaves were initiated during the 'spring phase' of growth. During summer dormancy, however, organogenesis was very intense and the upper leaves, the flower, the roots, the daughter bulbs with their tunic primordia, including the main daughter bulb with all its scales, were initiated in that period. The fleshy scales of the other daughter bulbs were initiated during autumnal growth and winter dormancy. Organogenesis in the daughter bulbs is particularly active in the winter months when the growth of all shoot parts stops completely (November, December and January). The interrelations of all these processes of growth and development are indeed remarkable.

A negative feature in the vegetative reproduction of *T.fosteriana* is a tendency of its daughter bulbs towards rejuvenation. The daughter bulbs of the adult individuals are usually not carried away by stolons, but remain inside the scales of the mother bulb. After their isolation they also initiate one or two daughter bulbs, which form stolons growing sideways and somewhat downwards or sometimes even upwards. They carry the daughter bulbs for a distance of 1-40 cm away from the mother bulb, but the main daughter bulb remains where it was. A plant which develops from an axillary bud is strongly rejuvenated, usually does not flower

Table 23
Time of initiation of the organs in *T. fosteriana*

Organs	March	April	May	June	July	August	Sept.	Oct.	Nov.	Dec.	Jan.	Febr.
Leaves	1st leaf	2nd leaf		3rd leaf	4th leaf							
Flower					tepals and stamens		pistil					
Roots					initiation							
Daughter bulbs in the axil of:												
the first scale				initiation	tunic			1st fl. sc.*	2nd fl. sc.	3rd fl. sc.		
the second scale				initiation	tunic				1st and 2nd fl. sc.	3rd fl. sc.	4th fl. sc.	
the third scale						initiation	tunic		1st fl. sc.	2nd fl. sc. and 3rd fl. sc.		
main daughter bulb					initiation		tunic 1st fl.sc. 2nd fl. sc. 3rd fl. sc. 4th fl. sc.					

and has only one leaf. It has spent much energy in growing the long stolons and for this reason forms smaller bulbs. The same happens in the next and following years. It is practically impossible to select from a great number of plants individuals forming hardly any stolons, but producing a nest of daughter bulbs in the same way as cultivated tulips do. By propagating such plants different varieties can be obtained. Very likely the varieties Red Emperor, Madame Lefeber and others were obtained in Holland through selection among plants received as bulbs from the USSR. For example, we obtained 'Materinskaya Slava' ('Mother's Glory') simply by selecting a plant which had a beautiful flower and could be well reproduced vegetatively, but with a weak tendency to stolon formation.

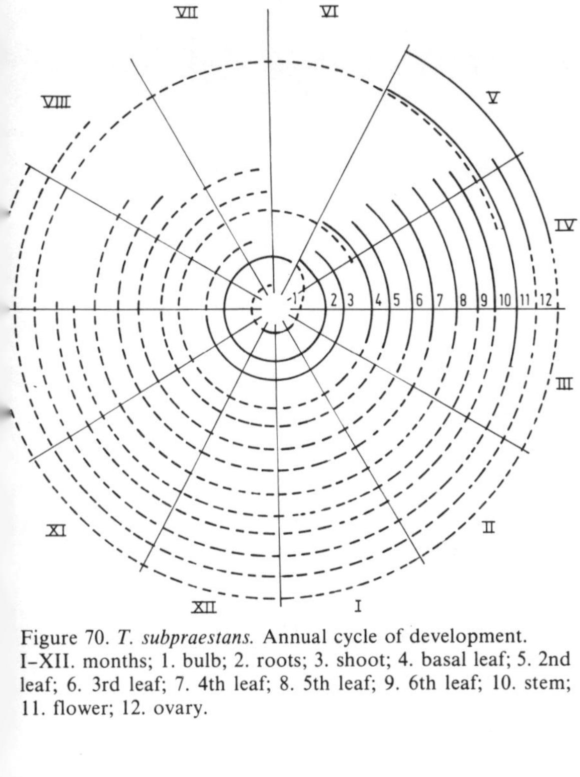

Figure 70. *T. subpraestans.* Annual cycle of development. I–XII. months; 1. bulb; 2. roots; 3. shoot; 4. basal leaf; 5. 2nd leaf; 6. 3rd leaf; 7. 4th leaf; 8. 5th leaf; 9. 6th leaf; 10. stem; 11. flower; 12. ovary.

TULIPA SUBPRAESTANS VVED.

Initiation and growth of the main daughter bulb

figs. 70, 71

The main daughter bulb was initiated early in July at the base of the new shoot of the young bulb which had just become separated from its mother bulb. The first six weeks the primordium did not grow. Between mid August and mid September it enlarged slightly and then started to grow steadily till mid November. This period coincided with the beginning of root activity in the mother bulb. During winter dormancy, from mid November till February, the young bulb did not grow. In the first half of February it began to grow vigorously, stopped between mid February and mid March and then resumed growth at an accelerated rate. Early in April it slowed down again to recommence with new vigour in early May. Development of the bulb was

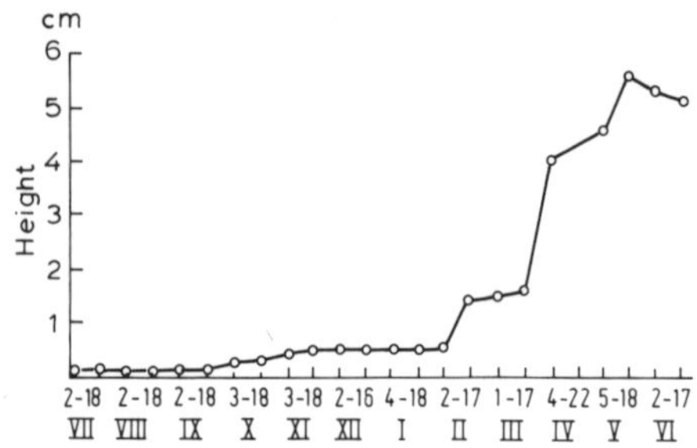

Figure 71. *T. subpraestans*. Growth and development of the main daughter bulb.

completed by mid May after which it slightly diminished in size owing to shrinkage of the tunic. In that period the bulb is severed from the mother bulb.

The initiation of the scales of this bulb occurred with the following time intervals: the tunic primordium appeared in mid August, that of the first fleshy scale a month later, that of the second fleshy scale two weeks after the first one with the third fleshy scale following almost directly after it. All the scales of the main daughter bulb were initiated by the time the mother bulb took root before winter dormancy. Thus, organogenesis of the main daughter bulb takes place during the summer dormant period of the mother bulb.

The latent period of the main daughter bulb, during which it is firmly attached to the mother bulb, was from mid July until the middle or the end of May of the next year and after losing contact with the mother bulb it lived for another year. It took 6 to 8 weeks after separation from the mother bulb before roots were initiated and 4½, or sometimes even 5 months, before they were fixed in the soil, usually in mid October. The life span of the main daughter bulb from initiation to senescence averaged 20½ months.

The roots
fig. 70

Root initiation started in mid July simultaneously with the initiation of the main daughter bulb. The roots remained concealed inside the basal plate till the first autumn rains in the latter half of October and on emerging from the basal plate became immediately active. The roots die in the middle of May or early in June of the next year. The entire period between initiation and senescence is 10 months, including the 3 months of latent growth and development inside the basal plate.

The shoot
fig. 70

The primordium of the next year's shoot was initiated in the main daughter bulb during flowering of the mother bulb in the last decade of April. For 2½ months it neither increased in size nor initiated organs. Towards the end of June or early in July the first leaf was initiated and 2 or 3 days later the second leaf opposite the first. Not earlier than mid August did a third leaf primordium become noticeable, soon followed by a fourth.

The apical flower was initiated in mid August. At that time there could also be observed in the axils of the two upper leaves the primordia of two or more flowering shoots on each of which a leaf was initiated towards the end of that month. The flowers of the lateral shoots were initiated early in September. Although these flowers appear later than the apical flower, they develop almost at the same pace and often open at the same time.

In the first half of March the red shoot tip emerged and the leaves unfolded soon after. Senescence of the shoot began around the end of May when the seeds ripen and the main daughter bulb at the base of the shoot becomes an independent organism.

The principal organogenetic processes inside the bulb of *T. subpraestans* were completed before the end of summer dormancy, in the period when the plant has no functioning roots or leaves and its contact with the outer environment is at a minimum.

The leaves and the stem
fig. 72

The basal leaf, initiated early in July, and the almost immediately following second leaf grew but little till mid August, during which period a third and soon after a fourth leaf were initiated. In the latter half of August the basal leaf and second leaf began to make some extension growth and by the second half of September all four leaves grew vigorously. At the same time the fifth, sixth and following leaves were initiated on the lateral shoots, which had appeared earlier in the leaf axils of the main axis. The higher a leaf is situated on the axis the sooner it begins to grow although its final increment will be proportionally less.

In the first days of October growth slowed down somewhat but it accelerated again around mid October. The reason for this was that when the plants were ready to strike root, they could not do so because of the hot and dry climate of Tashkent. Part of the food reserves in the mother bulb were depleted by the scales of the main daughter bulb which had begun to grow. Besides, the lateral shoots with their leaf primordia were initiated in the same period, which apparently took much of the plant's energy and affected subsequent growth of the leaves.

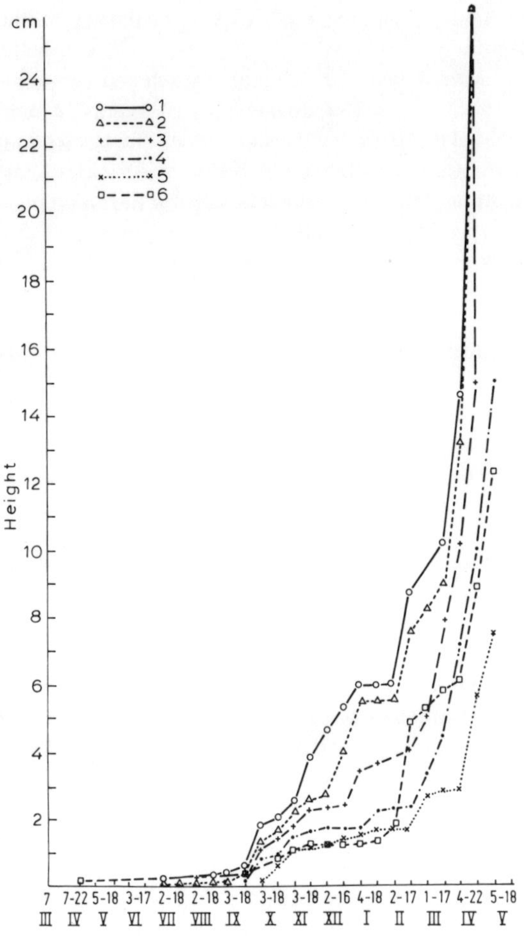

Figure 72. *T. subpraestans.* Growth of the stem and leaves. 1. basal leaf; 2. 2nd leaf; 3. 3rd leaf; 4. 4th leaf; 5. 5th leaf; 6. stem.

Rapid growth of all leaves was resumed in the latter part of October, but the fourth and the fifth leaf stopped growth in the first half of November and the second and third leaves in the latter part of that month. Only the basal leaf grew steadily till the beginning of January, when it stopped growth for a month and then continued to grow till the end of the vegetation period.

The upper leaves reacted differently to the environmental conditions. In mid November, the period when leaf growth stops, the second leaf slightly decreased its growth rate, its resting period being only two weeks, while that of the third leaf was a month and that of the fourth leaf 1½ month. The growth of the fifth leaf was resumed earlier, but slowed down towards the beginning of December to continue at a very slow rate during the next two months.

After the resting period in November the second leaf grew more actively than the basal leaf. In the beginning of January both leaves entered winter dormancy after which period they resumed growth at the same rate.

The third leaf, which completed the resting period two weeks later, continued growth at a decreasing rate till March. The fourth leaf extended the resting period till early in January and interrupted its growth again from the middle of that month till mid February. After this all leaves of the rosette grew vigorously, particularly in April, but with the exception of the fifth leaf, which from March till April hardly grew at all. This was apparently because the flower and the more vigorous lower leaves intercepted most of the water and nutrients. In early April it also started to grow rapidly, but it did not attain the size of the lower leaves.

The stem of the reproductive shoot, initiated in April, grew very slowly until mid September, when it had initiated all its organs. Between mid September and mid November it grew intensively and then entered winter dormancy, discontinuing visible growth till mid January. In the second half of January growth was resumed at a slow rate, but suddenly speeded up in the first half of February. Probably due to the cold weather conditions, growth of the stem and the basal leaf was retarded in the latter part of February.

The most active growing period of the four lower leaves began in mid March; that of the stem early in April before that of the fifth leaf.

In 1954, when these observations were made, growing conditions were rather unfavourable. However, the description of the interrelations of the growth of the organs of the reproductive shoot, as given above, is on the whole correct, because it was based on the study of ample material.

The flower
fig. 73

As a rule the stem of *T.subpraestans* is branched and carries, besides the flower of the main axis, lateral flowers at the top of the shoots growing from its leaf axils. The flower primordium of the main stem was initiated in mid August. Two weeks later the primordia of the outer tepals became clearly visible. Early in September the tepals began to grow and the primordia of the stamens appeared in the same sequence as the tepals. The elliptic bulges of the outer set came first, soon followed by those of the inner stamens. The pistil was initiated almost simultaneously with the latter.

From the first days of September till mid November the tepals developed rapidly. The anthers of the stamens also grew, but unlike the tepals, they slightly slowed down growth in mid October and stopped altogether by mid November. The pistil developed but slowly in comparison with the other organs and was still of a minute size in the beginning of November when it started winter dormancy earlier than the other flower organs did.

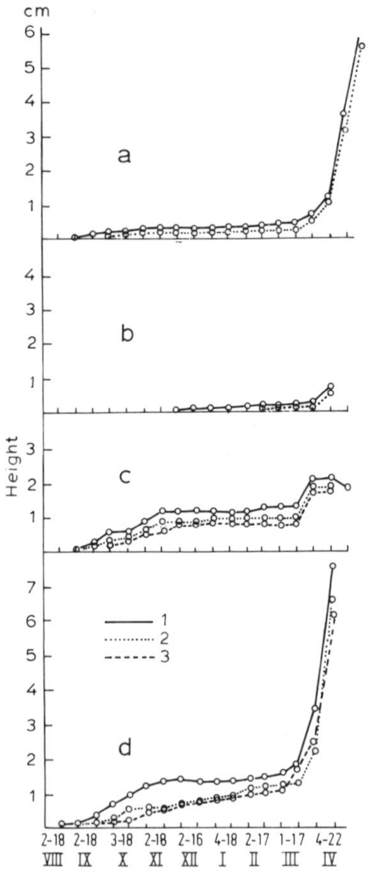

Figure 73. *T. subpraestans.* Initiation and growth of the flower organs of the inflorescence.
1. terminal flower; 2. second flower; 3. third flower; a. ovary; b. filaments; c. anthers; d. tepals.

Dormancy of the tepals was 3½ months and only in the latter half of that period was some slight growth observed. Dormancy of the anthers took 4 months, from mid November to mid March, but in the beginning of February they increased somewhat in length. Dormancy of the pistil lasted from November to mid March.

The filaments appeared as tiny triangles early in December and began to grow feebly in the second half of February when it became a little warmer.

All flower parts broke winter dormancy in the latter half of March. In the course of two weeks the anthers had attained their maximum length, but the stamens were in the first half of April still elongating, owing to the rapidly growing filaments. The tepals, which started to grow two weeks earlier than the anthers, grew most actively during the latter half of March and early in April. The pistil, which terminated winter dormancy together with the anthers, grew slowly but steadily between mid March and mid April and it was only after the flower unfolded that it began to grow more rapidly, forming a mature ovary by the beginning of May.

The anthers were first to stop growth and development, followed by the tepals and the stamens, which terminated this period at the same time. The last to stop growing was the pistil because the ovary develops after pollination. The lateral flowers showed the same pattern of growth and development as the flowers of the main axis.

The daughter bulbs
fig. 74

T. subpraestans, besides producing daughter bulbs in the axils of the fleshy scales, initiates one in the axil of the tunic, like the cultivars. We were able to study these processes in spite of the unfavourable weather conditions of 1954 because the initiation of the daughter bulbs depends largely on the conditions of the previous vegetation period.

A smallish meristematic protuberance appeared in the axil of the tunic in the latter half of April and at the same time another one became visible in the axil of the first fleshy scale. A third was initiated in the axil of the second fleshy scale early in May. In mid September, 4½ months later, formative tissue swelled up in the axil of the third fleshy scale.

The bud of the tunic began to grow directly after its initiation and had attained 2 mm after two weeks. It continued to grow slowly and measured 3 mm by mid September; then growth increased and a month later it was 6 mm long. After that growth slowed down to increase again in mid January. By the middle of March the daughter bulb of the tunic was 10 mm high, continued to grow actively and had reached 32 mm in June.

The daughter bulb of the first fleshy scale, initiated simultaneously with that of the tunic, hardly made any extension growth before mid August. During the following two months it grew slowly to 3.5 mm by mid October, 5.0 mm by mid January and 7.0 mm by mid

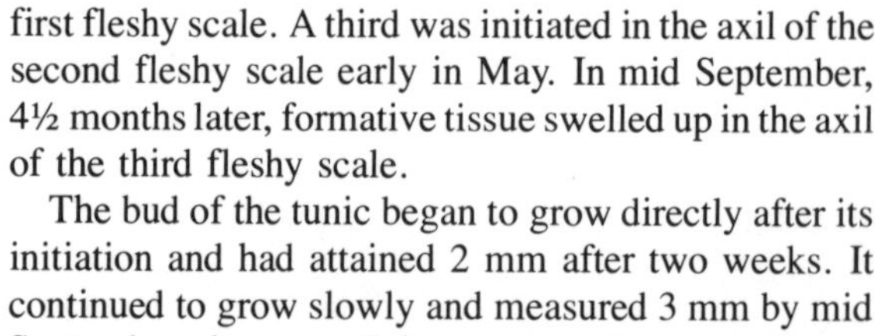

Figure 74. *T. subpraestans.* Growth of the daughter bulbs in one mother bulb.
1. daughter bulb in the axil of the tunic; 2. daughter bulb in the axil of the 1st fleshy scale; 3. daughter bulb in the axil of the 2nd fleshy scale.

April, after which its height increased rapidly to 30 mm in June.

The daughter bulb of the second fleshy scale, initiated in early May, began to grow noticeably in mid September. It measured 2.0 mm at the beginning of October when growth stopped. Early in February it grew 0.1 mm and eventually attained 7.0 mm by the first of March. Further growth depended on the weather conditions in spring. With sufficient moisture and rising temperatures this daughter bulb soon became as large as the main daughter bulb.

The daughter bulbs in the axils of the third and fourth fleshy scales developed in the same way. They were initiated in mid September, approximately 5 months after the first daughter bulb and did not have a latent period growth, but began to grow immediately, attaining 5.0 to 7.0 mm in the first month. Like the earlier initiated daughter bulbs they hardly grew during winter, but their active period of growth began somewhat earlier in the beginning of March. Provided the external conditions are favourable, they may reach the size of the main daughter bulb.

TULIPA KOROLKOWII REGEL

The next year's new shoot was initiated in the main daughter bulb in April during or after flowering of the mother bulb (fig.75). In the latter half of May of the next year it had produced a mature capsule. The shoot lived for ca. 13 months and the long interval between initiation of some of the organs is typical of its development. The basal leaf was initiated soon after the appearance of the shoot primordium in the last decade of April, while the second leaf was initiated after mid August. The third leaf primordium appeared simultaneously with that of the flower during the last week of August when the stem began to develop. Except for the basal leaf all the organs of the shoot were initiated within a fortnight.

The root primordia on the basal plate appeared early in June. By October the roots had completed their development and were breaking through the epidermis of the basal plate to penetrate into the soil with the first rains, early in November. They died off in mid April of the next year.

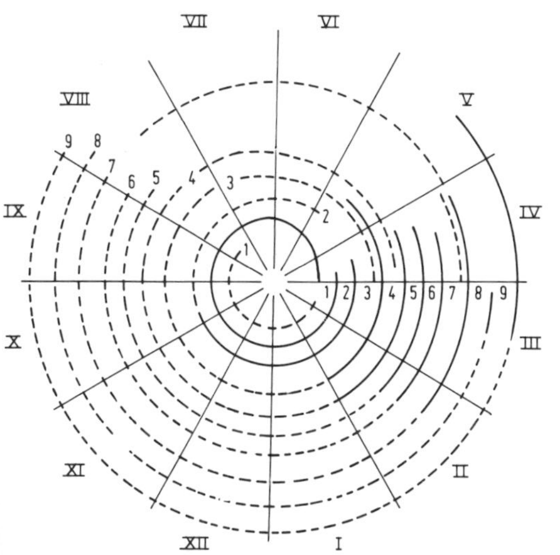

Figure 75. *T. korolkowii.* Annual cycle of development. I–XII. months; 1. bulb; 2. roots; 3. shoot; 4. basal leaf; 5. 2nd leaf; 6. upper leaf; 7. stem; 8. flower; 9. ovary.

Initiation and growth of the main daughter bulb
fig. 76

The main daughter bulb was initiated at the base of the young reproductive shoot in the beginning of September. The mother bulb died in mid April and the period between initiation of the main daughter bulb and its separation from the mother bulb was 7½ months. As it lived for another year as an independent organism its total lifespan was ca. 19½ months.

Because of the late initiation the tunic primordium was not more than 0.5 mm in the beginning of September. Two weeks later it had grown to 1.8 mm. During this period the first fleshy scale was initiated. It was 1.7 mm long by the beginning of October and nearly surpassed the tunic whose growth had slowed down. Between mid October and November both the tunic and the first fleshy scale grew slowly. During this period the second fleshy scale was initiated and till the beginning of February of the next year the tunic and both scales gained only ca. 0.5 mm a month. From mid February they began to grow more rapidly, in particular during the latter half of March, by which time the main daughter bulb had reached 19 mm. During flowering of the mother bulb growth of the tunic and fleshy scales slowed down but accelerated again early in April.

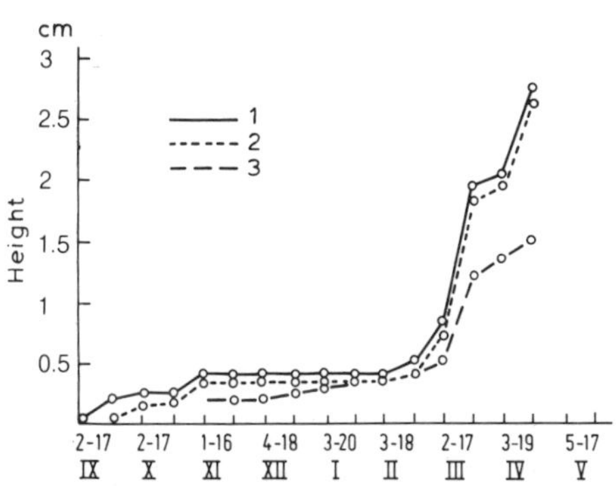

Figure 76. *T. korolkowii.* Growth of the main daughter bulb. 1. tunic; 2. 1st fleshy scale; 3. 2nd fleshy scale.

Towards the end of the growing period the tunic, including its elongated tip, was 26 mm long, the first fleshy scale 25 mm and the second fleshy scale 12 mm. This completed their development.

We see that the initiation of a new scale somewhat retards the growth of the preceding one. When a new scale had nearly overtaken the preceding one, both continued growth at the same rate. Only the second fleshy scale, which slowed down its extension growth during flowering, did not resume its previous growth rate, whereas the tunic and the first fleshy scale continued at the same uniform rate as before.

Growth of the shoot

The shoot and its first leaf for the next year were initiated in the apex of the young main daughter bulb in the period when it stopped growing (fig.76,77) and in the 4 months following no further organs were initiated. The basal leaf began to grow slowly in mid August. Early in September the second and the third leaf were initiated and the lowest stem internode began to make extension growth.
All leaves and the stem grew actively till mid October when the roots began to function. Leaf growth was retarded considerably in the latter half of October, because there was no precipitation, and growth stopped in the first half of November. The stem continued to grow. Rain began to fall in the latter part of November, activating root growth, by which the development of all organs was resumed, but only for two weeks, because from December till February all leaves stopped growth for two months. The stem stopped growing in the latter half of December, then grew 20 mm in the first half of January. Two weeks before the leaves terminated the resting period the stem again stopped growing from mid January to March. From February onwards the leaves grew vigorously and under the favourable temperature conditions appeared aboveground. By mid March they completed growth, after which they remained green for approximately six weeks. Two weeks before the leaves had ceased growing the stem became active again, grew slowly in the first half of March, but made 4 cm extension growth during the following period.

The early termination of leaf development is typical of *T.korolkowii.*

The flower
fig.75:8 and 9)

The flower primordium appeared early in September on the stem apex. An examination in mid August had not yet shown any flower initials, but early in September the primordia of all the organs were already initiated (fig.78). The first stages of development only show a superficial similarity.

From September till the first days of October the tepals grew 8 mm after which growth fell abruptly, the increment during the two following weeks being no more than 1 mm. Between mid October and mid November the tepals did not grow at all. After the plant had taken root near the end of November they grew 2 — 3 mm and then stopped again for two months till the beginning of February. From February till mid March they grew another 3 mm. It appears that over the entire growing period from September till mid March they grew only 13 mm, attaining 15 mm by the end of March.

The stamens grew steadily till mid November. Dur-

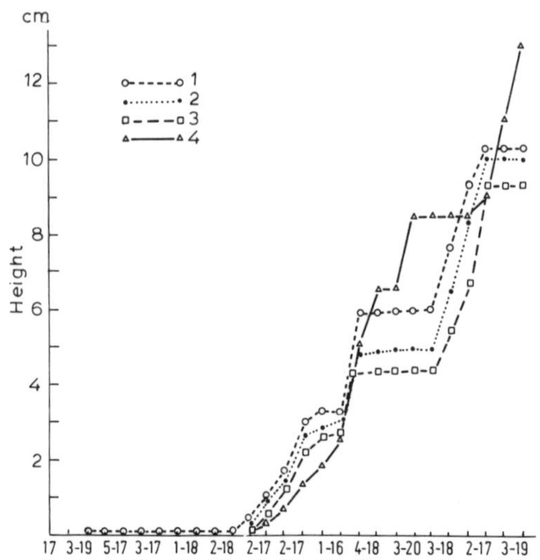

Figure 77. *T. korolkowii.* Growth of the leaves and the stem. 1. basal leaf; 2. second leaf; 3. upper leaf; 4. stem.

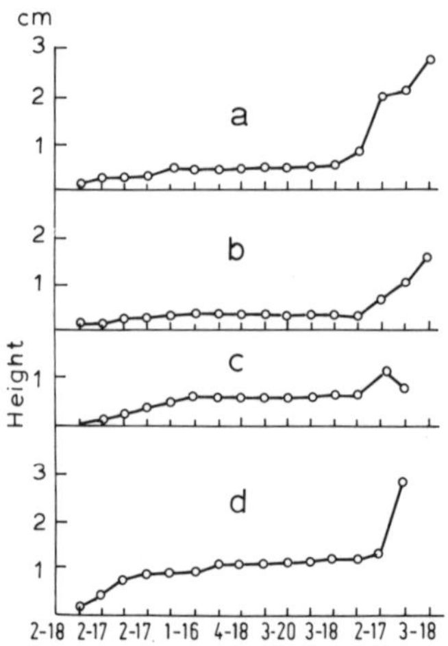

Figure 78. *T. korolkowii.* Growth of the flower organs. a. ovary; b. filaments; c. anthers; d. tepals.

ing this period the reduction division takes place in the anthers and the pollen hibernates as mononucleate pollen grains. From mid November till February the stamens did not grow. In the beginning of February some weak growth was observed and in the course of one month they made 2 mm extension growth. In the latter half of March they grew another 5 mm, half the length of the anthers, after which they dehisced during flowering and died.

The ovary, initiated together with the stamens early in September, grew slowly till the beginning of November, gaining no more than 4 mm. Then growth stopped for 3½ months to start again in mid February, at first slowly and from the beginning of March more actively. Between mid March and April the ovary hardly grew at all, but when the flower unfolded in the first half of April, it was growing rapidly, after which period its dimensional growth had terminated.

TULIPA BIFLORIFORMIS VVED.

The primordium of the new shoot was initiated in mid April in the main daughter bulb, which was then nearly an independent organism. Although its leaves had partly dried off, the mother bulb had not yet completed the vegetation period. By mid May, when the roots of the main daughter bulb were initiated, the basal leaf primordium of the new shoot was already visible.

The still flattish primordium of the next season's main daughter bulb became noticeable at the base of the central bud in mid June. By that time the initiated roots of the main daughter bulb were already connected with the main vascular system of the basal plate.

Early in June the second and the third leaf were initiated in the then still flattish shoot primordium. The latter began to make extension growth early in August. An examination in mid August showed the formation of two flower buds; a large one of the apical flower and a smaller one of a lateral flower. The tepals and stamens of the apical flower had already been initiated.

In the first half of August extension of the lower stem part, between the innermost bulb scale and the basal leaf of the new shoot, was observed. The ovary of the apical flower was initiated early in September.

Early in October, provided there is sufficient moisture and the soil temperatures are not too high, the bulb takes root after which the new main daughter bulb, initiated at the base of the new shoot in the axil of the last scale, starts to develop. The primordial tunic and the first fleshy scale were at that time already discernible and we observed a swelling of the second fleshy scale primordium behind the first fleshy scale. At the same time the mother bulb initiated a daughter bulb in the axil of the first fleshy scale.

In mid October the primordial bulge of a collateral daughter bulb appeared next to the new main daughter bulb. This completed the organogenesis in *T.bifloriformis.*

The scales of the daughter bulb in the axil of the first fleshy scale were initiated early in December and those of the collateral daughter bulb late in October.

The number of leaves and bulb scales of *T.bifloriformis* is limited. The reproductive shoot always has two leaves. The adult bulb, however, has three modified leaves, a tunic and two fleshy scales, a deviation from the usually distichous arrangement typical of Liliaceae.

Growth of the leaves and the stem
figs. 79, 80

In all species the basal leaf differs from the upper leaves in its morphological structure and functions. It resembles a tunic in serving as a cover for the upper leaves and it protects the reproductive shoot both underground and after emergence, against unfavourable external conditions. However, this is no longer the case when it assumes its main function as an assimilatory organ. The tunic on the other hand serves at first as food storage organ and after its depletion as a protective cover for the fleshy scales. The differences in growth and development of these two organs are evident.

The stem primordium of the main daughter bulb appeared in mid April and a month later its basal leaf was initiated, followed 3 months later by the second leaf. During this interval the first leaf, covering the bud, hardly made extension growth. Two weeks after

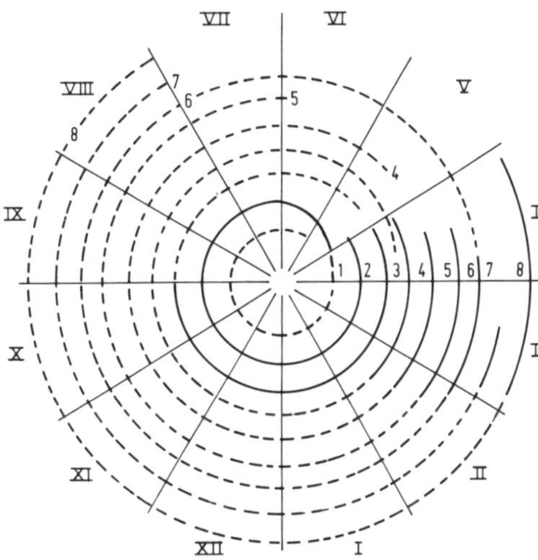

Figure 79. *T. bifloriformis.* Annual cycle of development. I–XII. months; 1. bulb; 2. roots; 3. shoot; 4. basal leaf; 5. second leaf; 6. stem; 7. flower; 8. ovary.

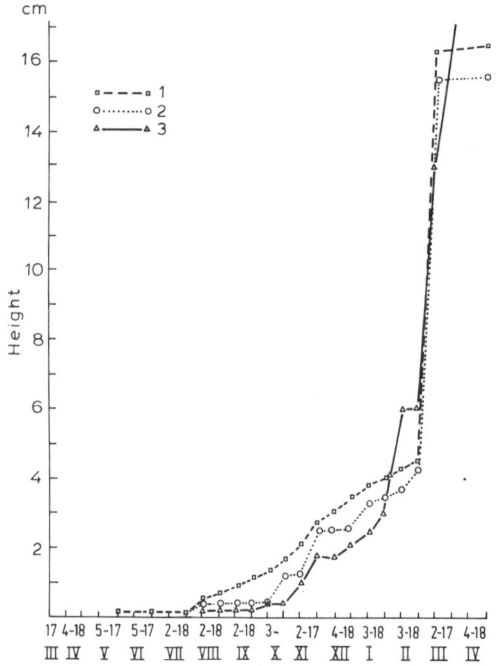

Figure 80. *T. bifloriformis.* Growth of the leaves and the stem. 1. basal leaf; 2. upper leaf; 3. stem.

initiation of the second leaf, the first leaf had reached 5 mm, the second 4 mm. The first leaf continued to grow even during summer dormancy when contact with the environment is at a minimum. During August and September it grew to 14 mm. The second leaf stopped in the first days of August and until October its increment was negligible.

The stem only grew 2 mm in the period between initiation and mid September. During the latter half of September it gained 2 mm, after which growth stopped again for two weeks. The basal leaf grew rapidly to ca. 4.5 cm between October and mid February. Up to that period active growth over 7 months averaged ca. 6 mm per month, but during the following period of maximal growth its increment was 11.8 cm.

The growth pattern of the second leaf was quite different in being strongly affected by growth of the other stem organs and the external conditions. Active growth, starting early in October, was retarded when the stem began to grow in mid October. During the first decade of November the growth was restored and both organs grew actively. Their growth was again stopped in mid November owing to the unfavourable conditions which may have reduced root activity. The stem resumed active growth early in December and the second leaf in the middle of that month. Growth of the latter slowed down considerably early in January and by February it had only enlarged 4 mm, while the stem gained 3 cm from mid January to February and in the middle of February stopped growth for two weeks until the bud emerged from the soil. Between mid February and early March the leaves attained their maximum size and did not grow any more till the end of the vegetation period around the end of April.

The flower
figs. 79:7 and 8, 81)

Since an examination of the bud early in August already revealed the primordia of the tepals, the stamens and the ovary, flower initiation took place late in July. The organs did not enlarge till mid September when the tepals and the stamens began to develop. The tepals grew till mid November and the stamens till the beginning of that month, by which time they were fully formed and the anthers contained pollen mother cells. Neither the stamens nor the pollen grains showed further development till February. The pistil began to grow later and only for two weeks but between mid October and February it enlarged.

A radical change in the growth and development pattern of all the flower organs occurred early in February. At that time the reduction division took place in the anthers and the chromosomes could be observed. There were 24 bivalents during diakinese and 24 separated pairs in the metaphase, showing the Tashkent form of*T. bifloriformis* to be a tetraploid. When all food reserves are mobilized for feeding the reproductive organs, growth of organs like the stem is arrested (fig.80). However, from the beginning of February the tepals, the stamens and the ovary began to grow again intensively. In the latter half of that month the tepals stopped growing for two weeks while all the other organs kept growing vigorously.

From the beginning of March, in the period when the flower opens, the tepals again grew rapidly. After de-

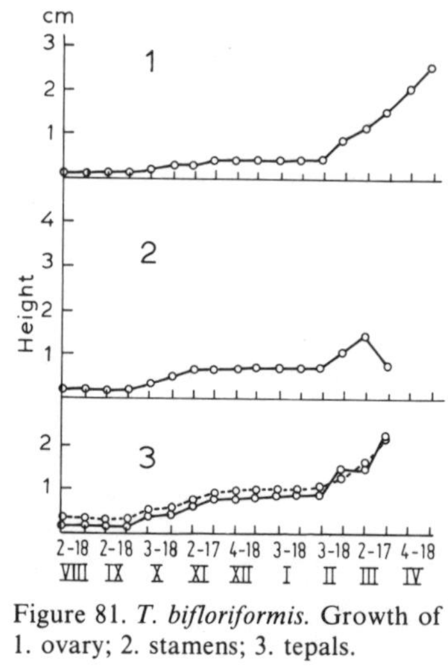

Figure 81. *T. bifloriformis.* Growth of the flower organs. 1. ovary; 2. stamens; 3. tepals.

hiscence of the anthers, the stamens shriveled and dried up. Two weeks later the tepals also stopped growing, but the ovary continued till the beginning of May. It matured in mid May.

The Tashkent form of *T. bifloriformis* produces fine capsules, containing numerous viable seeds, and has a fairly good capacity for vegetative reproduction. Although it forms only one daughter bulb in the axil of the first fleshy scale, it initiates at the base of the main daughter bulb a collateral daughter bulb that grows into a large bulb.

In the old Tashkent gardens, along the banks of the irrigation canals and under trees, one often sees large numbers of *T.bifloriformis* that have reproduced vegetatively. As a rule only about ten out of many hundreds of plants growing there bear flowers, while the others reproduce vegetatively.

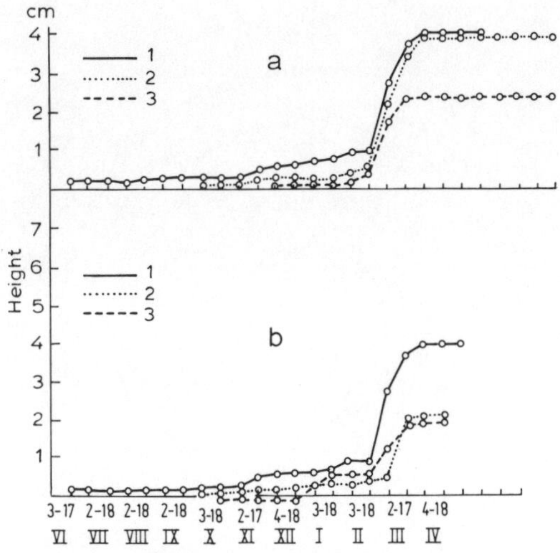

Figure 82. *T. bifloriformis.* Growth of the main daughter bulb and two other daughter bulbs.
a. main daughter bulb; 1. tunic; 2. 1st fleshy scale; 3. 2nd fleshy scale.
b. comparison of the growth of: 1. the main daughter bulb; 2. the daughter bulb in the axil of the first fleshy scale; 3. a collateral daughter bulb next to the main daughter bulb.

Initiation and growth of the main daughter bulb (figs.79.1 and 82)

The main daughter bulb was initiated in mid June and 3 months later still consisted of a ca. 2 mm high primordial tunic with a meristematic protuberance in the centre. The first fleshy scale was initiated in the beginning of October and the tunic began to grow early in November. In two weeks it grew to 6 mm.

In the first month the first fleshy scale slowly began to make some lateral growth. By mid November, when also the second fleshy scale had been initiated, it measured ca. 4 mm. In the meantime, probably because of low soil temperatures, growth of the tunic was retarded. It had reached 10 mm by the first of February after which it stopped growing. In that period also the stem and leaves of the mother bulb showed a decrease in their active growth, owing to the morning frosts which came after a long spell of warm spring weather. The innermost organs of the plant, the flowering parts and the inner scales of the main daughter bulb, protected from the direct influence of the environment, continued their growth.

The first fleshy scale resumed extension growth in the latter half of January and was 6 mm long by mid February. The second fleshy scale began to grow one month after initiation and measured 2 mm by mid December, when its growth stopped, apparently because its nutrients were intercepted by the collateral daughter bulb which had started to grow actively. When the tunic stopped growing, early in February, the second fleshy scale started to make extension growth. Active growth of the tunic and the scales was observed in the latter half of February, but when the flowering period began in March growth slowed down.

The second fleshy scale was the first to stop growing, in mid March, and was then much shorter (2 cm) than the tunic and the first fleshy scale. Growth of the tunic and the first fleshy scale stopped towards the end of March, after which for approximately 6 weeks new food reserves were accumulated and translocated from the tunic to the fleshy scales. After senescence of the leaves and the roots of the mother bulb, the main daughter bulb entered the dormant period.

The bulb tunic of the main daughter bulb of *T.bifloriformis* was initiated later than in the other species studied, but 3½ months earlier than its scales and the following daughter bulbs, which are initiated in the autumn period when the meristems of the basal plate are most active.

The initiation of the other daughter bulbs, which took place during the development of the main daughter bulb, proceeded in the following order: the daughter bulb in the axil of the first fleshy scale was initiated early in October simultaneously with the first fleshy scale of the main daughter bulb (fig.82b); the collateral daughter bulb next to the main daughter bulb followed in mid October and already after a month had three scale primordia; towards the end of November or early in December the daughter bulb in the axil of the second fleshy scale was initiated, including its first and second fleshy scale primordia. This completed the initiation of the organs of vegetative reproduction.

During the subsequent growth and development the young bulbs repeated the growth pattern of the main daughter bulb, although as a rule they did not reach its size and stopped growth two or three weeks earlier. The daughter bulb in the axil of the first fleshy scale entered the period of maximum growth two weeks later than

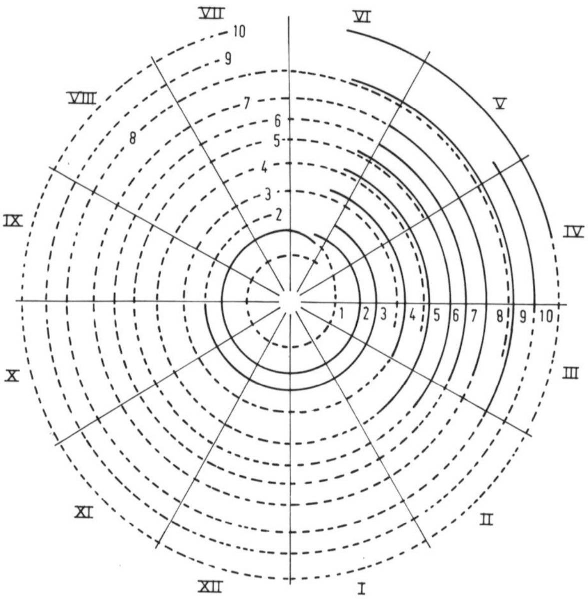

Figure 83. 'Jubilee'. Annual cycle of development.
I–XII. months; 1. bulb; 2. roots; 3. shoot; 4. basal leaf; 5. second leaf; 6. third leaf; 7. fourth leaf; 8. stem; 9. flower; 10. ovary.

the main daughter bulb and its collateral bulb.

The main daughter bulb of *T. bifloriformis* lives for 10½ months inside the mother bulb and another year as an independent organism. Its tunic, initiated in mid June dries up towards the end of the vegetation period, early in May of the next year. Its life span is 10½ months. The first fleshy scale, initiated early in October lives 19 months. During the latter months its food reserves are depleted. The second fleshy scale, initiated two weeks after the first one lives 18½ months. In January and February of the first year, before the vegetation period, it already functions as a sink receiving surplus starch and other nutrients from the scales of the mother bulb. During the last three months, from February to May of the next year, its food reserves are depleted.

THE DARWIN TULIP 'JUBILEE'

The shoot tips of this cultivar usually appear aboveground early in March, sometimes even near the end of February. When the leaves begin to unfold by mid March the main daughter bulb, which is still developing inside the mother bulb, initiates a shoot. The primordium of its basal leaf is then clearly visible. The second leaf is initiated in the first days of May, the third early in June and a fourth around the middle of that month. The lower internodes of the stem start to make extension growth early in September but flower initiation begins in mid July. Simultaneously with the initiation of the third leaf a meristematic protuberance appears at the base of the young shoot. It will develop into the new main daughter bulb which for nearly 12 months forms part of the mother bulb (fig.83).

Growth of the main daughter bulb
fig.84

The crescent-shaped primordium of the tunic was as a rule initiated in mid June. The formative tissues swelling up in its centre were the primordia of the first and second fleshy scales, which became clearly noticeable

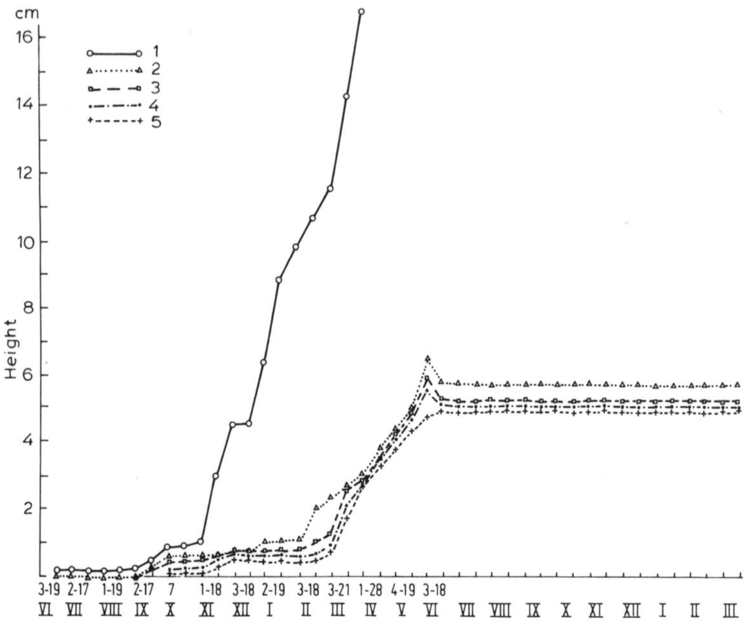

Figure 84. 'Jubilee'. Growth of the main daughter bulb.
1. tunic; 2. first fleshy scale; 3. second fleshy scale; 4. third fleshy scale; 5. fourth fleshy scale.

in the first days of August and September respectively. The third and fourth scales were initiated early in October. We see that all the organs are initiated during summer dormancy from mid June to October.

During the first two months the growth of the tunic was barely perceptible, but from mid August it gradually accelerated. When it had grown to 9 mm, in the beginning of October, growth stopped for a month because, as with the other species studied, lack of moisture in the soil had inactivated the root system. With the first autumn rains early in November the mother bulb took root and the tunic of the main daughter bulb started to grow actively, attaining 4.5 cm in December, which was almost the normal bulb length. In the first days of December growth was retarded, evidently because of a sharp drop in the air and soil temperatures. In the latter half of December growth accelerated and continued without interruption till the end of the vegetation period, because the tunic of the main daughter bulb of 'Jubilee' usually forms a leaf blade. It appears that this tunic is the broadened sheath of a green leaf and not a scale (figs.84,85).

The first fleshy scale, initiated soon after the tunic, showed during the first 4½ months a similar growth pattern as the tunic. However, from the first days of November, in the period of active growth of the laminate part of the tunic, its development was different. The first, second, third and fourth fleshy scales barely grew till February; the third and the fourth scales reached the size of the first two scales in the first half of November. Early in February the first fleshy scale broke winter dormancy and started to grow actively, gaining 1.3 cm in one month. The second scale started growing at the same time, but less rapidly; it gained 6 mm. In mid February the third and fourth scales followed and grew slowly till the beginning of March and thereafter more rapidly, suppressing to some extent the growth of the first scale. From mid March to June all scales grew rapidly and uniformly. Towards the end of the first decade of June, when the tunic dried up, all scales shrunk a little after which they did not change any more till the roots of the bulb began to grow in the autumn.

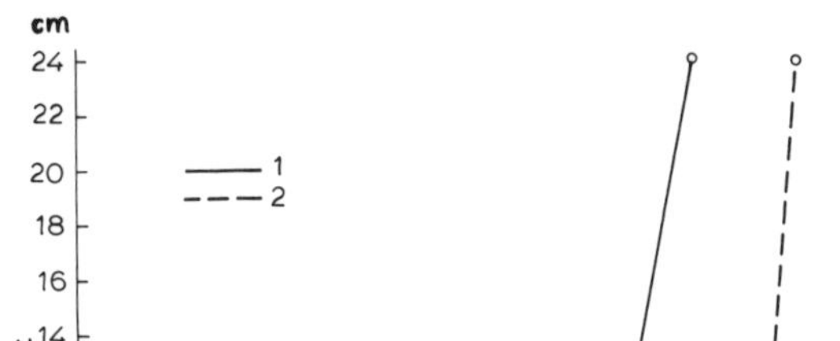

Figure 85. 'Jubilee'. Growth of the basal leaf, from initiation to March next year, and the stem, from August to May.
1. basal leaf; 2. stem.

In its second year towards the end of the vegetation period, in the latter half of June, the main daughter bulb becomes separated from the tunic and the basal plate of the mother bulb, which means that the cultivar 'Jubilee' lives for one year and two weeks on the food reserves of the mother bulb. After this period it exists another year as an independent organism so that its life span is two years and two or three weeks.

The roots were initiated in mid July of the second year during the summer dormant period. Depending on precipitation and soil humidity, the roots will break through the epidermis of the basal plate in October or November and when the bulb is planted they penetrate into the soil and begin to function. As with the wild species the root system of cultivated tulips begins to function in autumn and, provided the soil temperatures are high enough, remains active throughout the winter. The roots die off in the next year near the end of the vegetation period simultaneously with senescence of the leaves and the stem. Only the seed capsule takes somewhat longer to mature.

The shoot
fig. 83

After its initiation in mid March the basal leaf developed very slowly, but by the time the second leaf was initiated the provascular bundles had formed; the larger bundles in the centre and the smaller ones towards the periphery. Between mid May and mid June it grew 1 mm, after which no further growth was observed. In August it started growing rapidly and had reached 1 cm at the end of that month. During the first half of September growth had stopped again, but in the latter half of that month its size increased threefold. No growth was observed in October owing to the low temperatures and lack of moisture. However, the smaller leaves in the centre of the bud were less affected by the external conditions than the basal leaf which shields the bud.

We have noticed that tulip roots grow rapidly if watered early in autumn when the air and the soil temperatures are higher, but later on the soil becomes so hot that they do not grow any more and the entire zone of growth of the roots becomes thickened. Even under the most favourable conditions root growth is not resumed. The plants do not die but remain small and weak and form smaller and fewer daughter bulbs than normally.

When the temperature dropped and the rainfalls began, early in November, the growth rate of the basal leaf increased sharply. By mid November it had gained another 2 cm and emerged from the bulb. With a further lowering of the temperatures from mid November to mid December growth gradually slowed down but did not come to a standstill; it had grown 7 mm. Between mid December and mid January growth accelerated again to 1.5 cm a month. After mid January the basal leaf grew rapidly and steadily for 3½ months, reaching its final length by the beginning of flowering.

The second, third and fourth leaves developed almost in the same way as the basal leaf, only more gradually since they were initiated later and grew to a smaller size (fig.85).

The primordium of the stem consisted of a flattish zone of small meristem cells and was barely perceptible. It was therefore difficult to determine when the stem was initiated. We consider initiation to begin at the moment when the bulge of the first leaf primordium is observed, since from then on the stem functions as a primordial axis.

During initiation of the leaves and the flower the stem did not make any extension growth. This started in mid August, ca. two weeks after flower initiation, when all the organs became visible to the naked eye. Autumnal growth was completed around mid October and till the beginning of December the stem did not grow. From December to mid March the monthly growth increment was 5 mm and between mid March and April 20 it had reached 64 cm. During that period the extension growth of three days was as much as the total increment in the preceding year. After flowering the pedicel grew another 5 cm before the capsule reached maturation.

In the early stages the lowest internode of the stem, between the upper scale of the bulb and the basal leaf of the shoot, makes the extension growth. During subterranean growth the central internode grows very little, but the pedicel is active although it grows slower than the lowest internode. When all the leaves have appeared above ground the lowest internode practically stops growing, but the other internodes and in particular the pedicel grow vigorously (fig.85).

The flower
fig. 86

The flower bud was initiated around mid July, only one month after the main daughter bulb separated from the mother bulb. Initially the tepals grew slowly to 1 cm by mid November. At that time the bulb took root and growth accelerated somewhat but slowed down again

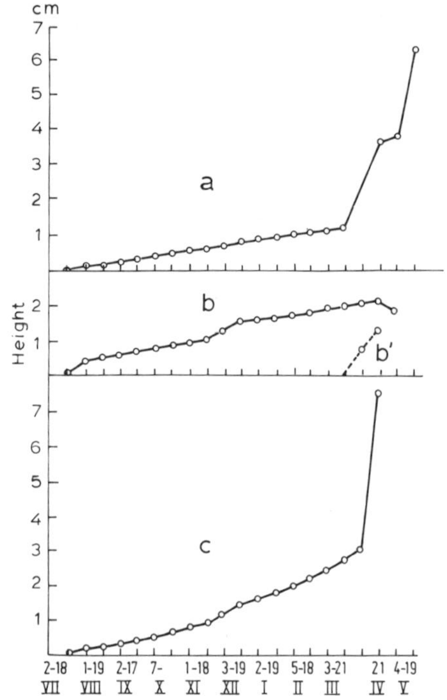

Figure 86. 'Jubilee'. Growth of the flower organs. a. ovary; b. stamens; b[1]. filaments; c. tepals.

towards mid December. However from mid February to April growth increased steadily, particularly during the last two weeks when the increment was much more than in the preceding period.

The stamens, initiated almost simultaneously with the tepals, grew 5 mm in the first two weeks and continued to grow at a slower rate, gaining another 5 mm by mid November. Then, like the tepals, they grew more rapidly, but they continued to grow at a uniform rate till the flower unfolded. During the last month their filaments grew to 1.3 cm. After the anthers had shed their pollen, the size of the stamens diminished radically.

The ovary, initiated in mid July, grew without interruption and more or less uniformly till mid March of the next year, after which its growth rate showed a sharp increase. It grew 3.7 cm in one month (cf. the 1.2 cm of the preceding 8 months). In mid April growth stopped for two weeks, but in May during the last two weeks after flowering it gained another 2.5 cm.

Growth and development of the axillary buds
(fig.87)

The buds of the daughter bulbs that develop in the axils of the scales are first order buds of the shortened bulb axis. They may develop into lateral shoots or accumulate food reserves in their scales (leaves) and after separation from the mother bulb continue life independently, because they are the organs of vegetative reproduction.

The daughter bulb buds in the axil of the tunic (figs 87, 92, 93. The first daughter bulb bud initiated by 'Jubilee' was the one in the axil of the tunic of the main daughter bulb, when the latter still formed part of the mother bulb. Initiation occurs very early, in the beginning of February, when a tunic primordium appears which begins to grow actively. At the beginning of May it measured 15 mm and, although this sometimes concludes its development, we observed in most cases a second growing stage in the latter half of July, when it grew another 10 mm.

The first fleshy scale, initiated during flowering of the mother bulb in the beginning of May, grew actively and was by mid July as large as the tunic. At this stage the first phase of development of the daughter bulb in the tunic axil was completed and the scales did not make further growth.

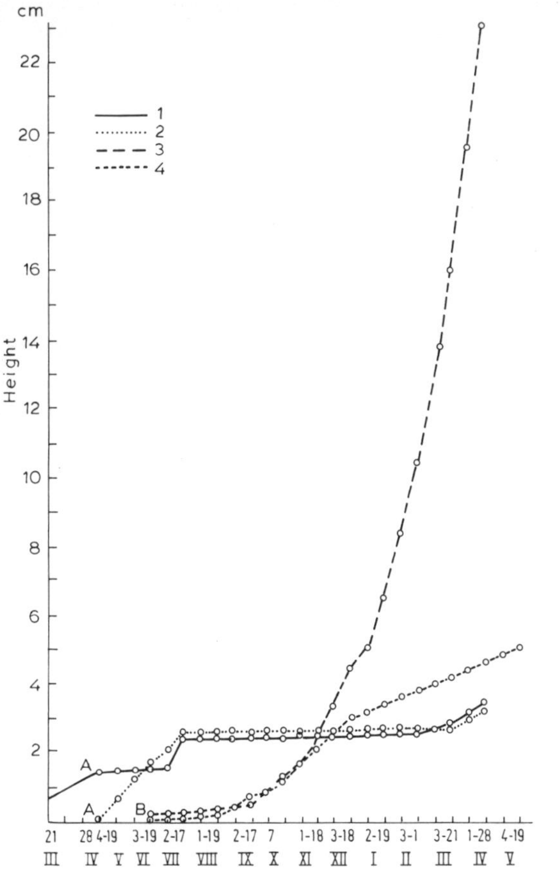

Figure 87. 'Jubilee'. Growth and development of the daughter bulb buds in the axil of the tunic.
A. first phase of development; B. second phase. 1. tunic; 2. first fleshy scale; 3. shoot (second phase); 4. tunic of new main daughter bub at the base of the shoot (second phase).

Rather suddenly, in mid July, there appeared in the apical growing point of this bud the primordia of two new organs, a shoot and the tunic of its main daughter bulb. For a long time their growth rates ran parallel. From the moment of initiation to the beginning of August the organs hardly increased in size, but in the following 6 weeks they grew 6-8 mm and by mid November they were 20 mm. Then the shoot began to grow vigorously while the new tunic developed steadily, but at a much slower rate. Growth was not affected by the winter dormant period because the organs received nutrients from the mother bulb and absorbed the accumulated reserves of the first two scales. Towards winter these scales, which still sheathed the base of the shoot and the new main daughter bulb, were depleted and had become thin and membranous. In July the three leaves of the central shoot were already discernible and in the beginning of September all the flowering parts.

In the period when the daughter bulb bud in the tunic axil was in its first phase of development and still depended on the mother plant, other daughter bulbs were initiated in the axils of its scales. Early in July we observed in the axil of its tunic a second phase daughter bulb and in the axil of the first fleshy scale the primordium of still another daughter bulb.

Besides these daughter bulbs, frequently a sister daughter bulb was initiated next to the daughter bulb in the axil of the first phase tunic. It was usually formed at the side of the mother bulb and was attached to the vascular bundle of the first daughter bulb in the tunic axil, which concealed the sister bulb.

As a rule the main daughter bulb of the tunic's daughter bulb, sitting at the base of the second phase generative shoot, had but a small number of scales, viz. a tunic and 2 or at most 3 fleshy scales. The remnants of the first phase scales could be found at its base as dry membranes.

In 10 out of 24 cases we found a well-developed daughter bulb behind the tunic of the first phase daughter bulb and quite as often a daughter bulb in the axil of its first fleshy scale. Sister bulbs were found only in three cases.

Nearly 50% of the large 'Jubilee' bulbs produced a flowering shoot from the tunic daughter bulb and in 10

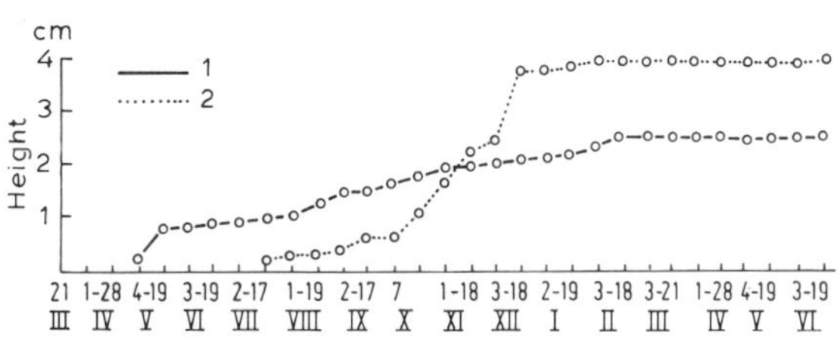

Figure 88. 'Jubilee'. Growth and development of the daughter bulb buds in the axil of the first fleshy scale.
1. tunic (first phase); 2. tunic (second phase).

out of 24 cases this shoot had a main daughter bulb at the base. Since the shoot was initiated in the second phase it is merely an axillary shoot with a flower and vegetative reproduction is in this case realized by the daughter bulbs in the axils of the first phase scales of the daughter bulb. These scales are homologous with the bud scales of other plants.

Thus, we see that there may arise from the axil of the tunic four organs: a shoot and 3 bulbs. In weak mother bulbs there may be formed only one large bulb.

The daughter bulb buds in the axil of the first fleshy scale (figs. 88, 92, 93). The tunic of this daughter bulb became discernible as a yellowish meristematic bulge in the first days of May. Towards mid May it had reached 8 mm in length but in the following two months it grew only another 2 mm. It could then already be observed that a next organ was being initiated. However, this was not a first phase fleshy scale (cf. the bud in the axil of the tunic), because further development showed that the first phase of this daughter bulb ended in most cases with the formation of a tunic. Probably because of its later initiation this next scale, observed in mid July, proved to be a tunic too; in this case already a second phase tunic. Simultaneously with the latter a second phase fleshy scale began to swell up, followed by two more scales in mid September.

In 9 out of the 24 plants studied that year the first phase was limited to the initiation of two scales, a tunic and a fleshy scale, and in the other 15 plants only one scale appeared, which developed as a tunic or rather as a fleshy scale without a green leaf blade. It accumulated starch and grew to 2.5 cm in mid February, but in December it was outgrown by the fleshy scale of the second phase daughter bulb, which by the middle of December had reached its maximum length of 4 cm.

Around mid November we noticed behind the first phase tunic a daughter bulb, consisting of two scales. One month later it had already formed four scales. In mid December we also found a sister daughter bulb (a serial bulb) of the daughter bulb of the first fleshy scale on the side of the second scale. It consisted of two scales and a meristematic bulge.

In a plant, of which the daughter bulb in the axil of the first fleshy scale, had developed in the first phase a tunic and a fleshy scale, we noticed in mid January, besides the bud behind the tunic, another bud primordium behind the first fleshy scale. Its tunic and two fleshy scales were already discernible.

In one case we observed that a bulb, dissected during the first phase in mid January, had developed, besides two scales, a flowering shoot of the second phase. It had a daughter bulb bud in the axil of the tunic and a daughter bulb bud in the axil of its first fleshy scale serving as a future main daughter bulb.

Of the 24 bulbs examined 3 daughter bulb buds in the axil of the first fleshy scale had not reached the second phase; one of them dried up in the initial stage.

We usually found the following organs in the axil of the first fleshy scale: a fully grown daughter bulb, developed from the second phase apical growing point; a bud in the axil of the first phase tunic; a bud in the axil of the first phase fleshy scale; a serial bud of the basal bud, on the same vascular bundle, or even a cluster of buds, initiated a few millimeters below the base of the scale.

The daughter bulb buds in the axil of the second fleshy scale (figs. 89, 93). The daughter bulb bud in the axil of the second fleshy scale of the main daughter bulb was initiated early in May. We observed first the bulge of a tunic primordium which was immediately followed by that of the first fleshy scale. After growing actively for two weeks the tunic measured 7 mm and then stopped growth for 6 weeks. It was resumed during the first half of August when it gained 2 mm. During the next 3 months it grew another 3 mm after which growth of the tunic terminated. It had grown to 12 mm in the course of 6½ months after initiation. The one 'true' fleshy scale had started to grow slowly early in August, 3 months after initiation, and reached its final length of 9 mm by mid January of the next year.

In June a second phase bud primordium appeared as a yellowish green meristematic bulge in the centre of the first phase bud. Early in July the first scale was noticed and two more scales early in November.

We see that the bud of the daughter bulb in the axil of the second fleshy scale consists in the first phase of two scales and in the second phase of five scales, a tunic and four fleshy scales.

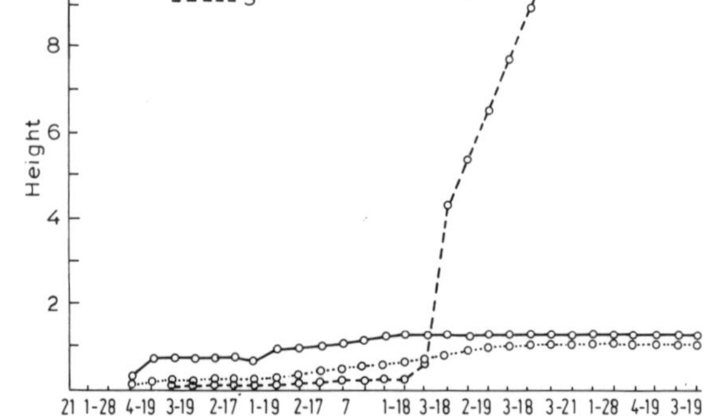

Figure 89. 'Jubilee'. Growth and development of daughter bulb buds in the axil of the second fleshy scale.
1. tunic (first phase); 2. first fleshy scale (first phase); 3. tunic which developed leaf blade (second phase).

The rather early initiated tunic of the second phase grew very slowly and was only 3 mm by mid November when its growth rate sharply increased. It grew 5 mm in two weeks. In the first half of December growth accelerated and it increased 3.5 cm. After that growth continued at a slightly reduced rate till mid March when it was 11.6 cm long and had developed a green leaf blade at the apex.

The first fleshy scale of the second phase did not make any extension growth before November when it began to grow slowly. It was 4 mm long in the beginning of December, after which growth stopped till mid January. By mid February it had grown to 1.2 cm and continued growing steadily. The other second phase primordia had begun to grow at a uniform rate simultaneously with the active growth of the tunic and the first fleshy scale.

Only one of the 24 pairs of plants we studied that year initiated a shoot in the axil of the second fleshy scale. In the other plants only bulbs were initiated in the apex and nearly all the second phase buds developed a green leaf blade at the tunic apex during its active growing period in December. It was difficult to observe this in the early stages, because it was then not possible to distinguish a leaf blade from a sheath.

In mid November we found a sister bulb (a serial bulb) in the axil of the second fleshy scale. At that time another bud had appeared in the axil of the first phase tunic. It had two scales, a tunic and a fleshy scale, and by mid December there were three scales. In mid January a plant which had developed two scales in the first growing phase, initiated, besides the daughter bulb bud in the tunic axil, a primordium in the axil of the one fleshy scale. In February this primordium initiated a tunic and a fleshy scale.

As in the axils of the tunic and the first fleshy scale, we may find in the axil of the second fleshy scale of 'Jubilee' four daughter bulbs. However as a rule there were fewer daughter bulbs, because for example a sister bulb occurred rarely.

The daughter bulb buds in the axil of the third fleshy scale (figs.90, 92, 93). Early in May the tunic of a daughter bulb was initiated in the third fleshy scale axil of the main daughter bulb. It grew 2 mm in the first two weeks and then stopped extension growth for 3 months. Late in August growth was resumed and in September it grew 10 mm. Further growth continued at a reduced rate until the latter half of January when it had reached 15 mm. This completed the first phase of development, because initiation of a fleshy scale was observed in only one of the 24 plants, early in June. It showed the same growth pattern as the tunic. This scale is a food store for the second phase buds, including serial buds, during their period of active growth. Moreover two plants did not form any first phase buds at all in the axil of the third fleshy scale, but initiated the daughter bulbs in the second phase.

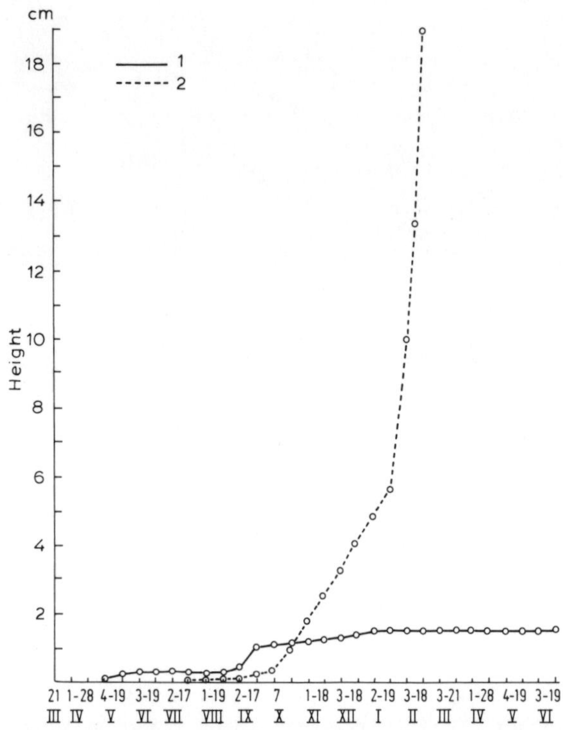

Figure 90. 'Jubilee'. Growth and development of daughter bulb buds in the axil of the third fleshy scale.
1. tunic (the only development during the first phase); 2. the shoot (second phase).

The second phase tunic appeared in mid July, started growth in mid September and had reached a month later 4 mm. Growth continued steadily and accelerated after mid January. By the end of February it was 19 cm long and had formed a green leaf blade, the fleshy scales which had followed remaining far behind in development. Around mid January for example the tunic had grown to 4.7 cm, the lst fleshy scale to 0.47 cm, the 2nd fleshy scale to 0.37 cm, the 3rd scale to 0.26 cm and the fourth to 0.15 cm. Although the fleshy scales continued to grow they did not reach the size of the tunic.

Like the buds in the axils of the outer fleshy scales of the main daughter bulb, the buds initiated in the axil of the third fleshy scale can initiate serial buds, for example next to the daughter bulb of the first phase tunic and next to the first phase fleshy scale (observed in the one case where this scale was initiated). Besides we noticed in mid January a 2 x 2 mm large primordium in the axil of the second phase tunic.

Thus, we may also find four daughter bulb buds in the axil of the third fleshy scale of the main daughter bulb. To these may be added the large bud behind the tunic of the main bud of that scale.

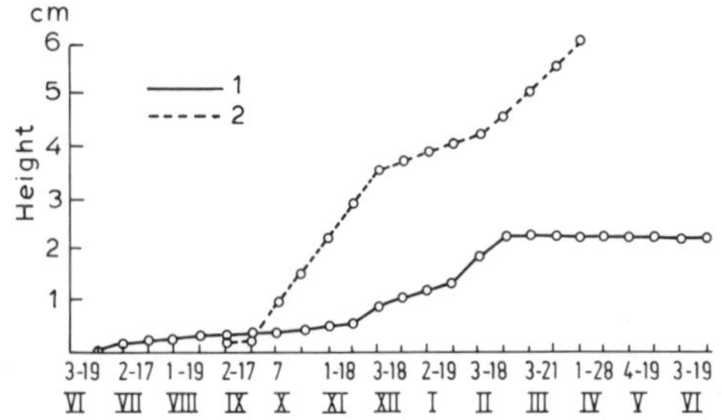

Figure 91. 'Jubilee'. Growth and development of daughter bulb buds in the axil of the fourth fleshy scale.
1. tunic, the only scale developing during the first phase; 2. tunic, initiated during the second phase.

The daughter bulb buds in the axil of the fourth fleshy scale (figs.91, 92, 93). The growth and development of the buds in the axil of the fourth fleshy scale was different from that in the outer scale axils. In the first phase only a tunic was initiated, in mid June. By mid November it was 5 mm long and growth accelerated. In the beginning of December it was 9 mm after which it slowed down, but from mid January to mid February it grew to 22 mm. Then growth ceased and the stored nutrients were translocated to the second phase bud.

We failed to observe the initiation of the second phase bud, but in the first days of September noticed a 3 mm long tunic and a 1 mm long fleshy scale with a 0.1 mm high primordium in its centre. They developed rapidly and by the beginning of December the second phase tunic was 3.5 cm. At that time four more scales were initiated which shows that the daughter bulb in the axil of the fourth fleshy scale may consist of a tunic and five fleshy scales.

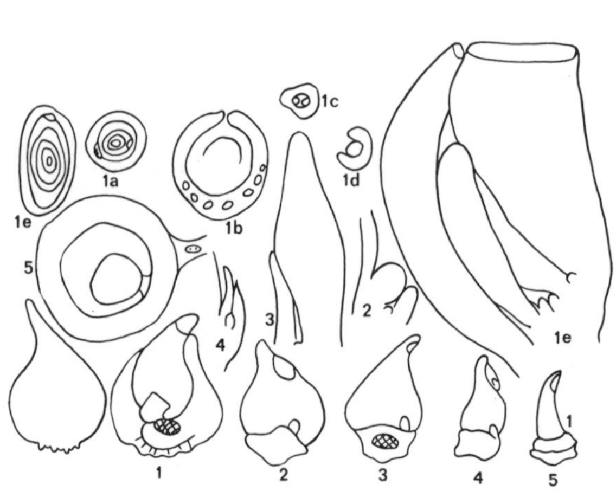

Figure 92. 'Jubilee'. Growth and development of daughter bulbs in the beginning of July. All scales of the first phase of development are initiated.
1. the daughter bulb in the axil of the tunic; 2. daughter bulbs in the axil of the first fleshy scale; 3. the daughter bulb in the axil of the second fleshy scale; 4. the daughter bulb in the axil of the third fleshy scale; 5. the apical bud. All scales of the first phase.

Between the scales of the second phase bulb we found late in April, when the tunic had reached its final size, primordia of the following dimensions: in the axil of the tunic, 2.5 x 0.5 mm; in the axil of the first fleshy scale, 2.0 x 0.7 mm; in the axil of the 2nd fleshy scale, 1.0 x 0.8 mm. This time there was no daughter bulb in the axil of the first phase tunic, but in previous studies with 'Jubilee' we had found one which had also a serial bulb.

In bulbs which had five scales a daughter bulb was initiated in the axil of the fifth scale directly at the beginning of the second phase. We did see neither the initiation of scales of the first phase nor buds in the axil of this scale.

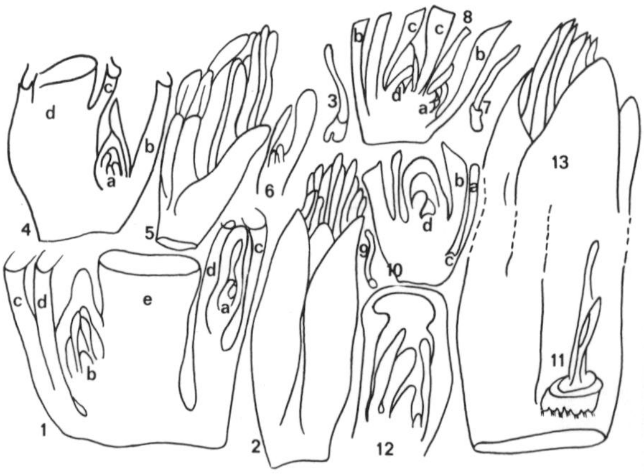

Figure 93. 'Jubilee'. Growth and development of the daughter bulbs in mid December.
1. The formation of daughter bulbs in the axil of the tunic, longitudinal section;
a and b the daughter bulbs initiated behind the first phase scales c and d resp.; e. the base of the generative shoot.
2. Flower bud showing the anthers (the microspore mother cells are at this stage in the profase of meiosis).
3. Daughter bulb bud of the first fleshy scale with a serial bud.
4. Longitudinal section of a mother bulb showing the daughter bulbs in the axil of its first fleshy scale; a. the daughter bulb initiated behind the first phase tunic (b); c. the first phase fleshy scale; d. the base of the generative shoot.
5. Flower bud of the generative shoot (4d).
6. Longitudinal section of a serial bulb next to the bulb in the axil of the first fleshy scale.
7. Daughter bulb bud in the axil of the second fleshy scale.
8. Longitudinal section of a mother bulb showing daughter bulbs in the axil of the second fleshy scale; a. daughter bulb in the axil of the first phase tunic (b); c. the basal parts of two green leaves of the second phase; d. bulb consisting of two fleshy scales (second phase).
9. Daughter bulb bud in the axil of the third fleshy scale.
10. Longitudinal section of a mother bulb showing daughter bulbs in the axil of its third fleshy scale; a. one scale of the first phase; b. basal part of a green leaf (the second phase tunic); c. the primordium of a daughter bulb between a and b; d. bulb consisting of two fleshy scales (second phase).
11. Basal plate with central bud and main daughter bulb.
12. Longitudinal section of main daughter bulb; all organs initiated in the second phase of development.
13. Flower bud of the central shoot.

A vigorous bulb of 'Jubilee' with six fleshy scales may form in the course of one year as many as 21 daughter bulbs of varying age and size and at different positions inside the bulb. As a rule they are less numerous owing to the development of reproductive shoots by some of these buds, as was frequently observed (Botschantzeva, 1954).

The Single Early tulip 'Duc van Tol Yellow'
fig. 94

The varieties of the Duc van Tol group flower in the same period as the wild species and it was therefore interesting to compare their annual cycles.

In Tashkent 'Duc van Tol Yellow' usually begins to flower late in March. There is at that time no indication of new organs being initiated in the main daughter bulb. However, at the end of the flowering period, late in April, the primordium of the first leaf for the next year became discernible in the apex of the main daughter bulb. Some bulbs may even show, besides the shoot primordium, a basal leaf primordium in mid April. Soon after initiation it acquires the typical shape of a developing leaf, which is somewhat raised on one side. Six weeks later, in the beginning of June, we noticed the crescent-shaped primordium of the second leaf, which is clasped for ¾ of its height by the base of the first leaf on the opposite side of the shoot. The third leaf was initiated in mid July and all following flower organs towards the end of that month. A bulb primordium was already initiated simultaneously with the second leaf. In the meantime the main daughter bulb had become separated from the mother bulb and its roots were initiated in mid July.

Growth of the leaves and the stem
fig. 94

With the initiation of the first leaf the stem, which is then still a flattened meristematic protuberance, begins to function. The leaves, irrespective of their time of initiation, hardly grew till mid August, showing that their extension growth does not depend on the age of the young leaf, but on the cyclic processes inside the bulb. During this period the bulb as a rule lies in dry soil or on a shelf in a store. In mid August all three leaves started to grow actively till mid September by which time the two lower leaves acquired pubescent tips. In September growth of the basal leaf slowed down but speeded up again in the beginning of October. A similar fluctuation was observed in mid October. The other two leaves, however, grew steadily throughout that period.

At the beginning of the winter dormant period, in

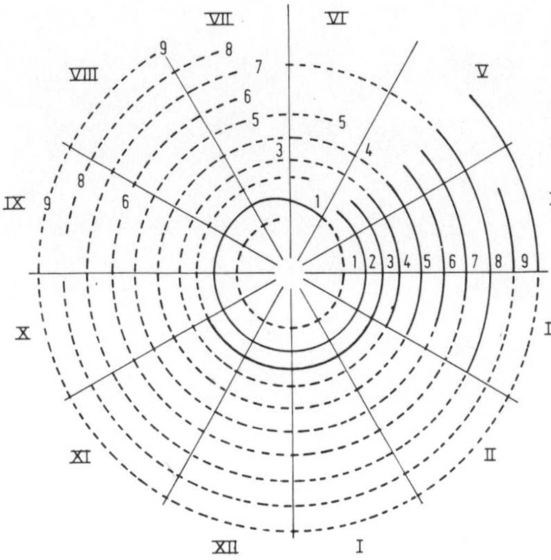

Figure 94. 'Duc van Tol Yellow'. Annual cycle of development. I–XII months; 1. bulb; 2. roots; 3. shoot; 4. basal leaf; 5. second leaf; 6. top leaf; 7. stem; 8. flower; 9. ovary.

mid November, all three leaves stopped growth; the two lower leaves till mid December and the top leaf till mid January. When the lower leaves resumed growth they gained 5-6 mm in two weeks, but growth slowed down in January owing to low soil temperatures and by mid January they had only grown 4 mm. The basal leaf grew actively from mid January onwards, reaching its maximum growth in the latter half of February. The second leaf did not grow from mid January, after which growth started again and also reached its maximum in the latter half of February. Early in March both lower leaves completely stopped for a month, apparently because of the actively growing flower organs, but in April growth was resumed. Towards the end of flowering and for some days thereafter growth of the leaves at first accelerated but subsequently stopped, because of the increasing dryness of the soil and high air temperatures.

Growth of the top leaf proceeded somewhat differently, presumably because of its shorter distance from the apical bud of the flowering shoot. Until the first half of November its growth pattern did not differ from that of the lower leaves, but the following winter dormant period was much longer and lasted till mid January. After growth was resumed it nearly stopped again early in February, but during the latter half of that month, the active growing period of the two lower leaves, it grew vigorously. Early in March, when the lower leaves had stopped growing the top leaf continued growth at a lower rate. From the latter half of March, when growth increased again, it gained 3 cm in two weeks, while the lower leaves did not grow at all, probably because the top leaf intercepted the flow of nutrients to the developing flower or because it used

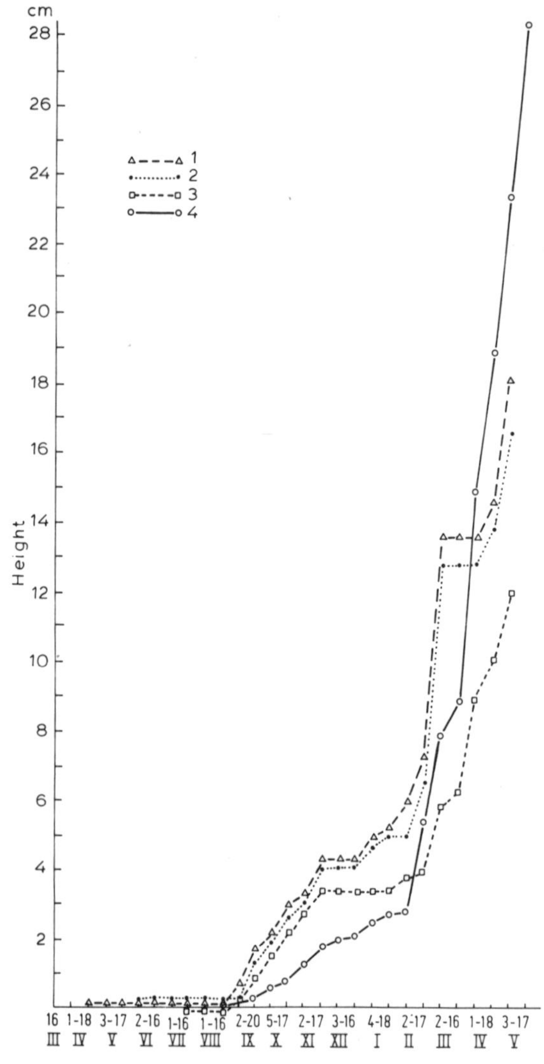

Figure 95. 'Duc van Tol Yellow'. Growth of the leaves and the stem.
1. basal leaf; 2. second leaf; 3. top leaf; 4. stem.

assimilatory products which were translocated from the lower leaves to the flower. In the first half of April growth of the top leaf decreased to less than ⅓ of its previous growth. By mid April growth of all the leaves accelerated again. Growth terminated in the first days of May. The above observations show that the growth rhythm of the leaves does not depend entirely on the external conditions, but is also determined endogenously.

The stem, initiated in mid April, started extension growth simultaneously with the leaves in the beginning of August. During the first month it gained only 2 mm in the lowest internode between the upper bulb scale and the basal leaf. In mid September its growth rate increased but remained below that of the leaves. When the bulb took root, in the first half of October, growth slowed down to increase again in the first half of November. Until mid December, the stem made hardly any growth at all. During the latter half of December growth was resumed and continued throughout January, during which period it gained merely 5 mm. In February it grew 5 cm, which was about twice the increment over the previous 5½ months. When the flower organs began to grow actively, during the first two weeks of March, growth fell off slightly but was restored again by mid March and continued till mid May.

It appears that stem growth proceeds without interruption in the following periods: when the bud emerges above ground; during flowering and during active growth of the ovary.

Initiation and growth of the flower
figs.94, 96

Floral initiation started late in July with the appearance of the three crescent-shaped primordia of the outer tepals. They were followed by the broader and shorter primordia of the inner tepals, which were initiated simultaneously with the stamens. In the early stages it is difficult to see which of the stamen primordia belong to the inner and which to the outer whorl, but somewhat later on this becomes quite clear. The ovary was initiated directly after the stamens.

By the beginning of September the tepals were only 2 mm long, but they grew 6 mm in the 2 following weeks. During that period the veining of the tepals and the pubescence at their tips became discernible. The inner tepals with emarginate tops are shorter than the outer ones and the longer stamens are standing opposite the shorter inner tepals and vice versa. Also the anthers with the protruding connectives had become clearly visible. The thin, peg-shaped, pistil has an overlapping stigma which had not yet developed papillae.

The tepals slightly slowed down their growth in the latter half of September, growing only 2 mm. In October they grew 6 mm after which extension growth was arrested from November till mid January. Then they grew another 3 mm and practically stopped growth, only 1 mm, till the end of February. Active growth, starting in March, continued till the end of the flowering period.

The anthers began to grow and develop at the same time as the tepals, but soon overtook them. Their period of active growth falls in the first half of September. During the latter half of that month, when the loculi were formed, growth was retarded. They continued to grow slowly during the first two weeks of October. After meiosis had taken place in the microspore mother cells, in mid October, the autumn period of growth of the anthers terminated by the end of that

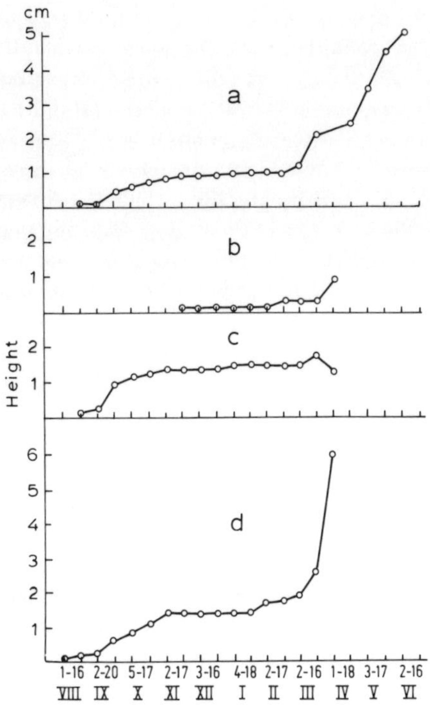

Figure 96. 'Duc van Tol Yellow'. Growth of the flower organs. a. ovary; b. filaments; c. anthers; d. tepals.

month. Early in November tetrads were still found in the anthers. The pollen hibernates as mononuclear grains. From November till March the anthers hardly grew at all. Their 'spring phase' of growth begins in the first half of March and generally ends two weeks before the flower opens, after which the anthers dehisce and subsequently dry up.

The lengthening of the stamens, just before opening of the flower, is caused by elongation of the filaments. It was only in mid September that the filaments became noticeable. They did not develop for 2½ months and retained their low triangular shape. During the first half of February they grew 3 mm and then stopped again for a month. Before opening of the flower, in the latter half of March, they reached their final length. After flowering the stamens fell off together with the tepals.

The three primordial bulges of the ovary, initiated almost simultaneously with the stamens in mid August, merged during the two following weeks to form a triangular-shaped ovary. Active growth began in September when it grew 5 mm. In the next two months it gained only 4 mm. From mid December till mid February it did not grow at all and from mid February to March it grew 2 mm. The active growing period fell in the first half of March when it was growing rapidly during and after flowering. After this period its growth gradually decreased and the ovary became yellow on reaching maturation.

Initiation and growth of the main daughter bulb

figs.94:1, 97d

The main daughter bulb of 'Duc van Tol Yellow' is initiated rather late. We observed the primordium of the bud around mid July. During the following 6 weeks it did not enlarge, but in the first half of September it grew 3 mm, then stopped and grew another 4 mm in October. From November till February there was practically no growth and by the beginning of the spring phase of growth it was only 8 mm long. During February it grew 5 mm but when the roots became active and the leaves unfolded, in March, it gained 2.9 cm. In April the bulb grew only 4 mm, probably because the flower of the mother bulb was growing actively during that month. In the beginning of May it slightly diminished in size owing to its shrinking tunic.

Initiation and growth of the other daughter bulbs

figs.97, a, b, c

Because of the unfavourable spring and autumn seasons of 1954 most of the daughter bulb buds failed to reach their normal size. However, we were able to observe their initiation and early development.

The 'Duc van Tol Yellow' plants did not form a bud in the axil of the tunic of the main daughter bulb. Early in

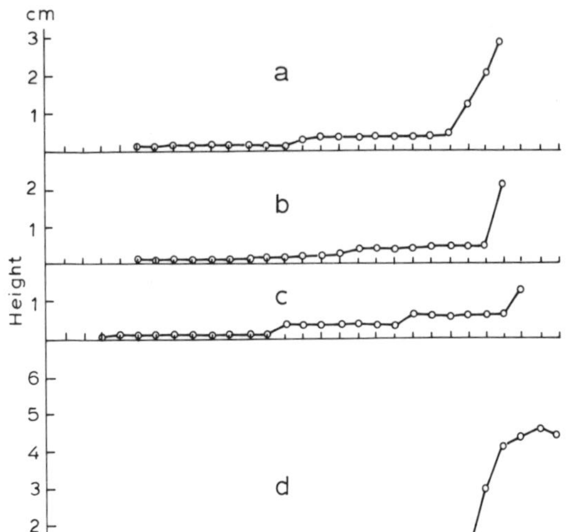

Figure 97. 'Duc van Tol Yellow'. Growth of the daughter bulbs. a. the bulb in the axil of the 3rd fleshy scale; b. the bulb in the axil of the 2nd fleshy scale; c. the bulb in the axil of the 1st fleshy scale; d. the main daughter bulb.

May a daughter bulb appeared in the axil of the first fleshy scale. During the first few months growth was barely noticeable. From mid September to October it grew 4 mm and then stopped for 3 months. Likewise during the first two weeks of January it grew 3 mm and stopped again till April, after which it restarted to grow more or less actively, depending on the condition of the plant. In plants with developing ovaries growth continued till mid April. In that period the starch from the tunic of the young bulb is translocated to the fleshy scales.

The daughter bulb bud in the axil of the second fleshy scale was also initiated in early May, when the main daughter bulb was still attached to the tunic and the basal plate of the mother bulb. In the following four months no growth was observed. In the first two weeks of September it grew 2 mm and then stopped till November after which growth was resumed till December. The small increment of only 2 mm during the autumn period was apparently due to the low precipitation in 1954, which prevented the mother bulbs taking root. No further growth was observed from December till mid March, the beginning of the active growing period, which was two weeks earlier than that of the daughter bulb in the axil of the first fleshy scale. In two weeks it gained 16 mm and after that terminated its extension growth.

The daughter bulb bud in the axil of the third fleshy scale, sitting on the same side of the bulb as the daughter bulb in the first fleshy scale axil, was initiated one month later than the first two bulbs. It skipped the long latent period and started to grow in mid June, making 2 mm extension growth in two weeks. Its summer dormant period lasted from July till October. In October it grew only 2 mm before entering the winter dormant period, lasting from the beginning of November till mid January. After that period it grew only 1 mm and stopped again in the first half of February. During the active period of spring growth, from mid February to the end of March, it reached its final length of 26 mm.

The outermost daughter bulb of 'Duc van Tol Yellow' was first to be initiated, successively followed by the bulbs in the axils of the scales higher up the basal plate. The last to be initiated was the new main daughter bulb. However, the closer to the centre of the basal plate the buds were initiated, the sooner the primordia started to grow. This was most obvious in the 'spring phase' of growth, commencing for the new main daughter bulb early in February, for the bulb in the axil of the third fleshy scale in mid February; for the bulb in the axil of the second fleshy scale in mid March and for the bulb in the axil of the first fleshy scale in the beginning of April.

CONCLUSIONS AND DISCUSSION

Bulb formation

The initiation of the main daughter bulbs of both the wild species and the cultivated varieties of the genus takes place in the summer dormant period and is spread over 2½ months from the beginning of June till mid August. The time of initiation varies from one species or variety to another, e.g. *T.bifloriformis* early in June; *T.kaufmanniana* in mid June; *T.fosteriana* early in July; *T.subpraestans* in mid July; *T.korolkowii* in mid August; 'Jubilee' early in June; 'Duc van Tol Yellow' in mid July.

Table 24 shows that the life span for wild species varies from 20 to 23 months. The active period from the moment the bulb becomes separated from the mother bulb till senescence was ca. 12 months for all species studied. The differences in life span between species are due to their unequally long primordial stages. We found that the early-flowering varieties live 22.5 months, the late-flowering ones 24.4 months.

Almost all wild species and cultivars initiate daughter bulbs in the axils of the fleshy scales, but initiation and growth proceed more slowly in the wild species. In *T.korolkowii* development of the daughter bulbs is often confined to the primordial stages.

Table 24a

Duration of life of the organs of the tulip (months)

Species or cultivar	Bulb			Roots			Shoot			Basal leaf		
	Primordium	Active period	Total	Primordium	Active period	Total	Primordium	Active period	Total	Primordium	Active period	Total
T. kaufmanniana	11.0	12.0	23.0	3.5	7.6	11.0	11.0	4.0	15.0	11.0	3.8	13.8
T. fosteriana	10.0	12.0	22.0	3.6	7.0	10.6	11.0	3.8	14.8	10.8	3.4	14.2
T. subpraestans	10.5	12.0	22.5	1.7	8.5	10.2	10.5	2.7	13.2	8.4	2.4	10.8
T. korolkowii	8.0	12.0	20.0	5.0	5.5	10.5	10.0	3.5	13.5	9.8	2.7	12.5
T. bifloriformis	10.5	12.0	22.5	4.5	6.8	11.3	10.0	2.2	12.2	9.1	2.0	11.1
'Jubilee'	12.4	12.2	24.6	2.6	8.0	10.6	11.5	3.4	14.9	11.5	4.0	15.5
'D. v. T. Yellow'	10.5	12.0	22.5	4.5	6.5	11.0	10.0	2.0	12.0	9.6	2.4	12.0

Many wild species do not develop a bud in the axil of the tunic. These buds were never found in *T. bifloriformis*, frequently in *T. fosteriana* and *T. kaufmanniana* and in nearly every plant of *T. subpraestans*. In cultivars of the Single Early group, e.g. 'D. v. T. Yellow', buds in the tunic axil rarely developed, but they were always initiated and developed in cultivars of the Darwin group.

The initiation of the daughter bulbs in wild species and in cultivars proceeds from the periphery to the centre of the bulb. Their growth and development can be devided into four consecutive processes: (1) initiation and primary growth of the meristem; (2) initiation and primary growth of the tunic; (3) a more protracted period of further organogenesis and weak growth; (4) rapid growth in spring without initiation of new organs. The later a daughter bulb is initiated, the sooner the first three processes are completed and the shorter the intervals between initiation of the various scales, in particular between that of the tunic and the first fleshy scale.

In the daughter bulb buds of wild species all the scales were as a rule initiated by the beginning of winter dormancy. The peripheral buds are the first to enter this period, but in spring the buds closer to the centre are the first to break dormancy.

We could distinguish two developmental phases of the daughter bulb buds of cultivars, particularly in those of the Darwin group. During the first phase organs are initiated in the scale axils of the main daughter bulb which is then still connected with the mother bulb and lives on the food reserves of the latter. The shoot of the mother bulb has at that time just emerged and assumes its assimilatory functions. With the beginning of summer dormancy the first phase of development is terminated and the second phase daughter bulbs are initiated in the axils of the scales of the first phase. The first phase tunic never completes development and, after its food reserves have been used up by the second phase buds, becomes thin and membranous like a depleted fleshy scale. The later a first phase daughter bulb is initiated, the fewer scales it has and consequently fewer second phase buds are initiated in their axils. However, the latter are formed more readily and have a larger number of scales.

A first phase bud in the axil of the tunic may initiate a second phase bud in the axil of its own tunic and another in the axil of its first fleshy scale, while its apical growing point often produces a reproductive shoot. Sometimes a main daughter bulb was initiated instead of a reproductive shoot and the second phase tunic bud had a serial bud.

As a result of the 'two-phase' vegetative reproduction one main daughter bulb of 'Jubilee' with six fleshy scales can produce in one year a cluster of 21 daughter bulbs of various sizes and ages. However, as a rule these numbers are smaller, because several buds may have formed a reproductive shoot. The formation of a shoot, instead of a daughter bulb, is a reversion to the habit of the remote ancestors of the genus, in a period when buds with a specialized function did not exist and a true shoot developed in the axils of their scales, which were at that time leaves. The exceptional cases in which shoot formation still occurs can explain how the organs of the tulip plant have evolved.

The roots

The roots are initiated during summer dormancy after the bulb has become separated from the tunic and the basal plate of the mother bulb. Of the species studied, the first to initiate roots was *T. bifloriformis* in mid May, followed by *T. korolkowii* early in June, *T. kaufmanniana* and 'Duc van Tol Yellow' in mid June, *T. fosteriana* early in July and finally *T. subpraestans* and 'Jubilee' in

Table 24b

Duration of life of the organs of the tulip (months)

Species or cultivar	Leaves: 2nd leaf: Primordium	Active period	Total	Leaves: Top leaf: Primordium	Active period	Total	Stem: Primordium	Active period	Total	Flower: Primordium	Active period	Total	Ovary: Primordium	Active period	Total
T. kaufmanniana	9.4	2.6	12.0	7.5	2.5	10.0	12.5	2.5	15.0	7.5	2.5	10.0	6.5	2.5	9.0
T. fosteriana	8.4	2.5	10.9	8.1	2.4	10.5	11.6	2.3	13.9	9.0	1.0	10.0	7.0	2.0	9.0
T. subpraestans	7.2	1.8	9.0	6.0	1.7	7.7	11.0	2.5	13.5	8.0	0.7	8.7	7.4	1.6	9.0
T. korolkowii	5.7	2.3	8.0	5.6	2.2	7.8	9.9	2.1	12.0	6.8	0.4	7.2	6.6	1.8	8.4
T. bifloriformis	—	—	—	2.5	8.1	10.6	10.4	1.7	12.1	7.1	0.5	7.6	6.1	1.8	7.9
'Jubilee'	9.1	2.9	12.0	9.3	2.7	12.0	11.5	3.5	15.0	8.5	1.3	9.8	9.3	2.3	11.6
'D. v. T. Yellow'	9.0	2.0	11.0	9.3	1.8	11.1	9.5	2.5	12.0	8.5	0.7	9.2	8.0	1.5	9.5

mid July. The earlier a species or a cultivar enters the dormancy period, the sooner the main daughter bulb becomes separated from the mother bulb and initiates roots. For the genus as a whole root initiation takes place between mid May and mid July.

Table 24 gives the life span of the roots which is for these species and cultivars from 10.2 to 11.3 months. There are two stages of growth and development. The initial stage, from initiation till penetration through the basal plate, and the active stage from emergence from the basal plate till senescence. In some species and cultivars, like *T.subpraestans* and 'Jubilee', the initial stage is rather short, 1.7 and 2.6 months resp.; in others it is longer, e.g. *T.korolkowii* and 'Duc van Tol Yellow', 5 and 4.5 months resp. The species and varieties with a longer initial stage of root formation have a comparatively shorter active period of root growth.

In the period between separation from the mother plant and root initiation the basal plate of the bulbs is at the optimal stage for experiments with grafts, transplantations etc., because it is then becoming detached from the enveloping tissues of the mother bulb and its meristems are to some extent stille capable of regeneration. At a later time all formative tissues have been used for root formation. The short-lived tulip meristem shows a marked polarity in its high specialization and short active periods.

The reproductive shoot

The time of initiation of the reproductive shoot depends on the species or variety, and is in *T.kaufmanniana* around the end of February, in *T.fosteriana* and 'Jubilee' in mid March, in *T.korolkowii* early in April, in *T.bifloriformis* and *T.subpraestans* in mid April and in 'Duc van Tol Yellow' in mid May. For the whole genus this period extends from the beginning to the end of the vegetation period of the shoot of the mother bulb. In this period the main daughter bulb is still connected with the mother bulb.

After initiation the stem priomordium, the axial organ, broadens slightly and begins to initiate organs. In 'Jubilee' the first leaf primordium was immediately initiated. In the other species studied this occurred after periods varying from several days in *T.fosteriana* and *T.korolkowii,* 10 days in 'Duc van Tol Yellow', 15 days in *T.kaufmanniana* to one month in *T.bifloriformis* or even more than two months in *T.subpraestans.* During the following phase the stem of the shoot develops in the same way as the basal plate of the bulb by initiating the leaves from the periphery to the centre; they are followed by the flower organs. This period of organogenesis was 70 days in *T.subpraestans,* 80 days in 'Duc van Tol Yellow', 110 days in *T.bifloriformis,* 120 days in 'Jubilee', 150 days in *T.kaufmanniana* and *T.korolkowii* and 170 days in *T.fosteriana.* During this period the stem does not make any extension growth. It begins the next phase of development only after all parts of the flower have been initiated. During that phase, in the autumn and winter, the stem and the initiated organs gradually start growing. Growth of the stem may be arrested in December and January, but early in February it starts active growth, which continues till after flowering when the foliage starts to wither.

The interrelations in the development of the organs is most clearly exemplified by the stem and the flower. We found that until flower initiation the stem repeats the growth pattern of the basal plate. After flower formation it begins to make extension growth. Removal of the flower bud interrupts stem growth, which is stimulated by flower development (Botschantzeva, 1952). Active stem growth in the autumn starts in the section between the bulb and the basal leaf. However, it is not until February or March, depending on the species and external conditions, that the basal internode carries the leaf rosette up to soil surface. The upper part, the pedicel, continues growth longer than any other part of the stem. The development of the stem in the species studied is shown in figs. 66, 72, 77, 80, 85 and 95.

The leaves

As with the stem we can distinguish three phases of development and growth of the leaves: (1) from initiation to the beginning of growth; (2) a phase of slow growth from autumn till the end of winter; (3) active growth starting in February and continuing till the leaf tops become yellow. Successive leaf formation frequently shows a shorter time interval between initiation of the first and second leaves than between that of the second and third leaves (table 25). In case a fourth leaf is formed the interval between initiation of the third and the fourth leaf is also comparatively shorter, showing a tendency of the leaves to be initiated in pairs. Later on the distances between the leaves increase because of the elongating internodes and the tendency is then less apparent.

The more rapidly the leaves grow during the first growing phase, the slower stem growth and conversely. However, during the active growing phase the plants are sufficiently vigorous to ensure a rapid increment of both leaf and stem.

The later the leaf is initiated, the shorter is its first phase of development (without noticeable extension growth) and the sooner it becomes functional. The shorter life span of the upper leaves is mainly due to the shorter first phase of development and the resulting accelerated growth rates.

Table 25
Number of days between initiation of tulip organs

Species or Cultivar	Stem and 1st leaf	1st leaf and 2nd leaf	2nd leaf and 3rd leaf	3rd leaf and 4th leaf	3rd or 4th leaf and flower	Flower and ovary	Shoot and bulb	1st leaf and bulb	Shoot and roots	Roots and bulb	Shoot and ovary
T. kaufmanniana	15	60	60	—	30	0	105	90	105	0	165
T. fosteriana	10	20	75	15	0	60	135	135	105	30	165
T. subpraestans	75	5	40	5	5	15	90	15	85	0	130
T. korolkowii	10	130	10	—	0	10	135	125	60	75	150
T. bifloriformis	30	40	—	—	35	30	50	20	30	15	110
'Jubilee'	0	45	30	45	0	0	75	75	115	45	120
'D. v. T. Yellow'	10	30	35	—	0	15	75	65	45	15	95

The flower

The flowering parts are usually initiated between July and September during the latter half of the summer dormant period. The period of initiation varies in different species and cultivars. *T.fosteriana* and 'Jubilee' initiated their flowers in July, *T.bifloriformis, T.kaufmanniana* and 'Duc van Tol Yellow' early in August, *T.subpraestans* in mid August and finally *T.korolkowii* in the beginning of September.

In *T.bifloriformis, T.fosteriana, T.subpraestans* and 'Duc van Tol Yellow' the duration of the latent period of growth after initiation was different. In *T.fosteriana* it was two months (from June to September); in *T.bifloriformis* six weeks (from August to mid September); in *T.subpraestans* two weeks (from mid August to September); in 'Duc van Tol Yellow' one month from August to September).

There may be distinguished four phases of growth and development of the flowering parts: 1) the initiation of the organs, followed by a period of latent growth; 2) growth during the autumn; 3) a winter dormant period and 4) an active spring phase of growth. A winter dormant period was observed for: *T.subpraestans* (4½ months), *T.bifloriformis* (3½ months), *T.fosteriana* (3 months). In 'Duc van Tol Yellow' the anthers stopped growth for 4 months, the pistil for 3 months, while the tepals broke winter dormancy because of a thaw and resumed growth.

T.kaufmanniana, T.korolkowii and 'Jubilee' did not show a latent period in the development of the flowering parts after their initiation. Since 'Jubilee' initiated the flower in mid July, *T.kaufmanniana* early in August and *T.korolkowii* early in September, this could not be explained by a late initiation. In the last two species we could dinstinguish 3 phases of growth and development, viz. a period of slow growth during the autumn, a winter dormant period and an active 'spring phase' of growth. In winter the flowering parts of *T.kaufmanniana* completely stop growth for 3½ months, those of *T.korolkowii* for 4 months.

'Jubilee' had only two phases of growth and development. A period of gradually accelerating growth in autumn and winter and the active 'spring phase' of growth. We could not observe any interruption in the growth of its flowering parts during winter under the Tashkent conditions.

Typical of every species is the time the reduction division occurs and the growth pattern of its tepals. In *T.fosteriana* for example the pubescent tip of the tepals is the first to develop. In *T.korolkowii* the midrib of the tepals keeps growing actively for a rather long time, while the reflexing parts of the tepals start development much later. The stomata begin to develop on the claw of the tepals.

In all species the early initiated filaments begin to grow shortly before opening of the flowers.

The life-form

As was mentioned in chapter 7 tulips can be classified by their life-form as 'geoxerophytic ephemeroids'. They are perennial plants that develop quickly, flower and bear fruit early in spring and enter the hot summer dormant period when the air temperatures rise and dryness increases. The dormant period of a living organism does not imply complete cessation of all activity since that would mean senescence. Dormancy is a necessary condition which arose in the past and without it further life and normal development are not possible. During that period the organism survives the maximum permissible severance from its environment.

Among the perennial plants with a wide distribution in Middle Asia are many ephemeroids. Their annual cycle includes two dormant periods: the winter or cold dormant period, which is normal for plants in a temperate climate, and the dry or hot summer dormant period, caused by high temperatures and excessively low precipitations. They are represented by species of the genera: *Crocus, Eremurus, Ferula, Iris, Rheum, Rindera, Tulipa* and many others. These plants usually do not form a rosette of leaves in the autumn.

The green leaves and the roots of the tulips die in the summer dormant period and consequently the plant does not assimilate; nor is there any uptake of water and dissolved nutrients. Respiration is at a minimum because the outer epidermis of the bulb scales has no stomata. The plant, i.e. the bulb, seems to have broken all contact with the environment because it only undergoes the influence of the temperature. Nevertheless, the life processes in the plant are drastically reorganized during this period. All the organs of the next year's shoot and the roots are being initiated and begin to develop.

The growth of the vegetative buds has not been visible and in the now accepted terminology it is referred to as 'embryonic'. The plants start to root with the first autumn rains and the vegetative buds of the aboveground organs grow actively under a protective cover of soil and dead leaves. For growth and development in the autumn and winter the buds require low temperatures, otherwise these buds, in particular the flower buds, may die. We have tested this by raising plants in a glasshouse (Botschantzeva, 1952).

If the bulb by-passes the summer dormant period, it is practically impossible to induce any growth, because it is only in this period that the food reserves which have accumulated are converted and translocated and the roots and buds are initiated and begin to develop.

The dormant period, inherent to the prevailing climate, is in the life of our plants a necessary condition, which evolved in the past. A distinction should be made, however, between obligate dormancy, without which the plants cannot continue development, and incidental periods of forced dormancy caused by unfavourable external conditions. It is during the obligate summer dormant period that food reserves accumulate in the bulb scales and are converted into other substances, which enable the axillary buds to develop. In the following cold autumn and winter period with its precipitations the buds will pass through the last stages of development. The influence of light during the autumn (Lysenko, 1942) is excluded here, since the buds are hidden inside the bulb which is covered by a layer of soil of at least 10 cm thick. In spring the plant will pass through the required light-controlled processes by using the leaves that have grown from the apical bud.

The described development is apparently common in all bulbous plants with a sympodial branching habit. Typical of this development is that a lateral first order bud, which does not form green leaves, completes its development with the aid of the leaves of the apical floriferous shoot and, when starting the dormant period, initiates its own floriferous shoot in the apex. During that period the food reserves stored in special organs, the scales, are again converted and enable the second order buds to pass through their further stages of development.

CHAPTER ELEVEN

Changes in tulips caused by environmental influences

INTRODUCTION

It is possible to create new formae of plants by steering plant development in seedling populations of hybrids or wild species. Positive results, however, are more readily obtained from hybrid seedlings. To establish the conditions that will lead to changes in the expression of the heriditary properties an understanding ot the processes ot growth and development is essential. As in particular the wild species show a high stability with respect to changes of these characters their new environmental conditions have to be maintained to ensure a normal plant development.

We have seen in the preceding chapter that the ephemerous life form of *Tulipa* is typical of the group of perennial plants, growing under the severe climatic conditions of Middle Asia with a pronounced summer dormant period. With lowering temperatures and the first rains in the autumn the bulb will form numerous young roots, and if the conditions are favourable, the central bud already appears in October above the apex of the tunic. At this stage the young shoot of the adult bulb has all its leaves. The basal leaf, differing morphologically from the upper leaves, covers the bud. The pedicel, which is then only a few millimeters long, usually carries one flower of which the various organs are already discernible. The outer tepals can be clearly distinguished from the inner ones by their nervature and their pubescent upper surface. The stamens are only for ¾ of their length covered by the tepals. The filaments still retain their initial triangular shape, whereas the anthers are already well developed and clearly show the locules and the connective with the typical protuberance above the anthers, a character which is rather inconspicuous in the fully developed flower. Also in multiflowered species all these organs are already present and in the large flowers of some species the pollen mother cells are at the stage of reduction division.

MULTIFLOWERED SPECIES

In the Liliaceae the inflorescence with two, three or more flowers is more frequent than the single, terminal, flower and it is doubtless a more primitive character. The single flower must have developed as an adaptation to more severe conditions. Of the subfamily *Lilioideae* most *Fritillaria* and *Lilium* species have multiflowered stems. Other multiflowered genera of this subfamily are: *Bellevalia, Hyacinthella, Hyacinthus, Leopoldia, Lloydia, Muscari, Muscarimia, Ornithogalum* and *Scilla* (Flora of the USSR, 4, 1935).

Among the Middle Asian tulips the naturally occurring multiflowered species are predominantly *Eriostemones*. Taliev (1930) referred to the tulips as 'deprived relations' of the lilies, but nevertheless drew attention to the predominating multiflowered habit in section *Eriostemones*. Whenever multiflowered tulip species are cited in the literature, mention is always made of the *T. bifloriformis* plant with 11 flowers, found by Popov near Darbaza station on the Tashkent railway. Most *Eriostemones* species of Middle Asia can produce in their native habitats 2 or 3 flowers on a stem.

Vvedenskiĭ gives the following data on numbers of flowers of *Eriostemones* with a Middle Asian distribution (Flora of the USSR, 4, 1935):

Species	Number of flowers
T. biebersteiniana	1(5)*
T. biflora	1 or 2(6)
T. bifloriformis	(1) 2 or 3 (to 11)
T. buhseana	(1) 2 or 3 (6)
T. dasystemon	1
T. dasystemonoides	1(2)
T. orthyioides	1
T. sogdiana	1(2)
T. tarda	1-8
T. turcomanica	1 (2 or 3)
T. turkestanica	1-7

*Numbers in brackets relate to exceptional cases.

With the exception of *T. orthyioides,* we grew all these species from bulbs collected in their native habitats. In cultivation they retain their multiflowered habit and under improved conditions for growing produce even larger numbers of flowers.

We did not find multiflowered individuals among

plants, collected in nature, belonging to sections *Lophophyllon, Orithyia, Spiranthera* and *Tulipanum.* Of the section *Leiostemones,* the richest in species and including plants of decorative value, many are multiflowered. The section comprises several natural groups. Vvedenskiĭ mentions for example that *T.praestans* has 1 or 2 flowers. This species has for long been a favourite with horticulturists because of its pure bright flower colour. It forms a natural group with *T.subpraestans*, which like the former, grows in the Pamir-Alay, but it is not multiflowered.

The multiflorous, or rather diflorous, habit is frequent in a group of closely related *Leiostemones* species with predominantly yellow, but not very striking flowers of medium size. They are:

T.anisophylla 1(2) fl.
T.ferganica 1(2) fl.
T.korshinskyi 1(2) fl.
T.platystemon 2 fl.
T.tetraphylla 1 or 2(4) fl.

We grew *T.ferganica* and *T.tetraphylla* and some individuals of these species had multiflorous stems.

MULTIFLOWERED CULTIVARS

Tyutyunnikov (1932), who for many years cultivated tulips and bred new varieties, had drawn attention to their multiflowered habit. A bulb of the variety 'Rucckiĭ Bogatyr' ('Russian Giant'), raised by him, produced a stem with 2 flowers. In 1924 another individual was found to have 4 flowers on the stem. Later on it appeared that the older, sturdy, bulbs gave multiflowered stems and that weaker plants which flowered for the first time bore single flowers. In the period 1923-1931 Tyutyunnikov observed the multiflowered habit in 10 out of 112 cultivars, including 'Russian Giant' (Double Early tulip). They were the following:

Single Early tulips

'Couleur Cardinal'	— one plant, 2 fl. (1928)
'Hobbema'[1]	— one plant, 2 fl. and one plant, 4 fl. (1928)
'Proserpine'	— one plant, 2 fl. (1926, 1931)

Double Early tulips

'Boule de Neige'	— one plant, 2 fl. (1931)
Double Red[2]	— one plant, 2 fl. (1923, 1924)
Double White[2]	— several plants, 2 or 3 fl. (1926)
'Leuchtfeuer'[3]	— one plant, 3 fl. (1930)

Breeder tulips

'Felicitas'	— one plant, 3 fl. (1930)

Darwin tulips

'Nora Ware'	— one plant, 2 fl. (1931)

Of the 30 Darwin varieties, which he had received from Van der Schoot,[4] only one plant of 'Nora Ware' bore two flowers on a stem. However, among the seedlings he grew from these cultivars individuals with from 2-4 flowers were found every year. Also seedlings raised from 'Russian Giant' produced at first flowering multiflowered individuals.

Hall (1929) mentions the commercially available cultivar 'Monsieur S. Mottet', which gives multiflowered plants. However, he understands by 'multiflowered' that one or more stems may be produced by a large bulb to which a daughter bulb is still attached. This is a different type of multiflowered habit from the one occurring in the wild species and cultivars of Tyutyunnikov. When referring to multiflowered species Hall (1940) and many other authors mention for example *T.turkestanica,* sensu Taliev (1930), which is cited as *T.bifloriformis* Vved. in the 'Flora of the USSR'.

THE FIRST TYPE OF MULTIFLOWERED PLANTS

There can be distinguished 3 types of multiflowered plants. They occur in wild species in the natural environment or are the result of certain methods of cultiva-

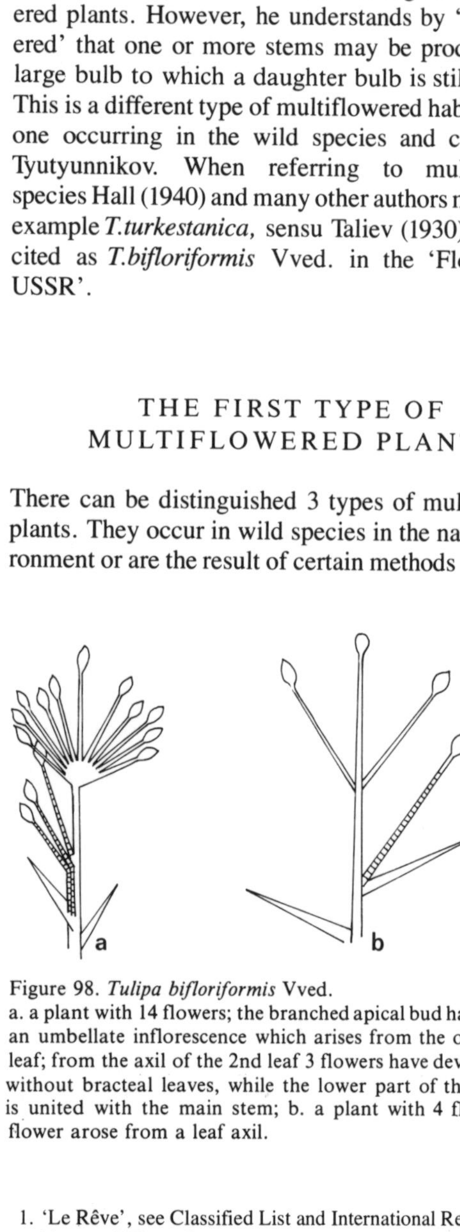

Figure 98. *Tulipa bifloriformis* Vved.
a. a plant with 14 flowers; the branched apical bud has produced an umbellate inflorescence which arises from the one bracteal leaf; from the axil of the 2nd leaf 3 flowers have developed also without bracteal leaves, while the lower part of their pedicels is united with the main stem; b. a plant with 4 flowers; one flower arose from a leaf axil.

1. 'Le Rêve', see Classified List and International Register of Tulip Names, 1971. (editor)
2. Cultivar name lost.
3. 'Vuurbaak'. (editor)
4. Bulbgrowers at Hillegom, Holland. (editor)

tion. The first type, typical of *Eriostemones,* includes plants which bear several terminal flowers on long pedicels. The inflorescence, which arises just above the upper leaf, has only one subtending bracteal leaf (fig.98). The flowers may be arranged in corymbs consisting of one-, two- or three-tier umbels. All these variations can be found among individuals of the same species. The inflorescence develops through branching of the apical bud and was only found in the *Eriostemones.* It was called the *'Eriostemones* type' or the 'first type', because it was the first to be mentioned in the literature of *Tulipa.* Neither the flowers of the terminal umbel nor the lateral flowers on the first order shoots, which develop in the axil of the second leaf, carry bracteal leaves (fig.98). The absence of bracteal leaves and the umbellate flower arrangement at the top of the stem are important characters in distinguishing this type of multiflowered habit from the others. The extreme reduction of certain organs of some *Eriostemones* is shown not only by the loss of leaves on the main stem (adult plants mostly have but two leaves), but also by the absence of bracteal leaves on the pedicels of the terminal flower and the flowers of the first order shoots. However, this does not prove the primitive evolutionary stage, claimed by Taliev (1930). Had the terminal flowers retained their bracteal leaves, the inflorescence would have been of the *Fritillaria* type, with a branched stem of which the serial first order branches have leaves. The smaller dimensions of the plants and the smaller number of leaves are adaptations to the severe conditions of a continental climate.

THE SECOND TYPE OF MULTIFLOWERED PLANTS

The second type of multiflowered plants, the *'Leiostemones* type', mostly occurs in species belonging to this section. It differs from the first type by the development of lateral flowers on leafy shoots, which are formed in the axils of the green leaves of the stem (fig.99). The flowering branches of the *Eriostemones* species do not bear any leaves, but the axillary shoots of multiflorous *Leiostemones* have as a rule one rather small bracteal leaf below the flower. Sometimes there are even 2 or 3 leaves and in some cultivars one can find nearly as many leaves on the lateral shoots as on the main stem, e.g. in 'Russian Giant' (fig. 100).

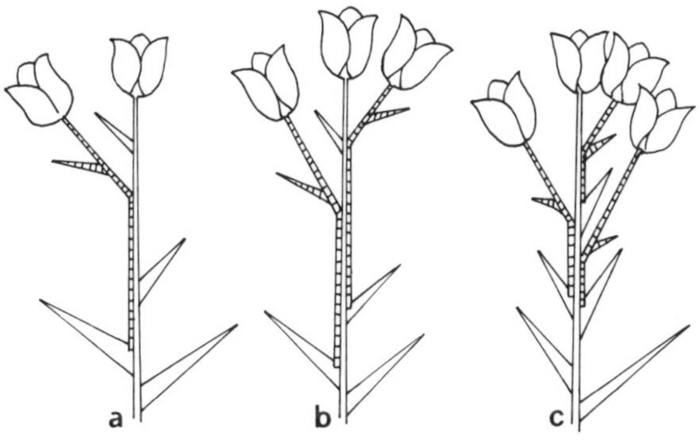

Figure 99. *Tulipa praestans* Hoog
a. a plant with 2 flowers; the 1st order branch arising from the 2nd leaf axil; b. a plant with 3 flowers; the 1st order branches arising from the 2nd and 3rd leaf axils; c. a plant with 4 flowers; the 1st order branches arising from the 3rd, 4th and 5th leaf axils; the 5th leaf of the main stem, bearing a first order branch in its axil, is the bracteal leaf of the terminal flower.

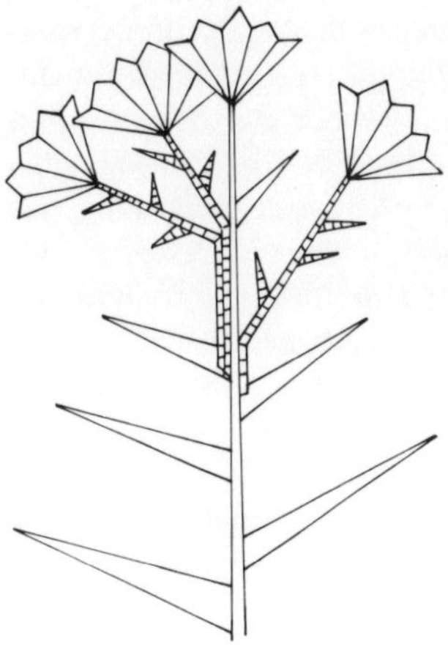

Figure 100. 'Russian Giant' a plant with 4 flowers; from the axils of the 4th and 5th leaf arise 1st order branches with one flower and 3 leaves; two branches have developed side by side from the axil of the 5th leaf and are intergrown with the main stem up to the first leaf above the place of their initiation.

Further development may lead to the formation of second order shoots with leaves and a flower from the buds in the leaf axils of the first order shoots. There may even develop several flowering second order shoots in one leaf axil. Third order branches were found in *T.subpraestans.* They usually bear one leaf and a flower and the higher the order of branching, the smaller the leaf. The rather low plants of *T.subpraestans* are so strongly branched that, when in flower, they look like patches of red balls lying on the ground.

The second multiflowered condition is sometimes also found in *Eriostemones* (fig.98b) when more than one bud has developed in one leaf axil. They are either serial or collateral buds. This tendency to form collateral buds can often be observed in both wild species and cultivars, e.g. *T.praestans, T.subpraestans,* 'Kathleen Parlow', 'Mother's Day', 'Murillo', 'Russian Giant' and many others.

THE THIRD TYPE OF MULTIFLOWERED PLANTS

The third type results from an accelerated development of the daughter bulbs. Under natural conditions the main daughter bulb bud remains dormant in the year after initiation and as a rule develops roots the next spring, but when the mother bulb receives abundant nutrients, the main daughter bulb already starts growing in the year of its initiation (second phase of development) and may form a shoot with leaves and a flower. Although the multiflowered *Eriostemones* type is never found among *Leiostemones,* there are cases where a type of branching of the inflorescence, resembling the *Eriostemones* type, occurs also in *Leiostemones,* e.g. in 'Elephant', 'Kathleen Parlow', 'Proserpine' and 'La Tulipe Noire'. Then a shoot develops from an axillary bud of the top leaf of the main stem and the terminal flower seems to have lost its bracteal leaf, which is however always present and subtends the lateral flower. It is quite easy to distinguish the terminal flower of the main stem from that of the first order branch, even when the stem and the lateral branch have intergrown. Following the coalesced parts down to the first node, we find that the first order branch is the one closest to the median vein of the leaf on that node. This branch usually bears a bracteal leaf, whereas the main axis, which is farther removed from the stem leaf, does not have a bracteal leaf on its terminal pedicel.

METHODS TO INDUCE THE MULTIFLOWERED HABIT

Experiments carried out in the botanical garden of the Academy of Sciences of the Uzbek SSR have shown that certain methods of cultivation can induce many Middle Asian tulip species to produce multiflowered forms. For this purpose adult bulbs, collected by expeditions, were planted on beds in September or October. The soil, the usual cultivated and irrigated cerozem of Tashkent, was mixed with sand and leaf mold at the rate of resp. 15 and 16 kg per 4.5 m². The nutrient level was raised by applying 30 g ammonium nitrate per m² before planting and 30 g with the first autumn rains. With thawing in spring 50 g ammonium nitrate was given simultaneously with loosening of the soil and at flowering the plants were fed once more.

After wilting of the flowers the ovaries were removed with a razor blade or a sharp knife, leaving the pedicels intact, to direct all soluble substances, accumulated by the assimilating leaves and stems for feeding the capsule and its seeds, to the bulb. Removal of the flower bud does not produce this effect because then the internodes stop growing and the leaves of the rosette will shade one another. The early removal of the ovaries results in an increased bulb size. With the beginning of summer dormancy the bulbs are lifted and stored in boxes until the time they are planted again on beds, which are prepared as described above.

The replanted bulbs received four dressings of ammonium nitrate and after flowering the ovaries were removed again. The beds were watered during periods without precipitation in the autumn and during the vegetation period in spring. We believe that watering in the autumn is very important, because the bulb must be firmly rooted to complete the stages of development that precede flowering.

THE EFFECTS OF MANURIAL TREATMENTS AND REMOVAL OF THE OVARY

Some tulip species and cultivars easily responded to the treatments and produced multiflowered individuals already in the second year. There remain, however, groups of cultivars and species which never give multiflowered plants (Botschantzeva, 1953, 1954). Although both the cultivars and the wild species respond to the conditions of cultivation, they do so in different ways. The effect bears in the first place on plant size.

Plant size

T.greigii Regel – Certain species can be very sensitive to improved growing conditions and in cultivation their flowers often become very large. By comparing the description of *T.greigii* growing in the natural habitats with those of plants cultivated in different areas of the Soviet Union, we found that the conditions of cultivation strongly influenced plant size.

In the 'Flora of the USSR', 1935, the following data are given for wild-growing individuals: stem 20-40 cm; number of leaves 3; basal leaf 3.5-14 cm broad; tepals 3-10 cm long. Plants of *T.greigii* grown in Leningrad (Saakov, 1950) were 25-40 cm high and had 4 leaves. The inner tepals were 5.6-8.0 cm long and 2.5-3.0 cm broad; the outer tepals 5.0-7.5 cm long and 2.3-3.0 cm broad.

Table 26 gives the measurements we took in 1952 on plants of *T.greigii*, that were growing in the hills near Darbaza station on the Tashkent railway. Because of the humid spring, conditions had been favourable for growth. The following data are derived from table 26:

Plant height: av. 18.1 cm; max. 24.5 cm; min. 13.0 cm.

Leaves: 4, rarely 3; av. 9.6 × 4.1 cm; max. length 13.6 cm; min. length 7.0 cm.

Outer tepals: av. 4.5 × 3.4 cm; min. 3.0 × 2.1 cm; max. length 5.9 cm; max. width 4.5 cm.

Inner tepals: av. 4.9 × 3.6 cm; min. 3.2 × 2.0 cm; max. length 6.2 cm; max. width 4.8 cm.

On April 2 of the same year we also measured a number of *T.greigii* plants that had been brought as bulbs from the hills near Darbaza station and flowered in the first year of cultivation. The following data are

Table 26

T. greigii. Measurements of plants growing in their natural habitat (Darbaza gully, Kelesskiy rayon)

Plant height (cm)	Number of leaves	Length and width of the leaves (cm)				Size of the tepals (cm)		Number of flowering parts
		Basal	Second	Third	Top	Outer	Inner	
18.0	4	10.0 × 4.0	10.0 × 2.8	8.5 × 1.5	7.2 × 0.9	5.4 × 3.6	5.2 × 3.8	6 + 6 + 3
14.5	4	13.6 × 5.6	13.0 × 3.3	12.0 × 1.7	10.5 × 1.2	5.4 × 4.5	5.4 × 4.8	6 + 6 + 3
20.0	4	9.0 × 4.6	9.5 × 3.2	7.0 × 1.4	7.0 × 0.8	5.6 × 3.8	6.3 × 3.9	6 + 6 + 3
18.5	4	7.5 × 4.5	5.5 × 3.1	6.5 × 1.7	6.3 × 0.9	4.3 × 3.6	4.6 × 3.4	6 + 6 + 3
20.5	3	7.0 × 3.6	6.3 × 1.6	5.5 × 0.9		5.5 × 4.3	3.6 × 3.3	6 + 6 + 3
14.0	3	7.0 × 3.0	6.7 × 1.7	6.5 × 0.7		4.7 × 3.2	4.6 × 3.3	6 + 6 + 3
13.0	4	7.0 × 3.9	6.0 × 2.3	4.5 × 1.0	4.0 × 0.6	3.0 × 2.1	3.2 × 2.0	6 + 6 + 3
19.0	4	10.0 × 4.2	9.6 × 2.5	7.7 × 1.1	5.7 × 0.7	5.9 × 3.8	6.2 × 3.3	6 + 6 + 3
22.0	4	10.5 × 3.8	10.0 × 2.6	10.0 × 1.6	7.0 × 0.8	5.2 × 4.1	5.5 × 3.6	6 + 6 + 3
21.5	4	12.0 × 5.2	12.0 × 3.4	9.0 × 1.3	6.7 × 0.8	4.7 × 3.2	4.5 × 3.4	6 + 6 + 3
24.5	4	13.0 × 4.7	12.0 × 3.9	9.5 × 1.5	7.5 × 0.9	5.5 × 3.9	5.9 × 4.3	6 + 6 + 3
13.5	4	9.0 × 13.6	8.0 × 1.9	6.5 × 1.0	5.2 × 0.5	4.5 × 2.8	4.5 × 3.1	6 + 6 + 3
19.0	4	11.5 × 3.7	9.0 × 2.2	8.0 × 1.0	6.7 × 0.7	4.5 × 3.0	4.8 × 3.2	6 + 6 + 3
16.5	4	8.0 × 3.6	7.0 × 2.6	4.3 × 0.9	4.5 × 0.7	3.7 × 2.7	4.3 × 3.6	6 + 6 + 3
17.0	4	9.5 × 3.3	9.0 × 2.4	8.0 × 1.0	6.3 × 0.7	4.5 × 3.2	4.5 × 3.4	6 + 6 + 3
av. 18.1		9.6 × 4.1	8.9 × 2.6	5.6 × 1.2	6.5 × 0.8	4.5 × 3.4	4.9 × 3.6	

Table 27

T. greigii. Measurements of plants in the first year of flowering after transplantation from the natural habitat

Plant height (cm)	Number of leaves	Length and width of the leaves (cm)				Size of the tepals (cm)		Number of flowering parts
		Basal	Second	Third	Top	Outer	Inner	
19.0	4	12.0 × 5.0	8.0 × 5.0	8.0 × 3.0	5.0 × 1.0	6.0 × 6.0	7.0 × 6.0	6 + 6 + 3
14.0	4	11.0 × 7.0	10.0 × 6.0	8.0 × 5.0	4.0 × 2.0	5.0 × 5.0	7.0 × 5.0	6 + 6 + 3
19.0	4	13.0 × 6.0	11.0 × 4.0	7.0 × 1.0	7.0 × 1.0	7.0 × 4.0	9.0 × 4.0	6 + 6 + 3
12.0	4	13.0 × 3.0	11.0 × 4.0	8.0 × 1.0	5.0 × 1.0	4.0 × 4.0	5.0 × 4.0	6 + 6 + 3
18.0	4	10.0 × 5.0	9.0 × 4.0	7.0 × 2.0	5.0 × 1.0	4.0 × 4.0	5.0 × 4.0	6 + 6 + 3
av. 16.4		11.8 × 5.4	9.8 × 4.2	7.4 × 2.6	5.2 × 1.2	5.2 × 4.6	6.6 × 4.6	6 + 6 + 3

derived from table 27:

Plant height: av. 16.4 cm; max. 19.0 cm; min. 12.0 cm.

Leaves: 4; av. 11.8 × 5.4 cm; max. length 13.0 cm; min. length 10.0 cm.

Outer tepals: av. 5.2 × 4.6 cm; min. 4.0 × 4.0 cm; max. length 7.0 cm.

Inner tepals: av. 6.6 × 4.6 cm; min. 5.0 × 4.0 cm; max. length 9.0 cm; max. width 6.0 cm.

The various parts were on the whole somewhat larger than those of the plants growing in their natural environment.

In the experiments where the nutrient level was raised by extra dressings of nitrogen and the ovaries were removed to divert the flow of assimilates, we found a marked change in plant dimensions (table 28). The plants were somewhat dwarfed, because they were grown on plots that were exposed to abundant sunshine and strong easterly winds. The following data are derived from table 28:

Plant height: av. 25.2 cm; max. 45 cm; min. 10 cm.

Leaves: 4 (5); av. of basal leaf 19.1x10.5 cm; max. 32x16 cm; min. length 13 cm; min. width 6 cm.

Outer tepals: av. 10.7x8.5 cm; max. length 14.0 cm; max. width 11.0 cm; min. length 8.0 cm; min. width 6.0 cm.

Inner tepals: av. 10.9x8.2 cm; max. 16.0x10.0 cm; min. length 8.0 cm; min. width 6.0 cm.

A comparison of the data in tables 26, 27 and 28 shows that the smallest flower measured is much larger than the large flowers in the natural habitat and in the first year transplants. It also appears that the shortest basal leaf had grown to the size of the longest basal leaf found in the plants of the natural habitat and in the transplanted individuals.

T. micheliana Hoog – The description of this species in the 'Flora of the USSR', 1935, includes the following numerical data: stem 15-40 cm; number of leaves 4-5; basal leaf 2-6 cm broad; tepals 2.5-10 cm long. Saakov (1950) mentions for *T. micheliana* grown in Leningrad: plant height 25-35 cm; number of leaves 4;

Table 28

T. greigii. Measurements of single-flowered plants after manurial treatments (Tashkent)

Plant height (cm)	Number of leaves	Length and width of the leaves (cm)					Tepals (cm)		Number of flowering parts
		Basal	Second	Third	Fourth	Fifth	Outer	Inner	
28	5	18 × 11	20 × 9	21 × 6	13 × 4	13 × 4	12 × 7	14 × 9	6+6+3
19	4	21 × 13	17 × 9	18 × 3	13 × 3	—	10 × 7	11 × 7	6+6+3
24	4	16 × 6	15 × 6	12 × 4	11 × 4	—	9 × 9	10 × 7	6+6+3
18	5	18 × 10	17 × 9	17 × 3	13 × 3	12 × 3	11 × 11	10 × 9	6+6+3
12	4	14 × 6	14 × 5	13 × 4	11 × 4	—	9 × 7	9 × 9	6+6+3
24	4	18 × 11	16 × 7	15 × 3	11 × 3	—	9 × 9	10 × 7	8+7+3
26	4	20 × 8	17 × 5	17 × 3	15 × 3	—	13 × 10	12 × 7	6+6+3
31	4	18 × 11	22 × 8	18 × 5	18 × 3	—	11 × 9	13 × 8	6+6+3
31	4	19 × 13	19 × 9	16 × 4	11 × 3	—	8 × 8	8 × 7	6+6+3
24	4	19 × 8	14 × 6	16 × 4	14 × 3	—	11 × 6	9 × 6	6+6+3
10	4	14 × 10	15 × 7	14 × 4	9 × 3	—	10 × 9	9 × 9	6+6+3
10	4	13 × 9	12 × 9	10 × 4	5 × 2	—	10 × 10	10 × 9	6+6+3
45	4	32 × 16	30 × 11	26 × 10	25 × 4	—	14 × 9	16 × 10	6+6+3
38	4	20 × 12	22 × 12	20 × 10	16 × 4	—	10 × 10	10 × 9	6+6+3
38	4	26 × 13	26 × 11	23 × 7	22 × 4	—	13 × 8	13 × 8	6+6+3
av. 25,2		19.1 × 10.5	18.1 × 8.2	17.1 × 4.9	13.8 × 3.3	12.5 × 3.5	10.7 × 8.5	10.9 × 8.2	6+6+3

Table 29

T. micheliana. Measurements of single-flowered plants after manurial treatments (Tashkent)

Plant height (cm)	Number of leaves	Length and width of the leaves (cm)					Tepals (cm)		Number of flowering parts
		Basal	Second	Third	Fourth	Fifth	Outer	Inner	
24	4	24 × 14	25 × 10	22 × 9	22 × 5	—	15 × 8	14 × 8	6+6+3
19	5	24 × 14	24 × 11	20 × 7	19 × 4	10 × 4	11 × 9	11 × 9	6+6+3
18	4	26 × 12	26 × 10	20 × 4	18 × 4	—	11 × 8	12 × 8	6+6+3
29	5	30 × 14	27 × 13	25 × 3	23 × 6	20 × 4	14 × 7	14 × 6	6+6+3
26	4	24 × 17	22 × 12	23 × 9	20 × 6	—	14 × 8	14 × 8	6+6+4
28	5	27 × 13	23 × 10	20 × 8	23 × 6	18 × 4	14 × 8	13 × 6	6+6+3
16	4	23 × 9	17 × 7	16 × 5	15 × 4	—	10 × 5	11 × 8	6+6+3
17	5	23 × 11	23 × 9	21 × 7	16 × 5	15 × 4	10 × 7	9 × 6	6+6+3
22	5	22 × 16	21 × 13	21 × 9	18 × 5	19 × 3	11 × 7	12 × 7	7+6+4
12	4	19 × 11	19 × 8	18 × 5	15 × 4	—	9 × 6	9 × 6	6+6+3
22	3	27 × 14	27 × 12	20 × 5	—	—	13 × 8	14 × 7	6+6+3
28	5	24 × 18	28 × 15	26 × 10	22 × 8	23 × 4	16 × 9	15 × 8	6+6+3
27	6*	25 × 15	24 × 10	22 × 6	25 × 6	21 × 5	14 × 8	13 × 9	8+6+3
23	4	21 × 16	25 × 12	21 × 9	18 × 7	—	14 × 8	14 × 8	5+6+3
27	5	27 × 20	26 × 15	24 × 8	16 × 9	21 × 8	14 × 9	15 × 8	6+6+3
26	4	29 × 18	25 × 8	21 × 7	24 × 4	—	13 × 7	14 × 7	6+6+3
24	4	27 × 12	26 × 10	24 × 7	23 × 5	—	13 × 8	13 × 6	6+6+3
24	4	24 × 16	24 × 12	22 × 8	18 × 4	—	13 × 8	13 × 6	6+6+3
26	4	24 × 20	24 × 14	22 × 10	18 × 4	—	13 × 7	12 × 7	6+6+3
27	5	30 × 20	31 × 14	26 × 7	21 × 5	—	16 × 8	13 × 8	6+6+3
av. 23,2		25 × 15	24.4 × 11.3	21.7 × 7.4	19.7 × 5.3	18.2 × 4.5	12.9 × 7.5	12.7 × 7.3	

*Size of a 6th leaf: 23 × 3 cm.

outer tepals 5-7 cm long and 2.3-3.0 cm broad; inner tepals 5.3-8.0 cm long and 2.2-2.9 cm broad.

A comparison of these data with those of our experiments in Tashkent (table 29) shows that the Tashkent plants are a little lower, but have larger leaves. The narrowest basal leaf was 1½ times, the broadest leaf more than 3 times as broad as the broadest leaf quoted in the 'Flora of the USSR'.

As regards size of the flowers, the average dimensions obtained in our experiments considerably exceed the Leningrad data and those given in the 'Flora of the USSR'. The average width of the tepals was nearly equal to the tepal length of the Leningrad plants. There were only minor deviations in the number of leaves of the main stem and in the flowering parts. Only 2 out of 20 plants measured had more than 6 tepals in the outer whorl, which cannot be explained by the higher nutrient level. We have observed similar deviations in

Table 30

T. ingens. Measurements of plants after manurial treatments (Tashkent)

Plant height (cm)	Number of leaves	Length and width of the leaves (cm)				Tepals (cm)		Number of flowering parts
		Basal	Second	Third	Fourth	Outer	Inner	
15	3	20 × 6	19 × 4	15 × 4	—	11 × 6	12 × 7	6+6+3
25	3	21 × 10	20 × 8	22 × 6	—	11 × 6	12 × 6	6+6+3
30	4	22 × 13	30 × 11	27 × 9	13 × 1	15 × 5	13 × 6	7+6+3
20	3	20 × 10	22 × 8	22 × 6	—	13 × 5	13 × 5	6+6+3
17	4	19 × 8	12 × 6	19 × 6	17 × 3	11 × 5	9 × 6	7+6+3
18	3	28 × 10	24 × 6	24 × 6	—	13 × 5	13 × 5	6+6+3
20	3	24 × 13	23 × 11	17 × 7	—	13 × 6	12 × 7	8+7+4
18	3	24 × 13	22 × 9	20 × 7	—	14 × 7	12 × 6	6+7+4
15	4	22 × 10	22 × 9	18 × 5	20 × 4	12 × 6	11 × 6	6+6+3
35	3	22 × 11	22 × 9	22 × 8	—	15 × 6	14 × 7	6+6+3
27	5*	22 × 11	24 × 8	27 × 4	24 × 3	13 × 6	11 × 5	6+6+3
18	3	16 × 11	19 × 9	19 × 7	—	13 × 6	12 × 6	6+5+3
15	4	16 × 10	18 × 11	16 × 10	20 × 5	12 × 7	12 × 6	6+6+3
24	4	22 × 11	24 × 8	24 × 4	24 × 3	13 × 5	13 × 6	6+6+3
20	4	22 × 10	27 × 6	25 × 3	25 × 2	13 × 5	13 × 5	6+6+3
av. 21.1		21.3 × 10.8	22.5 × 8.2	21.1 × 6.1	20.4 × 3.0	12.7 × 5.6	12.1 × 5.2	

* Size of a fifth leaf: 24 × 4 cm.

tulips grown from seed at first flowering and in consecutive years.

T. ingens Hoog – The data on number of leaves given in the 'Flora of the USSR', 1935, and by Saakov (1950), for Leningrad, were confirmed by those of our observations (table 30), although our plants were on the whole a little shorter. It is, however, not clear what the authors understand by 'plant height' or 'stem length'. Do they refer to the whole stem, from the bulb to the top, or only to the aboveground part, which begins approximately one centimeter below the node at the basal leaf?

The statement 'abundant flowering in the open' (Saakov 1950, p.281) does not apply to a plant which produces only one flower. When transplanted from Middle Asia to the different conditions of Leningrad, the plants may fail to flower, because they have not completed the required stages of development. It may also be that of a certain seedling population only part of the individuals had reached the adult phase. In those cases it is only possible to state the percentage of flowering individuals.

In the 'Flora of the USSR', 1935, Vvedenskiĭ mentions for *T.ingens:* basal leaf 2-5-(9) cm broad; tepals 4-11 cm long. The author gives only the width of the leaves, because in his opinion their length is not a diagnostic character. We are however convinced that the length : width ratio of the leaves is for many species a distinguishing character. Saakov (1950) gives for the plants grown in Leningrad the following dimensions: basal leaf 6 cm broad and 29-32 cm long; outer tepals 3-4 cm broad and 6-8 cm long; inner tepals 3.1-4.3 cm broad and 6.2-8.4 cm long.

The following data, obtained in our Tashkent experiments, are derived from table 30:

Basal leaf: av. 21.3 × 10.8 cm; max. length 28 cm; max. width 13 cm; min. length 16 cm; min. width 6 cm.

Outer tepals: av. 12.7 × 5.6 cm; max. length 15 cm; max. width 7 cm.

Inner tepals: av. 12.1 × 5.2 cm; max. length 14 cm; max. width 7 cm.

The basal leaves were somewhat shorter than those of the plants grown in Leningrad, but much broader than those of plants growing in their natural habitats. We also found an increase in the size of the flower organs of the plants grown at a raised nutrient level. There were no great deviations in the numbers of flower organs. Two of the 15 plants measured had 7 tepals and the normal number of stamens and carpels. One plant had 8 tepals, 7 stamens and 4 carpels and another plant had the normal number of tepals and carpels but 5 stamens.

T.fosteriana Hoog ex Irv. – In 'Flora of the USSR', 1935, the following data are cited for *T.fosteriana*: stem 15-35 cm; basal leaf 3-15 cm broad; tepals 4.5-12.5 cm long. Saakov (1950) took the following measurements of plants grown from bulbs in the open in Leningrad: plant height 20-40 cm; basal leaf 12-14 cm long and 6 cm broad; outer tepals 8 × 3.8 cm; inner tepals 8.5 × 4.0 cm.

The data derived from table 31, giving the measurements of the plants grown in Tashkent at a raised nutrient level, are:

Plant height: 23-44 cm.

Basal leaf: av. 22.3 × 13.7 cm; max. length 30 cm; max. width 18 cm.

Table 31

T. fosteriana. Measurements of plants after manurial treatments (Tashkent)

Plant height (cm)	Number of leaves	Length and width of the leaves (cm)					Tepals (cm)		Number of flowering parts	Leaves of daughter bulbs (cm)	
		Basal	Second	Third	Fourth	Fifth	Outer	Inner		First	Second
23	4	18 × 12	19 × 11	18 × 7	16 × 6	—	12 × 7	11 × 7	6 + 6 + 3	—	—
25	3	18 × 11	18 × 9	15 × 6	—	—	11 × 5	11 × 5	5 + 5 + 3	—	—
27	4	19 × 16	18 × 11	17 × 7	14 × 7	—	14 × 6	12 × 6	8 + 6 + 4	—	—
43	4	24 × 16	24 × 12	20 × 8	15 × 7	—	14 × 7	14 × 7	6 + 6 + 3	8 × 2	—
40	5	24 × 14	23 × 10	21 × 10	19 × 7	19 × 3	12 × 5	10 × 7	6 + 6 + 3	—	—
30	3	22 × 15	22 × 12	19 × 10	—	—	12 × 7	12 × 6	6 + 6 + 3	—	—
34	3	24 × 16	27 × 12	23 × 9	—	—	15 × 6	14 × 6	6 + 6 + 4	20 × 2	—
38	4	22 × 18	26 × 9	22 × 6	20 × 6	—	14 × 6	14 × 6	6 + 6 + 3	15 × 3	—
28	4	18 × 12	19 × 11	19 × 7	14 × 5	—	14 × 5	14 × 5	6 + 6 + 3	5 × 1	—
30	4	22 × 10	24 × 10	21 × 8	10 × 6	—	12 × 6	15 × 5	6 + 6 + 3	20 × 6	18 × 2
34	3	25 × 10	20 × 8	15 × 5	—	—	12 × 6	11 × 6	6 + 6 + 3	16 × 4	—
33	3	24 × 16	24 × 11	21 × 7	—	—	13 × 6	12 × 5	6 + 6 + 3	15 × 3	—
31	4	24 × 16	25 × 12	23 × 7	22 × 6	—	13 × 7	12 × 6	6 + 6 + 3	17 × 8	16 × 2
35	3	21 × 16	22 × 10	18 × 8	—	—	15 × 7	15 × 7	6 + 6 + 4	—	—
44	3	30 × 15	31 × 11	27 × 9	—	—	15 × 7	15 × 7	6 + 6 + 3	25 × 3	22 × 2
28	3	22 × 14	21 × 10	22 × 7	—	—	14 × 7	12 × 7	6 + 6 + 3	—	—
33	3	29 × 13	27 × 9	21 × 5	—	—	12 × 6	13 × 6	6 + 6 + 3	20 × 2	—
27	3	21 × 13	20 × 10	14 × 5	—	—	13 × 6	12 × 7	6 + 6 + 3	11 × 2	—
30	3	21 × 11	20 × 8	15 × 6	—	—	13 × 6	12 × 7	6 + 6 + 3	14 × 2	6 × 1
33	5	18 × 14	19 × 10	19 × 8	16 × 4	16 × 4	12 × 6	10 × 5	6 + 6 + 3	14 × 3	—
av. 32.3		22.3×13.7	22,7×10.3	18.5× 7.2	16.2×6.0	17.5×3.5	13.1×6.2	12.5×6.1		15.4×3.2	15.5×1.8

Outer tepals: av. 13.1 × 6.2 cm; max. length 15 cm; max. width 7 cm.

Inner tepals: av. 12.5 × 6.1 cm; max. length 15 cm; max. width 7 cm.

Despite the considerable increase in size of all parts, including the flower, the numbers of the flowering parts did not show any significant changes. The minor deviations in either direction merely prove the stability of this character.

T.fosteriana has the largest flowers of all Middle Asian species. A comparison of the size of largest flowers of *T.ingens, T.micheliana* and *T.fosteriana* shows that the flowers of the first two species are on the average smaller than those of *T.fosteriana,* although there is no great difference (tables 29, 30 and 31).

Serdyukov (1952) supplied data on the increased dimensions after cultivation for many years of *T.eichleri* plants from natural habitats. However, he gave only the light dressings of nitrogenous fertilizers as are commonly used in field trials. Our experiences with annuals (Botschantzeva, 1935, 1951) and those of Bragilevskaya (1949) show, that in floriculture heavy nitrogen dressings are highly effective.

Number of flowers

Many species and cultivars become multiflowered when the plants are grown in highly fertile soils with sufficient moisture during the vegetation period. The temperatures during the long summer dormant period should be high enough and the young ovaries should have been removed in the preceding years. Without this treatment multiflowered plants occur but rarely and usually remain unnoticed. As already mentioned, we found among species and cultivars three types of multiflowered plants, which are typical of certain groups (Botschantzeva, 1953, 1954).

As a result of the treatments the following species produced a new multiflowered type:

Section and species	Treated plants	Multiflowered Plants 1951	1952
Eriostemones			
T. bifloriformis	100	75	75
T. turkestanica	40	37	37
Lophophyllon			
T. regelii	75	·2	2
Leiostemones			
T. ferganica	22	0	1

T. regelii, formerly distinguished as an *Eriostemones,* was placed in the monotypic section, *Lophophyllon,* by Vvedenskiĭ, because of the unusual shape of its one leaf. We found a plant with two flowers, caused by branching of the apical bud. It had no bracteal leaf. This feature places *T. regelii* close to the *Eriostemones* section where this type of multiflowered habit is the rule.

In 1952 one of the two plants of *T. ferganica* produced a new type of multiflowered inflorescence. This was interesting because all branches lacked bracteal leaves (the other individual was a multiflowered plant of the *Leiostemones* type). The occurrence of this plant with an *Eriostemones* type of inflorescence in a *Leiostemones* species brings these two sections somewhat closer together. Moreover, *T. ferganica* and the group of closely connected species around it differ in some characters from the other *Leiostemones*.

Second type multiflowered plants in wild species

In 1951 the second type of multiflowered plant was obtained in 5 species of section *Leiostemones* and in 1952 it appeared in 14 Middle Asian species. Of the latter, eleven were of section *Leiostemones*, two of section *Tulipanum* and one of section *Spiranthera* (table 32).

Of the one multiflowered *T. greigii* plant (fig. 101) the buds in the axils of the 2nd and 3rd leaf axils produced a lateral shoot. The one in the 2nd leaf axil remained underdeveloped and had only two small leaves and an abortive flower of which nevertheless all parts, except the pistil, were discernible. The bud in the third leaf axil developed into a shoot with a normal flower.

T. fosteriana (fig. 102) produced a strong first order lateral from the axil of its first leaf. It bore one leaf and a large flower of the same size and colour as the terminal flower.

Table 32

Species with a multiflowered inflorescence of the second type

Section and species	Number of plants treated	Number of multiflorous plants 1951	1952
Leiostemones			
T. fosteriana	500	1	3
T. ingens	30	0	1
T. praestans	15	11	11
T. subpraestans	63	22	39
T. greigii	400	1	11
T. micheliana	130	0	1
T. korolkowii	300	0	2
T. vvedenskyi	450	0	1
T. zenaidae	4	0	2
T. ferganica	20	0	1
T. hissarica	50	2	2
Spiranthera			
T. kaufmanniana	1000	0	27
Tulipanum			
T. kuschkensis	50	0	3
T. hoogiana	75	0	3

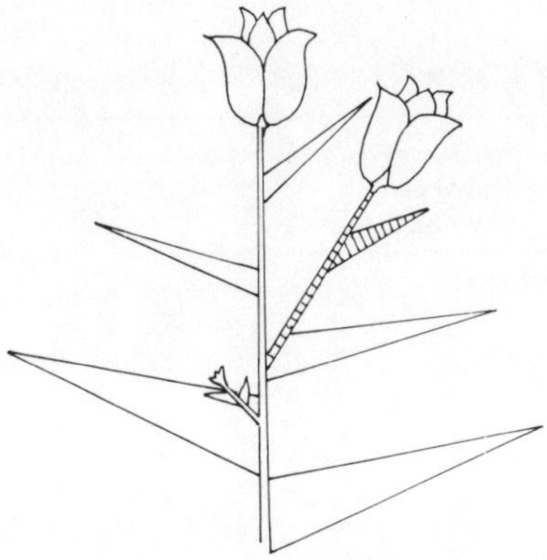

Figure 101. *Tulipa greigii* Regel, a plant with 2 flowers; in the 2nd leaf axil a shoot with two small leaves and an abortive flower.

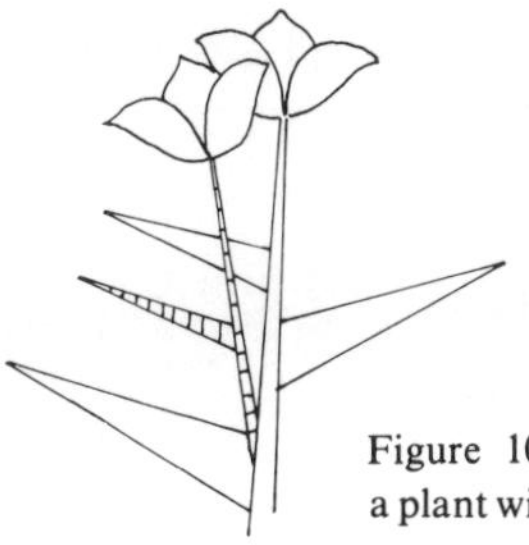

Figure 102. *Tulipa fosteriana* Hoog ex Irv., a plant with 2 flowers.

Twenty-two of the treated 63 *T. subpraestans* plants became multiflowered. As with the cultivars 'Proserpine' and 'La Tulipe Noire', not a single bud in the axil of the first leaf broke dormancy. The buds in the axil of the 2nd leaf developed in 12 plants, in the axil of the 3rd leaf in 20 plants, in the axil of the 4th leaf in 8 plants and in the axil of the 5th leaf in five plants.

The tendency of *T. praestans* to form branched inflorescences with many flowers has been noticed by several researchers (Tyutyunikov, 1932; Hall, 1940; Botschantzeva, 1953). The transplantation of 61 plants from their natural habitat to the experimental plots was sufficient to obtain at least 14 plants with branched inflorescences.

As with *T. subpraestans* the buds in the first leaf axils of *T. praestans* (fig. 99) remained dormant. The highest number of developed buds was found in the second leaf axil, viz. in 13 out of 14 multiflowered plants, whereas *T. subpraestans* produced the highest number in the third leaf axil. In 9 plants of *T. praestans* the bud in the third leaf axil developed; in 8 plants the bud in the second leaf axil and in a few plants also the buds in the fourth and fifth leaf axils. Unfortunately the inflorescences of the latter were nearly all removed by unauthorized persons. The remaining one is shown in fig. 99c.

Table 33a

T. greigii (multiflowered): Measurements of plants after manurial treatments (1952)

Plant height (cm)	Number of flowers	Number of leaves on the stem	Length and width of the leaves (cm)							
			1st	2nd	3rd	4th	5th	6th	7th	8th
32	4	6	27 × 16	26 × 14	22 × 10	23 × 6	22 × 4	18 × 4	—	—
16	2	6	18 × 11	19 × 8	13 × 5	17 × 3	13 × 3	12 × 3	—	—
31	3	4	25 × 15	22 × 10	21 × 7	20 × 5	—	—	—	—
34	2	5	31 × 21	26 × 14	25 × 7	23 × 5	14 × 2	—	—	—
34	2	6	27 × 23	28 × 14	24 × 10	25 × 8	24 × 5	24 × 3	—	—
23	3	6	24 × 10	24 × 9	22 × 7	19 × 2	8 × 1	4 × 1	—	—
35	2	6	24 × 15	25 × 12	19 × 10	19 × 7	18 × 4	8 × 2	—	—
29	2	8	28 × 16	29 × 13	27 × 7	25 × 4	21 × 6	22 × 3	25 × 3	20 × 2
25	2	6	19 × 13	17 × 12	15 × 7	14 × 5	12 × 3	12 × 5	—	—
36	2	6	32 × 10	28 × 19	22 × 6	24 × 5	24 × 3	19 × 2	—	—
33	2	4	22 × 13	22 × 10	17 × 5	19 × 3	—	—	—	—
av. 29.8		5.7	25.3 × 14.8	24.2 × 11.3	20.7 × 7.5	20.7 × 4.8	16.2 × 3.5	14.9 × 2.9		

Table 33b

T. greigii (multiflowered). Measurements of plants after manurial treatments (1952)

Plant height (cm)	Terminal flower: Number of flower organs	Terminal flower: Length and width of the tepals (cm)		Number of branches	Node of initiation no.	Lateral branches of the inflorescence: Length of the laterals	Length and width of the leaves (cm)				Lateral flowers: Number of flower organs	Length and width of the tepals (cm)	
		Outer	Inner				1st	2nd	3rd	4th		Outer	Inner
31,5	5+5+3	12 × 8	12 × 8	3	3	18	10 × 2	—	—	—	5+5+3	9 × 6	8 × 5
					4	23	11 × 2	—	—	—	5+6+3	9 × 6	9 × 6
					5	18	8 × 1	—	—	—	5+5+3	6 × 4	6 × 4
16	5+5+3	6 × 4	7 × 5	1	3	20	10 × 2	—	—	—	5+5+3	5 × 4	6 × 4
31	6+6+4	10 × 7	11 × 8	2	1	16	5 × 1	—	—	—	6+6+3	6 × 6	5 × 4
					2	10	5 × 2	—	—	—	6+6+2	8 × 6	7 × 6
34	6+6+3	6 × 6	7 × 6	1	1	24	4 × 2	—	—	—	6+6+3	5 × 5	5 × 6
34	6+6+3	11 × 8	12 × 8	1	2	30	14 × 6	10 × 3	8 × 3	—	6+6+3	9 × 6	9 × 5
23	6+6+3	12 × 7	13 × 8	2	4	18	4 × 1	3 × 2	3 × 1	2×1	6+6+3	9 × 7	9 × 5
					5	12	3 × 1	2 × 1	—	—	6+6+3	7 × 5	7 × 5
35	6+6+3	10 × 5	12 × 6	1	2	29	8 × 3	—	—	—	6+6+3	9 × 6	8 × 5
29	6+6+3	10 × 6	12 × 6	1	1	29	13 × 3	10 × 2	—	—	6+6+2	10 × 4	9 × 5
25	6+6+3	9 × 9	11 × 7	1	3	20	13 × 2	6 × 3	—	—	4+4+4	10 × 6	9 × 6
36	6+6+2	8 × 8	8 × 7	1	2	24	24 × 5	19 × 2	—	—	6+6+4	8 × 7	7 × 7
33	6+6+3	9 × 9	9 × 8	1	1	24	2 × 1	2 × 1	2 × 1	—	5+5+2	8 × 4	8 × 4
av. 29.8		9.1× 7.0	10.4× 7.0			21	9.0× 2.3	7.4× 2.0	4.3× 1.7	—		7.9× 5.5	7.5× 5.1

Table 34a

Measurements of seven tulip species after manurial treatments

Species	Plant height (cm)	Number of flowers	Number of leaves on the stem	Length and width of the leaves (cm)				
				1st	2nd	3rd	4th	5th
T. micheliana	32	2	4	26 × 19	25 × 19	22 × 29	20 × 18	—
T. hoogiana	10	2	4	22 × 11	20 × 5	12 × 3	12 × 4	—
T. hoogiana	15	2	4	25 × 8	19 × 3	19 × 3	19 × 3	—
T. kuschkensis	32	2	5	25 × 10	24 × 10	24 × 8	22 × 8	13 × 4
T. ingens	30	2	4	22 × 12	24 × 8	22 × 8	5 × 1	—
T. fosteriana	33	2	5	22 × 20	18 × 10	22 × 8	14 × 7	14 × 6
T. fosteriana	34	2	4	22 × 18	21 × 12	16 × 8	10 × 5	—
T. kaufmanniana	25	3	3	20 × 8	17 × 6	17 × 4	—	—
T. kaufmanniana	40	5	4	16 × 18	16 × 6	15 × 6	12 × 4	—
T. korolkowii	37	3	3	20 × 18	17 × 7	17 × 8	—	—
T. korolkowii	16	2	4	20 × 3	20 × 3	21 × 2	20 × 1	—

Comparison of plant size of the single- and multiflowered plants

The multiflowered *T.greigii* plants, which we obtained in 1952, were all measured (tables 33a, 33b).. Of the other species we measured only a few individuals (tables 33a, 33b). A comparison of the measurements of the multiflowered *T. greigii* and the other species with those of the treated uniflorous species (tables 28, 29, 30 and 31) shows that the size of the flowers and the green parts of the multiflowered plants had not diminished. The multiflowered plants were much larger than those from the natural habitats, described in the 'Flora of the USSR', and also larger than the plants we measured in the fields (table 26) and the plants grown in Leningrad (Saakov).

The data on *T.greigii* do not permit a conclusion with respect to the node at which most frequently a lateral develops. Table 33 gives the same percentages for laterals produced at the first and the second node. Further it is shown that laterals are also produced in the third, fourth and fifth leaf axils. The buds in the upper leaf axils are only initiated when there is a bud in a lower leaf axil.

Double flowers

There were no double flowers observed in either the treated single-flowered plants or the treated multiflowered ones.

We arrived at the conclusion that branching of the stem and the multiflowered habit are generic characters which existed in the not very remote past. In their present habitats the aerial parts lost the branching habit owing to xerophytisation of their present environment or because the plants were forced to migrate to more arid regions. When these plants receive sufficient water and nutrients their former latent ability to form branches becomes manifest again. The branching habit in the genus *Tulipa* is an ancient character like those found by Tarakanov (1952) in certain heterocarpic solonchak plants. This author showed how the different adaptations of morphologically different seeds to varying conditions of germination had evolved as a response to environmental conditions in the past and that when a new character developed the old one was retained.

The formation of double flowers, however, presupposes entirely different changes in the organism, viz. a change in the direction of the development of certain organs and their transformation into others. It results from a severe disorganisation in the organism which disrupts the interrelations between the developing flower organs and may be regarded as a pathological phenomenon. Double flowers are usually found in plants of hybrid origin.

Second type of multiflowered plants in cultivars

In 1951 multiflowered plants of the second type were obtained in 9 and in 1952 in 22 cultivars (table 35). All

Table 34b

Measurements of seven tulip species after manurial treatments

Species	Terminal flower: Number of flower organs	Terminal flower: Length and width of the tepals (cm) Outer	Terminal flower: Inner	Lateral branches of the inflorescence: Number of branches	Node of initiation no.	Length of the branches (cm)	Length and width of the leaves (cm)	Lateral flowers: Number of flower organs	Lateral flowers: Length and width of the tepals (cm) Outer	Lateral flowers: Inner
T. micheliana	6+6+3	13×5	13×6	1	3	13	5×2	4+6+3	5×4	5×3
T. hoogiana	6+6+3	6×3	6×3	1	2	8	11×4	6+6+3	6×4	6×3
T. hoogiana	6+6+3	7×3	7×3	1	2	12	16×4	6+6+3	7×4	7×3
T. kuschkensis	6+6+3	10×3	9×4	1	3	16	15×4	6+6+3	10×4	9×4
T. ingens	6+6+3	14×5	13×6	1	2	18	4×1	6+6+3	7×4	7×4
T. fosteriana	6+6+3	13×8	11×5	1	3	25	14×6	6+6+3	10×8	10×8
T. fosteriana	6+6+3	11×6	11×6	1	1	16	10×5	6+6+2	5×5	5×6
T. kaufmanniana	6+6+3	9×3	8×3	2	3	16	8×1	6+6+3	7×2	5×2
					3	16	2×1	6+6+3	3×1	3×1
T. kaufmanniana	6+5+3	12×3	11×4	4	1	12	8×2	5+5+3	9×3	8×3
					2	20	7×2	6+6+3	7×3	5×3
					2	25	13×3	5+4+3	7×2	7×3
					3	10	8×1	6+6+3	6×3	6×3
T. korolkowii	5+5+3	8×3	6×3	2	1	19	17×2	5+5+3	10×3	6×3
					2	15	5×1	5+5+3	3×3	2×2
T. korolkowii	6+6+3	5×6	5×6	1	1	12	12×1	6+6+3	4×5	3×5

Table 35

Second type of multiflowered plants in cultivars

Group and cultivar	1951 Number of bulbs planted	1951 Multiflowered plants	1952 Number of bulbs planted	1952 Multiflowered plants
Single Early:				
'Keizerskroon'	200	0	not counted	2
'Proserpine'	98	20	not counted	8
Double Early:				
'Mr. van der Hoef'	7	1	13	1
'Murillo'	30	10	not counted	15
Double Late:				
'Russian Giant'	4	1	8	4
Breeder:				
'Gold Lake'	4	0	8	1
'La Danee'	3	0	10	1
'Medea'	8	2	9	3
'Orange Delight'	7	0	16	1
'Golden 60' (from Adler)	1	0	3	1
Red with yellow base 55 (from Leningrad)	5	0	10	1
Cottage:				
'Arethusa'	7	1	11	0
'Golden Harvest'	10	0	24	4
'Maréchal Victor'	4	0	6	2
'Mother's Day'	2	0	2	1
Lily-flowered:				
'Solfatare'	4	0	6	1
Darwin:				
'Elephant'	532	29	not counted	90
'Kathleen Parlow'	175	26	not counted	55
'La Tulipe Noire'	270	28	not counted	34
'Philip Snowden'	1	0	17	1
'Princess Elizabeth'	1	0	1	1
'Violet' (from Leningrad)	2	0	6	1

the 1951 multiflowered plants also produced multiflowered individuals in 1952. The second type was found in nearly all cultivar groups, although in some groups they were more numerous than in others.

The buds from which the lateral flowers developed did not always arise in the same leaf axils. In 'Kathleen Parlow' these buds developed predominantly in the axil of the first leaf (fig. 103a); in 14 plants (nearly 50%). Eight plants produced a lateral flower in the second leaf axil and 7 produced one in the third. This cultivar formed also second order branches in the inflorescence (fig. 103b).

In 'Proserpine' we observed bud development mainly in the second leaf axil and none at all in the first leaf axil.

The cultivar 'La Tulipe Noire', like 'Kathleen Parlow', did not develop a single bud in the axil of the green leaf.[5] Three plants had a lateral in the second leaf axil, 15 plants had one in the third leaf axil and 3 plants had one in the fourth leaf axil.

5. There seems to be an error in the text here. Presumably 'green' should be 'first' and reference should have been made to 'Proserpine' instead of 'Kathleen Parlow'. (editor)

In 'Elephant' the distribution pattern of the buds that developed was as follows: in the first leaf axil (7 plants); in the second leaf axil (3 plants); in the third leaf axil (15 plants) and in the fourth leaf axil (5 plants). This cultivar may form buds in all leaf axils, but they were found most frequently in the third.

When experimenting with larger numbers of bulbs we found that one cultivar formed the buds mainly at the first node and another at the second node. In two cultivars the buds arose mainly at the third node. Cultivars like 'Arethusa' and 'Medea', of which we used only a small number of bulbs, formed buds at the third node. In 'Russian Giant', which has many stem leaves, only the buds at the fourth and the fifth node developed into laterals.

Number of leaves and axillary buds

Most adult plants bear from 2 to 6 leaves on the main stem. Exceptions are *T.regelii* (section *Lophophyllon*) with only one leaf, and some cultivars with more than 6 leaves, e.g. 'Russian Giant'.

It should be noted that some species and cultivars

develop 2 or even 3 buds in one leaf axil, e.g. *T.bifloriformis, T.subpraestans,* 'Kathleen Parlow', 'La Tulipe Noire', 'Proserpine' and 'Russian Giant'. One bulb can thus produce a whole bunch of flowers (figs.104, 105). The buds are collateral buds which are arranged side by side in the leaf axil and never one above the other (fig.103c). This type of branching was also observed in species of the section *Eriostemones,* e.g. in *T.bifloriformis,* from Sukhumi, after three years of cultivation (fig.98).

As a rule all the species and cultivars studied produced first order branches bearing one leaf and a flower. However, in some individuals of the cultivars 'Elephant' and 'Kathleen Parlow' the first order branches growing from the first leaf axil had 2 leaves, and in 'Elephant' one of the first order branches had 3 leaves.

In 'Russian Giant' the first order branches from the fourth and fifth leaf axils had 3 leaves. In the fifth leaf axil had developed two collateral buds, each forming 3 leaves and a flower (fig.100). This cultivar has a tendency to increase the number of leaves on the main stem and the number of flowering parts.

The abortive flowering branch in the second leaf axil of *T.greigii* had 2 nodes and two tiny leaves (fig.101). In 1959 we also found multiflowered *T.greigii* in the Karatau mountains.

An increase in the number of leaves on the first order branches enlarges the assimilating surface of the plants. Besides, second order branches may be formed in the leaf axils of the first order branches, e.g. in 'Kathleen Parlow' (fig.103) and *T.subpraestans* (fig.104).

We may conclude that by raising at the critical phases in the life of the plant the humidity and nutrient level of the substrate and by removing the ovary the growth rhythm can be changed, provided the temperatures during bud initiation are not too low. The resulting morphological changes are the formation of more and larger leaves and increased numbers of flowers. Usually single-flowered tulips will produce inflorescences with first and second order lateral branches.

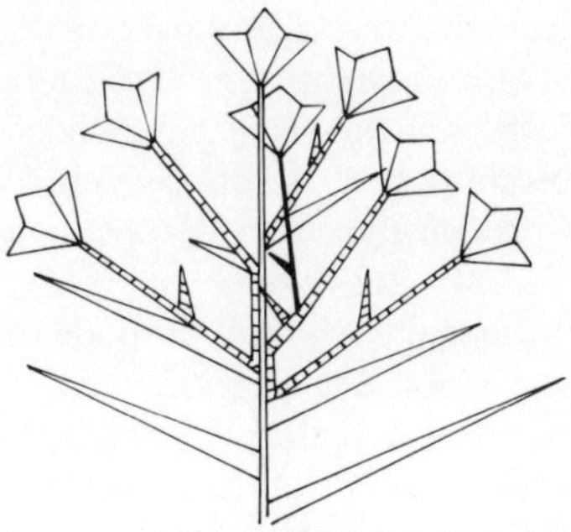

Figure 104. *Tulipa subpraestans* Vved.
A multiflowered plant with seven flowers, branched at the 3rd, 4th and 5th nodes; two serial buds developed in the 3rd and the 4th leaf axil and one bud in the 5th; a 2nd order branch arose in the axil of the only leaf on one of the 1st order branches from the 3rd leaf axil.

Concaulescence

Concaulescence of the main axis and a first order branch is frequently observed in tulip cultivars or in species that have become multiflowered in cultivation. 'La Tulipe Noire' (fig.105) is a good example. In extreme cases the coalesced stem part is rounded and the flowers of the axial branch and the first order branch have merged into a double flower with 12 tepals, 12 stamens and an ovary of 6 carpels. As a rule the stems coalesce to a short distance below the flowers,

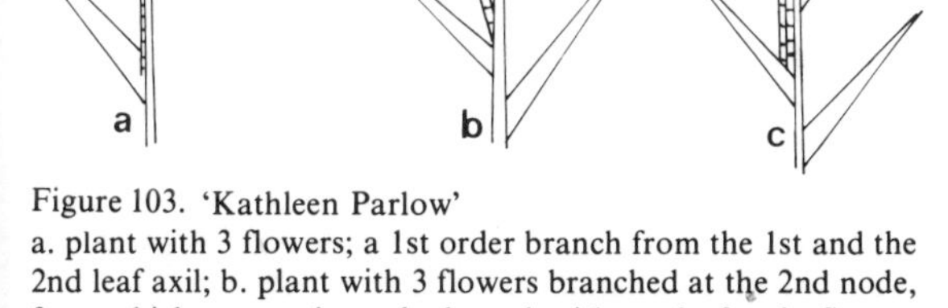

Figure 103. 'Kathleen Parlow'
a. plant with 3 flowers; a 1st order branch from the 1st and the 2nd leaf axil; b. plant with 3 flowers branched at the 2nd node, from which arose a 1st order branch with one leaf and a flower, which developed in the leaf axil a 2nd order branch, also with one leaf and a flower; c. a plant with 5 flowers branched at the 2nd and the 3rd node; at each node have developed 2 collateral buds which produced a branch with a leaf and a terminal flower.

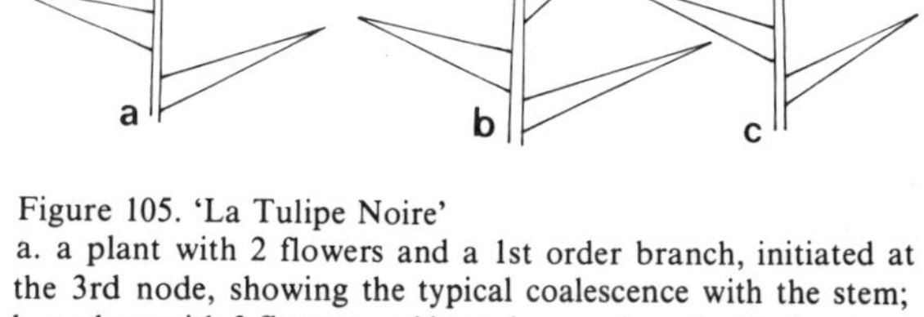

Figure 105. 'La Tulipe Noire'
a. a plant with 2 flowers and a 1st order branch, initiated at the 3rd node, showing the typical coalescence with the stem; b. a plant with 3 flowers and branches, coalesced with the stem, arising from the 3rd and the 4th leaf axil; c. a plant with a double flower resulting from complete coalescence of a first order branch with the main stem; the flower has 12 tepals, 12 stamens and an ovary with a six-lobed stigma.

which have short pedicels and form a terminal cluster. In some cases the stems grow together for only a short distance. 'Proserpine' showed the same pronounced tendency to form coalesced stems as 'La Tulipe Noire', but we observed this phenomenon also in other varieties. In 'Kathleen Parlow' (fig.103) the first order branches coalesce with the main axis from the node of initiation to the first node above it (fig.103a) and a second order branch showed a similar development (fig.103b). Other examples are 'Elephant', 'Russian Giant' (fig.100) and *T. subpraestans* (fig.104).

In 'Elephant' the first order branches which were initiated in the axil of the first leaf only coalesced with the main axis for the length of the leaf sheath. This habit was also observed in 'Arethusa', 'Kathleen Parlow', 'Medea', *T.fosteriana* and *T.greigii.* It results in luxuriant bunches of flowers which are more loosely arranged than the compact clusters of 'La Tulipe Noire'.

Concaulescence of the main stem and a first order branch often resembles fasciation, because the parts have in most cases not completely fused and are usually somewhat flattened. Rather, the parts seem to have grown together to some extent as a result of the treatments. The central bud, initiated in the autumn, remains inside the bulb till early spring and the shoot with the leaves and flowers is tightly packed in the innermost bulb scale with the result that certain parts grow together. Hence, the larger the innermost bulb scale, the smaller are the chances of coalescence and a closely branched inflorescence.

Aerial bulbils

The number of species and cultivars with branched stems becomes much larger if we include those which form aerial bulbils that develop. This occurs as a rule in the axil of the basal leaf (table 36).

In the 1951 experiment the cultivar 'Kathleen Parlow' yielded many plants which formed first order branches in the axils of the upper leaves and also one plant that had formed a bulbil in the axil of its basal leaf. Despite the large number of plants, the cultivar 'Jubilee' had in 1951 and 1952 not produced plants with any first order branches from the upper leaf axils, but one individual had produced a bulbil in the basal leaf axil. The branching habit of this cultivar resulted from premature development of this bulbil.

In 1952 we observed the formation of bulbils in the basal leaf axils of plants with and without branched inflorescences. Nine of the 14 species and cultivars examined (table 36) had initiated a bulbil in the axil of the basal leaf and only in two cases were they also initiated in axils of the upper leaves. A plant of 'Maréchal Victor' had one in the axils of its basal leaf and second leaf.

One plant of *T. kaufmanniana* initiated a rather large bulbil in the axil of its basal leaf and another bud in the axil of the second leaf developed into a shoot, which had one small, fleshy, leaf and a partially developed flower. From the third leaf axil rose merely a truncated green stump resembling neither a bulbil nor a shoot (fig.106).

In 1951 one plant of *T. fosteriana* formed a first order lateral with a large flower in the axil of its basal leaf and the main daughter bulb of this plant initiated the next year an aerial bulbil, also in the axil of its basal leaf. It appeared that where the mother bulb has formed a shoot, a daughter bulb may form an aerial bulbil next year.

The aerial bulbils are initiated on an adult shoot which develops from a bud which, after having reached its final stage of development, changes its

Table 36
Formation of bulbils in the axils of the leaves of the main stem

Cultivar or species	Number of plants with bulbils	2nd type multiflorous plants	Number of bulbils per plant	Node of initiation of the bulbils, no.
'Afterglow'	2	—	2	1
'Amidonette'	1	—	1	1
'Elephant'	2	—	1	1
'Jubilee'	5	—	1	1
'Kathleen Parlow'	1	+	1	1
'Keizerskroon'	4	+	1	1
'Maréchal Niel'	1	—	1	1
'Maréchal Victor'	4	+	1,2	1,2
'Miss Holland'	1	—	1	1
'Oranje Nassau'	1	—	1	1
'Venus'	1	—	1	1
T. fosteriana	1	+	1	1
T. kaufmanniana	1	+	3	1, 2 and 3

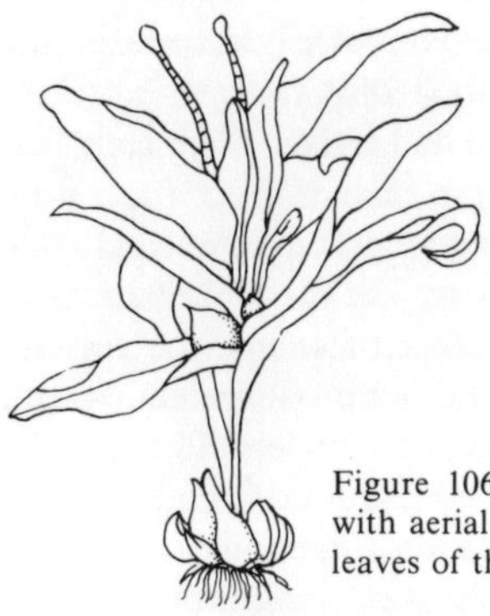

Figure 106. A plant of *T. kaufmanniana* with aerial bulbils in the axils of the stem leaves of the reproductive shoot.

growth pattern. The internodes of the shoot elongate and growth terminates with the appearance of the flower. The longest internode, between the bulb scales and the basal leaf, carries the bud up to soil surface. The central internodes may vary in length from one species to another, but they are never as short as those of the basal plate. The pedicel, between the top leaf and the flower, is nearly as long as the basal internode and sometimes even longer. It can be seen from the shape, size and texture of the leaves that the growth pattern of the stem of an adult plant gradually changes. Juvenile plants form only one sheathing leaf and fleshy scales but the shoot of an adult plant forms the green leaves. A bulbil may appear in the axil of an upper leaf when this has passed through all stages of development in contrast to the shoot which has not yet completed development. The bulbil is therefore at least one stage in its phasal development behind the shoot and, if planted, will flower in the next year. It follows that the initiation of a bulbil leads to a rejuvenation which detracts one stage of development. The phenomenon is similar to that occurring in onions, of which Avakyan (1948) described the formation of aerial bulbils. However, there is some difference, because in our case there is no complete rejuvenation.

It may be concluded from the example of *T. kaufmanniana* (fig.106), that the lower on the stem the bulbil is initiated, the more radical is the rejuvenation. The lowest bulbil, in the axil of the basal leaf, will not flower before the following year. The bulbil on the second node had only one thick, fleshy, leaf with the shape and function of a storage organ, while the other leaves remained green and the shoot terminated in a partially developed flower. In the axil of the third leaf only the stem part of a branch developed.

The branched inflorescences of the second type result from the development of lateral shoots in the leaf axils of the main stem. The aerial bulbils appearing in those leaf axils are also somewhat modified shoots.

The results of our experiments with tulips confirm Shakhov's (1952) conclusions that plant ontogenesis comprises a succession of developmental stages and, that with the transition from one stage to another, development may under certain conditions revert to that of an earlier form, adapted to different ecological conditions. The branched aboveground shoot of *Tulipa* is a more ancient character but, xerophytisation of the environment led to the present predominance of the uniaxial single-flowered habit.

Third type of multiflowered plant or cultivar type

The third type multiflowered plant is entirely different from the first and second types because it results from an accelerated development of daughter bulbs in the axils of the scales of the mother bulb. These scales do not contain chlorophyll. The phenomenon is frequently observed in cultivars; hence the alternative name 'cultivar type'.

As in the other multiflowered types the habit is induced by large applications of ammonium nitrate and removal of the ovary to redirect the flow of assimilates to the daughter bulbs when the assimilating surface of the plant is at its largest. In the first year only the size of the daughter bulbs and the amount of starch in the scales increase but not their number, since they were initiated during the summer dormant period of the previous year and developed at the expense of the reserves stored in preceding seasons.

In the next year the newly initiated buds grow vigorously and complete in one year two phases of their development owing to the large quantities of nutrients which have accumulated in the bulb scales.

A good example of the third multiflowered habit was provided by the Darwin tulip 'Jubilee'. If well-tended in the preceding year, particularly after removal of the ovary, and given extra dressings of ammonium nitrate in spring and in the autumn when the roots start growing, one bulb will produce a group of sprouts. The shape and length:width ratio of these sprouts already show that several floriferous shoots will be formed.

As a rule one bulb developed 6 sprouts of which 3 or 4 could be expected to flower. The usual number of flowers was 2, in some cases 3, and only one percent of the plants treated produced 4 flowers. The number of single-flowered plants was equal to the number of four-flowered plants.

With the unfolding of the leaves in spring it was often possible to predict simultaneous flowering of the main daughter bulb, the daughter bulb in the axil of the tunic and the apical bud of the mother bulb. Adult bulbs of 'Jubilee' have 6 scales, including the tunic. The buds, initiated in their axils may under favourable conditions grow to a considerable size and sometimes become as large as the main daughter bulb.

As a rule daughter bulbs do not form any roots and

leaves while still connected with the mother bulb. However, some daughter bulbs, which were initiated in the second phase, may develop leaves and roots in the year of their formation, depending on the available nutrient supply and the growing conditions in the preceding year. The completion of the second phase of development during the year of initiation results in the third type multiflowered plants. The usually delayed development of the daughter bulbs in comparison with the central bud is caused by an insufficient supply of nutrients and the prevailing temperatures during the summer dormant period. This nutrient deficiency is easily demonstrated by checking the absorbed amounts of nutrients which are stored in the fleshy scales of the bulb.

During summer dormancy the processes taking place inside the bulb are thoroughly reorganized. The amount of water in the bulb diminishes because of the high soil temperatures and the central bud begins to develop slowly at the expense of the food reserves accumulated in the scales. The food reserves in the innermost scale, which envelops the central bud and the bud of the main daughter bulb, are used first, and then those in the next scale and so on towards the periphery of the bulb. In every scale the quantity of starch diminishes first in the subepidermal layer, lying against the inner epidermis, and then in the underlying cell layers. As this process penetrates deeper into the scale, the transformation and translocation of starch are no longer evenly spread throughout the layer, but become more and more localized. Usually there are areas of empty cells, which are shaped like truncated pyramids with the bases turned inside the bulb. These areas, consisting of parenchymatous tissue, lie between the vascular bundles of the scale, which remain for a long time surrounded by cells filled with starch.

When roughly one third of the starch in the innermost scale has been dissolved, the same process starts in the next scale. The slow transformation of starch and consequent depletion of the scales do not show on the outside of the bulb until the beginning of September, when the basal plate starts swelling up, the tunic bursts and the growing central bud and the main daughter bulb are filling up the empty space resulting from the depletion of the scales.

The apical bud of the mother bulb and the bud of its main daughter bulb are the first to become active and they consume the bulk of the stored food reserves, after which the bud in the axil of the second scale starts growing, followed by the one in the axil of the third scale. A cross section of the mother bulb shows that those daughter bulbs which develop closest to the periphery are the smallest in this period. A large bulb contains enough food reserves for the initial stages of development of the apical bud, the bud of the main daughter bulb and buds lying close to the periphery of the bulb. In bulbs of cultivars, which have accumulated limited amounts of food reserves, the nutrients are intercepted by the apical bud and the main daughter bulb bud, which are ahead of the other buds and suppress their growth. With ample food reserves and after a long summer dormant period, the main daughter bulb bud overtakes the apical bud and also initiates several leaves and a flower. In some wild species, however, the daughter bulb buds do not develop although the mother bulb may contain a surplus of food reserves.

At a later stage the innermost daughter bulb buds develop more slowly than the peripheral ones, because the smaller inner scales contain insufficient food reserves which are practically used up by the growing central sprout before these buds are large enough to compete with it. The daughter bulbs in the axils of the peripheral scales develop in a different way. They start somewhat later but can draw upon abundant food reserves. Moreover, soon after they have become active, the mother bulb strikes root and the daughter bulbs take up nutrients from the environment. The initially weakly developed peripheral buds then quickly overtake the buds in the inner scale axils. In particular the bud in the tunic axil gets far ahead. After the tunic of the mother bulb, which is rather delicate in most cultivars, has become thin and membranous, this daughter bulb comes into direct contact with the soil and forms roots. The peripheral daughter bulbs of 'Jubilee' only formed generative shoots when strong bulbs were planted in a rich substrate. In a poor soil they may sometimes produce a leafy stem but no generative shoot (fig.107d). The tunics of the daughter bulb buds in the axils of the inner scales developed into a leaf. These buds were in a less favourable position than the others, but they were nevertheless further developed

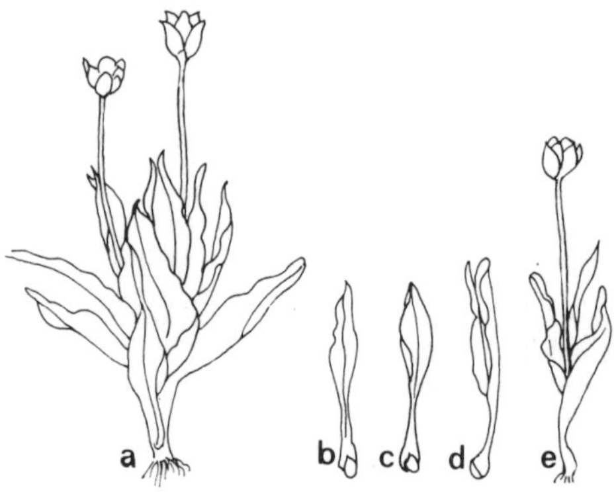

Figure 107. A multiflowered plant of 'Jubilee'
a. right, the apical shoot; left, the main daughter bulb, which developed a flowering shoot; b and c daughter bulbs from the 2nd and the 3rd fleshy scale axil with one leaf and a serial bud; d. the daughter bulb from the 3rd scale axil with several leaves and a sister daughter bulb bud; the flower did not develop; e. the daughter bulb from the tunic axil, which developed a stem with leaves and a flower.

than they would have been under the usual conditions of cultivation, because of the manurial treatments resulting in a larger accumulation of food reserves and the formation of a more vigorous root system of the mother bulb.

Serial bud initiation, as described for the second type of multiflowered plants, was also found in the third type. In this case the additional bud in the scale axil never sprouts in the year of initiation and only forms a small or medium-sized daughter bulb (fig.107b,c), because it is initiated much later than its sister bud and the main daughter bulb bud and is a stage behind in its phasal development.

Of the cultivars grown in our experimental plots many produced tunic daughter bulbs that flowered. They were found among plants of the second multiflowered type in various groups of cultivars:

SINGLE EARLY
'Duc van Tol Scarlet'
'Duc van Tol White maxima'
'Duc van Tol Yellow maxima'
'Keizerskroon'
'Proserpine'

DOUBLE EARLY
'Murillo'
'Mr. van Tubergen'

DARWIN
'Elephant'
'Jubilee'
'Kathleen Parlow'
'La Tulipe Noire'

Species with flowering tunic daughter bulbs were: *T. subpraestans* (frequently), *T. greigii* (one plant) and a natural hybrid of *T. ingens* x *T. fosteriana* (one plant).

Bulbs with double tunics

We found in 'Jubilee' and a number of other cultivars main daughter bulbs with two tunics, formed by the bases of two sheathing leaves (table 37, fig.108). The bulbs with two tunics were the most vigorous and had often initiated collateral daughter bulbs in the axil of one or of both tunics. Among the bulbs studied we only very rarely found collateral bulbs in the axils of the other scales. The double tunic of vigorous bulbs is formed by basal leaves or, more frequently, by the basal part of sheathing leaves.

Some bulbs of 'Jubilee' developed, instead of the main daughter bulb, another reproductive shoot. It was initiated at the same time as the reproductive shoot in the axil of the tunic but somewhat later than the apical shoot and developed accordingly.

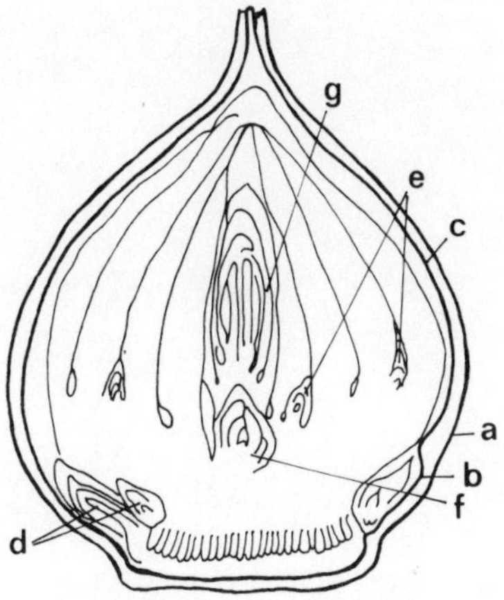

Figure 108. A bulb of 'Jubilee' with a double tunic.
a. the 1st tunic; b. the axillary bud; c. the 2nd tunic; d. collateral buds in the axil of the 2nd tunic; e. daughter bulb buds in the axils of the fleshy scales; f. the bud of the main daughter bulb; g. the apical bud with the flower and primordial green leaves.

In some cases a serial bud arose at the base of the reproductive shoot of the tunic axil in November or December. It was initiated a little lower at the same vascular bundle. We noticed in vigorous bulbs the initiation of such buds in the axils of the other scales as well. They often occurred in cultivars and less frequently in species and were then initiated later (fig.110).

The methods we applied to obtain multiflowered plants also affected the capacity for vegetative propagation. Of some species we already obtained individuals producing large numbers of daughter bulbs, which offers possibilities for the selection of new cultivars. The third type of multiflowered plants, observed in cultivars, again confirmed our conclusion that all the buds in the adult bulb are initially of the same age with

Table 37
Cultivars with double tunics

Cultivar	Number of vigorous bulbs		
		Bulbs with two tunics	
	Total	Number	%
'Bin 1'	27	1	3.7
'Elephant'	50	20	40.0
'Golden Harvest'	25	2	8.0
'G.W. Leak'	8	1	12.5
'Jubilee'	50	25	25.0
'Kathleen Parlow'	50	10	20.0
'Medea'	25	2	8.0
'Orange Delight'	23	2	8.7
'Proserpine'	11	2	18.2
'Venus'	11	1	9.2
Red with yellow base	18	2	11.0
'Golden Yellow'	3	2	66.7

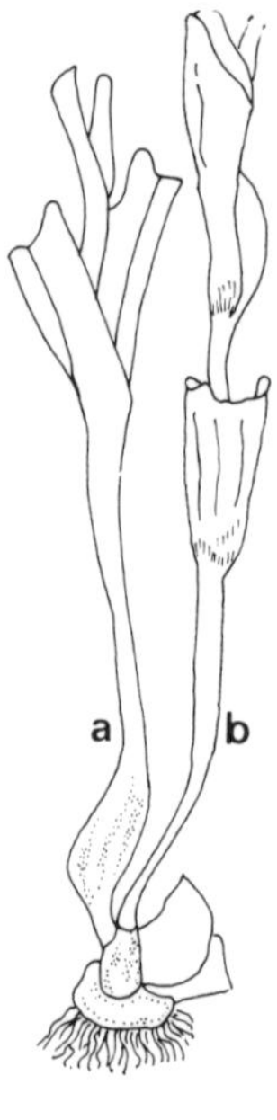

Figure 109. A plant of 'Jubilee' showing the accelerated development of the main daughter bulb.
a. the main shoot; b. the shoot of the main daughter bulb with a sheathing leaf, which is broadened at the base.

respect to phasal development. Because of their dormancy they do not form either roots or leaves and are unable to undergo the phasic changes and thus remain one phase behind the central bud. However, when the mother bulb has accumulated a surplus of nutrients the daughter bulbs overtake the central bud and can flower simultaneously with it.

Development of daughter bulbs and the external conditions

In vigorous bulbs the daughter bulbs are initiated before the beginning of summer dormancy. This may even occur before the final stages of vernalization are completed and long before the influence of the light stimulus becomes effective.

We followed bud initiation and development in vigorous bulbs of the following species and cultivars,

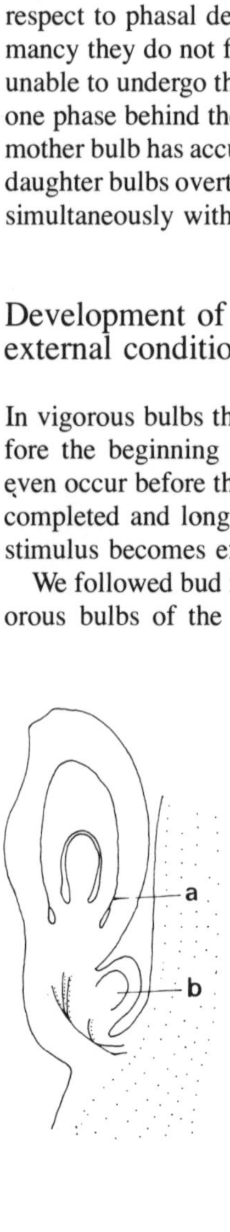

Figure 110. A serial bud in the axil of the first scale of a bulb of *T. fosteriana*, in March.
a. the main bud; b. the serial bud.

grown in Tashkent: *T.bifloriformis, T.fosteriana, T.kaufmanniana, T.korolkowii, T.subpraestans,* 'Duc van Tol Yellow' and 'Jubilee'. In most species it is a slow process, whereas bud initiation and development in a cultivar like 'Jubilee' is a more intensive process; in 'Duc van Tol Yellow' it is somewhat protracted.

In Tashkent summer dormancy starts around mid May and the late-flowering cultivar 'Jubilee' is at that time still growing. Dissection of a large bulb, which still has its green aerial parts, shows that buds are initiated in the axils of the bulb scales. They are at this stage undifferentiated meristems, while the apical bud still consists of a meristematic protuberance and two semicircular swellings. The bud of the main daughter bulb is not yet discernible.

Early in June two scales and a leaf primordium with a meristematic protuberance in the centre are visible in the axil of the tunic. The bud in the axil of the first fleshy scale has formed one scale and a meristematic protuberance; the one in the axil of the second fleshy scale is merely a meristematic protuberance. By that time the apical bud has already formed 3 leaf primordia and the provascular system is just discernible in the two peripheral swellings of the apical meristem; the bud of the main daughter bulb consists only of a flattened, still undifferentiated meristematic tissue.

The second stage of summer dormancy begins in the month of July, when the buds initiated by the large bulbs of cultivars start growing. The apical bud in the axil of the tunic, which consists of two white scales, usually develops into a shoot, and buds are initiated in the scale axils. One of these, at the base of the shoot, will be a main daughter bulb. The other near the periphery usually does not make any further growth. The buds in the axils of the fleshy scales of the mother bulb consist of one or two (the bud closest to the periphery of the bulb) white, fleshy scales in the first phase of development. In the second phase some of these buds may start forming a shoot from the apical growing point, but as a rule they develop a sheathing leaf. The broadened, basal part of this leaf, which arises from the apex of the bud, will form the tunic of the bulb of the next year.

New second phase daughter bulbs are initiated in the axils of the scales of the first phase buds. The bud of the main daughter bulb is the last to be initiated when the bulb enters the second stage of the summer dormant period. In large bulbs it usually begins to develop a sheathing leaf, typical of second phase daughter bulbs. It overtakes the fleshy scales of the first phase of all the lower seated buds. If, as sometimes happens, the sheathing leaf of the main daughter bulb is initiated before the second stage of summer dormancy, another one may be initiated below it.

By the beginning of the low temperature period the apical growing points of the axillary buds are sufficiently developed to be receptive to the required vernalization stimulus. However, the axillary buds of the second phase of development, the serial buds and the collateral buds have had no time to initiate their organs or are just beginning to do this and are then in their development one stage behind the first phase buds.

Thus, there occúr in vigorous bulbs, buds at three different stages of development: 1) buds which have completed all the stages required to flower in the next season; 2) buds which have completed the formation of their scales and have receptive growing points; 3) buds of the second phase, still in the process of formation, which have not yet initiated scale primordia and are not yet ready to respond to stimuli from the environment. In the first case the bulbs flower at the same time as the apical shoot of the mother bulb; in the second case the bulbs flower in the first year after having become separated from the mother bulb; in the last case they flower in the third year after initiation. The latter include daughter bulbs which developed from second phase buds, which are axillary buds of first phase axillary buds of the mother bulb and consequently its grandchildren. The serial and collateral buds are sister buds of the axillary buds and direct progeny of first generation daughter bulbs of the mother bulb. We therefore find in this case daughter bulbs of two generations (taking permission to use the term 'generation' with respect to vegetatively reproduced individuals).

Because of the accelerated growth processes of the buds in vigorous bulbs and their correspondingly advanced morphological development at the time the mother bulb forms roots, these buds can respond to the necessary vernalization stimulus. The relation between nutrition of the mother bulb and receptiveness to vernalization is evident.

In weaker bulbs bud initiation is somewhat delayed and their primordial growth during summer dormancy is very slow. When the roots start to grow in the autumn the buds are not sufficiently developed to respond to the autumnal conditions.

Since the bulbs develop roots in the autumn when their apical buds grow actively underground, tulips may be called 'winter' plants. However, they do not form a rosette of leaves in the autumn and when their growing points are passing through the stage of vernalization the influence of light is excluded, because the bulbs lie sufficiently deep in the soil and the growing buds are protected by living scales and dead membranes of preceding years. Likewise, the growing embryos in swelling tulip seeds have not yet acquired a green colour when the seeds undergo vernalization (Botschantzeva, 1951). Thus, for tulips the complex of conditions necessary for vernalization of 'winter' plants does not include the influence of light in the autumn, which Lysenko (1952) regards as essential for vernalization in 'winter' plants. As a 'winter' plant the tulip somewhat differs from wheat or rye. Many perennial Mediterranean plants, e.g. *Anemone coronaria, Cyclamen ponticum* and *Sternbergia lutea,* belong to 'winter' plants of the wheat type. Other perennial plants, like *Ferula, Leontice* and *Tulipa,* are Middle-Asian ephemeroids, which during the autumn develop underground without any exposure to sunlight, and their shoots do not appear aboveground until spring.

Under Tashkent conditions tulips usually start flowering in the open towards the second decade of March and finish flowering in the first decade of May; the flowering period lasts only two months at most.

At flowering the adult plants have several green stem leaves of which the basal leaf is the largest and broadest. The higher up the stem, the shorter and narrower the leaves (tables 26-31). Their size, shape, colour and arrangement on the stem vary from one species or cultivar to another. They may form a whorl at the base of the peduncle (*T.tetraphylla*), or ascend the stem almost up to the flower (*T.maximowitzii*). Their shape can be nearly round (some forms of *T.kaufmanniana),* broadly lanceolate (*T.praestans*) or grass-like (*T.maximowiczii*). The leaf surface can be smooth or have protruding ribs (*T.regelii*); be spotted (*T.greigii*) or striped (*T.micheliana*) by anthocyanin; glaucous (*T.albertii*) or pubescent (*T.vvedenskyi*).

These morphologically different green leaves all share one essential function of being the principal assimilatory organs and receptors of the photoperiodic light stimulus. The shoot with the terminal flower has completed all its stages of development and as the leaves do not merely have an assimilatory function, the question may be raised for which buds they function as light receptors. We believe that the green leaves of the adult plants are primarily active in completing the development of the ovary and in perceiving the light stimulus for the bud of the main daughter bulb, initiated in the preceding summer dormant period. It appears somewhat later than the apical bud, of which it is a sister bud. Thus, the bud of the main daughter bulb, after having passed through the vernalization stage in the preceding autumn and winter period, receives the required light stimulus with the aid of the leaves that developed from the apical bud of the mother bulb and reaches the reproductive phase. During the next summer dormant period the main daughter bulb, which developed from that bud, initiates a flower in its apical bud. As a rule this apical bud completes its stages of development in the absence of any green leaves of its own and even before the main daughter bulb separates from the mother bulb.

The same processes take place in the bud in the tunic axil of the mother bulb, which was initiated simultaneously with the meristem of the apical bud towards the end of the vegetation period, before the summer dormant period of the previous year. The bulb from this bud will flower at the same time as the apical bud of the mother bulb.

If initiated by a bulb that has stored sufficient food reserves, other axillary buds may react in the same way when their development is accelerated and their formation has proceeded far enough to enable them to respond to the autumn and winter conditions.

The early development of the main daughter bulb or of any other daughter bulbs results in multiflorous plants of the third type or 'cultivar' type (Botschantzeva, 1954). Although it occurs as a rule in cultivars, we observed this phenomenon also in species.

With the accelerated development of the bud in the axil of the innermost scale, which should normally form the main daughter bulb, the function of the latter is taken over by the daughter bulb in the axil of the next fleshy scale. Frequently the shoot, which is produced instead of the main daughter bulb, has its own main daughter bulb developing from a serial bud at its base. In these cases two main daughter bulbs are formed and in such bulbs the flowering daughter bulb in the axil of the tunic often forms a main daughter bulb too, which is initiated as a collateral bud or as a daughter bulb bud of its tunic. We may therefore obtain from one mother bulb 2 or 3 and in some cases even more main daughter bulbs, all of a different age and at different stages of development.

In the bulbs of the plants grown at a raised nutrient level, the axillary buds are early initiated and grow actively till the summer dormant period, during which they complete the second growing stage. The apical growing points of these buds are receptive to the vernalization stimulus required for flowering and the bulbs will flower simultaneously with the apical bud of the mother bulb.

The daughter bulb in the axil of the tunic

The bud in the axil of the tunic is in well-nurtured bulbs of cultivars the first to be initiated. Its initiation and growth were rarely observed in species and then only in very vigorous bulbs. Table 38 gives the percentages of individuals with bulbs from the tunic axil that flowered simultaneously with the apical bud of the mother bulb after the plants had received manurial treatments in previous years.

In some groups many cultivars responded to the given treatment. Of the Lily-flowered group both cultivars showed accelerated flowering of tunic bulbs, but the total number of individuals was not very high. A high percentage of cultivars with accelerated flowering of the tunic bulb was found in the Double Early group, 73%; in some of these cultivars 62% of the individuals.

Although the bulbs of the Single Early cultivars were not very large, nearly 100% flowered in the first year, but only 25% of the cultivars showed accelerated flowering of tunic bulbs. Within the cultivars the maximum number with flowering tunic bulbs was ca. 25%. It was also found that the daughter bulbs in the axils of the peripheral scales may only partially develop.

Under our conditions the bulbs of Single Early tulips remain rather small, with the exception of cultivars like 'Keizerskroon' and 'Proserpine', which we have cultivated for more than 15 years. The latter were not included in this experiment, because they were not replanted in 1951.

The rather weak development of the Single Early tulips may be due to the construction of the beds. The beds were a little raised with ditches running between

Table 38
Third type of multiflowered cultivars after manurial treatment

Group	Number of cultivars: Total	Number of cultivars: With flowering daughter bulbs from the tunic	Number of cultivars: With flowering main daughter bulbs	With flowering daughter bulbs from the tunic %	Flowering daughter bulbs in the cultivars %
Single Early	24	7	0	25	25
Double Early	11	8	0	73	62
Double Late	2	1	0	50	25
Mendel	21	7	0	33	50
Breeder	18	13	1	72	40
Bizarre	3	1	1	33	13
Parrot	4	3	0	75	20
Lily-flowered	2	2	0	100	22
Triumph	29	5	0	17	30
Cottage	24	15	5	63	60
Darwin	44	23	4	52	33
Total	182	85	11	47	34

them for watering by the infiltration method, but they dried out quickly. On the south side of the beds the plants started flowering 3-5 days earlier than on the north side. As the beds became too hot the plants soon completed the vegetation period. In the deeper lying beds the watering had made the soil very firm and in 1951 and 1952 the tulips on these beds flowered better than those on the raised beds.

The Cottage Group produced the lowest number of cultivars with earlier flowering tunic bulbs, 17.6%. Of the Darwin group 63% of the cultivars showed this tendency.

In 1952 ca. 75% of the Double Early, Breeder and Parrot cultivars produced plants with earlier flowering tunic bulbs, but the percentages of individuals showing this habit varied widely, namely from 62% in the Double Early cultivars to 20% in the Parrot group.

Flowering of tunic daughter bulbs in species

The species used in our experiments, also included individuals with tunic daughter bulbs that flowered simultaneously with the apical bud of the mother bulb. Owing to the weak development of the buds in wild species, their number was considerably lower than in the cultivars (table 39).

Because a good representative population of *T. kaufmanniana* was not available for this experiment the figures are very likely an underestimation. In other plantings of 1952 the tendency to produce third type multiflorous plants was much more pronounced.

Vegetative reproduction in *T. korolkowii* and *T. micheliana* is practically nil. Their daughter bulbs are, if present, weakly developed and they do not produce flowering daughter bulbs. By raising the nutrient level we obtained bulbs of *T. micheliana,* weighing 130 grams, but they did not form daughter bulbs. We found that shortening of the daylight periods has a strong modifying influence on the morphology of this species and hope to select forms that are capable of vegetative reproduction.

Some forms of *T. fosteriana* appear to propagate rather well vegetatively, but the periods the daughter bulbs require for reaching the adult phase vary for different mother bulbs. This tendency may explain the comparatively low percentage of flowering daughter bulbs in the axils of the tunics.

T. praestans and *T. subpraestans* showed the highest percentages of third type multiflowered plants. They are highly modifiable species and their response to the improved conditions is equal to that of the cultivars.

The development of the daughter bulbs in cultivars

Simultaneous flowering of the main daughter bulb with the apical bud of the mother bulb occurred in 'Jubilee' in most of the treated plants. It was also found in other Darwin tulips, e.g. 'Elephant' (in several plants), 'Turner' (in one plant), 'Clara Butt' (in one plant) and in the Cottage tulips: 'Bin I', 'Golden Harvest', 'Maréchal Victor', 'Miss Holland' and 'Mother's Day' (in one plant of each cultivar).

One individual of 'Krösschell'[6] produced, besides the main daughter bulb, 3 flowering shoots in other scale axils, each of which had formed its own main daughter bulb. In the Breeder tulip 'Lucifer' the tunic daughter bulb and the main daughter bulb flowered at the same time as the apical bud of the mother bulb. In vigorous bulbs the main daughter bulb and other

6. 'Fancy', see Classified List and International Register of of Tulip Names, Hillegom, 1971. (editor)

Table 39

Third type of multiflowered plants in species after manurial treatments

Species	Number of bulbs planted	With flowering daughter bulbs in the axil of the tunic		With flowering main daughter bulbs
		Number	%	
T. kaufmanniana	46	1	2	1
T. greigii	33	5	15	0
T. ingens	18	1	5	0
T. fosteriana	74	3	4	0
T. subpraestans	28	9	32	0
T. praestans	15	4	27	0
T. turkestanica	22	11	50	0
T. micheliana	20	0	0	0
T. hoogiana	75	2	3	0
T. korolkowii	300	0	0	0
T. ferganica	20	1	5	0

daughter bulbs may flower before separation from the mother bulb (table 38).

The absence of an adequate nutrient supply during the initial growth of the daughter bulbs of wild species leads in most cases to their premature senescence. This has become a positive factor in the struggle for survival of tulips growing under poor soil conditions and insufficient humidity. Of the many tulips inhabiting deserts and semideserts even juvenile plants hardly ever multiply vegetatively.

To understand the problems of vegetative reproduction it is necessary to find out what happens during the formation of daughter bulbs by an adult bulb. They either remain from the start behind the apical bud of the mother bulb, because their phasal development is not completed, or there occurs a rejuvenation. In a previous study (Botschantzeva, 1953) we considered that the daughter bulbs are one stage in phasal development behind the apical growing point. Further studies, with plants of different age classes of the process of initiation of daughter bulbs in the axils of the bulb scales and aerial leaves, and the formation of aerial bulbils in onions, as described by Avakyan (1948), led to the conclusion, that in all these cases the organs are remaining behind in phasal development. However, the formation of aerial bulbils, initiated at the apex of the flowering scape of the onion plant is a complete rejuvenation and development recommences as if from seed, whereas the initiation of the buds in the axils of bulb scales is only a partial rejuvenation, resulting in their being one or two stages behind in phasal development. These conclusions are based on the following observations: the simultaneous initiation of the meristems of daughter bulb buds and of the apical bud in the summer dormant period after completion of the vernalization and light-receptive stages; the simultaneous flowering of the daughter bulbs, in the lower bulb scales of vigorous bulbs, and the apical buds, e.g. in 'Jubilee', 'Krösschel' and 'Turner'; the many cases of simultaneous flowering of the daughter bulb in the tunic axil, seated at the base of the basal plate and in this respect the youngest bulb, with the apical bud of the mother bulb, the oldest bulb. It follows that all daughter bulb buds, initiated in the same season as the apical bud, are at the same stage of phasal development as the apical bud. The reason that the axillary buds, even in vigorous bulbs, are normally partially rejuvenated and lag behind the apical bud is due to the limited amount of available nutrients at the moment of initiation. The transformation of starch into sugars, taking place in the scales, begins only in the summer dormant period and the amounts are not sufficient to last till root formation.

The daughter bulbs of cultivars, in particular those of vigorous bulbs, are initiated in the axils of first order buds seated in the scale axils of the mother bulb. These second order buds always form basal scales. They undergo the low temperature period together with the apical growing points of the first order buds and are protected by the scales of the mother bulb with which they are closely connected. If sufficiently developed, the second order buds will perceive the light stimulus through the aerial parts of the mother bulb. Towards the end of the vegetation period the bulbs that developed from the first order buds with in their scale axils the second order buds, become separated from the mother bulb. The first order daughter bulbs will enter one more phase of development, that of organogenesis. After completion of the last stages of development they initiate the apical bud and somewhat later the main daughter bulb.

The second phase of organ initiation can be observed in all adult bulbs of wild species and cultivars. In strong bulbs the earlier initiated daughter bulb buds

Table 40
Development of sheathing leaves by daughter bulbs of wild species

Species	Total number of bulbs planted	Number of plants forming daughter bulbs with sheathing leaves — Number of d.b's with sheathing leaves per plant: 1	2	3 or more	Percentage of plants forming daughter bulbs with sheathing leaves
T. kaufmanniana	46	5	2	0	15.2
T. greigii	33	5	4	0	27.3
T. ingens	18	1	3	0	22.2
T. fosteriana	74	12	12	5	39.2
T. subpraestans	28	15	10	3	100.0
T. praestans	15	10	2	0	80.0
T. turkestanica	22	22	0	0	100.0
T. micheliana	20	1	0	0	5.0
T. hoogiana	75	10	1	0	14.7
T. korolkowii	300	1	0	0	0.3
T. ferganica	20	1	0	0	5.0

Table 41

Development of sheathing leaves by daughter bulbs of cultivars

Groups of cultivars	Number of cultivars	Number of plants forming daughter bulbs with sheathing leaves — Number of d.b's with sheathing leaves per plant: 1	2	3, or more
Single Early	24	12	11	5
Double Early	13	3	3	—
Double Late	2	—	—	—
Mendel	21	5	5	2
Breeder	18	11	11	4
Bizarre	3	—	—	—
Parrot	4	—	—	—
Lily-flowered	2	1	1	1
Triumph	29	6	6	6
Cottage	24	12	12	7
Darwin	44	24	17	21

usually sprout during initiation of the apical bud and start their second phase of development. In weaker bulbs and in many wild species initiation of these daughter bulb buds has then only just set in.

The sprouting of the daughter bulb buds, which had been second order buds, but became first order buds after the first order daughter bulb became separated from the mother bulb, means that a new series of daughter bulb buds is initiated in the axils of their scales, while their apical bud forms either a bulb or a reproductive shoot.

Frequently the new bulb from the second phase, arising from the apical growing point, is morphologically different from the bud from which it developed. The tunic of this new second phase bulb is formed by the broadened sheath of a green leaf with a more or less developed blade (tables 40, 41). It also has more fleshy scales than the bud from which it developed. The larger the nest of bulbs, the more numerous the bulbs with a sheathing leaf instead of the normal basal tunic. In some species this feature is more pronounced than in others.

SUMMARY

1. The results of experiments aimed at steering plant development have shown that the conditions of cultivation exert a strong influence on the development of the tulip plant and considerably increase the size of all plant parts of the wild species.

2. Applications of nitrogenous fertilizers, watering and removal of the ovary resulted in increased numbers of flowers obtained from one bulb, both in cultivars and wild species. There occurred 3 types of multiflowered plants.

3. The first type, the *Eriostemones* type, develops through branching of the apical growing point of the stem.

4. The second type, the *Leiostemones* type, results from the development of the usually dormant or abortive buds in the axils of the aerial leaves. Plants were obtained with first, second and third order branches, bearing leaves and terminating in a flower.

5. The third type of multiflowered plant, the cultivar type, results form an accelerated development of the axillary buds which flower simultaneously with the apical bud.

6. The initiation of daughter bulbs in an adult bulb proceeds in two phases. The first phase begins before the dormant period; the second begins during that period. The first phase lasts till the vernalization stage and the light-perceptive stages are completed, when all buds being initiated are vegetative ones. The second phase occurs in the latter half of the summer dormant period when the earlier initiated buds begin to develop and new bulbs or reproductive shoots are initiated by their apical meristems and new buds are initiated in the axils of their scales. At that time the apical bud of the mother bulb develops the reproductive shoot. Somewhat later the bud of the main daughter bulb is initiated, which in vigorous bulbs may even occur before the reproductive shoot is formed. The buds initiated in the second phase by the apical meristems of the axillary buds differ morphologically from the first phase buds. Their tunics form the basal parts of the sheathing leaves which later develop an aerial green blade. The number of scales of these buds is always higher than that of the buds from which they developed.

7. In weak bulbs of cultivars and in the bulbs of most wild species the beginning of bud initiation is delayed until the plant enters dormancy and sprouting of the buds with initiation of sheathing leaves does not occur.

8. Heavy dressings of nitrogenous fertilizers and additional watering, combined with removal of the young ovary to redirect the flow of assimilates to the bulb, resulted in the formation of branched inflorescences and an accelerated flowering of daughter bulbs. The branched stem and the multiflowered inflorescence are ancient characters of the genus. They were lost owing to xerophytisation of its present environment, or because of migration to more arid habitats. During this process the genus acquired under the influence of the changing environmental conditions the new uniflorous habit. However, in evolving this new type of growth and development the tulips did not completely lose the old one.

Bibliography

Abraham, A. 1939. Chromosome structure and the mechanism of mitosis and meiosis. *Ann. Bot.* 3 (11): 545–568.

Akhverdov, A.A., 1954. Biologiya nekotorykh dekorativnykh geofitov flory Armenii (Biology of certain ornamental geophytes of the Armenian flora). *Avt. ref. diss. sois. step. kand. biol. nauk. Bot. Inst. Arm. SSR.* Erevan.

Akhverdov, A.A. & N.V. Mirsoeva. 1950. Novyĭ vid tyul'pana iz yuzhnoĭ Armenii, *T. sosnowskyi* (A new tulip species from southern Armenia, *T. sosnowskyi*). *Trud. bot. Inst. AN Arm. SSR.* 7: 31–32.

Alferov, V.A. 1934. K razvitiyu tvsetochnykh lukovits (The development of flower bulbs).*Sovietskie subtropiki* 1, 2.

——, 1956. *Lukovitsnye tsvetochnye rasteniya* (Flower bulbs). Moskva, Sel'khozgiz.

Alkiperova, S.G. 1940. Novyĭ sposob vegetativnogo razmnozheniya nartsisov (A new method of vegetative propagation of *Narcissus*). *Sovietskie subtropiki* 11–12.

Anderson, E. 1949. *Introgressive Hybridization.* New York, John Wiley.

Artyushenko, Z.T. 1948. O morfologicheskoĭ sushchnosti nizhneĭ zavyazi nekotorykh predstaviteleĭ sem. Caprifoliaceae (The morphological essence of the inferior gynoecium of certain representatives of the Caprifoliaceae). *Bot. Zh.* 33 (2): 202–212.

Arutyunova, L.G. 1940. Morfologiya khromozom nekotorykh vidov klevera (Chromosome morphology of certain clover species). *C.R. Acad. Sci. URSS* 27 (8): 825–831.

Avakyan, A.A. 1948. Nekotorye voprosy individual'nogo razvitiya rasteniĭ (Some problems of individual plant development). *Agrobiologiya* 2.

——, 1948. Stadiĭnye protsessy i tak nazyvaemye gormony tsveteniya (Phasal development and the so-called flowering hormones). *Agrobiologiya* 2.

——, 1950. Tsvetenie dvukhletnykh kul'tur v pervom godu zhizni (Flowering of cutivated biennials in the first year of life). *Agrobiologiya* 2.

Avdulov, N.P. 1931. Kariosistematicheskoe issledovanie sem. Gramineae (Karyosystematic investigations of Gramineae). *Bull. appl. Bot. Gen. Pl.-Breed.* 26 (5), suppl.

——, 1933, Dopolnitel'nye karyologicheskie dannye k sistematike zlakov (Kariologische Ergänzungsdaten zur Systematik der Gramineen). *Bull. appl. Bot. Gen. Pl.-Breed. ser.* 2, no. 2: 131–136.

——, 1933. O sverkhkomplektnykh khromozomakh u kukuruzy (On the additional chromosomes in Maize). *Bull. appl. Bot. Gen. Pl-Breed. ser.* 2, no. 2: 101–130.

Avdulov, N.P. & N.N. Titova. 1933. Sverkhkomplektnye khromozomy u *Paspalum stoloniferum* Bosco (Additional chromosomes in *Paspalum stoloniferum* Bosco). *Bull. appl. Bot. Gen. Pl.-Breed.* ser. 2, no. 2: 165–172.

Bailey, L.H. 1947. *The Standard Cyclopedia of Horticuture.* New York, The Macmillan Co.

Baker, J.G. 1874. Revision of the genera and species of *Tulipeae. J. Linn. Soc. Bot.* 14: 275–296.

Baranova, P.A. & V.L. Astaurova (ed.). 1956. *Poliploidiya, Sbornik stateĭ* (Collected articles on polyploidy). Moskva.

Barth, T., K. Weinhausen & L. Steffen. 1954. *Die Kultur der Blumenzwiebeln und -knollen, Ihre Vermehrung, Anzucht und Behandlung.* Hamburg.

Bazilevskaya, N.A. 1957. Dekorativnoe tsvetovodstvo SSSR (Floriculture in the USSR). *Sad i ogorod* 11.

Beal, J. 1939. Cytological studies in relation to the classification of the genus *Calochortus. Bot. Gaz.,* 100: 528–547.

Beklemeshev, E.A., 1889. O kul'ture lukovits na Kavkaze (Bulb growing in the Caucasus) *Vest. Imp. Ross. Obshch. Sad. S. Petersburg* no. 1.

Belling, J. 1926. The iron-acetocarmine method of fixing and staining chromosomes. *Biol. Bull.* 50 (2): 160–162.

Belyaeva, V. 1935. Promyshlennoe razvedenie tsvetochnykh lukovits v Anglii i Gollandii (Commercial cultivation of bulbs in England and The Netherlands). *Sovietskie subtropiki* 12.

Berg, L.S. 1924. Zakonomernost' v obrazovanii organicheskikh form (Regularities in the evolution of organisms). *Bull. appl. Bot. Gen. Pl.-Breed.* 14 (5): 19–68.

Bergner, A.D., 1943. Geographical distribution of chromosome types in *Datura Metel. Am. J. Bot.* 30: 431.

Blumenthal, I.H. 1937. *Geobotanicheskiĭ ocherk zapadnoĭ okonechnosti Zailiĭskogo Alatau i Chuiliĭskikh gor* (Geobotanical survey of the western spurs of the Zailiiskiy Alatau and the Chuiliiskie mountains). Leningrad, L.G.U.

Bochantsev, V.P. & A.Ya. Butkov. 1937. *Dikorastushchie dekorativnye rasteniya Uzbekistana* (Wild plants of decorative value from Uzbekistan). Tashkent, Uzgiz.

Bochantseva, Z.P. 1933. K biologii tsveteniya Guayuly, *Parthenium argentatum* Gray (The flower biology of *Parthenium argentatum* Gray). *Trud. sred.-aziat. Gos. Un. Ser.* 8b, 15.

——, 1934. K metodiki tsitologicheskogo issledovaniya rasteniĭ (On the methods of cytological investigations of plants). *Sotsial. Nauka i Technika* 3–4.

——, 1937. Spermiogenez u tyul'panov i ego izuchenie dlya tseleĭ kariosistematiki (Spermiogenesis in tulips and its importance for karyosystematic studies). *Byull. sred.-aziat. Gos. Un.* 22: 329–336.

——, 1945. Preodoleni neskreshchivaemosti pri mezhrodovoĭ gibridizatsii v semeĭstve makavykh, Papaveraceae (Overcoming cross-incompatibility in the intergeneric hybridisation of Papaveraceae).*Byull. AN Uz. SSR* 5.

——, 1946, Miksoploidiya u nekotorykh rasteniĭ (Mixiploidy in certain plants). *Byull. AN Uz. SSR* 12.

——, 1951. K. voprosu o prorastanii semyan tyul'panov (Germination of tulip seeds). *Trud. bot. Sada AN Uz. SSR* 2.

——, 1952, K. vosprosu o stadiĭnom razvitii mnogoletnikh rasteniĭ (Phasal development in perennials). *Trud. Inst. Bot. AN Uz. SSR* 1.

——, 1952. Mnogotsvetkovost' tyul'panov, poluchennaya eksperimental'no (The induced multiflowered habit in tulips). *Trud. bot. Sada AN Uz. SSR* 3.

——, 1953. Vystupleniya uchastnikov soveshchaniya predstaviteleĭ botanicheskikh sadov v 1952 g. v Moskve (The 1952 conference of representatives of botanical gardens in Moscow). *Byull. glav. bot. Sada* 15: 176–178.

——, 1954. Mnogotsvetkovost' tyul'panov (The multiflorous habit of tulips). *Trud. bot. Sada AN Uz. SSR* 4.

——, 1954. Novyĭ vid tyul'pana iz Zapadnogo Tyan'-Shanya (A new tulip species from the western Tien Shan). *Bot. Mat. Gerb. Inst. Bot. AN SSR* 14: 3.

——, 1954. Tyul'pany i ikh kul'tura v Tashkente (Tulips and their cultivation in Tashkent). Ref. rab. AN Uz. SSR, vnedr. narod. Khoz. Tashkent.

——, 1955. Prorastanie semyan tyul'panov, podvergnutykh vozdeĭstviyu ul'trazvuka (Germination of tulip seeds after ultrasonic wave treatment). *Dok. Akad. Nauk Uz SSR* 4.

——, 1956. Ontogenez tyul'panov (Ontogeny in tulips). *Trud. bot. Sada AN Uz. SSR* 5: 71–108.

——, 1957. Otdalennaya gibridizatsiya u tyul'panov (Remote hybridization in tulips). *Tez. Dokl. soobshch. yubil. nauch. ses, posvyashch. 40-let. Vel. Okt. sotsial. Revol. AN Uz. SSR.* Tashkent.

——, 1958. Otdalennaya gibridizatsiya u tyul'panov (Remote hybridization in tulips). *Tez. dokl. soveshch. otdal. gibrid. rast. zhivot.* Moskva.

——,1958. Lola. 'Fan va Turmush' 5.

Boissier, E. 1882. *Flora Orientalis*, V. Genève, Basel and Lyon.

Borodin, I.P. 1904. *Kurs anatomii rasteniĭ*, 3rd ed. Leningrad.

Botschantzev, see Bochantsev.

Botschantzeva, see Bochantseva.

Bragilevskaya, S.M. 1949. Primenenie udobrenii u tsvetovodstve (The application of fertilizers in floricuture). *Byull. 'Obmen opytov po ozeleneniyu gorodov Uz. SSR'* 1. Kiev.

Breslavets, L. 1933. Tsitologicheskie issledovaniya lubyanykh rasteniĭ, I: *Cannabis sativa*, konoplya (Cytological investigations of fibre plants I. *Cannabis sativa*, hemp). *Trud. Inst. novogo lubyannogo syr'ya* 1. Moskva.

Breslavets, L., G. Medvedeva & M. Magitt. 1933. Tsitologicheskie issledovaniya lubyanykh rasteniĭ, II: kendyr-, rami, kenaf, kanatnik, sap, bamiya (Cytological investigations of fibre plants, II: dogbane, ramie, hemp, abutilon, hibiscus). *Trud. Inst. novogo lubyanogo syr'ya* 2. Moskva.

Brittingham, W.H. 1934. Cytological studies on some genera of the Iridaceae. *Am. J. Bot.* 21 (2): 77–83.

Bruun, H.G. 1932. Cytological studies in *Primula*. Symb. Bot. Ups. 46: 123–129.

Bugaev, V.A. 1946. *Klimat Sredneĭ Azii i Kazakhstana* (The climate of Middle Asia and Kazakstan). Tashkent, AN Uz. SSR.

Burrell, P.C. 1939. Root tip method for difficult material. *Stain Technology* 14.

Buxbaum, F. 1936. Die Entwicklungslinien der *Lilioideae* II. Die Wurmbacoideae. *Bot. Arch.* 38: 367–390.

Bylda, A.Z. 1952. O godichnom tsikle razvitiya u plodovykh rasteniĭ (The annual cycle of development of flowering plants). *Priroda* 2: 106–107.

Chernova, N.M. 1950. Izmenchivost' tyul'pana Shrenka v Krymu (The variability of *Tulipa shrenkii* on the Crimea). *Byull. glav. bot Sada* 5: 102–106.

Crane, M.B. & W.J.C. Lawrence. 1952. *The genetics of garden plants.* London, Macmillan Co.

Cutting, Stephen G. 1954. Forcing bulbs in the home. *Gard. J. New York Bot. Gard.* 4 (6): 173–174.

Daeva, O.V. 1951. Ritm razvitiya nekotorykh sredneaziatskikh rasteniĭ v usloviyakh Moskvy (The growth rhythm of certain Middle Asian plants under Moscow conditions). *Trud. glav. bot. Sada AN SSSR* 2: 59–72.

——, 1958. Sredneaziatskie vidy luka i opyty ikh kul'tury v Moskve (Middle Asian onion species and cultivation trials in the Moscow area). *Byull. glav. bot. Sada AN SSSR* 31: 31–39.

——, 1959. Biomorfologicheskie tipy luka Sredneĭ Azii (Biomorphological types in Middle Asian onions). *Byull. glav. bot. Sada AN SSSR* 33: 73–78.

Danilevskaya, O.N. 1955. Razmnozhenie tyul'panov na Severe (Propagation of tulips in the North). Avt. ref. diss. sois. step. kand. biol. nauk. Leningrad.

——, 1956. *Kul'tura tyul'panov* (The cultivation of tulips). Leningrad, Lenizdat.

Dark, S.O.S. 1934. Chromosome studies in Scilleae II. *J. Genet.* 29(1): 85–98.

Darlington, C.D. 1927. The behaviour of polyploids. *Nature* 119: 390–391.

——, 1932. *Chromosomes and Plant Breeding.* London, Macmillan Co.

——, 1937. *Recent advances in cytology*, 2nd ed. London, Churchill.

Darlington, C.D. & E.K. Janaki Ammal. 1945. *Chromosome Atlas of Cultivated Plants.* London, George Allen & Unwin.

Darlington, C.D. & L.F. La Cour. 1942. *The handling of chromosomes.* London, George Allen & Unwin.

Darlington, C.D. & A.P. Wylie. 1955. *Chromosome Atlas of Flowering Plants.* London, George Allen & Unwin.

Darwin, Charles. 1950. *Sochineniya* 6–7 (Collected works 6–7). Moskva.

Delone, L.N. 1915. Sravnitel'no-kariologicheskoe issledovanie neskol'kikh vidov *Muscari* Mill. (Comparative karyological investigations of certain species of *Muscari* Mill.). *Zap. Kievsk. Obshch. Estestvoz.* Kiev.

——, 1915. Tetraploidnaya osob' *Muscari latifolium.* (Tetraploid individuals in *Muscari latifolium*) *Prot. Zased. Kievsk. Obshch. Estestvozispyt.* Kiev.

——, 1922. Sravnitel'no-kariologicheskoe issledovanie vidov *Muscari* Mill. i *Bellevalia* Lapeyr. (Comparative karyological investigations in *Muscari* Mill. and *Bellevalia* Lapeyr.) *Vest. Tiflis. bot. Sada* 2(1).

——, 1925. Khromozomy vidov *Ornithogalum* L. (The chromosomes of *Ornithogalum* L. species). *Trud. Gos. Timiryazev issl. Inst., otd. esp. evol. ser.* 1, otd. 2(3).

——, 1926. Khromozomnaya teoriya nasledstvennosti u nekotorykh lileĭnykh (Chromosomes and the theory of inheritance in certain Liliaceae). *Vest. Tiflis. bot. Sada ser.* 2(2).

——, 1926. Prevrashcheniya kletochnogo yadra v nasledstvennoĭ izmenchivosti (Changes of the cellular nucleus with inherited variability). *Trud. s/g botaniki* 1(1).

——, 1928. Primenenie kariologicheskogo analiza k resheniyu voprosov spetsial'noĭ sistematiki (The application of a karyological analysis in solving special systematical problems). *Sborn. im. S.G. Navashina.* Moskva.

Doroganevskaya, E.A. 1951. *O svyazi geograficheskogo rasprostraneniya rasteniĭ s ikh obmenom veshchestv* (The relation between the geographical distribution of plants and their metabolism). Moskva, AN SSSR.

Dyles, W.R. 1930. *Notes on tulip species.* London, Herbert Jenkins.

Efremova, L.D. 1951. Embriologicheskoe issledovanie dekorativnykh tyul'panov i ikh znachenie v sadovodstve (Embryological investigations of ornamental tulips and their horticultural importance). Avt. ref. diss. sois. uch. step. kand. biol. nauk. Moskva.

Ellenhorn, Ya. E. 1940. O razlichii khromomerov u allelomorfnykh sputnichnykh khromozom *Allium cepa* (Differences of the chromomeres of allelomorphic satellite-chromosomes of *Allium cepa*). *C. R. Acad. Sci. URSS* 27(4): 357–360.

——, 1940, Kriticheskiĭ peresmotr mitoza v svyazi s ucheniem o spiral'noĭ strukture khromozom (Critical review of the mitosis in relation to the study of the spiral structure of chromosomes). *Bot. Zh. SSSR* 25(3).

Emme, E.K. 1923. Rezul'tat tsitologicheskogo issledovaniya nekotorykh vidov *Aegilops* (Results of a cytological investigation of certain *Aegilops* species). *Russ. bot. Obshch.* 8.

——, 1924. Materialy k tsitologii yachmeneĭ (Notes on the cytology of barley). *Russ. bot. Obshch.* 9.

——, 1927. Materialy k tsitologii roda *Secale* L. (Zur Zytologie der Gattung *Secale* L.). *Bull. appl. Bot. Gen. Pl.-Breed.* 17(3): 73–100.

Emme, K.A. 1936. Triploidnye gibridy stoĭkikh k fytofore vidov *Solanum antipoviczii* Vis. sp. coll. (Triploid hybrids of *Solanum antipoviczii* Vis. sp. coll. with resistance to Phytophthora). *Biol. Zh.* 1(5).

Ernst, A. 1901. Beiträge zur Kenntnis der Entwicklung des Embryosackes und des Embryo (Polyembryonie) von *Tulipa gesneriana*. *Flora oder All. bot. Z.* 88: 37–77.

Esinovskaya, V.N. 1958. Tyul'pany v Estonskoĭ SSR (Tulips in the Estonian SSR). *Izv. AN Est. SSR* 7(4).

Everett, T.H. 1954. *The American Gardeners Book of Bulbs.* New York, Random House.

Favorskiĭ, M.V. 1940. A new method for de-staining slides stained with iron haematoxylin. *C.R. Acad. Sci. URSS* 26(5): 493–494.

——, 1940. Rôle of clove-oil in de-staining slides stained with gentian-violet after Newton and its possible substitutes. *C.R. Acad. Sci. URSS* 26(5): 491–492.

Fedchenko, B.A. 1915. *Rastitel'nost Turkestana* (The vegetation of Turkistan). Petrograd.

——, 1929–30. Krasivo tsvetushchie rasteniya Sredneĭ Azii (Decorative flowering plants of Asia Media). *Bull. appl. Bot. Gen. Pl.-Breed.* 23(5): 273–308.

——, 1932. Krasivo tsvetushchie rasteniya. (Plants with decorative flowers). *Rastitel'nye bogatstva SSSR* 3, *Lesomaterialy i dekorativnye rasteniya.*

——, 1935. *Species of wild flowers of the USSR.* Moscow, Lektechsyrio corp.

Fedchenko, B.A., M.G. Popov & B.K. Shishkin (ed.) 1932. *Flora of the Turkmen SSR.*

Fedtschenko, B.A. see Fedchenko, B.A.

Fedorov, A.A. 1947. Sluchaĭ anomalii v stroenii tsvetka tyul'pana (Occurrence of an irregular flower structure in a tulip). *Sovetskaya botanika* 5: 288–295.

——, 1956. O nekotorykh anomaliyakh u tyul'panov i o znachenie etikh anomaliĭ dlya poznaniya morfologicheskoĭ struktury (Irregularities in tulips and their importance for understanding the morphological structures). *Bot. Zh.* 41(5): 681–689.

——, 1959. O differentsiatsii konusa narastaniya (The differentiation in the apical meristem). *Agrobiologiya* 1.

Fernandes, A. 1930. Estudos nos cromosomas das Liliáces e Amarilidáceas. Bol. Soc. *Broteriana* 7: 3–110.

Filimonova, Z.N. 1958. Period letnego pokoya u dikikh lukov (The summer dormant period of wild onions). *Sborn. rab. asp. AN Uz. SSR, otd. biol. nauk* 2. Tashkent.

——, 1958. Morfologiya i biologiya seyantsev pervogo goda razvitiya u vidov roda *Allium* L. (The morphology and biology of first-year-seedlings of species of *Allium* L.). *Uzb. biol. Zh.* 6.

——, 1959, Morfologiya lukovitsy nekotorykh vidov roda *Allium* L. (The morphology of the bulbs of certain species of *Allium* L. *Uzb. biol. Zh.* 4.

Foerster, K. 1956. *Der Steingarten der sieben Jahreszeiten; eine Provinz aller baulichen und naturhaften Gartenkunst.* Berlin.

Frijlink, A. 1954. The tulip, I: Its early history. *Gard. J. New York Bot. Gard.* 4(1): 5–7.

——, 1954. The tulip, II: The development of the modern tulip. *op. cit.* 4 (2): 47–51.

Gabe, D.R. 1939. Nasledovanie pola u *Mercurialis annua* v svyazi s tsitoplasmaticheskoĭ teoriĭ nasledovaniya pola (Sex inheritance in *Mercurialis annua* in relation to the cytoplasmatic theory of sex inheritance). *C.R. Acad. Sci. URSS* 39 (5): 478–481.

Gaiser, L.O. 1926. A list of chromosome numbers in Angiosperms. *Genetica* 8: 401–484.

——, 1930. Chromosome numbers of Angiosperms II. Bibliogr. *Genetica* 6: 171–466.

——, 1930. Chromosome numbers in Angiosperms III. *Genetica* 12: 161–260.

——, 1933. Chromosome numbers in Angiosperms IV. Bibliogr. *Genetica* 10: 105–250.

Gerasimova, E.N. 1940. O razmerakh sputnikov khromozom (On the size of the satellites of the chromosomes). *Izv. AN SSSR, ser. biol. Nauk* 1: 45–55.

Gerasimova-Navashina, E.N. 1954. Razvitie zarodyshevogo meshka dvoĭnoe oplodotvorenie i vopros o proiskhozdenii pokrytosememnnykh (The development of the embryo-sac, the double fertilization and the problem of the origin of the Angiosperms). *Bot. Zh.* 39 (5): 654–680.

Gilli, A. 1954–56. Neue Liliaceen und Iridaceen aus Afghanistan. *Feddes Rep. spec. novi reg. veg.* 56: 143–149.

Glotov, V. 1939. Dimorfizm kariotipa u *Chrysanthemum coronarium* L. (Dimorphism of the karyotype of *Chrysanthemum coronarium* L.). *C.R. Acad. Sci. URSS* 24 (2): 199–204.

——, 1939. Deĭsve kolkhitsina iz *Colchicum umbrosum* Stev. na kamfarnyĭ bazilik (Effect of colchicine from *Colchicum umbrosum* Stev. on the camphor basil). *C.R. Acad. Sci. URSS* 24 (5): 502–504.

——, 1940. Amfidiploidnaya plodovitaya forma *Mentha piperita* L., poluchennaya posle obrabotki kolkhitsinom (An amphidiploid fertile form of *Mentha piperita* L., produced by colchicine-treatment). *C.R. Acad. Sci. URSS* 28 (5): 450–453.

Gobulinskiĭ, I.N. & M.S.K. Schloss, 1932. K voprosu o partenogeneze u *Humulus lupulus* L. (A contribution to the problem of parthenogenesis in *Humulus lupulus* L.). *C.R. Acad. Sci. URSS* 23 (5): 486–487.

Gollandskie 1947. *Gollandskie tsvetochnye lukovitsy. Illyustrirovannoe opisanie* (Dutch flowering bulbs. Illustrated descriptions). Haarlem, Central Bulb Committee of the Royal General Bulbgrower's Association.

Goodspeed, T.H., F.M. Ueber & P. Avery. 1935. Chromosome structure in *Lilium longiflorum. Un. Calif. Pub. Bot.* 18: 33–42.

Grey, C.N. 1938. *Hardy bulbs,* 3 vols. London, Williams and Norgate.

Griffith, D. 1936. Tulips. *U.S. Dep. Agric. Wash., Circ.* 372.

Grossheim, A.A. 1940. *Tulipa* L. in: *Flora Kavkaza* II, 2nd ed. Baku.

Guignard, L. 1900. L'appareil sexuel et le double fécondation dans les Tulipes. *Ann. Sci. Nat. ser.* 8, Bot. 11: 365–387.

Guilermond, A., G. Mangenot & L. Plantefol. 1933. *Traité de cytologie végétale.* Paris.

Gushchin, G.N. 1940. Itogi otbora dikikh lukovichnykh rasteniĭ na Denauskom opytom pole VNIPSS i ikh khozyaĭstvenno-dekoratyvnye kachestva (Results of a selection for decorative quality in wild bulbous plants on the Denauskiĭ experimental station). *Byull. kul't. sukh. Subtrop.* 4 (56).

Hagerup, O. 1928. Morphological and cytological studies of Bicornes. *Dansk. bot. Arch.* 6, No. 1.

——, 1932. Über Polyploidie in Beziehnung zu Klima, Oekologie und Phylogenie. Chromosomzahlen aus Timbuktu. *Hereditas* 16: 19–40.

——, 1933. Studies on polyploid ecotypes in *Vaccinium uliginosum* L. *Hereditas* 18: 122–128.

Hall, A.D. 1929. *The book of the tulip.* London.

——, 1931. Polyploidy in tulips. *Gard. Chron.* 90 (3): 30.

——, 1937. Polyploidy in *Tulipa. J. Linn. Soc. Bot.* 50: 481–489.

——, 1940. *The genus Tulipa.* London, R.H.S.

Hargrave, J., F.C. Thompson & J. Wood. 1941. Size as a factor in the chemical composition and morphological structure of tulip bulbs. *J. Pomology & Hort. Sci.* 18: 307–324.

Heilbron, O. 1929. Chromosome numbers and taxonomy. *Proc. Int. Congr. Pl. Sci. Ithaca.*

Heitz, E. 1926. Der Nachweis der Chromosomen. *Z. Bot.* 18: 625–681.

Hellier, A.G. 1956. Tulips with a difference. *Amateur Gardening* 73, no. 7340.

Hoog, J.M.C. 1902. *Tulipa wilsoniana* Hoog. *Gard. Chron. ser. III,* 31: 50.

——, 1902. *Tulipa micheliana* Hoog. *op. cit.* 31: 350–51.

——, 1902. *Tulipa nitida* Hoog. *op. cit.* 31: 350, fig. 119.

——, 1902. *Tulipa ingens* Hoog. *op. cit.* 32: 14.

——, 1903. *Tulipa praestans* Hoog. *op. cit.* 33: 364, fig. 126.

——, 1904. *Tulipa tubergeniana* Hoog. *op. cit.* 35: 358, fig. p. 368.

Hsu Siang. 1932. Structure of somatic chromosomes in *Lilium tigrinum. La Cellule* 41: 163–178.

Ingwersen, W. 1951. Using bulbs in the rock garden. *Gardens and Gardening.*

Il'in, M.M. 1937. K. proiskhozhdeniyu flory pustyn'Sredneĭ

Azii (On the origin of the Middle Asian desert flora). *Sovetskaya Botanika* 6: 95–109.
Irmisch, Thilo. 1850. *Zur Morphologie der monokotylischen Knollen- und Zwiebelgewächse.* Berlin.
Irving, W. 1906. *Tulipa fosteriana* Hoog ex. W. Irving. *Gard. Chron. ser. III,* 39: 322, fig. 130.
Ishikawa, M. 1916. A list of numbers of chromosomes. *Bot. Mag. Tokyo* 30: 404–448.
Johansen, D.A. 1950. *Plant embryology.* Waltham, Mass. Chronica Botanica Co.
Kakhidze, N.T. 1940. Khromomernoe stroenei khromozom pshenits v mitoze (Chromomere structure of mitotic chromosomes in wheat). *C.R. Acad. Sci. URSS* 26 (5): 468–470.
Kapinos, G.E. 1947. K teratologii tsvetka tyul'pana; *Tulipa gesneriana* L. (Teratology of the tulip flower; *Tulipa gesneriana* L.). *Dokl. AN Azerb. SSR* 3 (6).
——, 1949. Iz nablyudeniĭ po fenologii tyul'pana na apsherone (Phenological observations on the tulips of the Apsheron peninsula). *Byull. glav. bot. Sada AN SSSR* 4: 67–69.
——, 1951, K teratologii tsvetka tyul'pana, II. (Teratology of the tulip flower, II). *Izv. AN Azerb. SSR* 11.
——, 1952. Vyrashchivanie giatsintov i tyul'panov iz semyan (Growing hyacinths and tulips from seed). *Dokl. AN Azerb. SSR* 8 (11).
Karpechenko, G.D. 1922–23. Chisla khromosom i geneticheskie vzaimootnosheniya u kul'turnykh Cruciferae (The number of chromosomes and the genetic correlations of cultivated Cruciferae). *Bull. Appl. Bot. Gen. Pl.-Breed.* 13 (2): 1–14.
——, 1924–25. Kariologicheskiĭ ocherk roda *Trifolium* L. (Karyologische Studien über die Gattung *Trifolium L.). Bull. Appl. Bot Gen. Pl.-Breed.* 14 (1): 271–279.
——, 1927, Polyploidnye gıbrıdy *Raphanus sativus* L. I *Brassica oleracea* L. (Polyploid hybrids of *Raphanus sativus* x *Brassica oleracea*). *Bull. Appl. Bot. Gen. Pl.-Breed.* 17 (3): 305–410.
——, 1930. K. sintezu konstantnogo gibrida iz trekh vidov (The synthesis of a constant hybrid from three species). *Trud. Syez. Gen. Sel. Sem. plem. Zhiv.* 2.
——, 1935. *Teoriya otdalennoĭ gibridizatsii; teoreticheskie osnovy selektsii, I.* (The theory of remote hybridization; I. The theoretical basis for selection).
Kasparyan, A.S. 1934. Obzor rabot po poliploidii i amfidiploidii za poslednie gody (Review of recent works on polyploidy and amphidipoloidy). *Bull. Appl. Bot. Gen. Pl.-Breed. ser.* 2 (6): 205–222.
Kazaryan, V.O. 1952. *Stadiĭnost' razvitiya i stareniya odnoletnykh rasteniĭ* (Phasal development and senescence in annuals). Erevan, AN Arm. SSR.
Kermit, Olson. 1953. Want early spring color, plan your bulb planting now. *Minnesota Hort.* 81 (8).
Kichunov, N.I. 1905. *Luchshie lukovichnye rasteniya dlya grunta* (Better bulbs for outdoor planting). Sad. Bibl. St. Petersburg.
——, 1909. *Krasivo tsvetushchie gruntovye lukovichnye rasteniya* (Decorative bulbous plants for outdoor planting). Moskva.
——, 1935. *Mnogoletniki* (Perennials). Moskva, Leningrad.
Kimura Yojiro. 1954. Système et phylogénie des monocotyledones. *Rapp. Congr. Int. Bot. Paris* 2.
Kirchner, O., E. Loew & C. Schröter. 1914. *Lebensgeschichte der Blütenpflanzen Mitteleuropas.* Stuttgart.
Kiselev, G.E. 1950. Opyt tsvetovodov-moskvicheĭ (Experiments of Moscow floriculturists). *Moskva, Mosk. rabochiĭ.*
Klimochkina, L.V. 1940. Chromosome morphology in *Helianthus annuus* L. *C.R. Acad. Sci. URSS* 27 (6): 584–86.
Komarov, V.L. 1940. *Uchen'e o vide rasteniĭ.* (Study on plant speciation) Moskva & Leningrad, AN SSSR.
——, 1945–1953. *Izbrannye sochineniya* (Selected works). Moskva & Leningrad.
Komarov, V.L. (ed.) 1935. *Flora of The USSR, 4.* Leningrad, Izd. AN SSSR.
Konovalov, I.N. 1949. O prichinakh vozniknaveniya yavleniĭ prolifikatsii tsvetov s tochki zreniya stadiĭnogo razvitiya rasteniĭ (On the causes of proliferations in flowers as a phenomenon of phasal development). *C.R. Acad. Sci. URSS* 64 (4): 705–708.
Konovalov, I.N. & Z.T. Artyushenko. 1951. Morfologiya plodov tipa korobochka i yagoda (Morphology of the capsule and berry-type of fruits). *Sborn. Morf. Anat. Rast. ser.* 7 (2). Leningrad.
Korovin, E.P. 1928. Rod *Scaligera* DC. (Umbelliferae) i ego filogeniya (The genus Scaligera DC. and its phylogeny). *Trud. sred.-aziat. gos. Un.* 8.
——, 1934. *Rastitel'nost Sredneĭ Azii* (The vegetation of Asia Media). Tashkent, SAOGIZ.
——, 1949. *Osnovnye puti estestvennoĭ akklimatizatsii teplolyubivoĭ flory; Tsitrusovye i subtropicheskie kul'tury Uzbekistana* (Principles of the natural acclimatization of thermophyllous plants; Citrus fruit and the subtropical cultures of Uzbekistan). Tashkent, Izd. AN Uz. SSR.
Korovin, E.P. & S.A. Shuvalov. 1948. O biogennoĭ kompleksnosti pochvenno-rastitel'nogo pokrova v aridnoĭ zone (The biogenic complex of the soil and vegetation cover in the arid zone). *Byull. Mosk. Obshch. Ispyt. Prir. otd. biol.* 53 (1): 81–96.
Kozhevnikov, A.V. 1950. Vesna i osen' v zhizni rasteniĭ (Spring and autumn in plantlife). *Byull. Mosk. Obshch. Ispyt. Prir.* 24: 238.
Kozhin, A.E. 1950. Puti selektsii tsitrusovykh v novykh raĭonakh ikh kul'tury (Selection methods for Citrus fruit in the new areas of cultivation). *Selektsiya i semonovodstvo* 2.
Krasinskiĭ, N.I. 1937. *Metody uskorrenoĭ vygonki tsvetochnykh rasteniĭ* (Methods for forcing flowers). Moskva, Sel'khozgiz.
Krenke, N.I. 1928. *Khirurgiya rasteniĭ* (Plant surgery). Moskva, Novaya derevnya.
——, 1940. *Teoriya tsiklicheskogo stareniya i omolozheniya rasteniĭ* (The theory of cyclic senescence and rejuvenation in plants). Moskva, Sel'khozgiz.
Krevan, P. 1928. Eshche k voprosu o kul'ture giatsintov i tyul'panov v SSSR (More about the cultivation of hyacinths and tulips in the USSR). *Sad i ogorod* 10.
Kroker, V. 1950. *Rost rasteniĭ* (The growth of plants). Moskva, Il.
Kudryashov, S. 1930. Rastitel'nost raĭona gor Khabdun-Tau i Karach Tau (The vegetation of the Khabdun-Tau and Karacha Tau mountain region). *Trud. sred-aziat. gos. Inst. ser.* 8v (13). Tashkent.
Kul'tiasov, M.V. 1950. Problema stanovleniya zhiznennykh form u rasteniĭ (The problem of the origin of the life forms of plants). *Sborn. Probl. bot. Moskva & Leningrad, ANSSR.*
Kuperman, F.M. 1951. Morfologicheskiĭ metod issledovaniya na sluzhbu selektsii rasteniĭ (The morfological method for plant selection). *Selektsiya i semenovodstvo* 9.
——, 1953. *Biologicheskie osnovy kul'tury pshenitsy* (The biological principles of wheat cultivation). Moskva, MGU.
Kuperman, F.M., S.A. Dvoryakin, Z.P. Rostovtseva & E.I. Rzhanova. 1955. *Etapy formirovaniya organov plodonosheniya zlakov,* I (The developmental stages of the fruit organs in grasses, I). Moskva, MGU.
Kus'mina, N.E. 1927. O kromosomakh *Beta vulgaris* L. (On the chromosomes of *Beta vulgaris* L.). *Bull. appl. Bot. Gen. Pl.-Breed.* 17 (3): 241–252.
La Cour, L.F. 1931. Improvements in everyday technique in plant cytology. *J. roy. microsc. Soc.* 51: 119–126.
Lapin, P.I. 1936–38. Opyt upravleniya tsveteniem dekorativnykh rasteniĭ (Experiments on controlling flowering of ornamentals). Diss. sois. uch. step. kand. biol. nauk. Leningrad.
——, 1939. K. voprosu o rasshirenii perioda tsveteniya lukovichnykh rasteniĭ (On the extension of the flowering period in bulbous plants). *Zelenoe stroitel'stvo* 5–6. Leningrad.
——, 1956. Vesenniĭ pokaz tsvetov (Spring flower show). *Priroda* 8: 55: 58.
Lavrenko, E.M. 1957. Pamyati Valeriya Ivanovicha Talieva (In commemoration of V.I. Taliev). *Bot. Zh.* 42 (9): 1331–1337.
Lawrenko, see Lavrenko
Lebedev, D.V. 1957. Evolyutsionnye vzglyady Valeriya

Ivanovicha Talieva (The evolutionary views of V.I. Taliev). *Bot. Zh.* 42 (9): 1338–1353.

Levan, A. 1931. Cytological studies in *Allium*. A preliminary note. *Hereditas* 15: 347–356.

——, 1933. Cytological studies in *Allium*, II. *Hereditas* 16: 257–294.

Levier, E. 1884. Les Tulipes de l'Europe. *Bull. Soc. Sci. Nat. Neuchâtel* 4: 201–312.

Levitin, A.N. 1954. Dikorastushchie lukovichnye rasteniya Sredneĭ Azii i ikh ispol'zovanie v kul'ture (Wild bulbous plants from Middle Asia and their utilization). *Byull. glav. bot. Sada AN SSSR* 17: 22–31.

Levitin, A.N. & G.N. Gushchin. 1953. Kratkoe opisanie sortov massovogo i gruppovogo otbora iz dikorastushchikh lukovichnikh rasteniĭ Sredneĭ Azii (A short description of the varieties resulting from mass- and group selections in wild bulbous plants from Middle Asia). *Byull. Kul't. sukh. Subtrop.* 11–12. Dushanbe.

Levitskiĭ, G.A. 1924. *Material'nye osnovy nasledstvennosti.* (The essential principles of inheritance). Kiev.

——, 1929. O tsitologicheskom metode v sistematike (On cytological methods in sytematics). *Trud. vsesoyuz. cyezd. Gen. Sel. Semenov. plem. Zhivotnov.* 2. Leningrad.

——, 1931. Morfologiya khromozom (The morphology of chromosomes). *Bull. appl. Bot. Gen. Pl.-Breed.* 27 (1): 19–174.

——, 1931. Opyt tsitologicheskogo analiza fiksatsionnogo deĭstviya khrom-atset-formola i khrom-formola (A cytological analysis of the fixing action of chromacetformol and chromoic formalin). *Bul. appl. bot. Gen. Pl.-Breed* 27 (1): 175–186.

——, 1931. Morfologiya khromozom i ponyatie " kariotipa" v sistematike (Chromosome morphology and the 'karyotype' in systematics). *Bull. appl. Bot. Gen. Pl.-Breed.* 27 (1): 187–240.

——, 1933. Tsitologicheskiĭ metod v selektsii (The use of cytological data in selection). *Sotsial. Rast.* 5/7. Leningrad.

——, 1934. Novye fiksatsionnye smesi, proyavlyayushchie morfologiya khromozom (New fixing mixtures revealing the morphology of chromosomes). *C.R. Acad. Sci. URSS* 4: 143–145.

——, 1934. Fiksationnye izmeneniya tela khromozom (Fixation changes of the chromosome body). *C.R. Acad. Sci. URSS* 4: 222–224.

——, 1935. Tsitologicheskiĭ metod v selektsii (The use of cytological data in selection). *Teoret. osnov. sel.* I. Leningrad.

——, 1936. Ocherk geneticheskoĭ tsitologii (Principles of genetical cytology). *Posob. sel. I.* Moskva.

——, 1939. Tsitolocheskie osnovy evolyutsiya (The cytological principles of evolution). Priroda 5.

Levitskiĭ, G.A. & A. Araratyan. 1931. Preobrazovanie khromozom pod vliyaniem rentgenovskiĭ lucheĭ (Transformations of chromosomes under the influence of X-rays). *Bull. appl. Bot. Gen. Pl.-Breed.* 27 (1): 265–303.

Levitskiĭ, G.A. & G.K. Benetskaya. 1929. Tsitologicheskoe issledovaniye konstantno-promezhutochnykh rzhano-pshenichnykh gibridov (Cytological investigation of constant rye-wheat hybrids). *Spis. dokl. tez. vsesoyuz. cyezd. Gen. Sel.* Leningrad.

——, 1931. Tsitologiya psenichno-rzhannykh amfidiploidov (Cytology of *Triticale* amphidiploids). *Bull. appl. Bot. Gen. Pl.-Breed.* 27 (1): 241–264.

Levitskiĭ, G.A. & N.E. Kuz'mina. 1927. Kariologicheskiĭ metod v sistematike i filogenetike roda *Festuca* L. (The use of karyological evidence in the systematics and phylogeny of *Festuca* L.). *Bull. appl. Bot. Gen. Pl.-Breed.* 17 (3).

Levitskiĭ, G.A. & M.A. Sizova. 1934. O zakonomernostyakh v preobrazovaniakh khromozom, vyzyvaemykh X-luchami (On regularities in chromosome transformations induced by X-rays). *C.R. Acad. Sci. URSS* 4: 84–87.

Levitskiĭ, G.A., E.M. Shepeleva & N.N. Titova. 1937. Novye kariotipicheskie rasy, vyvedennye iz potomstva rentenezirovannykh rasteniĭ (New karyotypical varieties from progenies of plants treated with X-rays). *Sotsial. Rast.* 11. Leningrad.

Lipschitz, see Lipshits

Lipshits, S. Yu. 1932. Zametka o *Tulipa krauseana* Rgl (Notice sur la *Tulipa krauseana* Rgl). *Izv. bot. Sada AN SSSR* 30: 733–736.

Litovchenko, A.G. 1940. Znacheni trikotiledonii (trekhdol'nosti) v rastenievodstvo (Role of tricotyledons in the socialist plant industry). *C.R. Acad. Sci. URSS* 27 (8): 816–820.

Lodkina, M.M. 1957. Osobennosti razvitiya tychinok pshenitsy i lilii v svyazi s obscheĭ fiziologieĭ tsvetka (The development of the stamens of *Triticum* and *Lilium*). *Trud. bot. Inst. AN SSSR ser.* 7 (4): 323–377.

Lopott, M.S. & P.K. Ozolin. 1941. *Prakticheskie ukazaniya po gruntovomu tsvetovodstvu v Sredneĭ Azii* (A practical guide for outdoor floriculture in Middle Asia). Dushanbe.

Luchnik, Z..I. 1951. *Dekorativnye rasteniya Gornogo Altaya* (Decorative plants from the Altai mountains). Moskva, Sel'khozgiz.

Lyshenko, T.D. 1943. *Raboty v dni Velikoĭ Otechestvennoĭ Voĭny* (The scientific work in the days of the Great Patriotic War). Moskva. Ogiz-Sel'khozgiz.

——, 1948. *Agrobiologiya* (Agrobiology). Moskva.

L'vova, N.N. 1950. Nekotorye voprosy tsitologii oplodotvoreniya uzlakov (Cytological problems in the fertilization of grasses). *Selektsia i semenovodstvo* 9.

Lyubich, F.P. 1951. Raznokachestvennost plodov i semyan u rasteniĭ i ee znachenie v zhizni vida (Differences in fruit and seed quality and their importance for the survival of the species). *Agrobiologiya* 5.

Maevskiĭ, P.F. 1954. *Flora Sredneĭ polosy Evropeĭskoĭ chasti SSSR* 8-e ed. (Flora of the Central European Region of the USSR, 8th ed.). Moskva & Leningrad, Sel'khozgiz.

Maheshwari, P. 1954. *Embriologiya pokrytosemyannykh* (Embriology of the Angiosperms). Moskva.

Makarova, N.A. 1934. Nekotorye ekologicheskie osobennosti v anatomicheskom stroenii i fiziologii tyul'panov yugovostoka (Ecological data on the anatomical structure and the physiology of tulips with a south-eastern distribution). *Sovetskaya botanika* 3: 103–110.

Mamporiya, F.D. 1951. *Osobennosti vosproizvedeniya, rosta, razvitiya i formoobrazovanniya tsitrusovykh i nekotorykh drugikh pomerantsevykh* (On the reproduction, growth, development and training of citrus fruits). Tbilisi.

Markov, A.G. 1955. *Otechestvennye sorta dekorativnykh rasteniĭ* (Indigenous varieties of decorative plants). Moskva.

Mashinskiĭ, L.O. 1939. Promyshlennoe osvoenie kul'tury dikorastushchikh tyul'panov v sadovodstvakh Leningrada (The cultivation of wild tulips for horticultural production in the Leningrad area). *Zelenoe stroitel'stvo* 11–12.

Matsura, H. 1935. On karyo-ecotypes of *Fritillaria camschatkensis* (L.) Ker Gawl. *J. Fac. Sci. ser. Y,* 3 (5).

Maude, P.F. 1939. A list of chromosome numbers of species of British flowering plants. *New Phytol.* 38: 1–31.

Mezhdunarodnyĭ kodeks nomenklatury dlya kul'turnykh rasteniĭ, 1957 (International code of nomenclature of cultivated plants, 1957). Moskva & Leningrad, AN SSSR.

Michurin, I.V. 1933. *Vyvedeni novykh uluchshennykh sortov plodovykh i yagodnykh rasteniĭ* (Breeding for better varieties of fruit and berry crops). Moskva, Sel'khozgiz.

——, 1941. *Sochineniya* (Collected works). Moskva & Leningrad, Ogiz-Sel'khozgiz.

Migushova, E. 1954. Osobennosti sadovogo tyul'pana (Notes on the garden tulip). Aft. diss. sois. uchen. step. kand. biol. nauk. Leningrad.

——, 1954. Osobennosti biologii sadovogo tyul'pana (Notes on the biology of the garden tulip). Aft. diss. sois. uchen. step. kand. biol. nauk. Leningrad.

Minina, E.G. 1952. *Smeshchenie pola u rasteniĭ vozdeĭstviem faktorov vneshneĭ sredy* (Changes in sex expression caused by environmental factors). Moskva, AN SSSR.

Mishchenko, P. 1912. Dikie vidy *Tulipa* i *Scilla* Kavkaza, Kryma i Sredneĭ Azii kak material dlya kul'tury (Wild *Tulipa* and *Scilla* species from the Caucasus, the Crimea and Middle Asia, suitable for cultivation). *Bull. appl. Bot. Gen. Pl.-Breed.* 5 (2): 37–59.

Modilevskiĭ, Ya. S. 1910. Ob anomal'nykh sluchayakh razvitiya zarodyshevogo meshka u vysshikh rasteniĭ (Cases of abnormal embryo-sac development of the higher plants). *Zap. Kievsk. Obshch. estest-ispytat.* 20.

——, 1929, Zinochiĭ gametofit u Angiosperma (Der weibliche Gametophyt der Angiospermen). *Ukr. bot. Zh.* 5: 5–40.

Mol, W.E. de 1925. Het celkundig-erfelijk onderzoek in dienst gesteld van de veredeling der hyacinthen, narcissen en tulpen. *Genetica* 7: 111–118.

——, 1927. Duplication and quadruplication of the generative nuclei of tulips. *Genetica* 9: 116.

——, 1928. Zusammenfassung der zytologischen und genetischen Ergebnisse des Versuchs der Duplizierung und Quadruplizierung von Sexualkernen bei Hyazinthen und Tulpen. *Z. indukt. Abst.* 48: 145–148.

——, 1928. The originating of diploid and tetraploid pollen grains in Duc van Tol-tulips (*Tulipa suavolens* Roth.) dependent on the method of culture applied. *Genetica* 11: 119–212.

——, 1928. Producing at will of fertile diploid and tetraploid gametes in Duc van Tol, Scarlet (*Tulipa suavolens* Roth.). *Beibl. 15 Viertel.-jahrschr. naturf. Ges. Zürich* 73: 73–97.

——, 1933. Die Entstehungsweise anormaler Pollen-körner bei Hyazinthen, Tulpen und Narzissen. *Cytologia (Tokio)* 5: 31–65.

Molchanov, L.A. 1935. Klimaticheskie usloviya naibolee teplykh raĭonov Sredneĭ Azii (The climatical conditions in the hot regions of Asia Media). *Sb. 'Subtropischeskie kul'tury Sredneĭ Aziĭ'.* Tashkent, Gosizdat Uz. SSR.

Moor, N. 1956. Historic tulips. *Amat. Gard.* 73. no. 3743.

Morinaga, T. et al. 1929. Chromosome numbers of cultivated plants, II. *Bot. Mag. Tokyo* 43: 589–594.

Murzaev, E.M. 1957. *Srednyaya Aziya* (Asia Media). Moskva, Geografgiz.

Mytarev, I.M. 1929. K voprosu o kul'ture giatsintov i tyul'panov v SSSR (On the cultivation of hyacinths and tulips in the USSR). *Sad. i ogorod* 1.

Nakajama, G. 1931. The chromosome numbers in cultivated and wild Angiosperms. *Bot. Mag. Tokyo* 45.

Navashin, M.S. 1927. Ob izmenenii chisla i morfologicheskikh priznakov khromozom u mezhvidovykh gibridov (On the variation of the number and morphological characters of the chromosomes of interspecific hybrids). *Bull. appl. Bot. Gen. Pl.-Breed.* 17 (3): 121–150.

——, 1928. Morfologiya kletochnogo yadra u vidov *Crepis* v svyazi s voprosom vidoobrazovaniya (The morphology of the cellular nucleus of *Crepis* species in relation to species formation). *Sbor. im. S.G. Navashina.* Moskva.

——, 1933. Novye dannye po voprosu o samoproizvol'nykh mutatsiakh (New data on the origin of spontaneous mutations). *Biol. Zh.* 2 (2, 3): 111–115.

——, 1936. *Metodika tsitologicheskogo issledovaniya dlya selektsionnykh tseleĭ* (Methods of cytological investigation for selection purposes). Moskva, Sel'khozgiz.

Navashin, M.S. & N. Gerasimova. 1939. Poluchenie poliploidov iz list'ev obrabotannykh kolkhitsinom (Production of polyploids from leaves treated with colchicine). *C.R. Acad. Sci. URSS* 24 (9): 948–950.

Navashin, S.G. 1910. O nekotorykh yavleniakh pri sliyanii yader u zarodyshevom meshke posle oplodotvoreniya i zamechaniya o tipicheskom delenii yader u *Fritillaria tenella* (Some phenomena observed during the fusion of nuclei in the embryo sac after fertilization with notes on the typical nuclear division in *Fritillaria tenella*). *Prot. zased. Kievsk. Obshch. est.-ispyt.* Kiev.

——, 1911. Ob individual'nykh i vidovykh otlichiakh khromozom (Individual and specific differences in chromosomes). *Prot. zased. Kievsk. Obshch. est.-ispyt.* Kiev.

——, 1912, O dimorfizme yader v somaticheskikh kletkakh u *Galtonia candicans* (Dimorphism of the nuclei in somatic cells of *Galtonia candicans*). *Bull. Acad. Imp. Sci. St. Petersburg:* 373–385.

——, 1912. Novye nablyudeniya nad prevrashcheniem kletochnogo yadra u *Galtonia candicans* (New observations on changes of the cellular nucleus in *Galtonia candicans*). *Prot. Kievsk. Obshch. est.-ispyt.* Kiev.

——, 1913. Getero i idiokhromozomy u rasteniĭ v svyazi s voprosom ob opredelenii pola u zhivotnykh (Hetero- and idiochromosomes of plants in relation to sex determination in animals). *Prot. Kievsk. Obshch. est.-ispyt.* Kiev.

——, 1914. Zametka o chisle khromozom v kletochnom yadre u *Najas maior* (Notes on the chromosome number of the cellular nucleus in *Najas maior*). *Prot. Kievsk. Obshch. est.-ispyt.* Kiev.

——, 1914–16. O nekotorykh priznakakh vnutrenneĭ organizatsii khromozom (Some characters of the inner chromosome structure). *Sbor. stat. posvashch. K.A. Timiryazev.* Moskva.

——, 1921. Rezyume vozrazheniĭ na doklad L.N. Delone. *Russ. bot. Obshch.* 6.

——, 1936. Priroda i poryadok vozbuzhdeniya yadernogo i kletochnogo deleniya v meristeme koreshka i mitogennye luchi prof. A.G. Gurvicha (Natur und Gang der Erregung von Kern- und Zellteilung in Wurzelmeristem und die 'Mitogenen Strahlen' von Prof. A.G. Gurwitsch). *Biol. Zh.* 5 (2): 260–270.

——, 1936. O diminutsii khromatina *Ascaris megalocephala* (Uber die Chromatindimunition bei *Ascaris megalocephala*). *Biol. Zh.* 5 (2): 236–248.

——, 1951. *Izbrannye trudy*, I (Selected works, I). Moskva & Leningrad, AN SSSR.

Neporozhnyĭ, G.D. 1950. *Gladiolus.* Moskva, Sel'khozgiz.

Newton, W.C.F. 1927. Chromosome studies in *Tulipa* and some related genera. *J. Linn. Soc. Bot.* 47: 339–354.

Newton, W.C.F. & C.D. Darlington. 1927. Meiosis in a triploid tulip. *Nature* 120: 13.

——, 1929. Meiosis in polyploids, Part I. Triploid and pentaploid tulips; Part II. Aneuploid hyacinths. *J. Genet.* 21 (1): 1–15.

——, 1929. Meiosis in polyploid tulips. *J. Genet.* 19.

Nikitin, S.A. 1936. Peski Zapadnogo Pribalkhash'ya (The sands of the Pribalkhash area). *Trud. Pochv. Inst. AN SSSR* 11: 147–226.

Nikolaeva, A.G. 1920. Primenenie tsitologicheskogo metoda pri reshenii nekotorykh voprosov genetiki (Application of the cytological method for solving certain genetical problems). *Trud. 3rd Vseross. c"yezd. sel. semenov. v Saratov.* Saratov.

——, 1922–23. Tsitologicheskie issledovaniya roda *Triticum* Etude cytologique du genre *Triticum*). *Bull. appl. Bot. Gen. Pl.-Breed.* 13 (1): 31–44.

——, 1924. Materialy po tsitologii rzhano-pshenichnykh gibridov (A contribution to the cytology of *Triticale*). *Selekts. stan. Sel'khoz. Akad.* Moskva.

Nikolaeva, M.G. 1948. K biologii prorastaniya nekotorykh vivov *Ferula* (Germination of certain *Ferula* species). *Trud. bot. Inst. AN SSSR ser.* 4 (6): 218–228.

——, 1948. Zavisimost' prorastaniya semyan nekotorykh zontichnykh ot sostoyaniya pokrovov zarodysha; na primere vidov *Ferula* (Seed germination of some Umbelliferae, e.g. *Ferula* species, depending on the condition of the integuments). *Trud. bot. Inst. AN SSSR ser.* 4 (6): 229–241.

——, 1950. Znachenie stratifikatsii semyan dlya dal'neĭskogo razvitiya seyantsev evropeĭskogo bereskleta (The importance of stratification for the development of the seedlings of *Euonymus europaeus*). *C.R. Acad. Sci. URSS* 71 (5).

Nisbet, A.F.R. 1956. Bulb propagation in the Hebrides. *Comm. Grower:* 229–230.

Novikov, G.N. & M.G. Nikolaeva. 1940. O prorastanii semyan nekotorykh vidov *Ferula* (Seed germination in certain *Ferula* species). *Sovet. bot.* 5–6: 331–337.

Pavlov, N.V. 1940. Literaturnye istochniki po flore i rastitel'nosti Kazakhstana (Literature sources on the flora and vegetation of Kazakhstan). *Kazakh. fil. Trud.* 13, 19. Moskva; Leningrad, AN SSSR.

——, 1947. *Rastitel'nye resursy Yuzhnogo Kazakhstana.* The vegetational resources of southern Kazakhstan). Moskva; Leningrad.

——, 1959. *Rasskazy o dikikh tsvetakh Kazakhstana* (On the wild flowers of Kazakhstan) Alma-Ata, AN Kaz. SSR.

Petrov, D.F. 1942. Geneticheskoe i evolyutsionnoe znachenie razlichnykh tipov zarodyshevykh meshkov (Genetical and evolutionary importance of the different types of embryo-sacs). *Zh. Obshch. Biol.* 1–2.

Pobedimova, E.G. 1949. Novyĭ vid tyul'pana iz Sredneĭ Azii, *Tulipa neustruevae* (A new tulip species from Middle Asia, *Tulipa neustruevae Pob. sp. nov.*). *Bot. Mat. Gerb.* 11: 62–63.

Poddubnaya-Arnol'di, V.A. 1930. Embriologicheskiĭ metod v sistematike pokrytosemyannykh rasteniĭ (The embryological aspects of Angiosperm systematics). *Izv. Ass. Nauch.-issled. Inst.* 3 (1a). Moskva.

——, 1939. Embriogenez pri otdalennoĭ gibridizatsii v rode *Taraxacum* (Development of pollen and embryo-sac in interspecific hybrids of *Taraxacum*). *C.R. Acad. Sci. URSS* 24 (4): 374–377.

——, 1945. K morfologii khromozom tverdykh pshenits (A contribution to the chromosome morphology of hard wheat). *C.R. Acad. Sci. URSS* 48 (4): 278–280.

——, 1947. Uspekhi sovetskoĭ embriologii rasteniĭ (Progress of Soviet plant embryology). *Byull. Mosk. Obshch. Ispyt. Prir.* 52 (5): 81–94.

Popov, M.G. 1923. *Flora pestrotsvetnykh tolshch Krasnopeschannikovykh nizkogoriĭ* (Flora of the variegated red sandstone of the lower mountain belt). Tashkent.

——, 1925, Desyat'let raboty Sredneĭ Aziĭ (Ten years' work in Middle Asia). *Bull. Inst. Pédol. Géobot. Centr. Asian State Un.* 1.

——, 1927. Osnovnye cherty istorii razvitiya flory Sredneĭ Azii (Basic outlines of the history of the Middle Asian flora). *Byull. Centr. Asian State Un.* 15.

——, 1940. Opyt monografii *Eremostachys* Bunge (Contributions to a monography of *Eremostachys* Bunge). *Mem. Mosk. Obshch. ispyt. prir.* 19.

——, 1940. *Rastitel'nyi pokrov Kazakhstana*, 18 (The vegetation of Kazakstan, 18). Alma-Ata.Fil. Kaz. AN SSR.

——, 1950. O primenenii botaniko-geograficheskogo metoda v sistematike rasteniĭ (The application of the geographical method in plant systematics). *Sbor. bot. Moskva & Leningrad, AN SSSR.*

——, 1957. O vzaimootnosheniyakh i istorii rodov *Papaver* i *Roemeria* (The interrelations and history of the genera *Papaver* and *Roemeria*). *Bot. Zh.* 42 (9): 1389–1397.

Popov, V.I. 1937. Tyul'pany v SShA (Tulips in the USA). *Sov. Subtrop.* 8. Moskva.

Proskoryakov, E.I. 1926. Temperaturnye usloviya prorastaniya semyan vesennikh mnogoletnikov (Temperature conditions for the germination of spring perennials). *Izv. glav. bot. Sad. SSSR* 25: 1–22.

——, 1952. Prisposobleniya prorastayushchikh semyan rasteniĭ sredneaziatskoĭ flory (Adaptations in germinating seeds of Middle Asian plants). Aft. sois. uch. step. dokt. biol. nauk. Samarkand, UzGU.

Proskoryakov, E.I. & A.S. Dzhalilov. 1956. K biologii odnogo iz sredneaziatskikh tyul'panov (A contribution to the biology of a Middle Asian tulip species). *Trud. bot. Sada AN UzSSR* 5.

Prozina, M.N. 1923. Sravnitel'noe kariologicheskoe issledovanie I. Somaticheskoe delenie (Comparative karyological investigations I. Somatic division). *Russ. bot. Obshch.* Moskva.

——, 1937. Opyt postroeniya filogenicheskogo razvitiya roda *Eremurus* (Versuch einer Rekonstruktion der phylogenetischen Entwicklung von *Eremurus*). *Biol. Zh.* 3: 511–512.

Prozorovskiĭ, A.B. 1936. O bibliogicheskikh tipakh rasteniĭ pustyni (Biological types in desert plants). *Bot. Zh.* 21 (5).

Pylkov, N.A. 1915. *Komnatnaya kul'tura luchshikh lukovichnykh rasteniĭ. Prakticheskoe rukovodstvo* (The best bubous plants for room cultivation. A practical guide). Moskva, Sadov. bibl. Soĭkina.

Rabotnov, G.A. 1954. Raboty v oblasti izucheniya zhiznennogo tsikla mnogoletnikh travyanistykh rasteniĭ v estestvennykh tsenozakh (Studies on the life cycles of herbaceous perennials in natural populations). *Sborn. Vopr. Bot. 2 Moskva & Leningrad, AN SSSR.*

Regel, E.A. in: Vestnik Rossiĭskogo Obshchestva Sadovodstva. 1873, 1877, 1878, 1879, 1884. St. Petersburg.

——. In: *Gartenflora. Allgemeine Monatsschrift für deutsche, russische und schweizerische Garten- und Blumenkunde.*

——, 1862. *T. orphanidea* Boiss. ex Heldr., *op. cit.* 11: 309, t. 373.

——, 1873. *T. greigii* Rgl. *op. cit.* 22: 290–292, t. 773. Uebersicht der Arten, 22: 292–299.

——, 1874. *T. eichleri* Rgl, *op. cit.* 23: 193–194, t. 799 and *T. hageri* Heldr., 23: 97–99, t. 790.

——, 1875. *T. silvestris* L. var. *tricolor* Ldb., *op. cit.* 24: 99–100, t. 827.

——, 1877. *T. kaufmanniana* Rgl, *op. cit.* 26: 194–196, t. 906.

——, 1878. *T. triphylla* Rgl, *op. cit.* 27: 193–194, t. 942, *T. altaica* Pall., 194, t. 942, *T. kolpakowskiana* Rgl, 294–295, t. 951.

——, 1879. *T. iliensis* Rgl, *op. cit.* 28: 162–163, t. 975, t. 982. *T. kesselringii* Rgl, 34–35, t. 964.

——, 1881. *T. turkestanica* Rgl, *op. cit.* 30: 228–229, t. 1050.

——, 1882. *T. brachystemon* Rgl, *op. cit.* 31: 323, t. 1099.

——, 1884. *T. ostrowskiana* Rgl, *op. cit.* 33: 34, t. 1144. *T. triphylla* Rgl var. *Höltzeri* Rgl, 33: 34–35. t. 1144. *T. cuspidata* Rgl, syn. *T. elwesii* Rgl, 66–67, t. 1147.

——, 1886. *T. linifolia* Rgl, *op. cit.* 35: 622–623, t. 1235.

——, 1888. *T. leichtlinii* Rgl, *op. cit.* 37: 93–94.

——, 1889. *T. dammani* Rgl, *op. cit.* 38: 314, t. 1300. *T. batalinii* Rgl, 38: 506–507, t. 1307^2. *T. maximowiczii* Rgl, 38: 505–506, t. 1307^1.

——. In: *Acta Horti Petropolitani* St. Petersburg.

——, 1873. Enumeratio specierum hucusque cognitarum generis Tulipae. 2: 437–457.

——, 1874. Descriptiones plantarum novarum in regionibus Turkestanicis ad viris Fedchenko, Korolkow, Kuchakewicz et Krause collectis cum adho tationibus as plantas vivas in Horto botanico Petropolitano cultas. *op. cit* 2 (2).

——, 1875. *T. eichleri* Rgl, *op. cit.* 3: 156–157. *T. korolkowii* Rgl, 3: 295–296. *T. tetraphylla* Rgl, 3: 296. *T. turkestanica* Rgl, 3: 296–297.

——, 1877. *T. albertii* Rgl, *op. cit.* 5: 264–265. *T. kaufmanniana* Rgl, 5: 265. *T. kolpakowskiana* Rgl, 5: 226.

——, 1878. *T. triphylla* Rgl, 5: 636–637. *T. kesselringii* Rgl, 5: 637.

——, 1879. *T. iliensis* Rgl, *op. cit.* 6: 301. *T. behmiana* Rgl, 6: 505. *T. dasystemon* Rgl, 6: 507. *T. krauseana* Rgl, *T. thianschanica* Rgl, 6: 508.

——, 1883. *T. brachystemon* Rgl, *op. cit.* 8, fasc. 1: 279.

——, 1884. Descriptiones plantarum novarum et minus cognitarum. *op. cit.* 8, fasc. 9: 641–698. *T. lanata* Rgl, 8: 647–648, t. IV. *T. linifolia* Rgl, 8: 648–649, t. V, 1–2. *T. ostrowskiana* Rgl, 8: 649–650. *T. suavolens* Roth, 8: 650. *T. triphylla* Rgl var. *Hölzeri* Rgl, 8: 651. *T. thianschanica* Rgl, 8: 652, t. V, 3.

——, 1887. Descriptiones plantarum nonularum Horti imperialis Botanici in statu vivo examinatorum. *op. cit.* 10 (1): 366–377. *T. boettgeri* Rgl, 366. *T. leichtlinii* Rgl, 366–367.

——, 1889. Descriptiones et emendationes plantarum in horto imperiali botanico petropolitano culturam. *op. cit.* 10 (2): 687–698. *T. maximowiczii* Rgl, 687. *T. batalinii* Rgl, 688. *T. dammani* Rgl, 688.

——, 1876. *Turkestanskaya flora po materialam sobrannym O.A. Fedchenko, Karelinym i Kirilovym, Kail'barsom, Korol'kovym, Krauze, Kushakevichem, Potaninym, Semenovym, Severtso-*

vym, Sharngorstom, Shrenkom i dr. 6. St. Petersburg.

——, 1888. *Vecennie krasivotsvetushchie mnogoletnie i lukovichnye rasteniya, ikh soderzhanie i vospitanie v sadakh* (Decorative spring-flowering perennials and bulbous plants; their maintenance and cultivation). St. Petersburg.

Rodionov, A.I. 1950. Izuchenie stadii yarovizatsii u persika i peschanoĭ vishni (A study of the vernalization stages of peach and maraschino cherry). *Izv. AN SSSR ser. biol.* 2: 51–67.

Rollo Meyer, H. 1951. The Tulip. *The Studio, ser. Gardens and Gardening* 128–135.

Romanov, I.D. 1936. *K morfogenezu khlopchatnika. Voprosy tsitologii, embriologii i anatomii khlopchatnika* (Morphogenesis of the cotton plant. Cytology, embryology and anatomy). Tashkent, Soyuz NIKHI.

——, 1938. Novaya forma zarodyshevogo meshka Adoxa u *Tulipa tetraphylla* i *T. ostrowskiana* (A new Adoxa-type embryo-sac in *T. tetraphylla* and *T. ostrowskiana*). *C.R. Acad. Sci. URSS* 19 (1, 12): 113–115.

——, 1939. Dve novye formy zarodyshevogo meshka v rode *Tulipa* (Two new types of embryo-sacs of the genus *Tulipa*). *C.R. Acad. Sci. URSS* 22 (3): 139–141.

——, 1940. Izmenchivost' i napravlenie evolyutsii zarodyshevogo meshka v rode *Gagea* (Variability and evolution of the embryo-sac in the genus *Gagea*). Aft. diss. sois. uch. step. kand. biol. nauk. Tashkent.

——, 1944. Evolyutsia zarodyshevogo meshka tsvetkovykh rasteniĭ (The evolution of the embryo-sac in the flowering plants). Aft. diss. sois. uch. step. dokt. biol. nauk. Tashkent.

——, 1952. K. anatomii reliktovykh rasteniĭ. Anatomiya lista *Tulipa regelii* (A contribution to the anatomy of relict plants. The leaf of *Tulipa regelii*). *Mosk. Obshch. Ispyt. Prir., otd. biol.* 57 (5): 79–87.

Rozanov, M.A. 1933. Geneticheskie osnovy sovremmenoĭ sistematiki rasteniĭ (The genetic bases of modern systematics). *Uspech. sovrem. biol.* 2 (4, 5). Moskva.

——, 1940. O genotipicheskom razlichii ras *Rubus caesius* L. (Genotypic differences in races of *Rubus caesius* L.). *C.R. Acad. Sci URSS* 27 (6): 590–593.

Rubashevskiĭ, A.A. 1949. *Filosofskoe znachenie teoricheskogo nasledstva I.V. Michurina* (The philosophical concept of the theory of inheritance of I.V. Michurin). Moskva.

Rubtsov, L.I. 1939. Prodlenie zhizni srezannykh tsvetov (Lengthening the life of cut flowers). *Zelenoe stroitel'stvo* 7, 8.

Ryabova, T.I. 1956. Vegetativnoe razmnozhenie eremurusov (Vegetative propagation of *Eremurus*). *Izv. otd. estest. nauk. AN Tadzh. SSR* 14.

Rybin, V.A. 1936. Gibridy terna i alychi i problema proiskhozhdeniya kul'turnoĭ slivy (Hybrids between blackthorn and wild mirobalan and the origin of the cultivated plum). *Bull. appl. Bot. Gen. Pl.-Breed. ser.* 2 (10). Leningrad.

——, 1929. Tsitologicheskiĭ metod v plodovodstve i orgorodnichestve (The cytological method in fruit- and vegetable growing). *Dost. Persp. Obl. priklad. Bot. Gen. Sel.*

Saakov, S.G. 1935. Dikorastushchie tyul'pany Sredneĭ Azii i opyty ikh prakticheskogo ispol'zovaniya (Wild tulips from Middle Asia in utilization trials). *Za zelenyĭ Leningrad* 6.

——, 1950. Opyt ispol'zovaniya v tsvetovodstve dikikh tyul'panov Sredneĭ Azii (Floricultural trials with wild tulips of Middle Asia). *Trud. bot. Inst. AN SSSR ser.* 6 (1): 268–294.

——, 1955. Tsvetovodstvo v. SSSR i perspektivy ego razvitiya (Floriculture in the USSR and its future development). *Trud. bot. Inst. AN SSSR ser.* 6 (4): 53–81.

Sa'di, Musharrifu'din. 1957. *Gulistan.* Moskva.

Sakamura, T. 1918. Kurze Mitteilung über die Chromosomenzahlen und die Verwandtschaftsverhältnisse der *Triticum* Arten. *Bot. Mag. Tokyo* 32: 151–154.

Schischkin, see Shishkin

Schlechtendal, D.T.L. 1826. Ueber die Monstrosität der gemeinen Gartentulpen (*Tulipa gesneriana* L.). *Linnaea* 1: 595–603.

Selden, M.M. 1942. The ups and downs of tulip bulbs. *Nation. hort. Mag.* 21 (1). Washington.

Senjaninova-Korczagina, see Senyaninova-Korchagina

Senyaninova-Korchagina, M.V. 1931. Kariosistematicheskoe issledovanie roda *Aegilops* (A karyosystematical investigation of the genus *Aegilops*). *Bull. appl. Bot. Gen. Pl.-Breed.* 28.

——, 1931. Kariologicheskoe issledovanie voprosa proiskhozhdeniya *Vicia faba* (A karyological investigation of the origin of *Vicia faba*) *Bull. appl. Bot. Gen. Pl.-Breed.* 28.

Serdyukov, B.V. 1950. K voprosu akklimatizatsii i vvedeniya v kul'turu tsvetochnykh rasteniĭ dikoĭ flory Gruzii (The acclimatization and introduction to cultivation of wild flowering plants from the Georgian republic). *Vest. Tiflis. bot. Sad.* 59.

——, 1952. Vvedenie v kul'turu nekotorykh dikorastushchikh tsvetochnykh rasteniĭ Gruzii (Introduction to cultivation of certain wild-growing flowering plants from the Georgian republic). *Vest. Tiflis. bot. Sad.* 60.

Serebryakov, I.G. 1952. *Morfologiya vegetationykh organov vysshikh rasteniĭ* (Morphology of the vegetative organs of higher plants). Moskva.

Severtsov, A.N. 1939. *Morfologicheskie zakonomernosti evolyutsii* (Recurrent morphological patterns in the evolution). Moskva.

Shakhov, A.A. 1952. O tipakh razvitiya rasteniĭ v protsesse istoricheskogo prisposobleniya k zasushlivym usloviyam (Types of plant development in the past processes of adaptation to arid conditions). *Sbor. Rastenie i sreda.* Leningrad, AN SSSR.

Sharp, L.W. 1943. *Fundamentals of Cytology.* London, McGraw-Hill Book Co.

Shcherbakov, N.I. 1932. *Fitoklimaticheskie analogi lesnykh zon Uzbekistana i sootvetstvuyushchikh po shirote chasteĭ Tadzhikistana i.t.d.* (Phytoclimatological analogies of the forest belts of Uzbekistan and the corresponding latitudinal parts of Tadzhikistan etc.). Tashkent, TIZ Uz. SSR.

Shishkin, B.K. & A.I. Vvedenskiĭ (ed.) 1951. *Flora of the Kirghiz SSR* 3. Frunze.

Shreder, R.R., E.P. Korovin & A.I. Vvedenskiĭ. 1941. *Flora Uzbekistana.* Tashkent.

Shtamm, V.A. 1954. K biologii gusinogo luka (Biology of the bulb of *Gagea*). *Byull. glav. bot. Sad. AN SSR* 19: 90–95.

Silina, Z.M. 1952. Tyul'pany v Leningrade; biologiya i selektsiya. (Tulips in Leningrad; biology and selection). Aft. diss. sois. step. kand. biol. nauk. Leningrad.

——, 1952. Opyty po gibridizatsii tyul'panov (Hybridization experiments with tulips). *Trud. bot. Inst. AN SSSR ser.* 6 (2): 230–271.

——, 1953. Kul'tura tyul'panov v Lengradskoĭ oblasti (Cultivation of tulips in the Leningrad region). *Trud. bot. Inst. AN SSSR ser.* 6 (3): 5–146.

——, 1955. Opisanie sortov tyul'pana (Description of tulip cultivars). *Trud. bot. Inst. AN SSSR ser.* 6 (4): 336–342.

Simons, A.J. 1955. *Bulb growing.* London.

Sizova, M.A. 1939. Strukturnye izmeneniya khromozom v predelakh vida *Triticum durum* (Structural changes in chromosomes of *Triticum durum*). *C.R. Acad. Sci. URSS* 25 (1): 75–77.

Smirnova, V.A. 1954. O peredvizhenii yader pri mozaichnykh zabolevaniyakh nekotorykh odnodol'nykh (On the displacement of nuclei in some monocotyledons with mosaic disease). *Byull. Mosk. Obshch. Ispyt. Prir. otdel. biol.* 59 (5): 71–76.

Sokolov, V.S. 1952. *Alkaloidonosnye rasteniya SSSR* (Alkaloid containing plants of the USSR). Moskva & Leningrad, AN SSSR.

Sokolovskaya, A.N. & O.O. Strelkova. 1940. Kariologicheskoe issledovanie vysokogornoĭ flory Glavnogo Kavkazskogo khrebta i problema geograficheskogo raspredeleniya poliploidov (Karyological investigations of the high-mountain flora of the main Caucasian mountain range and the geographical distribution of polyploids). *C.R. Acad. Sci. URSS* 29 (5, 6).

Sorokina, O.N. 1928. O khromozomakh vidov *Aegilops* (The chromosomes of *Aegilops* species). *Bull. appl. Bot. Gen. Pl.-Breed.* 19 (2): 523–532.

——, 1936. Gibridizatsiya egilops s pshenitseĭ (Hybridization between *Aegilops* and *Triticum*). *Bull. appl. Bot. Gen. Pl.-Breed. ser.* 2 (6): 7–36.

Stewart, R.N. 1947. The morphology of somatic chromosomes in *Lilium*. *Am. J. Bot.* 34 (1): 9–26.

Stout, A.B. 1919. Tulip droppers. *J. Int. Gard. Club* 3 (3): 463–472. New York.

Sugiura, T.A. 1931. A list of chromosome numbers in Angiospermous plants. *Bot. Mag.Tokyo* 45: 353–355.

——, 1937. A list of chromosome numbers in Angiospermous plants III. *Bot. Mag. Tokyo* 51.

Sveshnikova, I.N. 1927. Kariologicheskiĭ ocherk roda *Vicia*. (Karyological studies on *Vicia*). *Bull. appl. Bot. Gen. Pl.-Breed.* 17 (3): 37–72.

——, 1929. *Vicia sativa* L. i *Vicia cracca* L. K voprosu o znachenii kariotipa v sistematike (The importance of the karyotypes of *Vicia sativa* L. and *V. cracca* L. for systematics) *Izv. sel.-khoz. Akad. Timiryazev.* 4.

——, 1947. Noviĭ metod izucheniya evolyustsiya vida (A new method for studying species evolution). *Byull. Mosk. Obshch. Ispyt. otd. biol.* 52 (3): 75–81.

——, 1948. Izuchenie mutatsionnoĭ izmenchivosti v svyazi s problemoĭ evolyutsii u rasteniĭ (The study of mutational variability in relation to plant evolution). *Byull. Mosk. Obshch. Ispyt. otd. biol.* 53 (1): 73–80.

Sveshnikova, I.N. & Yu. Belekhova. 1936. Problema translokatsii u gibridov (Translocation in interspecific hybrids). *Bull. appl. Bot. Gen. Pl.-Breed. ser.* 2 (9): 63–70.

——, 1936. *Izmenenie khromozom v rezul'tate mezhvidovoĭ gibridizatsii u Viki i ikh filogeneticheskoe znachenie* (Changes in chromosomes resulting from interspecific hybridization and their phylogenetical importance). Moskva, VASKHNIL.

Svetozarova, V.V. & E. Ya. Elenhorn. 1950. Protsess oplodotvoreniya u rasteniĭ (The process of fertilization in plants). *Selek. i Semenov.* 9.

Synge, Patrick M. (ed.) 1956. *Dictionary of Gardening of the Royal Horticutural Society*; 2nd ed. Oxford, Clarendon Press.

Täckholm, G. 1922. Zytologische Studien über Rosa. *Act. Hort. Berg.* 3: 97–381.

Takhtadzhian, A.L. 1948. Morfologicheskaya evolyutsiya pokrytosemyennykh (Morphological evolution of the Angiosperms). *Mosk. Obshch. Ispyt. Prir.* 301.

——, 1954. *Voprosy evolyutsionnoĭ morfologii rasteniĭ* (Evolutionary problems of plant morphology). Leningrad.

Taliev, V.I. 1930. Protsess vidoobrazovaniya v rode *Tulipa*, sektsii *Eriostemones* (Species formation in the section *Eriostemones* of the genus *Tulipa*). *Bull. appl. Bot. Gen. Pl.-Breed.* 24 (2): 57–122.

Tarakanov, K.N. 1952. Opyt istoriko-ekologicheskogo izucheniya geterokarpii u nekotorykh vidov solonchakovykh rasteniĭ (An emperical ecological study of heterocarpy in certain species of solonchak plants). *Sborn. Rastenie i sreda.* Leningrad, AN SSSR.

Taylor, W.R. 1926. Chromosome morphology in *Fritillaria, Alstroemeria, Sylphium* and other genera. *Am. J. Bot.* 13: 179–193.

Ternovskiĭ, M.F. 1936. Mezhvidovaya gibridizatsiya *Nicotiana* poluchenie poliploidnykh i gaploidnykh rasteniĭ (Interspecific hybridization in *Nicotiana* and the occurrence of polyploid and haploid individuals). *Bull. appl. Bot. Gen. Pl.-Breed. ser.* 2 (9): 125–130.

Terpilo, A.I. 1948. *Anatomicheskiĭ atlas lekarstvennykh rasteniĭ* (Anatomical atlas of medicinal plants). Charkow.

Thibodeau, Jean. 1956. Adventures with the species tulips. *Horticulture, Mass. Hort. Soc.* 34 (9): 450.

Timofeeva, L.S. 1956. Dinamika rosta i razvitiya sadovogo tyul'pana kak osnova ratsional'noĭ agrotekhniki (The dynamics of growth and development of the garden tulips as a basis for a rational cultivation and production). Aft. diss. sois. uchen. step. kand. sel.-khoz. nauk. Moskva.

Tischler, G. 1921/22. *Allgemeine Pflanzenkaryologie.* Berlin.

——, 1938. Pflanzliche Chromosomen-Zahlen IV. *Tab. Biol.* 16 (8): 162–218.

Transkovskiĭ, D.A. 1930. Metod tsitologicheskogo issledovaniya pyl'tsevykh trubok i ego perspektivy (Method and aspects of the cytological study of pollen tubes). *Trud. C"ezd. Gen. Sel. Semen. Zhivotn.* 2.

——, 1931. Zytologische Beobachtungen über die Entwicklung der Pollenschläuche einiger Angiospermen. *Planta, Arch. wiss. Bot.* 12 (1): 1–18.

Trofimov, T.T. 1952. Vegetativnoe razmnozhenie khokhlatki Galera plotnoĭ (Vegetative propagation of *Corydalis halleri* Willd.). *Byull. Mosk. Obshch. otd. biol. Ispyt. Prir.* 57 (3): 68–76.

——, 1954. O podsnezhnom roste i razvitii rannevesennikh rasteniĭ (The growth and development of early spring plants under snow cover). *Priroda* 1: 102–105.

Troll, W. 1954. *Praktische Einführung in die Pflanzenmorphologie*, I. Jena, Fischer.

Turbin, N.V. 1950. *Genetika s osnovami selektsii* (Genetics and the principles of selection). Moskva.

——, 1954. *O biologicheskoĭ roli chuzherodnogo doopyleniya* (The biological significance of pollination by foreign pollen). Leningrad, LGU.

——, 1954. *Voprosy biologii oplodotvoreniya* (Problems of fertilization). Leningrad, LGU.

Tutayuk, V. Kh. 1949. Obrazovanie mikro- i megaspor v odnom sporolistke u pokrytosemyannykh (The formation of micro- and megaspores on one sporophyll in the Angiosperms). *Bot. Zh.* 34 (3): 267–272.

——, 1952. Anatomo-morfologicheskiĭ analiz makhrovosti pokrytosemyannykh rasteniĭ (Anatomic-morphological analysis of double flowers in angiospermous plants). *Trud. bot. Inst. AN SSSR ser.* 7 (3): 293–403.

Tyutyunnikov, A.I. 1932. Mnogotvsetkovyĭ dernistyĭ tyul'pan (A multiflowered, in tufts growing, tulip). *Izv. bot. Sad. AN SSSR* 30 (5, 6): 737–742.

——, 1945. O tyul'pane dazistemon, *T. dasystemon* Rgl, i tyul'pane tarda, *T. tarda* Stapf. (Notes on *T. dasystemon* Rgl and *T. tarda* Stapf). *Sovetskaya botanika* 13 (5): 37.

Upcott, M.B. 1935. Chromosomes of the tulip in mitosis. *Nature* 135: 957–958.

——, 1936. The mechanics of mitosis in the pollen-tube of *Tulipa*. *Proc. Roy. Soc. B.* 121: 207–20.

——, 1937. The genetic structure of *Tulipa*, II. Structural Hybridity. *J. Genet.* 34 (2): 339–398.

Upcott, M. & L. La Cour. 1936. The genetic strcture of *Tulipa*, I. A chromosome survey. *J. Genet.* 33 (2): 237–254.

Upcott, M. & J. Philip. 1939. The genetic structure of *Tulipa*, IV. Balance, selection and fertility. *J. Genet.* 38 (1, 2): 92–123.

Ustinova, E.I. 1949. Ritm razvitiya konusa narastaniya v lukovitsakh proleski (*Scilla sibrica*) v osenne-zimnee vremya (Development rhythm of the apical growing point of *Scilla sibirica* in the autumn/winter period *C.R. Acad. Sci. URSS* 26 (5).

Vakar, B.A. 1930. Tsitologicheskoe issledovanie gibridov *Triticum persicum* Vav. s. drugimi vidami pshenitsy (A cytological investigation of the hybrids of *Triticum persicum* Vav. with other wheats). *Trud. C"ezd. Gen. Sel. Semen. plem. Zhivot.* 2.

——, 1935. Pshenitchno-pyreĭnye gibridy. Filogeneticheskoe issledovanie (*Triticum-Agropyron* hybrids. A fylogenetical study). *Bull. appl. Bot. Gen. Pl.-Breed. ser.* 2, no. 8: 121–161.

——, 1935. *Tsitologiya pshenichno-pyreĭnykh gibridov* (Cytology of *Triticum-Agropyron* hybrids). Omsk, Omsk. oblast. izd.

Val'nev, E.M. 1937. Tyul'pany Sredneĭ Azii (The tulips of Middle Asia). *Sots. nauk. techn.* 9. Omsk.

Vasilevskaya, V.K. 1936. Znachenie anatomicheskikh priznakov v cheshue lukovits sredne-aziatskikh tyul'panov dlya ikh sistematiki (Importance des caractères anatomiques dans la tunique des bulbes de tulipes de l'Asie Centrale pour leur systématique). *Trud. Sred. Az. Gos. Un. ser. 8b, Bot.* 33.

——, 1938. O znachenie anatomicheskikh koeffitsientov, kak priznake zasukhoustoichivosti rasteniĭ (Anatomical coefficients as characters of drought resistance in plants). 23 (4). Ashkhabad.

Vasil'ev, A.V. 1930. Perspektivy razvitiya promyshlennogo sadovodstva na yuge i v subtropikakh SSSR (Future developments of the industrial horticulture in the south and the sub-tropical regions of the USSR). *Subtropiki* 7, 12.

Verushkin, S.M. 1933. *Pshenichno-pyreĭnye gibridy* (*Triticum-Agropyron* hybrids). Moskva, Sel'khozgiz.

——, 1936. *Osnovnye zakonomernosti v otdalennoĭ gibridizatsii. Posobie po selektsii* I (Principles of remote hybridization; a basis for selection). Moskva.

Vil'son, E. 1938. Vliyanie klimata i usloviĭ na razvitie tsvetochnykh lukovichnykh rasteniĭ (The influence of climate and environment on the development of flower bulb plants). *Opyt zarubezhnykh subtropikov*, 7.

——, 1940. *Kletka i ee rol' v razvitii i nasledsvennosti*, 2. (The cell and its function in development and heredity, 2). Moskva, Leningrad.

Vorob'ev, A.I. 1934. Kratnye khromozomnye chisla u rodstvennykh form (Multiples of chromosome numbers in related forms). *Bull. appl. Bot. Gen. Pl.-Breed. ser.* 1 (6).

Vvedenskiĭ, A.I. 1935. Tulipae at Junones novae. *Byull. sred.-aziat. goss. Un.* 21: 147–152.

——, 1935. *Tulipa* L. in: *Flora of the USSR* 4: 246–280.

——, 1952. *Tulipa binutans* Vved. in: *Not. Syst. Herb. Inst. Bot. Acad. Sci. Uzbek.* 13: 33.

Wanscher, J.H. 1934. The basic chromosome numbers of the higher plants. *New Phytol.* 33: 101–125.

Warmke, H.E. 1935. A permanent root tip smear method. *Stain Techn.* 10 (3): 101–103.

Wiedmer, T. 1954. *Wild Tulpen Zauber.*

Woods, M.W. 1937. The nucleolus in *Tulipa*. *Am. J. Bot.* 24: 528–536.

Woods, M. & R. Bamford. 1937. Chromosome morphology and numbers in *Tulipa*. *Am. J. Bot.* 24 (4): 175–184.

Wright, Murray. 1957. A garden of Ten Million Bulbs. *Gard. J. New York Bot. Gard.* 7 (2): 42–43.

Wulf, E.V. 1944. *Istoricheskaya geografiya rasteniĭ* (Historical plant geography). Moskva, Leningrad.

Yakovlev, M.S. 1946. Odnodol'nost' v svete dannykh embriologii (Embryological aspects of the monocotyledonous habit). *Sovetskaya botanika* 14 (6): 351–363.

——, 1950. Struktura endosperma i zarodysha zlakov kak sistematicheskiĭ priznak (The structure of the endosperm and the embryo as systematical characters in grasses). *Trud. bot. Inst. AN SSSR ser.* 7 (1): 121–218.

Yakovleva, S.V. 1933. Kariologicheskie issledovaniya nekotorykh vidov *Salvia* (Karyological investigation of some species of the genus *Salvia*). *Bull. appl. Bot. Gen. Pl.-Breed. ser.* 1 (5).

Yanishevskiĭ, D.E. 1934. Iz zhizni tyul'panov na nizhneĭ Volge (About the tulips from the Lower Wolga). *Sovetskaya botanika* 3: 72–103.

Yarkova, A.S. 1939. Gipoteza zashchitnykh veshchestv v rastitel'nom organizme (A hypothesis with respect to the protective substances in a vegetative organism). *Sovetskaya botanika* 3: 9–16.

Yurtsev, V.N. 1955. Osobennosti novoobrazovaniĭ i stadiĭnogo razvitiya pridatochnykh pochek i pobegov u mnogoletnikh travyanistyikh i drevesnykh rasteniĭ (On the formation and phasal development of adventituous buds and shoots in herbaceous and woody perennials). *Bull. appl. Bot. Gen. Pl.-Breed.* 32 (1): 134–171.

Zabelin, N.A. 1948. Pestrolepestnost' tyul'panov v zavisimosti ot sorta i sredy razvedeniya (Variegated flowers in tulips depending on the variety and conditions of cultivation). *Trud. gos. Nikit. bot. Sad* 24 (3).

Zaitseva, E.N. 1956. Dikorastushchie vidy tyul'panov i ikh sadovye formy v kolektsii Glavnogo botanischeskogo Sada (Wild tulip species and their garden forms in the collection of the Main Botanic Garden). *Byull. glav. bot. Sad.* 26: 48–52.

——, 1957. Kollektsiya sadovykh form tyul'panov v Glavnom botanicheskom Sadu (The collection of tulip cultivars in the Main Botanic Garden). *Byull. glav. bot. Sad.* 27: 51–54.

——, 1958. *Tyul'pany* (Tulips). Moskva.

Zatvornitskiĭ, G.F. 1954. Stepnoĭ tyul'pan (The steppe tulip). *Byull. glav. bot. Sad. ANN SSSR* 18.

Zolotarev, I.I. 1896. *Flora sadovodstva, Illyustrirovannyĭ ukazatel' rasteni'* (Flora of horticultural plants, an illustrated guide). Moskva.

Zhadovskiĭ, A.E. 1924. Gormony kletochnykh deleniĭ (ranevye gormony i nekrogormony) i ikh rol' v razvitii normal'nykh i pridatochnykh zarodysheĭ u rasteniĭ (The hormones of cell division, wound hormones and necrohormones, and their role in the development of normal and additional plant embryos). *Usp. eksp. biol.* nos. 3–4.

——, 1924–25. Tipy razvitiya pyl'tsy i ikh filogeneticheskoe znachenie (Types of pollen development and their phylogenetic importance). *Bull. appl. Bot. Gen. Pl.-Breed.* 14 (5).

Index of plant names

The page references for the morphological and karyological descriptions of species and groups are printed in *italics.*

Subject index